International Conference on Space Science and Communication (IconSpace 2017)

AA006079

Journal of Physics: Conference Series Volume 852

Kuala Lumpur, Malaysia
3 - 5 May 2017

Editors:

Wayan Suparta
Mahamod Ismail
Mardina Abdullah

ISBN: 978-1-5108-4243-4
ISSN: 1742-6588

Printed from e-media with permission by:

Curran Associates, Inc.
57 Morehouse Lane
Red Hook, NY 12571

Some format issues inherent in the e-media version may also appear in this print version.

Copyright© (2017) by the Institute of Physics
All rights reserved. The material featured in this book is subject to
IOP copyright protection, unless otherwise indicated.

Printed by Curran Associates, Inc. (2017)

For permission requests, please contact the Institute of Physics
at the address below.

Institute of Physics
Dirac House, Temple Back
Bristol BS1 6BE UK

Phone: 44 1 17 929 7481
Fax: 44 1 17 920 0979

techtracking@iop.org

Additional copies of this publication are available from:

Curran Associates, Inc.
57 Morehouse Lane
Red Hook, NY 12571 USA
Phone: 845-758-0400
Fax: 845-758-2633
Email: curran@proceedings.com
Web: www.proceedings.com

Table of contents

Volume 852

2017 International Conference on Space Science and Communication

3–5 May 2017, Kuala Lumpur, Malaysia

Accepted papers received: 19 May 2017
Published online: 6 June 2017

Atmospheric and Magnetospheric Sciences

Geoscience and Remote Sensing

Satellite and Communication Technology

Interdisciplinary Space Science

Table of Content

Preface

2017 International Conference on Space Science and Communication "Space Science for Sustainability"

The present volume of the *Journal of Physics: Conference Series* represents contributions from participants of the 2017 International Conference on Space Science and Communication (IconSpace2017) held in Kuala Lumpur, Malaysia from May 3-5, 2017. The conference was organized by Space Science Centre (ANGKASA), Institute of Climate Change, Universiti Kebangsaan Malaysia (UKM) with a theme on "Space Science for Sustainability".

IconSpace2017 is the fifth series of conferences devoted to bringing researchers from around the world together to present and discuss their recent research results related to space science and communication, and also to provide an international platform for future research collaborations. This biennial international conference is an open forum where members in the field and others can meet in one place to discuss their current research findings.

The technical program of this conference includes four keynote speakers, invited speakers, and the presentation of papers and poster. The track of the session includes Astrophysics and Astronomy, Atmospheric and Magnetospheric Sciences, Geoscience and Remote Sensing, Satellite and Communication Technology, and Interdisciplinary Space Science. Apart from the main conference, there will be a special talk on "Space Exploration & Updates" on 5 May 2017. More than 100 scientists and engineers from various academic, government, and industrial institutions in Europe, Asia, Australia, Africa, and the Americas attended the conference. The papers for this conference were selected after a rigorous review process. The papers were all evaluated by international and local reviewers and at least two reviewers were required to evaluate each paper.

We should like to offer our thanks for the professionalism of the organizing committee, authors, reviewers, and volunteers deserve much credit for all the time for the successful run of the conference. We also wish to express our appreciation to all speakers and authors whose papers and presentations make the event a very exciting forum to add values to learn, discussion and exchanges of ideas, and to meet old or new people from different countries and interact with them. We would also like to express our sincerest gratitude to our sponsors: Malaysian Communications and Multimedia Commission (MCMC) for making this conference more valuable.

Wayan Suparta
The Editor

Content from this work may be used under the terms of the Creative Commons Attribution 3.0 licence. Any further distribution of this work must maintain attribution to the author(s) and the title of the work, journal citation and DOI.
Published under licence by IOP Publishing Ltd

IOP Conf. Series: Journal of Physics: Conf. Series **852** (2017) 011001 doi:10.1088/1742-6596/852/1/011001

Proceeding Editor

Wayan Suparta
Mahamod Ismail
Mardina Abdullah

Member of Editor

Mohammad Rashed Iqbal Faruque
Khairul Nizam Abdul Maulud
Mohd Shahrul Mohd Nadzir
Mandeep Singh Jit Singh
Zamri Zainal Abidin
Sabirin Abdullah
Teh Wai Leong
Md Firoz Khan
Tariqul Islam
Siti Aminah Bahari
Noridawaty Mat Daud

Papers

Astrophysics and Astronomy

Atmospheric and Magnetospheric Sciences

IOP Conf. Series: Journal of Physics: Conf. Series **852** (2017) 011001 doi:10.1088/1742-6596/852/1/011001

Geoscience and Remote Sensing

Satellite and Communication Technology

Interdisciplinary Space Science

IOP Conf. Series: Journal of Physics: Conf. Series **852** (2017) 011001 doi:10.1088/1742-6596/852/1/011001

IconSpace2017
IconSpace2017
2017 International
Conference on

IOP Conf. Series: Journal of Physics: Conf. Series **852** (2017) 011001 doi:10.1088/1742-6596/852/1/011001

IOP Conf. Series: Journal of Physics: Conf. Series **852** (2017) 011001 doi:10.1088/1742-6596/852/1/011001

IOP Conf. Series: Journal of Physics: Conf. Series **852** (2017) 011002 doi:10.1088/1742-6596/852/1/011002

Peer review statement

All papers published in this volume of *Journal of Physics: Conference Series* have been peer reviewed through processes administered by the proceedings Editors. Reviews were conducted by expert referees to the professional and scientific standards expected of a proceedings journal published by IOP Publishing.

Content from this work may be used under the terms of the Creative Commons Attribution 3.0 licence. Any further distribution of this work must maintain attribution to the author(s) and the title of the work, journal citation and DOI.
Published under licence by IOP Publishing Ltd

IOP Conf. Series: Journal of Physics: Conf. Series **852** (2017) 012001 doi:10.1088/1742-6596/852/1/012001

Surface charging of a crater near lunar terminator

A K Anuar

Department of Communication Engineering, Faculty of Electrical and Electronic Engineering, Universiti Tun Hussein Onn Malaysia, 86400 Parit Raja, Johor, MALAYSIA

E-mail: abul@uthm.edu.my

Abstract. Past lunar missions have shown the presence of dust particles in the lunar exosphere. These particles originate from lunar surface and are due to the charging of lunar surface by the solar wind and solar UV flux. Near the lunar terminator region, the low conductivity of the surface and small scale variations in surface topology could cause the surface to charge to different surface potentials. This paper simulates the variation of surface potential for a crater located in the lunar terminator regions using Spacecraft Plasma Interaction Software (SPIS). SPIS employs particle in cell method to simulate the motion of solar wind particles and photoelectrons. Lunar crater has been found to create mini-wake which affects both electron and ion density and causes small scale potential differences. Simulation results show potential difference of 300 V between sunlit area and shadowed area which creates suitable condition for dust levitation to occur.

1. Introduction

The transport of dust particles over the lunar atmosphere has been observed by astronauts during the Apollo missions and is believed to be the cause of *horizon glow* [1]. The first evidence of dust transport across the lunar surface was gathered by the Apollo 17's Lunar Ejecta and Meteorite (LEAM) experiment suggesting both vertical and horizontal dust transport particularly around the terminator region [2]. These lifted dust particles can be a major environmental hazard to any space activities on lunar surface. Major effects include reduced equipment lifetime due to adhesion and abrasion, reduction in optical visibility due to dust levitation, astronaut's health hazards and contamination of solar panels [3].

The transport of dust particle is believed to be due to the charging of the lunar surface by the ambient solar wind plasma and other charging process such as photoemission which create repelling electrostatic force between the dust particle and the surface [4]. It is therefore important to understand the charging process of the surface, in particular near the area where increase dust particle flux has been observed. Recently, a new particle in cell code has been developed with the capability of simulating surface charging and dusty plasma in the lunar environment [5, 6]. This paper presents simulation of the lunar surface charging over the lunar terminator region using the code.

2. Lunar surface charging

Surface interactions with solar wind plasma and UV radiation causes the lunar surface to become electrically charged with the process has been investigated since the early years of Apollo missions. In general, the charging of the lunar surface begins with the collection of incident solar wind particles

Content from this work may be used under the terms of the Creative Commons Attribution 3.0 licence. Any further distribution of this work must maintain attribution to the author(s) and the title of the work, journal citation and DOI.
Published under licence by IOP Publishing Ltd

and emission of secondary electrons and photoelectrons which will charge the surface to a floating potential. The current balance equation for the charging process is given by

$$I_e = I_i + I_{se} + I_{ph} \tag{1}$$

where I_e, I_i, I_{se}, I_{ph} are electron, ion, secondary electron and photoemission current respectively [4].

The resulting potential depends on which of these 4 main sources is dominant and is not consistent over the whole lunar surface because of the low surface conductivity with small scale variations have been observed and reported in many previous works (see for example [7]). On a large spatial scale, the sunlit side is charged to positive potential due to photoelectrons while the night side charges to negative potential as ions and solar UV flux are being screened by the lunar body.

In the presence of solar UV flux, incidence photons excite electrons on the surface, emitting them as low energy ($E_{ph} \sim 2\ eV$) photoelectrons. This process is highly dominant on the dayside of the lunar surface with the photocurrent exceeds the ambient electron current, thus charging the surface to positive potential. On the night side however, the absence of solar UV photons and the screening of solar wind protons mean the main charging current is the high energy electrons which are the first to enter the cavity due to their higher thermal speed. These high energy electrons charge the surface to high negative potential in order to attract more solar wind ions into the cavity, as well as preventing more electrons from reaching the surface. Figure 1 shows current collection and surface potential of a flat surface in the lunar terminator region. The absence of photoemission due to solar UV flux charges the surface to negative potential, as opposed to positive potential often observed in daylight region.

2.1. *Lunar terminator region*

Halfway between the sunlit and the night side region is the terminator region, where lunar surface is exposed to near parallel solar wind flow with very little or no solar UV flux. At this region, the main charging current comes from the ambient solar wind plasma but local surface topology such as crater and boulder could create localized regions of positive and negatively charged surfaces [7]. These localized regions are more prevalent in the terminator region because the almost parallel approach of solar wind and solar UV flux means a simple structure such as a boulder or a crater would have had a significantly shadowed region.

A crater at this region for example will have one side of the crater's rim obscured from both solar wind and solar UV flux, with the opposite with are accessible to both, creating a shadowed region. The shadowed region can be described as *mini wake* because it acts in a similar way as the lunar wake, i.e. it has depleted ion density and no photoemission [7]. As the lunar surface is full of craters and boulders, it is possible to have potential variations around these structures which would complicate the effort to understand the lunar exosphere. On a large scale however, lunar surface in the terminator region is charged to $\sim -70\ eV$ with the transition from positively charge dayside surface to negatively charge nightside surface happen inside the dayside region due to reduced solar wind and solar UV induced currents (ion and photoemission current) [7].

The potential developed on the lunar surface due to the charging processes creates a non-neutral layer of plasma sheath which screens either electrons or ions, creating a density gradient which in turn leads to electric field on the surface. On the night side, the electric field is directed towards the surface which accelerates ions onto the surface. The opposite happens on the dayside, where the field is directed outward and accelerates electrons towards the surface. The plasma sheath layer, assuming a Maxwellian velocity distribution for electrons can be described by

$$\Phi(z) = -\Phi_0\, exp\left(\frac{z}{\lambda_D}\right) \tag{2}$$

where Φ_0 is the potential bias of the surface, λ_D is the plasma Debye length $\left(\lambda_D = \sqrt{\frac{\epsilon_0 k_B T}{n_0 e^2}}\right)$ and z is the distance normal from the surface.

3. Simulation parameters

Spacecraft Plasma Interaction Software (SPIS) which employs particle-in-cell (PIC) method [9] has been upgraded to allow lunar surface charging to be investigated [5]. In addition to simulating the motion of solar wind particle, the code also allows simulation of photoemission originating of the surface which has been found to be one major charging current especially in the dayside region. Photoemission and photoelectrons are modelled using a similar model to the one proposed by [10]. The 3-dimensional and unstructured mesh employed by SPIS allows close representation of the lunar surface topology which is suited to this study.

SPIS is employed to model the lunar charging process using plasma parameters given by [8], where $n_e \sim n_i = 10^7\ \mathrm{m}^{-3}$ and $T_e \sim T_i = 10$ eV with Maxwellian velocity distribution and ion drift velocity of $v_{sw} \sim 400\ \mathrm{km \cdot s^{-1}}$. Simulation is first performed for a flat lunar surface over an area $6\lambda_D \times 6\lambda_D$ where $\lambda_D = 7.5$ m is the solar wind plasma Debye length. At terminator, it has been assumed that there is no photoemission from the surface due to very low exposure to solar UV. Figure 1 shows the current collection for ions and electrons and the resulting surface potential. Surface is found to have charged to around ~ -60 V after approximately 2 ms.

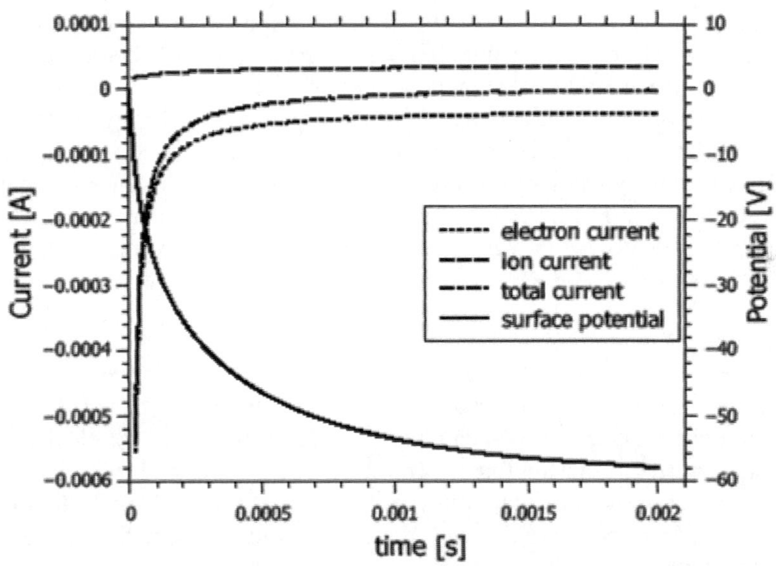

Figure 1. Current collection and surface potential for a surface near the terminator region. No photoemission in this region charges the surface to negative potential.

4. Surface charging near lunar terminator

Surface topography plays an important role in charging process and could produce localized potential difference [7]. Structures such as crater and boulder could disturb the natural solar wind flow and solar UV flux creating shadowed region. This shadowed region is charged to different potential than the

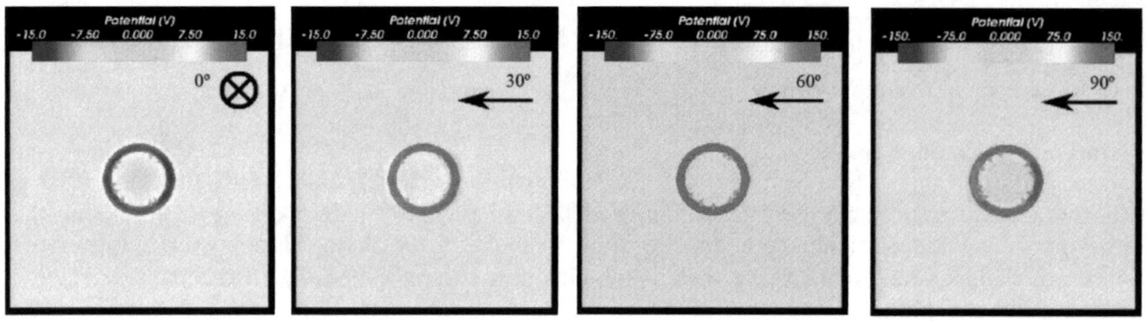

(a) 0° (b) 30° (c) 60° (d) 90°

Figure 2. Surface potential around a 1-m depth, 5-m diameter crater for (a) 0°, (b) 30°, (c) 60° and (d) 90° solar wind incidence angle.

illuminated region where the effects is more prevalent near the terminator region due to the low sun angle. In the simulation, charging process are observed for 4 solar wind and solar UV illumination angles which are 0°, 30°, 60° and 90°.

Figure 2 shows the potential of a 1-m deep, 5 meter in diameter crater for the 4 sun incidence angles. In Figure 2(a), the whole surface of the crater is exposed to the solar wind and solar flux. Potential difference is observed between the crater rim and the rest of the surface. In Figure 2 (b), opposite edges of the crater are charged to opposite polarity. The edge in blue is the ones in shadow which is charged to negative potential because solar wind photons are being obstructed by the crater's rim. The side facing the solar wind is charged to $\sim 9\,V$ while the one in shadowed is charged to -3 V. In Figure 2(c), the low solar angle has reduces the flux to the shadowed side even further with the surface is charged to ~ -70 V. In Figure 2(d), surface facing the solar wind is charged to positive potential while the one in shadow is charged to negative potential. A potential close to ~ -100 V can be seen on the shadowed side.

4.1 Electron and ion densities

Figure 3 shows the densities and potential across a crater located at 90° from the sub solar point. In Figure 3(a), the electron density near the crater is reduced to almost 3 order of magnitude smaller due to the strong electric field created by the negatively charged surface. Further electron depletion can be observed at the shadowed region inside the crater where density dropped to 10^2 m^{-3} which is due to the strong potential barrier. In comparison, ion density in the area outside the crater is almost constant over the surface with its density approximately the same as its ambient density ($n_i \sim 10^7$ m^{-3}), as shown in Figure 3(b). However, near the shadowed region inside the crater, the density is reduced to almost zero as most of the subsonic ions are completely obscured from the region. This causes the surface to charge to high negative potential as there is no ions to help lowering the magnitude of the surface potential. On the other hand, surface facing the shadowed region is charged to positive potential as ions are being collected by the surface. The surface and plasma potential are shown in Figure 3(c).

IOP Conf. Series: Journal of Physics: Conf. Series **852** (2017) 012001 doi:10.1088/1742-6596/852/1/012001

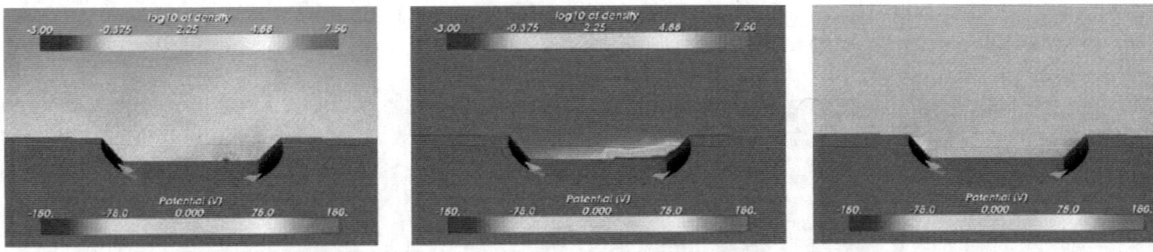

(a) electron density (b) ion density (c) surface and plasma potential

Figure 3. Plasma density and potential near crater at 90°.

4.2 Energetic solar wind plasma

Solar wind density and temperature are two parameters that could result in different potential level than the one observed in the preceding simulations. These two parameters vary over time and are highly dependent on solar activities. A simulation is performed for the case of crater near terminator ($\theta = 90°$) in a more energetic solar wind plasma. The average temperature of the solar wind is doubled from 10 eV to 20 eV and the remaining parameters of the simulation are retained.

Figure 4(a) and (b) show the surface and plasma potential for the crater after approximately 50 ms of simulation time. In Figure 4(a), the area outside the crater is charged to ~ -90 V while the base of the crater is charged to ~ -100 V. Having more energetic populations results in higher negative surface potential and is largely due to the increase in electron thermal energy. As for ions, since their motion are largely dominated by the subsonic flow, they would contribute approximately the same charging current to the surface under these conditions.

In Figure 4(a), the area outside the crater reaches its equilibrium potential after approximately 10 ms of simulation time whilst the one in shadow has yet to reach its equilibrium potential after 50 ms. The slow charging process of the shadowed surface is due to the lack of thermal ions reaching the surface. It is predicted that the surface will continue to charge to higher negative potential in order to attract more ion particles to the surface. It is also believed that an increase in solar wind speed would only result in a more negative patch since ions will have larger horizontal velocity component, and thus are more likely to be shielded by the crater's edge.

In addition to the high negative potential found on the shadowed crater's rim, a small patch of highly negative surface can be seen appearing near the top of the positively charged crater's rim. The formation of this patch is believed to be the result of screening of ions by the positively charged sun-facing crater's rim. The positive barrier shown in Figure 4(b) causes incoming ions to be deflected over the area which is indicated by the slightly high ion density near the patch (see Figure 4(c)). This patch also repels incoming electrons creating small region of reduced electron density as illustrated in Figure 4(d).

4.3 Bi-Maxwellian solar wind

The solar wind near lunar surface is more likely to contain both the fast and slow solar wind components. A simulation is performed for the case of crater for this type of plasma condition where the slow solar wind component is made of plasma with $n_s = 10^6 \mathrm{m}^{-3}$ and $T_s = 10$ eV whilst the fast solar wind is set to $n_f = 5 \times 10^6\ \mathrm{m}^{-3}$ and $T_f = 20$ eV. The streaming speed for both slow and fast solar wind are set to $400\ \mathrm{km \cdot s^{-1}}$ and $700\ \mathrm{km \cdot s^{-1}}$ respectively. Figure 5 shows the surface potential obtained from the simulation. From the figure, it is obvious that a similar pattern of surface potential can be observed near the lunar crater where the side in shadow is charged to negative potential whilst the one facing the solar wind is charged to positive potential. The simulation indicates an increase in the magnitude of the surface potential on both sides of the crater (shadowed and sunlit) but within the range anticipated by [8]. The simulation shows that there is around $200\ V$ difference between the side in shadow and the one facing the solar wind which could create strong electric field. The changes in surface potential also confirm the time varying nature of lunar surface potential with regards to the change in solar wind plasma properties.

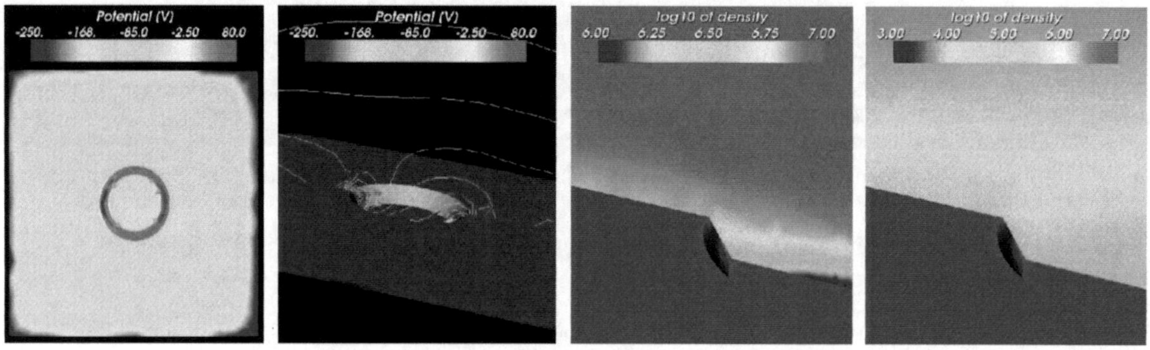

(a) (b) (c) (d)

Figure 4. Potential and species densities of a crater for TeV = 20 eV for (a) surface potential, (b) plasma potential, (c) ion density and (d) electron density.

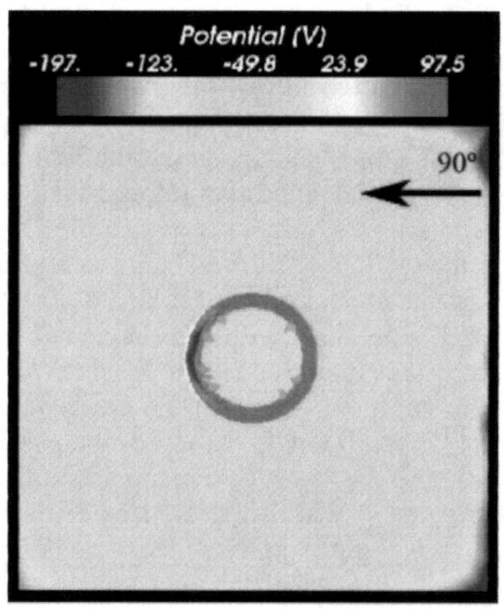

Figure 5. Surface potential near a crater for Bi-Maxmellian solar wind distribution.

5. Conclusion

Results obtained show the capability of SPIS in simulating the lunar surface charging process. Near the lunar terminator, the parallel approach of the solar wind and solar UV causes the surface to be negatively charged due to lack of photoemission current. Surface topology such as crater could charge the surface to different charging potentials with the solar wind facing surface is charged to positive potential while the shadowed surface is charged to negative potential. Potential difference of about 200 V can be observed and could increase in the presence of more energetic and bi Maxwellian solar wind plasma. This high negative potential has been shown to be able to levitate dust particles which could create a dusty lunar exosphere.

References

[1] Criswell D R 1972 Horizon-glow and motion of lunar dust *Abstracts of the Lunar and Planetary Science Conference* **3**163

[2] Berg O, Wolf H and Rhee 1976 j Lunar soil movement registered by the Apollo 17 cosmic dust experiment. *Interplanetary dust and zodiacal light* 233

[3] Christofoersen R, Lindsay J F, NobleS K *et al.* 2009 Dust Effects on Spacesuit Systems: Insight from Apollo Spacesuits, NASA/TP-2009-214786, Johnson Space Center, Houston, Texas

[4] Manka R H 1973 Plasma and Potential at the Lunar Surface. In R. J. L. Grard, editor, Photon and Particle Interactions with Surfaces in Space *Astrophysics and Space Science Library* **37** 347

[5] Anuar A K 2013 Honary F, Hapgood M and Roussel J -F 2013 Three dimensional simulation of dust charging and dusty plasma using SPIS *J. Geophys. Res.: Space Phys.* **118(**10) 6723

[6] Hess S L G and Sarrailh P, Mat_eo-V_elez J -C *et al.* 2015 New SPIS capabilities to simulate dust electrostatic charging, transport, and contamination of lunar probe *IEEE Trans. Plasma Sci.* **43**(9) 2799

[7] Farrell W M, Stubbs T J, Halekas J S *et al.* 2008 Loss of solar wind plasma neutrality and affect on surface potentials near the lunar terminator and shadowed polar regions *Geophys. Res. Lett.* **35**(5) L05105

[8] Colwell J E, Batiste S, Hor_anyi M, Robertson S and Sture s 2007 Lunar surface: Dust dynamics and regolith mechanics *Rev. Geophys*. **45**(2) RG2006

[9] Roussel J -F, Rogier F, Dufour G *et al* 2008 SPIS open-source code: Methods, capabilities, achievements, and prospects. *IEEE Trans. Plasma Sci.* **36**(5) 2360

[10] Poppe A and Horanyi M 2010 Simulations of the photoelectron sheath and dust levitation on the lunar surface *J. Geophys. Res*. **115** A08106

Limiting magnitudes and night sky brightness at the Langkawi National Observatory based on observations of standard stars

C -C Ngeow and S -C Luo

Graduate Institute of Astronomy, National Central University, Jhongli 32001, Taiwan

E-mail: cngeow@astro.ncu.edu.tw

Abstract. We present the estimation of limiting magnitudes and night sky brightness at the Langkawi National Observatory with observations of selected standard stars. We first derived the transformation equations using the standard stars in our CCD images, followed by photometric calibration of the 2MASS point sources located in these images. Based on the calibrated magnitudes, we estimated a limiting magnitude of V=16.6±0.1 mag can be reached with a 60 second exposure time. For the night sky brightness, our measurements give V_{sky}=19.0±0.1 mag arc-second^{-2} at the zenith direction. Our results can be used for planning the night time observations at the Langkawi National Observatory.

1. Introduction

The Langkawi National Observatory (hereafter LNO) is one of the research facility under the administration of ANGKASA (the Malaysian National Space Agency). Being located near the equator, LNO has an advantage to observe both of the northern and southern sky. LNO hosted several small-aperture telescopes, among them the 20-inch Ritchey-Chretien (RC) reflector telescope is the main telescope used for night time observations. This telescope has a focal length of 4115mm and hence a focal ratio of f/8.1, housing inside a 5m ObservaDOME. This telescope uses a Paramount-ME German equatorial mount for pointing and tracking celestial objects. A SBIG 1001E CCD Camera is equipped on the telescope. This CCD has 1024×1024 pixels, each pixels has a size of$24\mu m \times 24\mu m$, which produces a field-of-view of $20.1' \times 20.1'$ with a pixel scale of 1.2" per pixel. According to the vendor, this CCD can reach to quantum efficiency (Q.E.) of ~75% at 5500Å. The CCD has a gain of 2.2e^-/ADU and a typical readout noise of 15e^- RMS. Data taken from the LNO's telescope and CCD has been used in the study of, for example, the transit of a exoplanet HAT-P-13 [1]. In addition to CCD imaging, this 20-inch RC telescope has also been used in spectroscopic observations [2, 3, 4]. Recently, LNO has acquired an observatory code as O43 from the Minor Planet Center [5].

Both of the limiting magnitudes and night sky brightness are two important metrics to evaluate the observing capability of an observatory, as well as for the planning of night time imaging observations. For example, a celestial object with an apparent magnitude fainter than the limiting magnitude is impossible to be observed. A limiting magnitude of V=18.6±1.0 mag, inferred from the night sky brightness, has been measured at LNO in the past [6]. However, this measurement was done using a different combination of the telescope and detector – the Meade LX200GPS 8-inch Schmidt-Cassegrain telescope and a SBIG STV camera. Therefore, it is important to measure the limiting magnitudes and night sky brightness at LNO directly from the 20-inch RC telescope and the SBIG

Content from this work may be used under the terms of the Creative Commons Attribution 3.0 licence. Any further distribution of this work must maintain attribution to the author(s) and the title of the work, journal citation and DOI.
Published under licence by IOP Publishing Ltd

IOP Conf. Series: Journal of Physics: Conf. Series **852** (2017) 012002 doi:10.1088/1742-6596/852/1/012002

1001E CCD mentioned above. In this work, we present such measurements based on the observations of selected standard stars at the LNO. The observations were described in Section 2, followed by Section 3 on data reduction and analysis. In Section 4 we present our estimation of the limiting magnitudes and night sky brightness at LNO. A conclusion of this work is given in Section 5.

2. Observations

Our observation of standard stars at LNO was carried out on the night of 01 April, 2016. The night was clear and appeared to be photometric. The third quarter Moon was illuminated by ~45%, and rose around 01:44am local time. A number of standard stars were selected from [7], and they were observed at various air masses with standard broad-band BVRI filters. Our selected standard stars and the corresponding standard stars fields are:

- GD 108: GD 108, GD 108A, GD 108B, GD 108C, GD 108D
- SA104: SA104 350, SA104 364, SA104 456, SA104 457, SA104 460, SA104 461, SA104 470
- PG0918: PG0918+029, PG0918+029A, PG0918+029B, PG0918+029C, PG0918+029D
- PG1323: PG1323-086, PG1323-086A, PG1323-086B, PG1323-086C, PG1323-086D

Table 1 gives the list of observed standard stars fields and the observing log. In total, 48 images were acquired in BVRI filters for the selected standard stars in four different standard star fields.

Table 1. Observing Log for Standard Stars Observations at LNO on 01 April 2016.

Starting UT[a]	Airmass	Filter and Exposure time [b]	Standard Stars Fields
15:19	1.06	BVI: 30s; R: 15s	PG0918
15:30	1.14	BVI: 30s; R: 15s	SA104
15:39	1.06	B: 60s; VRI: 30s	GD108
15:51	1.26	B: 60s; VRI: 30s	PG1323
17:40	1.47	BVI: 30s; R: 15s	GD108
17:46	1.51	BVI: 30s; R: 15s	GD108
17:51	1.54	BVI: 30s; R: 15s	GD108
18:09	2.17	BVI: 60s; R: 30s	PG0918
18:30	1.05	B: 120s; V: 90s; RI: 30s	PG1323
18:42	1.07	BVRI: 30s	SA104
18:48	1.08	BVRI: 60s	SA104
18:58	1.06	BVRI: 60s	PG1323

[a] UT = Universal Time.

[b] Exposure time is in second. We reduced the exposure time in R-band because some standard stars were saturated if using the same exposure time as in V-band. Similarly, we increase the exposure time for some B-band images to increase the signal-to-noise ratio for the standard stars (as the Q.E. in B-band is generally lower than other longer wavelength bands).

The operating temperature of the CCD was set to -14°C during our observations because of the high and varying ambient temperature around 30°C. Therefore we employed the AUTODARK feature during the images acquisition. This feature will take a dark image with the same exposure time as the science image and subtracted from it, hence the dark current and the bias signal was immediately removed from the science image. For flat-field images, we took a series of dome flat images in BVRI filters in day-time. A set of dark and bias frames was also taken at the same time to remove the bias and dark signals in the dome flat images.

3. Data Reduction and Analysis

The reduction of the CCD images was done within the IRAF (Image Reduction and Analysis Facility, available at http://iraf.noao.edu/) environment. We combined the dome flat images to create master flat images in each BVRI filters, after removed the bias and dark signals in each of the dome flat images, using the flat combine task in IRAF. Since the bias and dark signals have already been subtracted in our science images (see Section 2), we only applied the master flat images with ccdproc task to remove flat field signatures in the science images. We also performed the astrometric refinement on the reduced images using the solve-field code available from the astrometry.net package [8,9], with a tweak order of 2. Stars in the images will then have correct equatorial coordinates (Right Ascension α and Declination δ, in epoch of J2000) after the astrometric refinements.

3.1. Deriving the Transformation Equations

To determine the limiting magnitudes and night sky brightness at LNO, it is necessary to derive the conversion, or transformation, between the instrumental magnitudes and standard (or calibrated) magnitudes. Standard stars in our reduced images can be used for such purpose, as the standard magnitudes of them are available in [7]. We first extracted the instrumental magnitudes of the standard stars listed in Table 1 using the phot task in IRAF. Radius of the aperture was set to 5.93 pixels (or 7", see [7]). A larger annulus was used to estimate the background counts for each standard stars and subtracted from the aperture counts.

In preparing for solving the transformation equations, we performed the following steps. These include: (a) created a text file linking the reduced images and the standard stars found in the images; (b) executed the mknobsfile task on the text file from previous step and the output magnitude files from the phot task to prepare the necessary observation files (these observation files contain the information of BVRI instrumental magnitudes and air masses for each standard stars in the reduced images); (c) edited the fnlandolt.dat and tlandolt.dat files copied from the IRAF installation and executed the mkconfig task to create the configuration file; and (d) prepare a data file containing the standard magnitudes and colors taken from [7] for our standard stars. Once all the necessary input files were ready, we executed the fitparams task (with 2.5σ iterative outliers rejection) in IRAF to solve the following transformation equations in each BVRI filters: $m_{instr}=m_{std}+ZP+\zeta X+\eta C$, where m_{instr} & X are instrumental magnitudes and air masses based on observations, m_{std} & C are standard magnitudes and colors adopted from [7], and ZP, ζ & η represent the CCD zero-point, the air mass-term & the color-term, respectively, that need to be solved. Note that the ZP is relative to 25 mag (the initial setting given in IRAF's apphot package). Our solved transformation equations are presented below:

$$b = B + 4.512(\pm 0.030) + 0.476(\pm 0.022)X - 0.180(\pm 0.015)(B - V), \sigma = 0.043, \quad (1)$$

$$v = V + 4.286(\pm 0.018) + 0.363(\pm 0.013)X + 0.035(\pm 0.010)(B - V), \sigma = 0.030, \quad (2)$$

$$r = R + 4.046(\pm 0.010) + 0.292(\pm 0.006)X + 0.079(\pm 0.010)(V - R), \sigma = 0.016, \quad (3)$$

$$i = I + 4.555(\pm 0.010) + 0.232(\pm 0.007)X - 0.024(\pm 0.006)(V - I), \sigma = 0.017. \quad (4)$$

The lower case letters, *bvri*, represent the instrumental magnitudes, while the upper case letters, BVRI, are standard magnitudes (and colors) adopted from [7]. σ is the standard deviation from the fits. Instrumental magnitudes of the standard stars can be re-calibrated with equation (1) - (4) to standard magnitudes using the invertfit task in the IRAF's photcal package. Figure 1 compares the V-band magnitudes and colors for the standard stars after re-calibration and the standard values given in [7]. Good agreements can be seen from this Figure that illustrates the validity of our derived transformation equations.

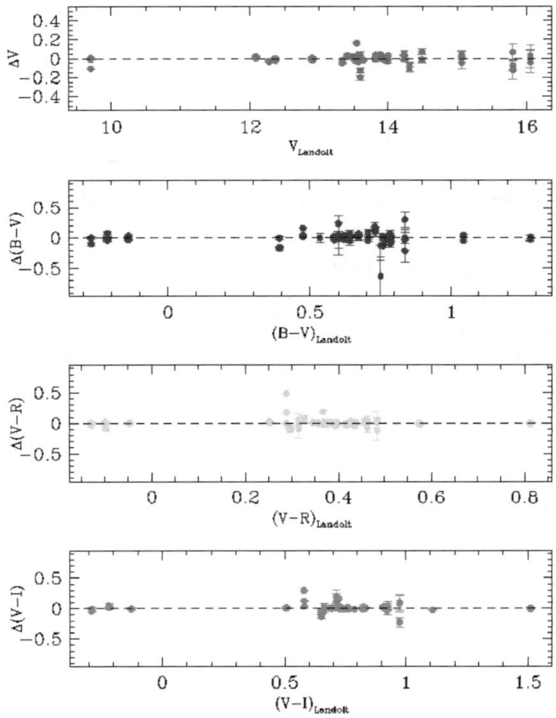

Figure 1. To evaluate the performance of our solved transformation equations, we compare the V-band magnitudes, the (B-V), (V-R) and (V-I) colors between our re-calibrated photometry, using equation (1) - (4), to the standard values taken from [7] as shown in the x-axes. The Δ in y-axes are the difference between them for the standard stars observed in this work. The dashed lines indicate Δ=0 and *not* the fit to the data. Results from this figure demonstrate that our solved transformation equations are reliable for a range of V-band magnitudes and colors.

3.2. Photometry for Stars in Standard Stars Fields

Stars located in our standard stars images, not just the standard stars, will be used to evaluate the limiting magnitudes at LNO. We first extracted the known stars or point sources that are expected to fall within our images from the 2MASS (The Two Micron All Sky Survey, [10]) All-Sky Point Source Catalog ([11]). The sky coordinates (α and δ in J2000) for these 2MASS sources were then converted to the pixel coordinates in (X,Y), using the wcs-rd2xy code from the astrometry.net package ([8,9]), and stored as coordinate files (.coo file). We also discarded 2MASS sources that located within 10 pixels from the edges of the images. With these coordinate files, aperture photometries were calculated with the phot task at the converted pixel locations of these 2MASS sources (a technique known as forced photometry), using a series of aperture sizes scaled with the seeing (or FWHM = Full-width at half maximum) of a given image. The later part is needed to calculate the aperture corrections using the mkapfile task. The calculated aperture corrections for each image was then applied to the aperture photometry based on an aperture radius of $3 \times FWHM$ of the images, and combined the BVRI photometry to a magnitude file with the mknobsfile task. Finally, these aperture-corrected instrumental magnitudes were calibrated to standard magnitudes using the results from previous sub-section by executing the invertfit task.

4. Estimation of Limiting Magnitudes and Night Sky Brightness at LNO

4.1. Estimating the Limiting Magnitudes

Limiting magnitudes from CCD images can be estimated using several methods (for examples, see [12,13]). In this work, we adopted the 5σ detection method demonstrated in [13]. This method is straight forward by plotting the magnitude errors versus the magnitudes, then the limiting magnitude at the 5σ detection level is the magnitudes where the magnitude error of 0.2 mag intersects with the data points in this plot. The BVRI-band calibrated magnitudes and their magnitude errors for the 2MASS sources in the standard stars images, based on the results from Section 3.2, were plotted in Figure 2(a), 2(b), 2(c) & 2(d). To find out the limiting magnitudes, we took an average of the calibrated magnitudes for those data points that have magnitude errors fall between 0.19 mag and 0.21 mag. Since images with 30 second and 60 second exposure time are common in all BVRI bands, we only determined the limiting magnitudes in these two sets of images. The estimated limiting magnitudes are presented in Table 2, where the errors represent the standard deviation of the means.

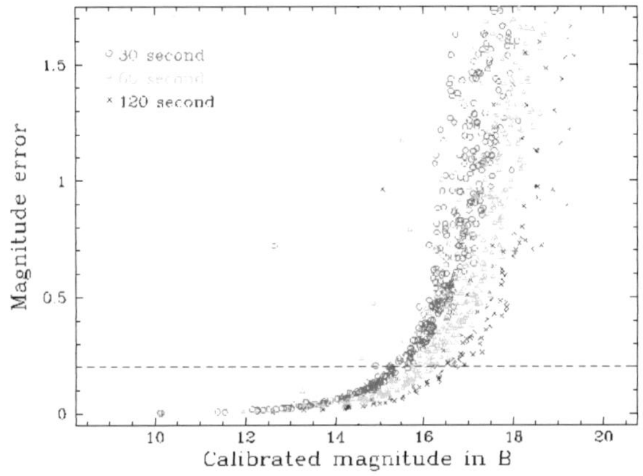

Figure 2(a). Plot of magnitude errors versus calibrated magnitudes in B-band for the 2MASS stars and point sources in our standard stars images. The horizontal dashed line indicates the magnitude error of 0.2 mag, corresponding to a signal-to-noise ratio of ~5.

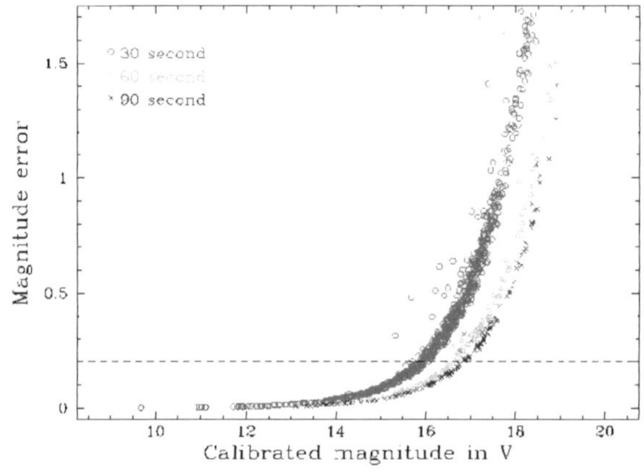

Figure 2(b). Plot of magnitude errors versus calibrated magnitudes in V-band for the 2MASS stars and point sources in our standard stars images. The horizontal dashed line indicates the magnitude error of 0.2 mag, corresponding to a signal-to-noise ratio of ~5.

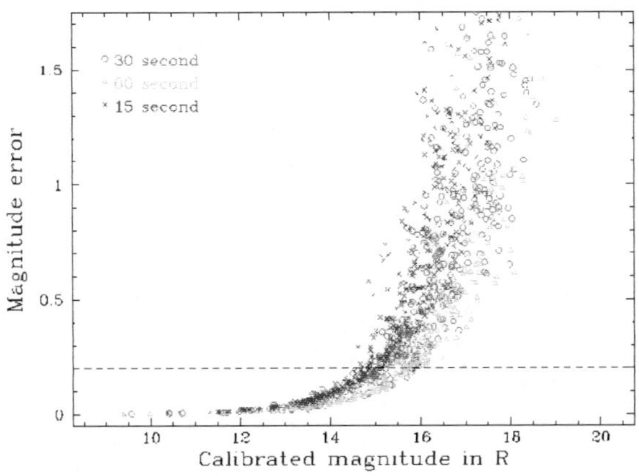

Figure 2(c). Plot of magnitude errors versus calibrated magnitudes in R-band for the 2MASS stars and point sources in our standard stars images. The horizontal dashed line indicates the magnitude error of 0.2 mag, corresponding to a signal-to-noise ratio of ~5.

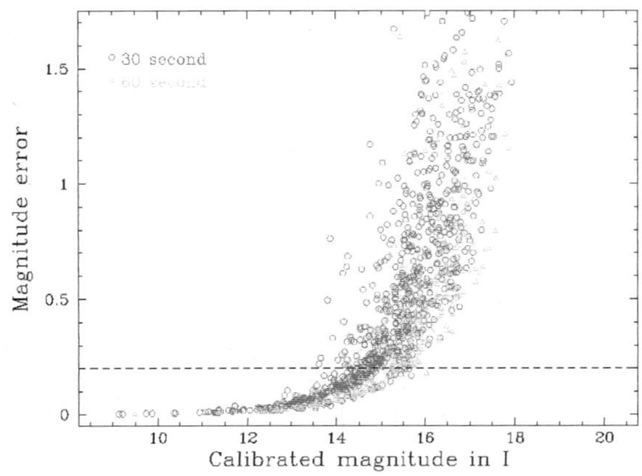

Figure 2(d). Plot of magnitude errors versus calibrated magnitudes in I-band for the 2MASS stars and point sources in our standard stars images. The horizontal dashed line indicates the magnitude error of 0.2 mag, corresponding to a signal-to-noise ratio of ~5.

Table 2. Estimated Limiting Magnitudes for the 5σ Detection at LNO.

Band	30 second exposure time [mag]	60 second exposure time [mag]
B	15.3±0.2	16.1±0.3
V	15.9±0.1	16.6±0.1
R	15.5±0.4	15.8±0.6
I	14.9±0.4	15.1±0.4

Our results revealed that the limiting magnitudes at LNO are ~15.4 mag and ~15.9 mag in exposure time of 30 and 60 second, respectively, regardless of the filters. Table 2 also shows that the best estimated limiting magnitude is in V-band, this is because the V-band data points in the magnitude errors versus magnitude plot do not exhibit large scatters as in other filters (see Figure 2). The larger scatters of data points seen in BRI bands are mainly due to the color terms, such as (B-V), (V-R) and (V-I), because magnitudes in these filters were calculated from the V-band magnitudes and colors. Therefore, uncertainties of the BRI-band magnitudes would be amplified and results a larger scatter toward fainter magnitudes. Because of this, we recommended using the V-band results (Table 2 and Figure 2) for the estimation of limiting magnitude when planning observations at LNO.

Figure 3 shows a small portion of the V-band image from one of our reduced standard stars images. On this Figure we also overlaid the locations of 2MASS sources with green circles, as well as labeled the calibrated V-band magnitudes on top of the circles. From this Figure we observed that sources brighter than ~17 mag are detectable, while sources with magnitudes between ~17 mag to ~18 mag are barely detected. The brightness of one source with V~20.6 mag in this Figure is almost same as the background noise. Hence, Figure 3 illustrates that our estimated limiting magnitudes are reliable.

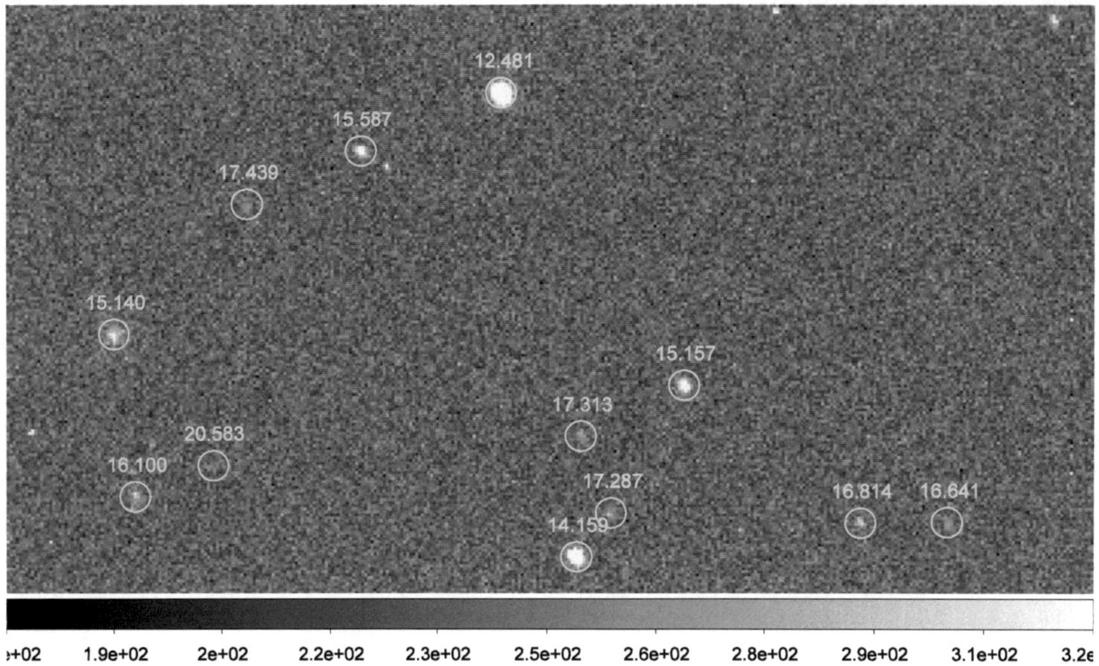

Figure 3. A small portion of one of the V-band image. Circles indicate the location of the 2MASS sources. The numbers above the circles are the calibrated V-band magnitudes for the given sources.

4.2. Estimating the Night Sky Brightness

Our standard stars images could also be used to estimate the night sky brightness at LNO. Since our images are relatively un-crowded, only a small fraction of the pixels will have high ADU counts due to illumination by stars light or cosmic rays. These high pixel values will be located at the (long) tail of the distribution (or histogram) for the count values of all pixels in an image, therefore the mode statistics (i.e. the peak of the distribution) is a good estimator of the night sky flux F_{sky} (expressed in ADU; on a single pixel) of the images. We employed the imstat task in IRAF to calculate the mode values or the F_{sky} for our standard star images. The instrumental night sky brightness is then expressed as (for examples, see [14,15]): $m_{sky}=Q-2.5\log(F_{sky}/A\ t_{exp})$, where Q=25.0 is the zero-point set in IRAF, t_{exp} is the exposure time (in second) of each images, and $A=1.44''^2$ is the area of a single pixel in the image. The unit of m_{sky} is mag arc-second^{-2}. We then calibrated these instrumental sky brightness to B_{sky}, V_{sky}, R_{sky} & I_{sky} using equation (1) - (4), assuming vanishing colors for the night sky.

Figure 4 shows the results of measured night sky brightness based on all of the standard stars images. The night sky brightness shows a clear trend that becoming brighter as air mass increases. Therefore, we fit a linear relation in the form of $M_{sky}=a+b(X-1)$ to the measurements shown in Figure 4, where M_{sky} represents the calibrated sky brightness in filter M. The measurements of night sky brightness are generally reported toward the zenith direction (for example, see [14] because usually it is the darkest part of the night sky. The intercepts of the linear relation, a, represent the night sky brightness at airmass of one (X=1), i.e. the zenith direction. Therefore, our final adopted night sky brightness at LNO is (measured at zenith): $B_{sky}=19.2\pm0.1$ mag arc-second^{-2}, $V_{sky}=19.0\pm0.1$ mag arc-second^{-2}, $R_{sky}=18.8\pm0.1$ mag arc-second^{-2} & $I_{sky}=18.1\pm0.1$ mag arc-second^{-2}. Our results in V- and R-band are consistent with the measurements made at the Weihai Observatory (WHO) of Shandong University ($V_{sky}^{WHO}\sim19.0$ mag~arc-second^{-2} & $R_{sky}^{WHO}\sim18.6$ mag arc-second^{-2} at the darkest night, [15].

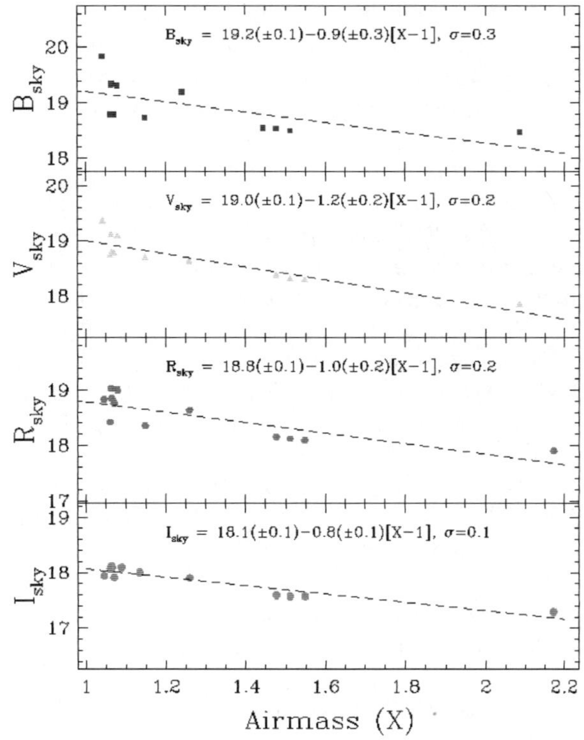

Figure 4. Based on our standard stars observations mentioned in Section 2, we measured the night sky brightness, in mag arc-second^{-2}, as a function of air mass in the BVRI bands. A trend is clearly seen from the data points such that the night sky brightness becomes brighter when air mass increases, therefore we fit a linear relation to the data points presented here. The dashed lines are fitted linear relation of the data points, with fitted results given at top of each panels. σ represents the dispersion of the linear fits. The night sky brightness in BVRI bands at zenith direction is represented by the intercepts of the fitted linear relations.

5. Conclusion

In this work, we measured the BVRI-band limiting magnitudes and night sky brightness at LNO. Our measurements were based on observations of selected standard stars fields taken on 01 April, 2016. This night represents typical observable nights at LNO, and hence our results can be used as guidance to plan the imaging observation at LNO with its 20-inch RC telescope. Our measured limiting magnitudes and night sky brightness were summarized in Table 2 and Figure 4, respectively. It is worth to point out that our results are not in disagreement with [6] – in fact both of the measurements on V-band night sky brightness are in good agreement. For the V-band limiting magnitudes, [6] inferred the limiting magnitude based on their sky brightness measurements using different telescope and instrument, while in our work we applied the 5σ detection method. Therefore, these different approaches could be the main reason behind a ~2 mag difference in the measured V-band limiting magnitudes.

Acknowledgments

This work was partly supported by the Taiwan's MOST grant 104-2112-M-008-012-MY3. We also thank the hospitality at LNO, especially the LNO staff Mohammad Redzuan Tahar and Karzaman Ahmad, during our stayed at the LNO. This publication makes use of data products from the Two Micron All Sky Survey, which is a joint project of the University of Massachusetts and the Infrared

Processing and Analysis Center/California Institute of Technology, funded by the National Aeronautics and Space Administration and the National Science Foundation. IRAF is distributed by the National Optical Astronomy Observatories, which are operated by the Association of Universities for Research in Astronomy, Inc., under cooperative agreement with the National Science Foundation.

References

[1] Szabo G M, Kiss L L, Benko J M *et al.* 2010 *Astronomy and Astrophysics* **523** A84
[2] Ahmad N and Zainuddin M Z 2009 *The Eighth Pacific Rim Conference on Stellar Astrophysics: A Tribute to Kam-Ching Leung* Astronomical Society of the Pacific Conference Series, vol 404, ed B Soonthornthum, S Komonjinda *et al.* (San Francisco: Astronomical Society of the Pacific) p 297
[3] Mohamad-Yob S-J, Gopir G K, Malasan H L and Anwar R 2009 *The Eighth Pacific Rim Conference on Stellar Astrophysics: A Tribute to Kam-Ching Leung* Astronomical Society of the Pacific Conference Series, vol 404, ed B Soonthornthum, S Komonjinda *et al.* (San Francisco: Astronomical Society of the Pacific) p 316
[4] Zainuddin M Z, Muhyidin M A, Ahmad N *et al.* 2011 *AIP Conference Proceedings* **1328** 56
[5] Loon C W, Zainuddin M Z, Ahmad N *et al.* 2015 *AIP Conference Proceedings* **1657** 050006
[6] Zainuddin M Z, Loon C W and Harun S 2010 *American Institute of Physics Conference Series* **1250** 51
[7] Landolt A U 2009 *The Astronomical Journal* **137** 4186
[8] Lang D, Hogg D W, Mierle K, Blanton M and Roweis S 2010 *The Astronomical Journal* **139** 1782
[9] Lang D, Hogg D W, Mierle K, Blanton M and Roweis S 2012 *Astrophysics Source Code Library* ascl:1208.001
[10] Skrutskie M F, Cutri R M, Stiening R *et al.* 2006 *The Astronomical Journal* **131** 1163
[11] Cutri R M, Skrutskie M F, van Dyk S *et al.* 2003 *VzieR On-line Data Catalog: II/246*
[12] Harris W E 1990 *Publications of the Astronomical Society of Pacific* **102** 949
[13] Gwyn S D J 2008 *Publications of the Astronomical Society of Pacific* **120** 212
[14] Duriscoe D M, Luginbuhl C B and Moore C A 2007 *Publications of the Astronomical Society of Pacific* **119** 192
[15] Guo D- F, Hu S- M, Chen X, Gao D- Y and Du J- J 2014 *Publications of the Astronomical Society of Pacific* **126** 496

IOP Conf. Series: Journal of Physics: Conf. Series **852** (2017) 012003 doi:10.1088/1742-6596/852/1/012003

Limb flares measurement from Langkawi National Observatory on 5th January 2016

F Kamarudin, M R Tahar and N R Saibaka

Langkawi National Observatory, National Space Agency, Empangan Bukit Malut, 07000 Langkawi

*E-mail : farahana@angkasa.gov.my

Abstract. Sun is the source of energy and has a lot of activity that will be influence the Earth. One of the solar activities is the limb flare or prominence. It release energy and expel outward the Sun surface. Langkawi National Observatory (LNO) had started photographic solar activity in Hα since 2008. LNO used the dedicated solar telescope to monitor and observed the solar activity daily. In 5th January 2016, LNO captured one limb flare at the south-west position of the Sun. The maximum height of the limb flare is 267,347 km and it increase gradually build up the velocity. Luckily, this eruption does not facing and directed towards Earth so there was no significant impact.

1. Introduction

The Sun is a giant ball of hot gasses held together by its immense gravity. It can occasionally release high energetic plasma from its magnetic fields in many forms like filaments, prominences, solar flares and coronal mass ejections (CME). Our Sun is filled by various activities and it occurs in the different layers of solar atmosphere like photosphere, chromosphere and corona. Some of the higher energy activities cause what we call space weather and it can affect and interrupt some technologies on Earth like communications, power grids and satellite navigations [1].

The chromospheric layer is one of the remarkable layers of our dynamics sun. Observers and researchers can see the developing of new active regions, plagues, prominences, flares and others features in this layer. Generally, there are two main spectral lines are used to observe the solar activity at chromosphere layer which is Hydrogen- Alpha (H-α) line at 3834 Å and Calcium-K line at 3868 Å [2]. For researchers, they detect prominences in the H-α band spectrum. A prominence is like a puff of material noticeable above the solar surface. They appear bright against the dark background of space in the solar limb due to the reduced emission loss thus it increase the brightness temperature in that line. But prominence is called as dark filament when it seen on the disk as it appears darker as compared to the more energetic surface of the chromosphere. [2][3].

A systematic observation of prominence has been carried out for almost 125 years from various parts of the world. Early observation of solar prominence is observed during the solar eclipse made by Muratori (1239) and Vassenius (1733) [4]. Beginning from 1860 the observations and study of prominences has been done with aided tools like photography and spectroscopy. In 1900's, Brocchino and Borocas had studied the relation prominence to the solar cycle [5].

Prominence will appear in variety of shapes and size. Sometimes it can persist almost few thousand kilometers to nearly radius of the Sun. It also can be seen by few hours, days or several month. Sometimes it can unexpectedly explode and reconstruct in same shape and position. Classification of prominences can be divided into 2 classes as below:-

Content from this work may be used under the terms of the Creative Commons Attribution 3.0 licence. Any further distribution of this work must maintain attribution to the author(s) and the title of the work, journal citation and DOI.
Published under licence by IOP Publishing Ltd

Table 1. Prominence classification

Class	Name	Description
1	Quiescent (long lived) prominence	a) Prominence or filament in or near active regions & in quiet regions b) Rising prominence
II	Active (flare associated, temporary) prominence	a) Limb flares b) Surges c) Sprays d) Loops

Usually the quiescent prominence will be form in active region, between two active regions or over polar coronal holes (called polar crown). The dynamics of prominence can be understood as a contribution of magnetic-force resistance to gravitational attraction of interior and solar surface [6]. This paper is organized as follow. In section 2, we will introduce the Langkawi National Observatory (LNO) Solar Observatory and our H-α solar observation procedures. The analysis of the solar limb detected is discussed in section 3. Meanwhile, in Section 4 we will discuss about our results. Finally, our work of this paper is summarized in the last section.

2. Langkawi National Observatory: H-alpha solar observation

In Malaysia, there are several H-α observation that is done by various parties. Most of these observations however are not continuous or conducted routinely. LNO is an exception as it is well equipped and staffed, can therefore carries out continuous H-α observation. LNO is situated north of Malaysia, on the Langkawi Island with latitude of 6.307147°N and longitude 99.781205°E [1]. The main focus of LNO is more on optical astronomy related researches and is one of the facilities under National Space Agency of Malaysia (ANGKASA). LNO is furnished with 3 solar observation systems to observe the sun in different wavelengths which are white light for monitoring the photospheric activities, H-α (6562.8Å) and Calcium K-line (3933.7Å) for monitoring the chromospheric activities. In this paper, we will be focusing on the output from the H-α system that recorded an active prominence or flare at the solar limb which happened on the 5th January 2016.

LNO uses digital images for all their solar observations. The imaging train consists of a Lunt 60mm f/8.3 dedicated solar telescope with an additional 50mm aperture double stack filter to increase the effective bandwidth down to <0.5 Å. The telescope's filter is an etalon design. Meanwhile, rear blocking filter used in tandem with this telescope is a Lunt's straight-thru B3400. With this blocking filter, the Sun's image is very bright and chromospheric activities can easily be seen despite the double stack setup. The telescope rides atop a Paramount ME tracking mount to follow the motion of the Sun throughout the day. Both etalon filters are easy to adjust to the preferred ultra-narrow H-α band of 6562.8Å. We used a SKYnyx 2-2 M CCD camera for recording the solar events in video format. The SKYnyx 2-2M is capable of recording lossless 8 & 12 bit monochrome images at 12 fps in full resolution of 2 megapixels. This capability makes it suitable for monitoring the sun activity in H-α due to the rapid changes of the solar chromosphere features like solar flares and active prominence. On our setup, the Sun disk diameter is around 1086 pixels. It will make 1 pixel equals to 1286 kilometer of the Sun.

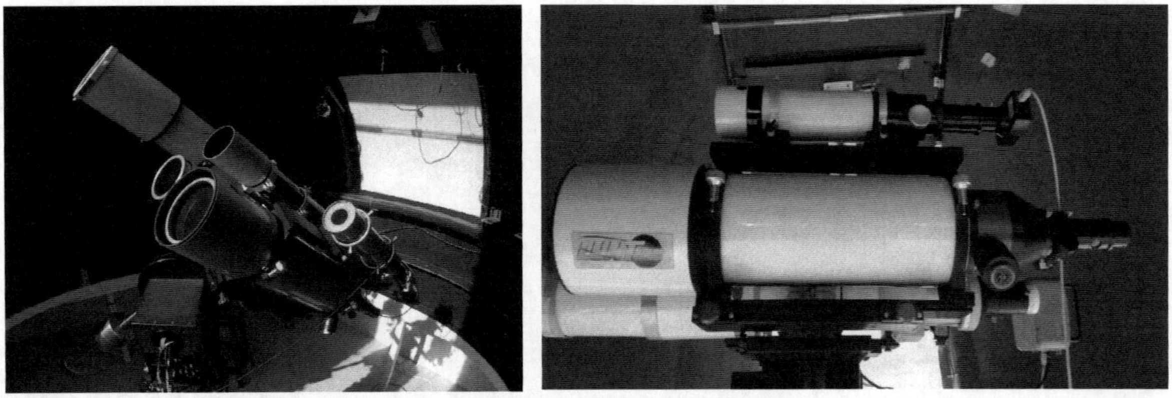

Figure 1. Instrumentation used for H-α imaging and data collection at LNO:- Double-stack solar telescope, SKYnyx 2-2M Video CCD Camera and Lunt Blocking Filter

H-α observations at LNO are performed daily whenever the weather permits (including holidays). Usually, we record H-α images between 9:30 AM until 4:00 PM, after the white light observation was completed. Observers will adjust the exposure time to fill 2 thirds of the CCD's well size to ensure it is well within the linearity range and corresponds to the atmospheric conditions at the time observation. The quality of atmospheric condition which involves seeing and sky condition is recorded along with every solar observation at LNO. They are rated ranging from 1 to 5, with 1 representing poor conditions and 5 for perfectly clear blue sky with very stable atmospheric turbulence [7]. Due to Malaysia's tropical weather, sometimes it is a bit difficult to get the best image even in good weather condition due to high humidity.

Firstly, observers need to ensure the H-α image is in focus. The sky condition is a contributing factor to determine the best focus where the spicules can be seen at the limb of the Sun in excellent sky conditions. Once good focus has been achieved, the observer will start to record the solar image at every 1 minute interval. Later, the data will be processed and viewed using Registax and observers will look for feature changes at the surface or limb of the Sun. The daily solar observing data from LNO are published on the LNO website and the additional data can be requested with an inquiry to LNO.

3. Limb flare analysis

Figure 2 shows the H-α image taken by LNO on 5th January 2016. During that day, a limb flare was observed at the south-west side of the Sun between 0235 UTC and 0423 UTC. This active prominence originates from AR2473 due to the same longitude position during this active region at the Sun surface. This sunspot or active region has been producing many eruptions ever since it appeared on the eastern solar disk on 21st December 2016.

IOP Conf. Series: Journal of Physics: Conf. Series **852** (2017) 012003 doi:10.1088/1742-6596/852/1/012003

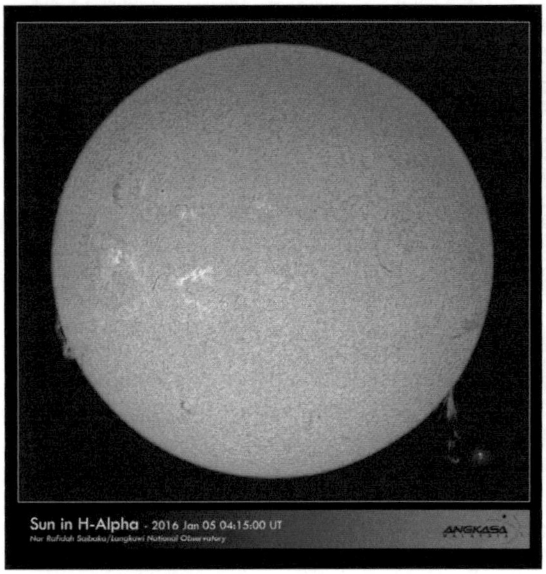

Figure 2. Limb Flare taken from LNO at 5th January 2016

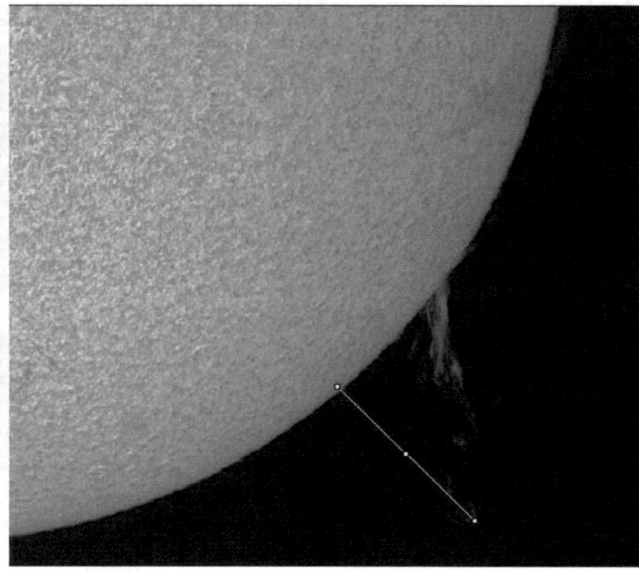

Figure 3. The maximum height identification and calculation

During the energy release process, the beginning height of this limb flare is 13,435 km and it reached the maximum height of 267,347 km as in Figure 3. It is calculated by multiplying the pixel along the height measurement of the burst with the image resolution (1286 km/pixel). It is due to unstable state of the magnetic fields at this active region.

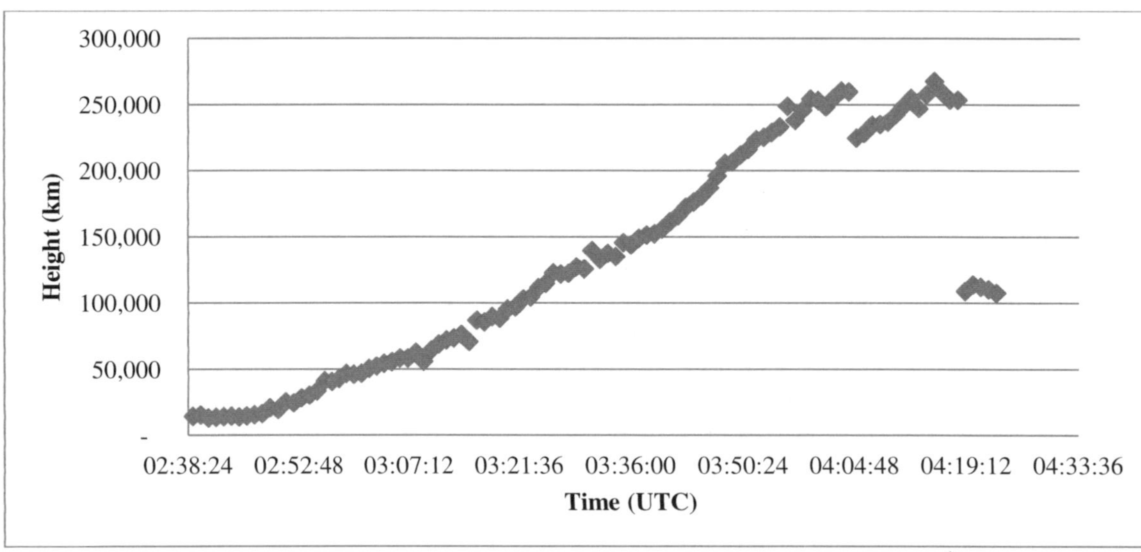

Figure 4. Limb flare height measurement in kilometre over time of observation on 5th January 2016

From the height measurement in the H-α wavelength as shown in Figure 4, the speed of the flare's height shows some slow gradual build-up. This is also reflected in the Figure 6. As the flare increased in height, the speed of height increment also increases slowly. Mostly this is due to the distance to the gravity pull that starts to weaken as it gets further away.

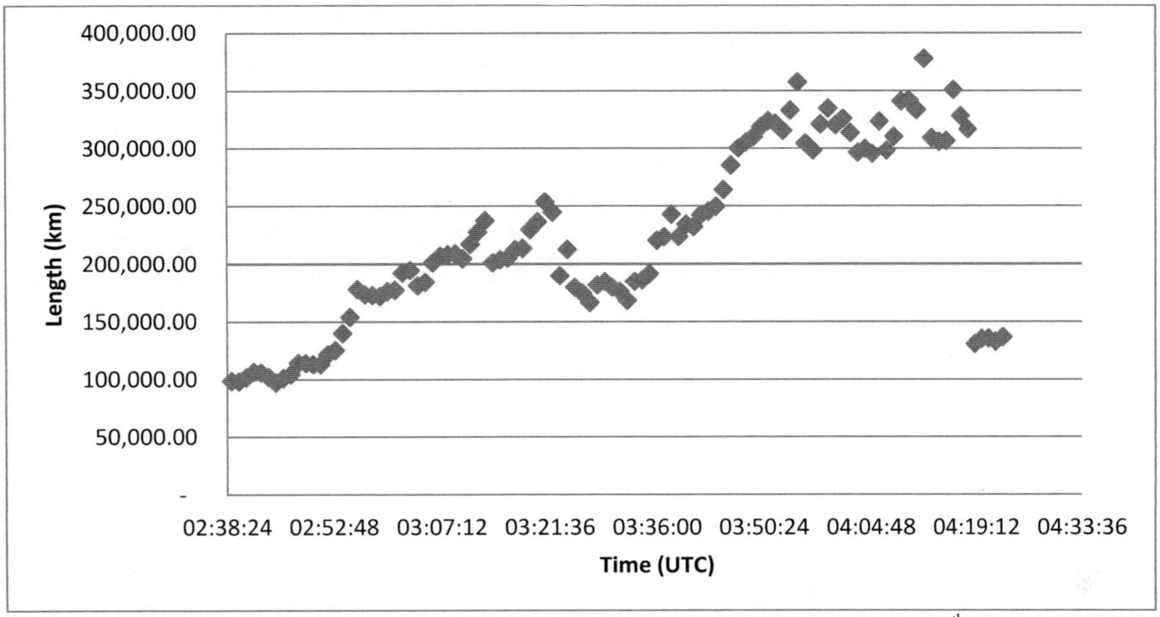

Figure 5. Limb flare length measurement in kilometre over time of observation on 5th January 2016

This beginning flare length is 74,781.79 km and it reached its maximum length of 377,636.93 km was shown in Figure 5. This shows that the flare detached from the Sun surface at the maximum point of its height and length.

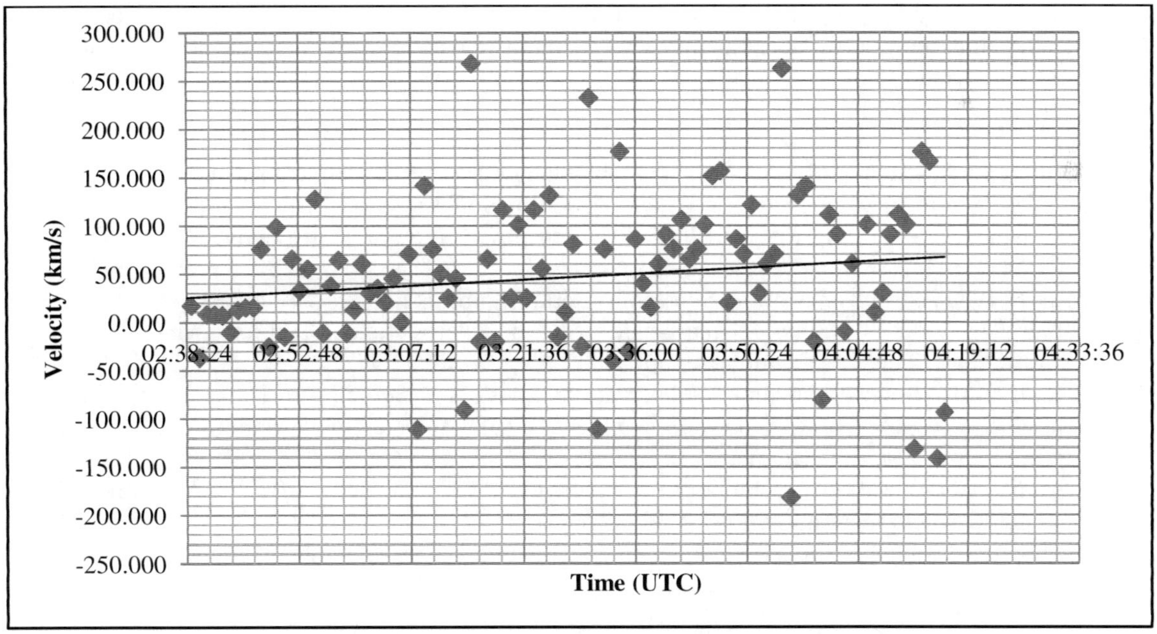

Figure 6. Shows the velocity (vertical) over time

Interestingly, the velocity of the flare along the length of the flare position decreases over time as it loses or starts to dissipate its energy. As evident in Figure 7, the overall pattern decreases over time but at the same time the fluctuations also increase as it get more unstable. This is the opposite of what is happening to the flare's vertical velocity which is increasing.

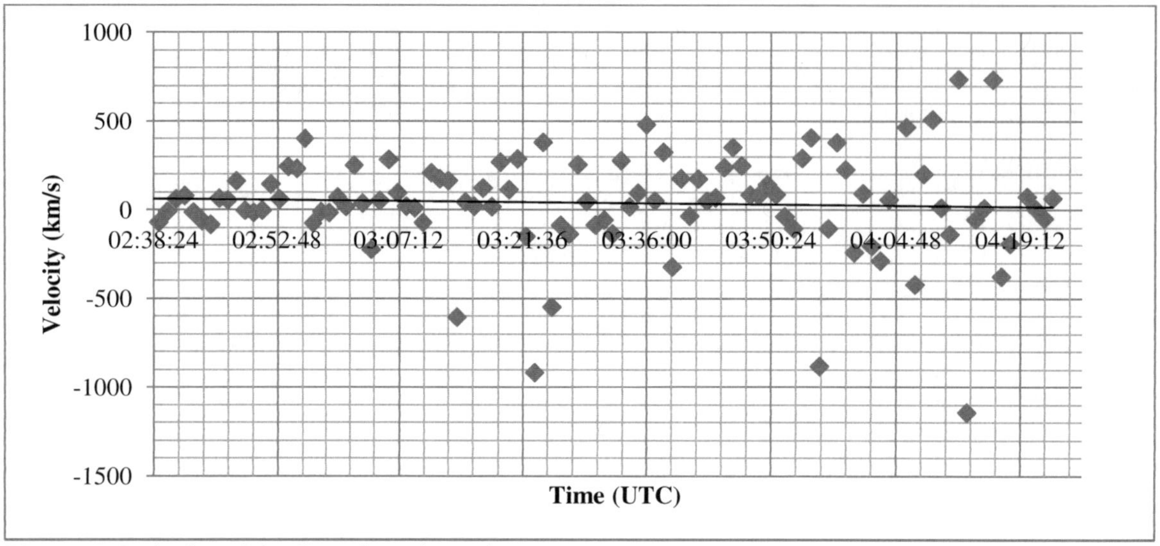

Figure 7. Velocity of the flare along its path.

4. Conclusion

From this paper, LNO is capable to carry out H-α observation of the Sun. This particular limb flare which is AR2473 was recorded on 5th January 2016. The maximum height of the flare is 267,347 km and the maximum length is 377,636.93 km. The speed of this flare eruption was slow gradually built up. This is due to gravity pull back to the Sun surface. As the flare erupts, the velocity is increased as it gets further from the Sun surface. Luckily, this eruption was not facing Earth and will not cause the space weather effect to Earth.

References

[1] Kamarudin F, Tahar M R, Saibaka N R and Padang L A L 2016 Relative sunspot number observed from 2013 to 2015 at Langkawi National Observatory *Adv. Sci. Lett.* **23**(2) 1285
[2] Bhatnagar A. & Livingston W, 2005 Fundamentals of Solar Astronomy, World Scientific Publishing (New Jersey: Hackensack)
[3] Zirin H 1989 Astrophysics of the Sun (New York: Cambridge University Press)
[4] Papics P I 2007 Solar Prominences, Solar Flares and Coronal Mass Ejections http://papics.eu/SolarPhenomena.pdf
[5] Parenti S 2014 Solar Prominences: Observations *Living Rev. Solar Phys.* 11
[6] Naufal L *et al.* 2016 Prominence measurement of total solar eclipse: March 9th 2016 Ternate Indonesia *J. Phys. Conf. Ser.* 771
[7] USSPS Code (2000) [Online] [Accessed 15th March 2016] Available from http://www.spaceweather.org/ISES/code/od/ussps.html

IOP Conf. Series: Journal of Physics: Conf. Series **852** (2017) 012004 doi:10.1088/1742-6596/852/1/012004

Comparison of relative sunspot numbers measured in Malaysia with international sunspot number calculated by SIDC-SILSO

N Rasmani[1], N S A Hamid[1*], F Kamarudin[2], W M A W M Kamil[1] and I Sarudin[1]

[1]School of Applied Physics, Faculty of Science and Technology, Universiti Kebangsaan Malaysia, 43600 UKM, Bangi, Selangor, Malaysia
[2]National Space Agency, 42700 Banting, Selangor, Malaysia

E-mail: shazana.ukm@gmail.com

Abstract. The accuracy of Langkawi National Observatory (LNO) relative sunspot number (RSN) data has been officially acknowledged when LNO has become one of the contributors to International Sunspot Number (ISN), R_{ISN} since June 2015. However, RSN from LNO (R_{LNO}) can only be obtained if the weather is favourable and the instruments are working well. These disadvantages can be solved if there are other observers across Malaysia that can contribute data of RSN to LNO. This research aims to compare the RSN collected by an amateur observer at National Planetarium, R_{NP} with R_{LNO} and R_{ISN}. 11 days data between the month of November 2014 and January 2015 are analysed to see the differences between R_{NP} to both R_{LNO} and R_{ISN}. Results show that, the value of R_{NP} is lower than R_{LNO} and R_{ISN}. The highest percentage difference is between R_{NP} and R_{ISN} which is due to the different method and instruments used in observing the sunspot. The correlation of relative sunspot number is high between R_{LNO} and R_{ISN} proving the reliability of LNO as contributor to ISN. The observational index is also determined in this study based on the ratio of correction factor of R_{NP} to R_{LNO}.

1. Introduction

Solar activities can affect the earth's magnetic field and affect our ground-based technology especially at high latitude country. Constant monitoring of solar activity including sunspots observation will help researchers to predict the behavior of the sun and provide an early warning to those who will be impacted by the space weather. Sunspot is the dark area that can be seen on the photosphere layer of the sun. The strong magnetic field is the main cause of low temperature of the sunspot area and thus, making the sunspot appears darker [1].

It can be said that sunspot is the subject of interest to many astronomers and solar physicists [2]. Sunspot observation is one of the oldest science existed in this world. Mankind has been observing the sun since the first telescope was built more than 400 years ago [3]. The data of sunspots has been collected ever since and it is the longest-running and still on-going observation [4]. The numbers of sunspots was observed and properly recorded in 1849 by R. Wolf who initiated the International Sunspot Number (ISN) [5]. The formula defined by R. Wolf to calculate the relative sunspot number (RSN) is based on the number of sunspots groups, g and total number of sunspots, f as follows:

$$R = k\,(10g+f) \qquad (1)$$

Content from this work may be used under the terms of the Creative Commons Attribution 3.0 licence. Any further distribution of this work must maintain attribution to the author(s) and the title of the work, journal citation and DOI.
Published under licence by IOP Publishing Ltd

The variable denoted by k is the correction factor of the observer. Although all observers uses the formula that is also known as Wolf formula to calculate RSN, there are still inconsistencies in the data collection due to different instrumentation used, local seeing conditions and also according to the observer skills [6]. Generally, sunspot observation is performed using the projection method. The observers draw or sketch the sunspots image projected from the eyepiece of the telescope. Then, they will manually count the number of sunspots and report it to the World Data Center (WDC).

The WDC for Sunspot Number originally based in Zurich Observatory but it was moved to Royal Observatory of Belgium, Brussel in 1981. This is lead to the foundation of Solar Influences Data Analysis Center (SIDC). Today, more than 80 observatories contribute daily sunspot numbers to SIDC. From the daily sunspot data, SIDC will calculate R_{ISN} and this relative sunspot number becomes the official index of sunspot number which is known as International Sunspot Number, ISN.

Langkawi National Observatory (LNO) officially became one of the contributors for sunspot numbers observation to WDC at Belgium since June 2015 [7]. Although in Malaysia there are quite many observatories, LNO is the only solar observatory that consistently performed solar observation. However, there are some disadvantages of this as LNO can only collect data if weather permits and if the instruments are working well. Therefore, we need more observers from Malaysia to take part in observing the sun and collecting sunspot numbers data. Amateur observers can support LNO's observation and provide a more complete RSN data from Malaysia.

In this research, we are comparing the data collected by an amateur observer at National Planetarium Observatory (NP) to LNO's data and ISN to see the percentage differences between the three sources. As we encourage amateur observers to support LNO's sunspot data, we introduce a method to calculate the observational index, β or known as correction factor ratio of NP data to LNO data derived from the Wolf formula.

2. Methodology

2.1 Solar Observation at National Planetarium, Kuala Lumpur

The observation was held at National Planetarium Observatory that located at Taman Tasik Perdana, Kuala Lumpur with coordinate 3°9' N, 101°41' E and 112 meter altitude above sea level. The main instrument that is used to observe the sun and sunspot is telescope Takahashi FS-102 that is mounted on Paramount ME German equatorial mount and covered with a Thousand Oak glass solar filter. The observation of sunspot here was done using digital images using DSLR camera Canon EOS 5D mark III. This camera has CMOS detector that is able to take 6 frames images per second. The setup of the instrumentation used at National Planetarium is shown in Figure 1.

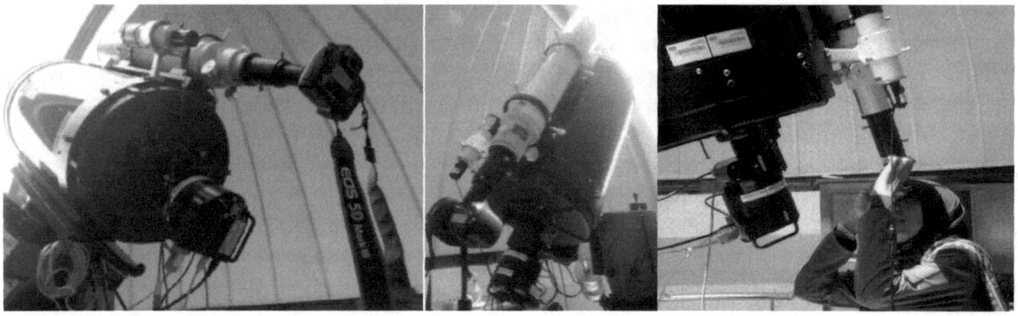

Figure 1. Instrumentation used for sunspot imaging with Camera EOS 5D mark III attached to Telescope Takahashi FS-102 NSVat National Planetarium, Kuala Lumpur.

Sunspot observations at National Planetarium were performed between November 2014 and January 2015 at time range 9.00 am until 12.00 pm. Sometimes the observation is repeated to ensure a

IOP Conf. Series: Journal of Physics: Conf. Series **852** (2017) 012004 doi:10.1088/1742-6596/852/1/012004

quality image is obtained. In those 3 months duration, there were only 11 days of data have been successfully collected. This is due to the rainy season at the end of the year in Malaysia.

The image of the sun was taken in 10 to 20 shots so that the best image can be selected to be analyzed as shown in Figure 2. The sunspots are manually counted by the observer by referring to the Zurich classification of sunspots and Wolf formula as mentioned before.

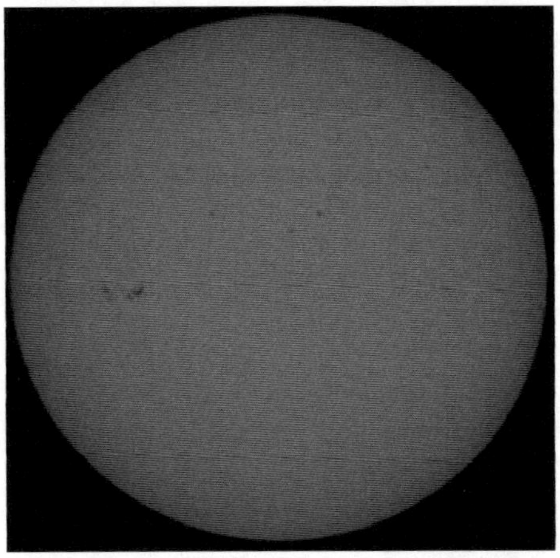

Figure 2. Example of a white light image taken by National Planetarium solar observation system on 16 December 2014.

2.2 Solar Observation at Langkawi National Observatory

Langkawi National Observatory (LNO) is situated at Langkawi Island, north of Peninsular Malaysia with latitude of 6.307°N and longitude of 99.781°E. This observatory has three solar observation systems to observe the sun in three different wavelengths. White light filter is used to observe sunspots, Hydrogen Alpha filter is used to observe chromosphere activity and Calcium K-Line filter is used to observe sunspots and plaques.

In this research, we focus on the images took using the white light filter. LNO used 152mm (6 inch) F/7.9 TMB telescope on a Paramount ME tracking mount and Baader ND5.0 solar filter film is used to achieve white light spectrum which filters 99.999% of sunlight. Besides that, the STL-4020M was chosen among many types of couple charged device (CCD) to take digital images of the sun as it fits the full solar disc when matched with TMB-152 telescope. This type of CCD also equipped with electronic shutter that is needed for very fast exposure. The instrumentation used in LNO is shown in Figure 3.

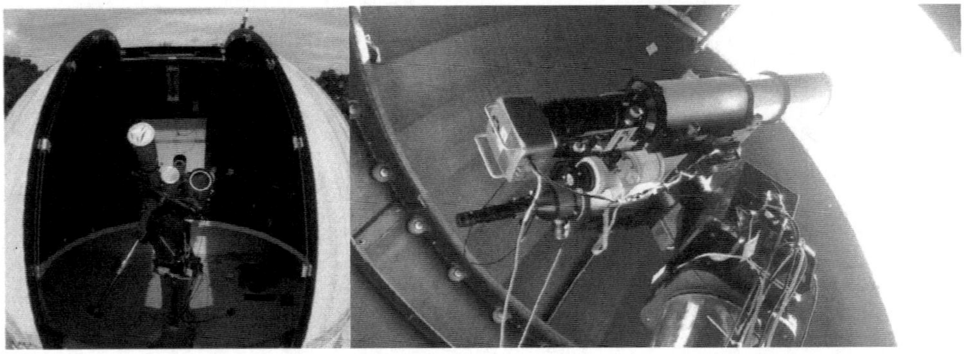

Figure 3. Instrumentation used for sunspot imaging and data collection at LNO: - Telescope TMB 152mm, SBIG STL4020 CCD Camera and Baader Solar Film.

IOP Conf. Series: Journal of Physics: Conf. Series **852** (2017) 012004 doi:10.1088/1742-6596/852/1/012004

Sunspot observation at LNO is performed daily only if weather permits between 9.30AM to 4.30PM local time [8]. It is preferable to take image of sunspot when the altitude of sun is higher and less affected by the atmosphere so there is less turbulence in the final image. The most crucial procedure in sun observation is making sure that the image is in focus as shown in Figure 4. Once good focus has been achieved, 50 white light full disk images are taken and the best image will be selected for further analysis as shown in Figure 5.

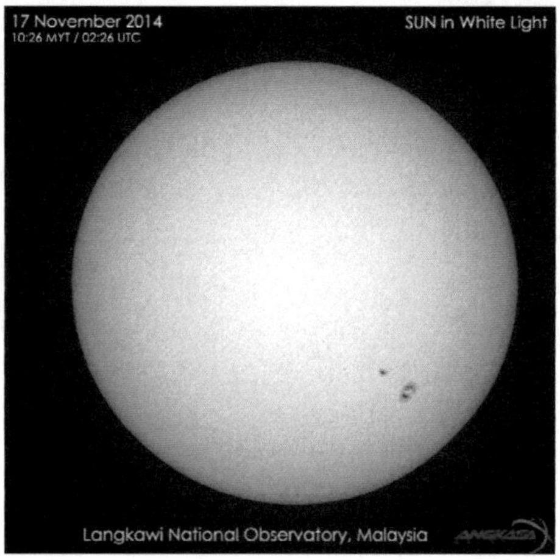

Figure 4. Example of a white light image taken by LNO solar observation system on 17 November 2014.

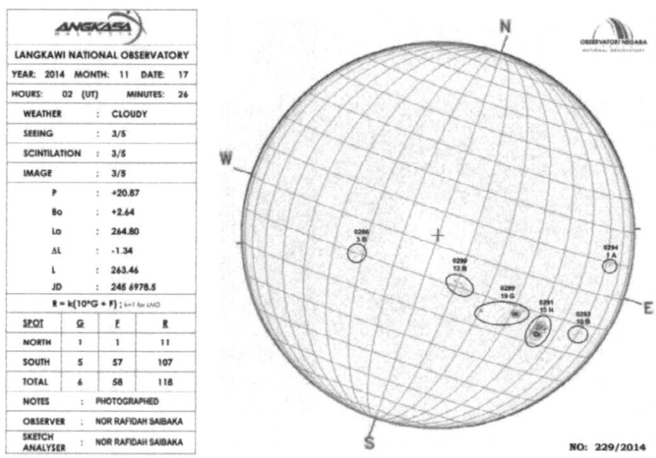

Figure 5. After analysis, a sketch and detailed sunspot count is produced for reference.

3. Results and Discussions

The RSN data collected from all the three sources are tabulated in Table 1. Unfortunately, there are 5 days missing data from LNO in December and early January due to weather condition. The daily sunspot number observed at National Planetarium, R_{NP} is represented in the red line, the sunspot number observed at LNO, R_{LNO} is green line and the International Sunspot Number, R_{ISN} is blue line in Figure 6. There are total of 11 days of RSN data compared in this study. One can observe that R_{ISN} exceeds both R_{LNO} and R_{NP} in the plot. However, we can perceive similar trends in RSN from all three sources. This indicates that R_{NP} obtained by amateur observer is in good agreement with the one observed by LNO and ISN.

Table 1. Relative sunspot numbers from National Planetarium, Langkawi National Observatory and International Sunspot Number.

Date	R_{NP}	R_{LNO}	R_{ISN}
05/11/2014	67	115	137
09/11/2014	44	76	69
17/11/2014	37	118	99
15/12/2014	122	-	166
16/12/2014	107	-	175
18/12/2014	89	-	154
19/12/2014	64	-	175
02/01/2015	96	-	122
04/01/2015	63	-	88
08/01/2015	87	117	99
15/01/2015	38	85	59

Figure 7 shows the scatter plot of R_{ISN} against R_{LNO}. Linear curve fitting was applied and the correlation coefficient of determination, R^2 are shown in the figure. R^2 represent the correction factor, k as mentioned in equation (1) [9]. The correction factor between R_{ISN} and R_{LNO} is quite high, 0.64 although there are missing data from LNO. This proves the reliability of LNO as one of contributor to ISN data.

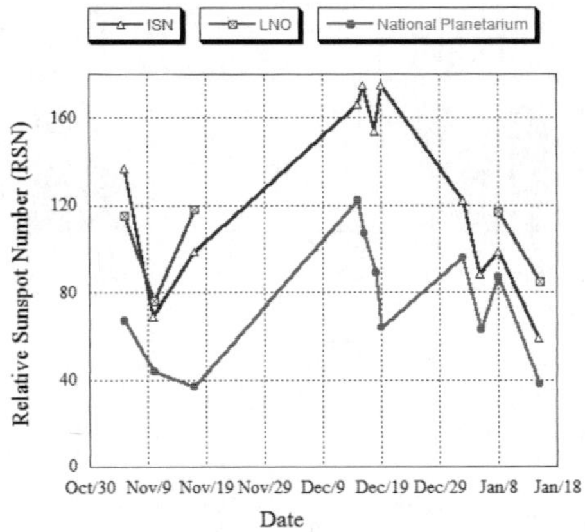

Figure 6. National Planetarium RSN, R_{NP} (red line) compared to Langkawi National Observatory RSN, R_{LNO} (green line) and International Sunspot Number R_{ISN} (blue line).

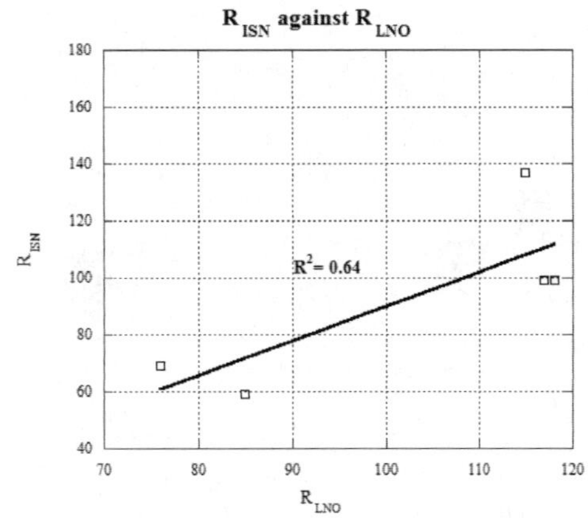

Figure 7. Relationship between the daily sunspot numbers observed at ISN (R_{ISN}) and LNO (R_{LNO}).

We further calculate the percentage differences of each day of the observation. The results obtained are plotted in Figure 8. The highest error between R_{LNO} and R_{ISN} is 27% as shown in Figure 8 (a) that is on Jan 15. The negative value indicates that value of R_{LNO} is higher than value of R_{ISN}. On the other hand, we found that the error between R_{NP} and R_{LNO} is much higher that is 80% as illustrated in Figure 8 (b). This high value could be due to the difference of type instruments used in capturing digital images of the sun as LNO uses CCD which takes better quality image than the DSLR camera used by National Planetarium. Therefore, the R_{LNO} calculated is higher since CCD produces images with higher resolution resulting in clearer images than the DSLR camera used by National Planetarium. Figure 8 (c) shown in the plot that the differences between R_{NP} and R_{ISN} is the highest, which is more than 100% caused by the different methods and instruments used in observing the sun. As mentioned earlier, most of the contributor for ISN used solar projection method to sketch the sunspots while National Planetarium uses digital images from DSLR camera to capture the images of sunspots. From these result, we can conclude that data obtained using solar projection method is more accurate than CCD and DSLR camera and data collection from CCD are more accurate than from DSLR camera.

We continue the analysis by calculating the observational index using equation (1) that is mentioned before. This equation is used to calculate the ratio of correction factor of R_{NP} to R_{LNO} as below:

$$\beta = \frac{k_{NP}}{1} = \frac{(10g_{LNO}+f_{LNO})}{(10g_{NP}+f_{NP})} \quad (2)$$

β is the ratio of correction factor or known as observational index. The best value of ratio is defined as 1 as correction factor of LNO, K_{LNO} as a contributor to ISN is also given as 1. It is shown in Figure 9 that the lowest value of this observational index is on 8th January 2015 which is 1.34 and the highest

IOP Conf. Series: Journal of Physics: Conf. Series **852** (2017) 012004 doi:10.1088/1742-6596/852/1/012004

ratio is on 17th November 2014 with the value of 3.19. This shows inconsistencies of the observational index of the 5 days data.

(a) (b) (c)

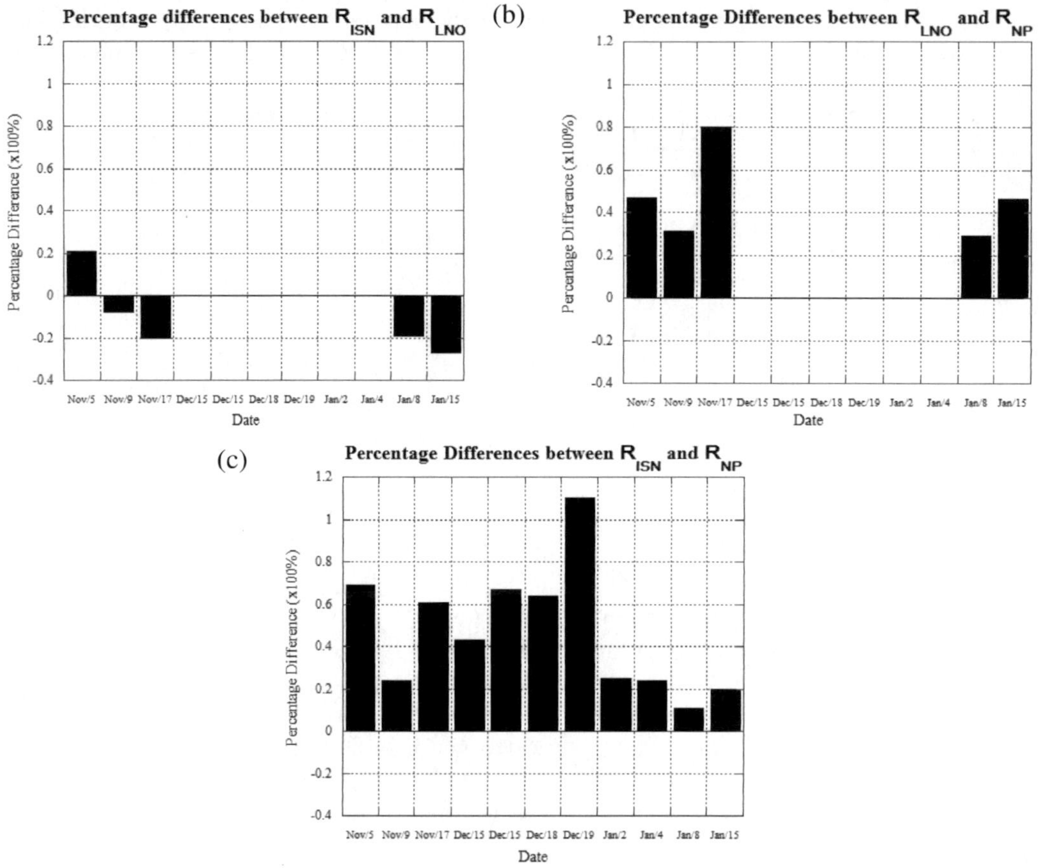

Figure 8. Percentage differences between (a) R_{ISN} and R_{LNO} (b) R_{LNO} and R_{NP} and (c) R_{ISN} and R_{NP} for 11 days.

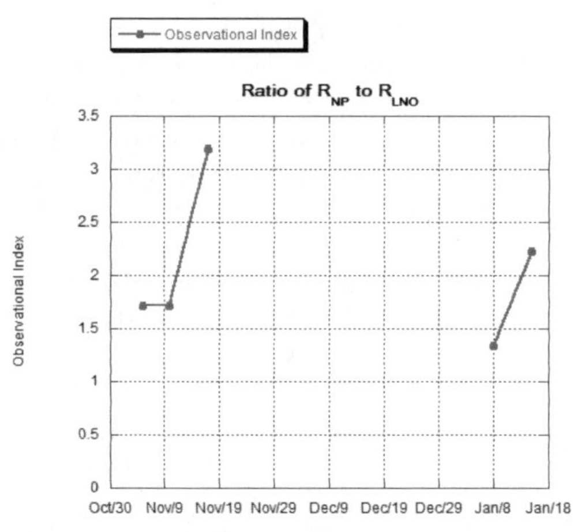

Figure 9. Ratio of correction factor of R_{NP} to R_{LNO} or known as observational index, β.

4. Conclusion

From the result discussed in this paper, LNO has proved its reliability as a dependable solar observatory in Malaysia as its data has high correlation with ISN. However, there is a high percentage difference of R_{NP} with R_{LNO} and R_{ISN}. The observational index, β is determined between R_{NP} and R_{LNO} using the Wolf formula. This index is important as it represents the ratio of correction factor from one source of data to R_{LNO}. We suggest that any studies to support LNO data done by other solar observers in Malaysia to use this method in determining their respective observational index to R_{LNO}.

Acknowledgements

The authors are grateful to SIDC and Langkawi National Observatory, National Space Agency for providing the RSN data that was used and referred in this paper. Special thanks to National Planetarium, Kuala Lumpur for letting us use their facilities and supporting this project. Financial support was provided by Universiti Kebangsaan Malaysia and Ministry of Education Malaysia through grants GUP-2016-016 and FRGS/1/2015/ST02/UKM/02/01.

References

[1] Koskinen H E J 2011 *Physics of Space Storms* ed J Mason and M Shardlow (Berlin: Springer) pp11 12
[2] Nguyen T T, Willis C P, Paddon D J, Nguyen S H and Nguyen H S 2006 *Fund. Inform.* **1** 105
[3] Bray R and Loughhead R 1965 *Sunspots* (New York: Wiley)
[4] Owens B 2013 Long-term research: Slow science *Nature* **495** 300
[5] Clette F, Svalgaard L, Vaquero J and Cliver E 2014 Revisiting the Sunspot Number *Space Sci. Rev.* **186** 35
[6] Vaquero J 2007 Historical sunspot observations: A review *Adv. Space Res* **40** 929
[7] Kamarudin F, Tahar M R, Saibaka N R and Padang L A L 2016 Relative sunspot number observed from 2013 to 2015 at Langkawi National Observatory *Adv. Sci. Lett.* **23**(2) 1285
[8] Radzi Z M, Kamarudin F, Hashim M H, Tahar M R, Saibaka N R, Ishak A N and Nurlisman Z K D 2017 Space weather monitoring using facilities in National Space Agency *Adv. Sci. Lett.* **23**(2) 1248
[9] Hathaway D 2015 The Solar Cycle *Living Rev. Sol. Phys.* 12

IOP Conf. Series: Journal of Physics: Conf. Series **852** (2017) 012005 doi:10.1088/1742-6596/852/1/012005

An analysis of heat wave trends using heat index in East Malaysia

W Suparta[1] and A N M Yatim[1]

[1]Space Science Centre (ANGKASA), Institute of Climate Change, Universiti Kebangsaan Malaysia, 43600 Bangi, Selangor Darul Ehsan, Malaysia

wayan@ukm.edu.my

Abstract. This paper aimed to investigate the heat wave trends in East Malaysia based on the National Weather Services (NWS) Heat Index. The heat index was calculated by using mean temperature and mean relative humidity on monthly basis for 5 meteorological stations in East Malaysia during the period 2008 to 2010. The trends for heat wave were estimated from Heat Index based on the least square regression analysis at each station level. Results showed that the heat wave trends are increasing at all stations. The highest heat index was occurred in Sandakan on July 2010 with heat index 35°C while the lowest heat index happened at Kuching in January 2009 with 27.3°C. From the heat wave observed, East Malaysia is still in caution categories or normal condition (27°C–32°C) and the extreme caution (32°C–41°C) was observed during southwest monsoon (May-July). The safety condition of heat waves in East Malaysia is possibly due to weak to moderate El Nino occurred during the period of observation.

1. Introduction

A heat wave event is a prolonged period of abnormally hot weather and anticipated to become more frequent, longer, and severe with a warming of the Earth's climate. There are many studies discuss about the effects of climate change on global temperatures and their distribution in space and time [1]. The rising in global mean temperatures because of climate change has high impact for the earth's ecosystems and other natural systems such as extreme climate events, including heat waves.

An extreme heat wave can affect to serious health concern, especially heat stress and heat stroke among the sick and the elderly. There were 200 cases have been reported by the Ministry of Health Malaysia (MOH) related to hot weather which is 52 cases of heat cramps (hyperthermia, spasms of large muscle), 126 cases of heat exhaustion (fatigue, collapse) and 22 cases of heat stroke (hot red or flushed dry skin, nausea or vomiting, coma) includes two (2) deaths reported as a result of heat stroke [2]. The heat waves event for long period not only gives an impact on human health but also can cause severe damage to the environment, vegetation and production. For example, the exceptional weather patterns of hot and dry conditions from El Nino phenomena had affected the Fresh Fruit Bunch (FFB) yield and Oil Extraction Rate (OER) and contributed to decreasing crude palm oil yield in 2009[3].

East Malaysia is located in the equatorial region, and the seasonal variation of the climate is usually described as dry and wet seasons throughout the year. The average of rainfall is 250 centimeters a year and the temperature ranges between 26°C to 29°C on an average while the average relative humidity from 70% to 80%. In this region, wet season (northeast monsoon) spans from the months of October to March and dry season (southwest monsoon) spans from the months of April to September are

Content from this work may be used under the terms of the Creative Commons Attribution 3.0 licence. Any further distribution of this work must maintain attribution to the author(s) and the title of the work, journal citation and DOI.
Published under licence by IOP Publishing Ltd

typically influencing. East Malaysia is also bare to the El Niño impact, which decreases rainfall during dry season. The climate shows noticeable impact on Malaysia by increasing sea levels and rainfall, flooding risks, severe droughts and extreme heat wave [4].

Recently, the heat wave trend and characteristic in Malaysia has not been extensively investigated. Until now, there is no standard acceptable definition of heat waves, but the heat wave vary relying on the length of consecutive days, the type of data temperature (minimum, average, maximum), the thresholds used to determine an extreme temperature, and whether humidity is taken into account[5]. Steadman [6, 7] defined the Heat Index (Hi) or known as the apparent temperature (Ta) might also be considered when studying the heat wave event. Steadman also translated combinations of air moisture and temperature and other factors such as wind speed and sun radiation into a single scale, calculated in the same units as air temperature [6]. Hi originally was developed to measure thermal comfort [8]but it has become a famous metric in environmental health since the U.S. National Weather Service (NWS) has associated different Hi ranges to environmental health threats [9]. Hi generated from temperature and relative humidity is more important than temperature for the human health because Hi shows how hot it exactly feels when relative humidity is combined with the actual air temperature. For example, if the air temperature is 34°C and the relative humidity is 75%, Hi is felt hot as 49°C [9] and the same effect is reached at just 32°C when the relative humidity is 90%.

In line with significant effects of Hi to our life and environment, this research focuses on the investigation of heat wave trends in East Malaysia based on the Hi defined from NWS for the period 2008-2010. This study is very important to further determining the occurrence of heat wave event in the future as an early warnings system especially during extreme weather in Malaysia.

2. Methodology

2.1. Data and location

The monthly mean temperature (°C) and mean relative humidity (%) was collected from Malaysian Meteorological Department (MetMalaysia) for the period 2008-2010 in order to calculate the heat index. In this study, we focused on East Malaysia because the beaches in East Malaysia are one of the most popular in Malaysia and become the tourist attraction every year. Economic strength and ecosystem stability by the evaporation of sea surface to inland in this coastal area is very important for the future sustainability. In addition, the selected station is close to the South China Sea and possible influence by the climate variability such as El Niño. Five (5) stations were chosen across East Malaysia and the details of the each station and locations are presented in Table 1 and Figure 1.

Table 1. Five selected stations in East Malaysia

No	Station	Longitude	Latitude
1	Kota Kinabalu (urban)	5° 55' 58.800" N	116° 02' 60.00" E
2	Sandakan (sub-urban)	5° 54' 0.000" N	118° 04' 1.200" E
3	Tawau (urban)	4° 16' 01.200" N	117°52' 59.60" E
4	Kuching (industrial)	1° 28' 58.800" N	110°19' 58.80" E
5	Bintulu (urban)	3° 12' 0.00" N	113°01' 58.70" E

2.2 Data Processing

This study chose the NWS method, which was developed by Rothfusz [10] based on the Steadman [6] for calculating Hi. The method was widely used in order to produce of weather warnings in real-life situations. NWS has modified this index for operational purposes such as specific ranges to health effects [14] and extreme heat warnings when needed. The calculation of Hi in this research is performed in agreement with equation:

$$Hi = -42.379 + 2.049(1.8C + 32) + 10.14R - 0.224(1.8C + 32)R - 6.83 \times 10^{-3}(1.8C + 32)^2 - 5.48\times10^{-2}R^2 + 1.22 \times 10^{-3}(1.8C + 32)^2R + 8.52 \times 10^{-4}(1.8C + 32)R^2 - 1.99\times 10^{-6}(1.8C + 32)^2R^2 \quad (1)$$

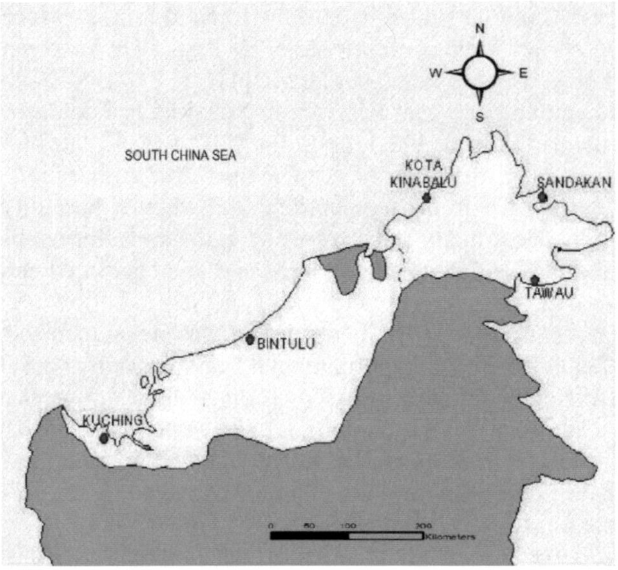

Figure 1. Locations of five selected stations in East Malaysia

where R stands for the relative humidity (%) and C is the temperature (°C). The original equation was used Fahrenheit (°F) [11] for temperature but in this study it was converted into Celsius (°C) unit. Based on this equation, it is can be used only when air temperature and humidity is higher than 26°C and 39%, respectively [9]. Then, the trends for each station were calculated using a least square regression analysis. The significance of trend was tested using *t*-test at 95% confidence level. The linear least square was applied to find the best fitting through the set of data. The linear least square fitting considers the deviation of a dependent variable, *y*, from its expected value.

$$y = (b_0 + b_1 x) + e \tag{2}$$

where x is the predictor variable while b_0 and b_1 are unknown parameters, representing the intercept and trend coefficient of the fitted straight line. From this equation, the observed trend was heat index value (y) with respect to time (annual basis) (the predictor variable, x). The values of b_0 and b_1 are estimated by minimizing the function of e. The least squares regression line aim to minimize these prediction errors. The error of prediction associated with x be e, which is

$$e = y - y_1 \tag{3}$$

A positive value of the trend coefficient indicates an increasing trend and negative value indicates otherwise. Then, the Hi value that obtained from this equation later will be classified into 4 categories based on the health risk defined in Table 2, while Figure 2 shows the overall methodology that has been conducted in this study.

Table 2. Health risks to various heat index categories adapted from NOAA (2009)

Heat Index (Hi)	Description
27 – 32°C	Caution: fatigue is possible with prolonged exposure and activity. Continuing activity could result in heat cramps.
32 – 41°C	Extreme caution: heat cramps and heat exhaustion are possible. Continuing activity could result in heat stroke.
41 – 54°C	Danger: heat cramps and heat exhaustion are likely; heat stroke is probable with continued activity.
> 54°C	Extreme danger: heat stroke is imminent.

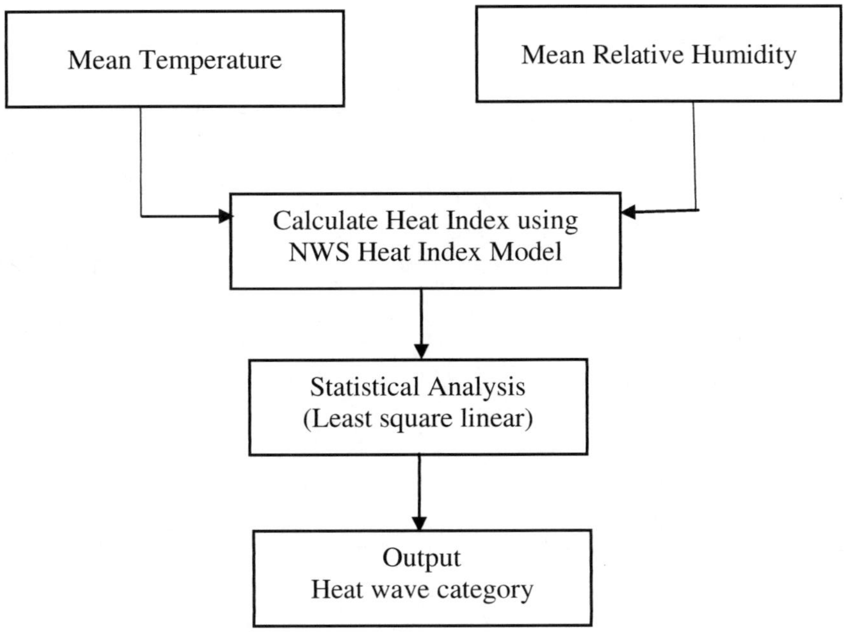

Figure 2. Summary of research methods

3. Result and Discussion

The variation of mean temperature, mean relative humidity, and the Hi value calculated using NWS model for all stations are presented in Figure 3. Based on the figure, overall temperature at all stations is still within normal range which is 26°C to 28°C [12]. The mean minimum temperature (25.8°C) was recorded at Kuching in January 2009 while the mean maximum temperature (29.4°C) was recorded at Kota Kinabalu in May 2010. From the figure, there is no slight difference in temperature recorded for the urban, suburban, and industrial area and no extreme temperature (>35°C) recorded during this period. The variation of mean temperature is slightly higher in the year 2009 and 2010 due to moderate El Nino event. El Nino events normally will change the climate patterns around the globe but the outcomes of each event are never exactly the same [13]. For the mean relative humidity, it showed in the range 68.2% to 91.2% with the minimum value was recorded at Kota Kinabalu in February 2008 and the maximum value was recorded at Bintulu in July 2010. For the Hi value, the minimum (27.3°C) are observed at Kuching in January 2009 while the maximum (35.0°C) has occurred at Sandakan in July 2010.

Based on the figure, the temperature at each station is in normal condition but the present of relative humidity have increased the heat index than the normal temperature [11]. The result shows that the heat index is increased due to the increment of the relative humidity. The high of relative humidity makes moist air feels hotter than dry air because it evaporates process less efficient. From the comparison results suggested that the peak values of the heat index mostly rising during May to July which has occurred during dry season (southwest monsoon). During this season, most areas experience monthly minimum rainfall and become drier or hotter climate conditions. From our analysis, it is interesting to note that all the stations showed increasing trends in heat index with the largest positive trends found in Sandakan followed by Kuching, Kota Kinabalu, Bintulu, and Tawau. This trend at the 95% confidence level is obtained from the linear equation as depicted in each figure. The highest trends magnitude happened in suburban area compared to another area is due to the effective difference between temperature and relative humidity.

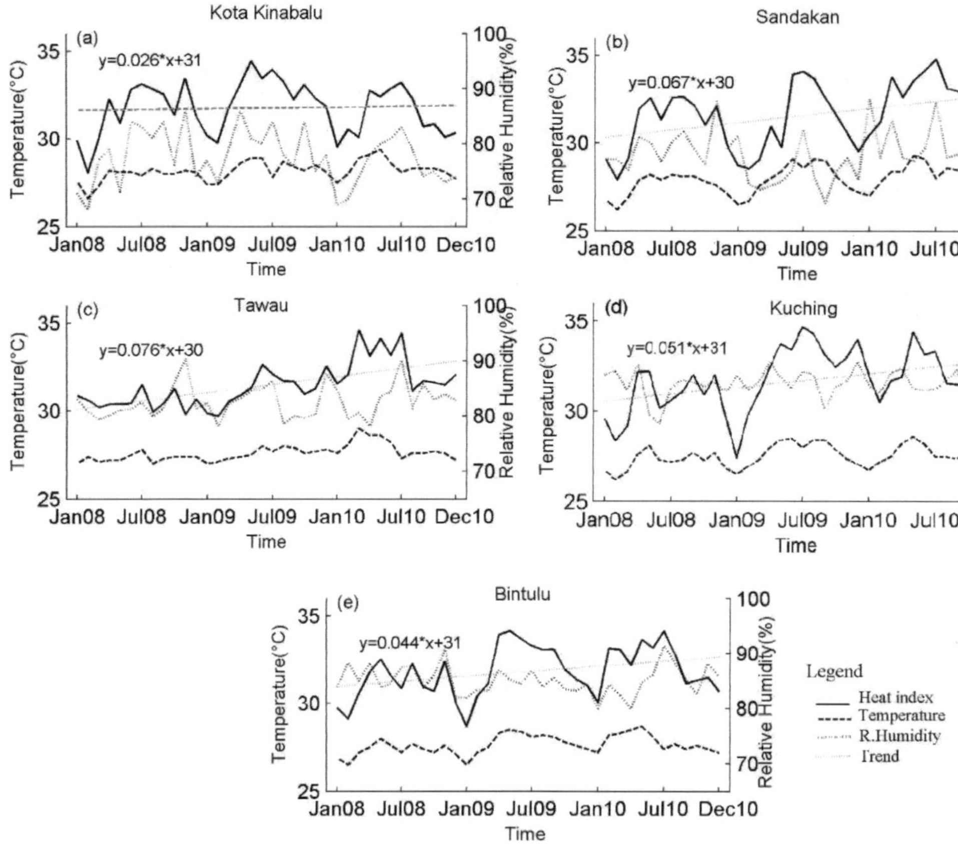

Figure 3. Variation of mean temperature, mean relative humidity, heat index and their trends for (a) Kota Kinabalu, (b) Sandakan, (c) Tawau, (d) Kuching, and (e) Bintulu

Figure 4 shows the minimum and maximum heat index for each station during the study period. Based on the figure, Sandakan experiences the highest value of heat index in the month of July 2010 when it crosses 35°C followed by Kuching 34.8°C in July 2009, Bintulu in June 2009 and Tawau in May 2010 with 34.6°C and last but not least is Kota Kinabalu with 34.4°C. The entire maximum heat index occurred from June 2009 till July 2010 which is during El Nino event. For minimum heat index, the lowest was recorded at Kuching in January 2009 with 27.3°C. Sandakan recorded 27.8°C in February 2009 while Kota Kinabalu recorded 28.0°C in February 2008 and Bintulu with 28.6°C in January 2009. Tawau showed the highest for minimum heat index in February 2009 with 29.4°C.

Table 3 showed the heat wave categories where temperature during this period are in caution categories (27°C-32°C) but it becomes an extreme caution categories (32°C-41°C). During the period 2008 until 2010, the heat wave in East Malaysia is not in the crucial condition except during dry season (southwest monsoon). This condition could be changed once temperature increases due to climate change. The combination of temperature and relative humidity is important to understand local heat wave exposure and its effect on human health directly. Fountain et al. [14] found few differences in human response to air relative humidity exposure between 60% and 90% of the temperature range 20°C-26°C. The relationship between heat waves and their effects on human health also depends on the exposure time to high temperatures, the rate of temperature increase, and people's climate adaptation capacity.

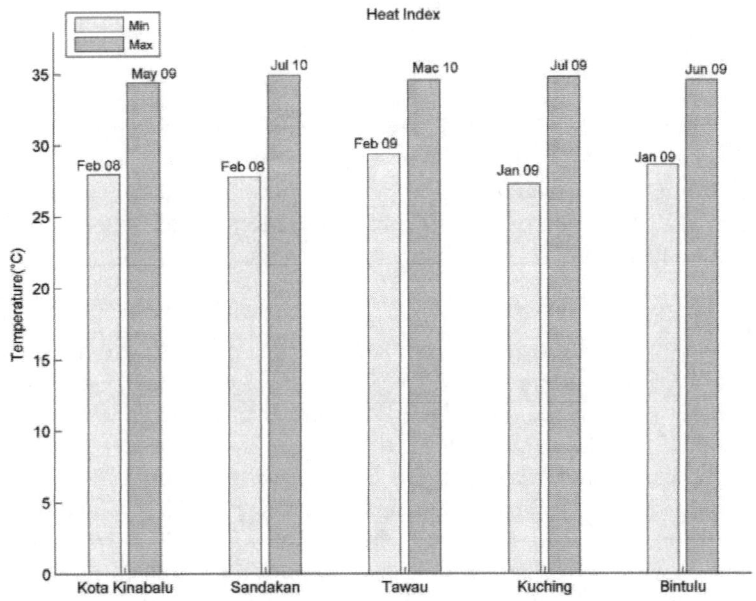

Figure 4. Minimum and maximum of heat index

Table 3. Categories of heat wave event

Station	Temperature (°C)	Humidity (%)	Heat Index (°C)	Time	Categories
K.Kinabalu	28.9	81.3	34.4	May-09	Extreme caution
	26.6	68.9	28.0	Feb-08	Caution
Sandakan	28.2	80.1	35.0	Jul-10	Extreme caution
	26.2	78.0	27.8	Feb-08	Caution
Tawau	29.0	80.5	34.6	Mac-10	Extreme caution
	26.8	82.5	29.4	Feb-09	Caution
Kuching	28.5	88.2	34.8	Jul-09	Extreme caution
	25.8	87.3	27.3	Jan-09	Caution
Bintulu	28.7	84.7	34.6	Jun-09	Extreme caution
	26.5	81.8	28.6	Jan-09	Caution

4. Conclusion

This paper successfully presents the analysis of heat wave trends in East Malaysia based the on the heat index value calculated using the mean surface temperature and mean relative humidity for the period 2008-2010. The mean temperature was observed with a normal condition in the range 26°C to 28°C and no extreme temperature was recorded during the study period for all stations. The results showed the lowest heat index was occurred at Kuching in January with 27.3°C and the highest index was occurred at Sandakan in July 2010 with 35.0°C. The maximum heat index happened during March-July which is during the dry season. The heat index was in caution category with the lowest (27°C–32°C) but it becomes extreme caution (32°C–41°C) during March-July. We can assume that the heat wave still in safe condition due to moderate El Nino event occurred during the study period. From our analysis, the frequency trend of the heat wave is found significantly increased over time due to global climate change events. The increased trend of heat waves needs to be anticipated to prevent illness and death happen. Thus, further research needs to be conducted to find exact definition and status of heat waves for the preparation of an accurate early warning system in Malaysia. The use of daily temperature and relative humidity and another parameter would possible to give more accurate information about the occurrence of heat wave events.

Acknowledgments

This paper is partly funded by Ministry of Science, Technology and Innovation Malaysia (MOSTI) under Flagship Programme: ZF-2014-016. We would like to thank the Malaysia Meteorological Department for providing the temperature and relative humidity data. The second author is PhD candidate and supported by the Faculty Science and Natural Resources, University of Malaysia Sabah, 88400, Kota Kinabalu, Sabah

References

[1] Diffenbaugh N S and Ashfaq M 2010 Intensification of hot extremes in the United States: Intensification of hot extremes *Geophys. Res. Lett.* **37** doi:10.1029/GL043888

[2] Ministry of Health Malaysia (MOH) 2016 *Planning division Health Informatics Centre* MOH/S/RAN/17.16 (AR) (AUGUST 2016)

[3] MPOB 2010 *Overview of the Malaysian Oil Palm industry* Report http://Bepi.Mpob.Gov.My/Images/Document/Overview202010_Final.pdf

[4] MMD 2009 *Climate change scenarios for Malaysian 2001-2090* Malaysian Metrological Department, Scientific Report, Petaling Jaya Malaysia pp 1-84

[5] Perkins S E, Alexander L V, and Nairn J R 2012 Increasing frequency, intensity and duration of observed global heat waves and warm spells *Geophys. Res. Lett.* **39** L20714 Doi: 10.1029/2012GL051120

[6] Steadman R G 1979 The assessment of sultriness. Part I: A temperature–humidity index based on human physiology and clothing science *J. Appl. Meteorol.* **18** 861

[7] Steadman R G 1984 A universal scale of apparent temperature *J. Clim. Appl. Meteor.* **23** 1674

[8] NOAA (National Oceanic and Atmospheric Administration) 2012 *Heat: A major killer.* Available: http://www.Nws.Noaa. Gov/Om/Heat/Index.Shtml [Accessed 12 October 2012].

[9] NWS (National Weather Service) 2011 *Meteorological conversions and calculations: Heat index calculator* Available http://www.Hpc.Ncep.Noaa.Gov/Html/Heatindex.Shtml [Accessed 12 October 2011]

[10] Rothfusz L P 1990 *The Heat index "Equation" (or, more than you ever wanted to know about heat index)* SR 90–23 Fort Worth, TX: National Oceanic and Atmospheric Administration, National Weather Service, Office of Meteorology

[11] NWS (National Weather Service) 1994 *Excessive heat watch, warning and advisory heat index criteria.* Regional operations manual letter E-5-94, Eastern Region, NWS, Bohemia, NY pp 3

[12] Malaysia's Second National Communication (NC2) 2000 *Vulnerability and adaptation assessment* Ministry of Natural Resources and Environment Malaysia

[13] FAO 2013 Climate impact on agriculture. *Understanding the drought impact of El Niño on the global agricultural areas* Rome http://www.fao.org/nr/ climpag

[14] Fountain M, Arens, Xu T, Bouman F and Massayuki O 1999 An investigation of thermal comfort at high humidifies *ASHRAE Transactions* **105** 1

IOP Conf. Series: Journal of Physics: Conf. Series **852** (2017) 012006 doi:10.1088/1742-6596/852/1/012006

A short review on the effects of aerosols on visibility impairment

Emetere Moses E[1] and Akinyemi ML[1]

[1]Department of Physics, Covenant University Canaan land, P.M.B 1023, Ota, Nigeria

Email: moses.emetere@covenantuniversity.edu.ng

Abstract. Ozone in the lower planetary boundary layer of the earth atmosphere is dangerous to people and vegetation, since it oxidizes natural tissue. The diminished in visibility is because of dispersing of sun based radiation by high convergences of anthropogenic aerosols. Visibility impairment is most prominent at high relative mugginess when the aerosols swell by the take-up of water to expand the cross sectional area for dispersing; this is the wonder known as haze. Haze has become a major air pollution challenge the aviation industry has to cope with in recent time. In this review, two major problems were spotted to be responsible for air disaster in any region of the world. While some developed countries had almost resolved the challenge of visibility impairment by seeking relevant solutions, most developing countries do not have a recovery plan. Therefore, the resolution of this major challenge may be the leverage for most developing nations to draw out a recovery plan.

1. Introduction

The impairment of visibility has a negative role in the aviation and communication industries. For example, the incessant cancellation of flight in airport and the attenuated communication signals which results to poor signal reception. Visibility impairment also has a positive role of reducing heat flux from the sun by the scattering and absorption of visible light [1]. Fine particulate matters from anthropogenic activities in form of suspended particles are dispersed into the lower atmosphere [2]. The transport of the suspended particulates as they relates with other particulates in the atmosphere may be very difficult to explain in clear terms.

There are lots of emphases on the beneficial nature of ozone (O3) in the upper troposphere and lower stratosphere. However, ozone in the lower planetary boundary layer of the earth atmosphere is dangerous to people and vegetation, since it oxidizes human tissues. O3 in the earth surface is created from the oxidation of CO (i.e. the reaction of CO with the hydroxyl radical (•OH)) and hydrocarbons by OH as a secondary reaction of NOx. In thickly populated districts with high discharges of NOx and hydrocarbons, fast O3 generation happens during daylight and result in a surface air contamination issue alluded to as smog [3]. Fuel burning and different exercises of our modern culture are responsible for its discharge into the air. The particular kind of air contamination related with diminished visibility is called "smog". The decrease in visibility is because of scrambling of sun powered radiation by high groupings of anthropogenic vaporizers as appeared in the figure beneath [3]. Dispersing of sun powered radiation by mist concentrates is the primary procedure constraining visibility in the troposphere. Without aerosols our visual range would be around 300 km, restricted by diffusing air atoms. Anthropogenic mist which concentrates in urban regions regularly lessen visibility by one order of magnitude relative to unpolluted conditions [4]. Debasement of visibility by anthropogenic aerosols is additionally a significant issue in numerous urban region of the developed

Content from this work may be used under the terms of the Creative Commons Attribution 3.0 licence. Any further distribution of this work must maintain attribution to the author(s) and the title of the work, journal citation and DOI.
Published under licence by IOP Publishing Ltd

world. The physics of haze is accomplished when the visibility decrease is most noteworthy at high relative humidity when the aerosols take-up water and increase its cross-sectional area for dissipating light.

2. The Physics of Visibility Impairment

The sun illuminates sight path (Figure 1), the atmosphere in the intervening path interferes with the light through the intervention of particulates-mostly aerosols and gases. The particulates affect visibility as shown in Figures 1a & b. The physics of visibility impairment is shown in Figure 1c. The incident multiple light strand from the sun are scattered by the particulates in the atmosphere. This occurrence leads to a diminishing contrast and subdues colours. The remaining absorbed light gives the scene a grayish cast which is seen as visibility impairment (Figure 1b). The collective participation of atmospheric particles and gases to the absorption and scattering are somewhat difficult to estimate. The extension of this major challenge by extension is responsible for signal attenuation in the atmosphere. Particles or gas do not scatter light uniformly in all directions (as shown in Figure 1c) because of its angle-dependent nature [5] and its size.

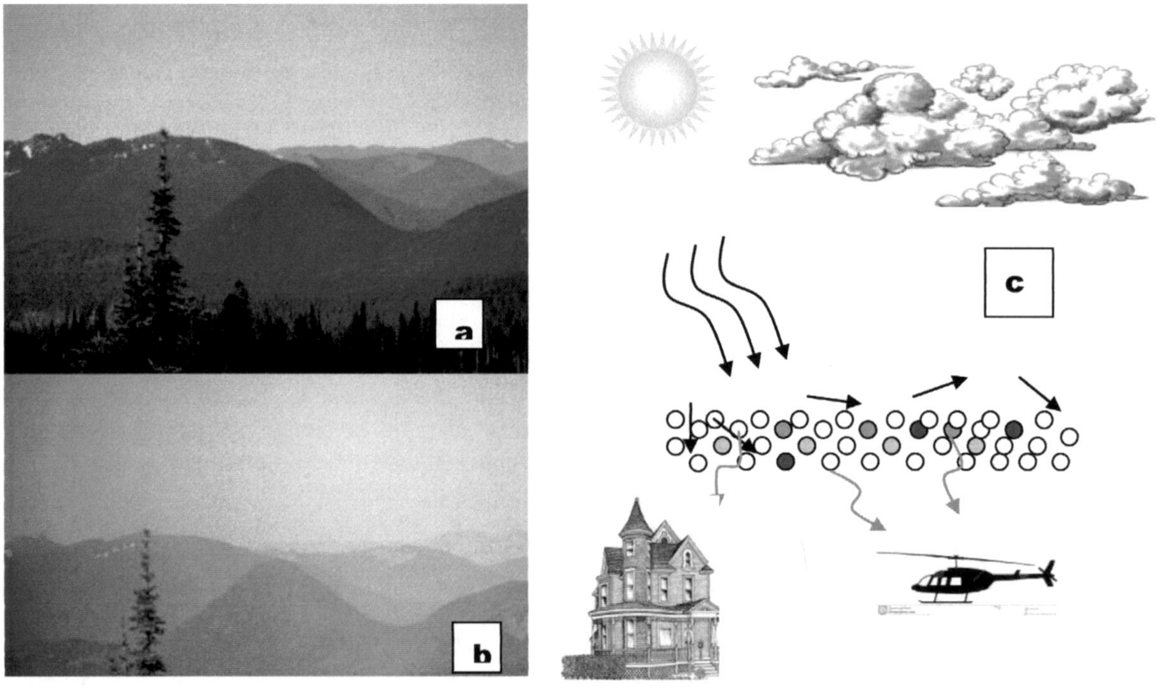

Figure 1. The physics of visibility impairment

The physics of visibility impairment is hinged on couple of physical quantities. For example, Horvath [6] showed that it dependent on the distance from the source of pollution. The minimum distance given was within the range of 50–100km away from the pollution source. This means that visibility impairment over a large geographical area is a result of a multi-faceted pollution sources that eject a large concentration of pollutants. By inference, a densely populated region like Lagos-Nigeria is expected to eject a large volume of anthropogenic air pollution that is expected to affect the visibility of the area. The second parameter that could affect the physics of visibility impairment is the wind speed [7, 8]. Based upon an in-depth study carried out on the wind stagnation parameters over Ota-Nigeria [9], it easy to infer that when there is high wind stagnation, a high visibility impairment is expected. The size of the atmospheric particles and gas over a geographical area can impact upon the visibility. Tsai et al. [10] observed that the visibility variation in Kaohsiung was as a result of the varying sizes of the PM10 (particulate matter of diameter, Dpo10 mm). The varying sizes per time simple means that there would be diverse scattering and absorption mechanism which increases the gray colouration of the particles and gas to increase the visibility impairment. As explained in this section, the number of participating particles and gas in atmosphere may be difficult to estimate. Scientifically, the rough estimation can be obtained from the calculation of scattering coefficient in the lower atmosphere. However, the influence of day and night over the scattering coefficient makes it dynamic in nature. For example, scattering coefficient is influenced by atmospheric temperature and chemical composition [11]. Hence, an increase in the scattering coefficient leads to a reduced visibility. This feat was achieved in UK as a result on the policy on the use of clean energy sources since 1973 [12].

3. Visibility Impairment And The Meteorological Influence

In southern Taiwan, Tsai [10] observed from forty-two years data set (1961–2003) that July which happens to be the hottest month with an average temperature of 29.3 1C has highest average visibility at 17.7 km. It was also observed that the visibility influenced by annual rainfall could increase averagely by 5.94%. The influence of rainy and non-rainy period is further expressed in Figure 2 below.

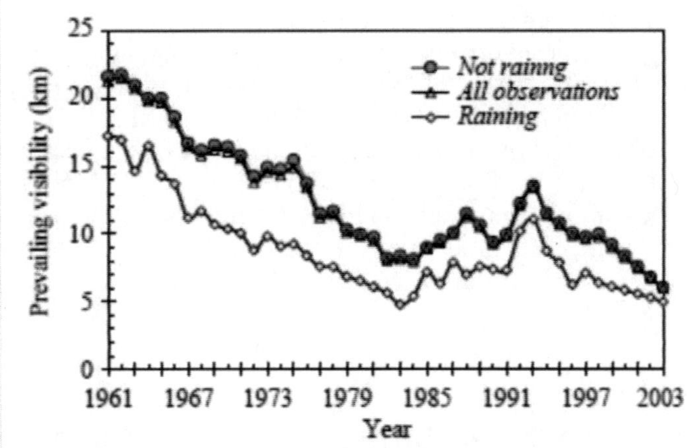

Figure 2. Mean annual visibilities in Tainan between 1961–2003 [10]

The meteorological influence on the lower atmosphere is enormous largely due to massive updrafts and downpour over a geographical location. It affects the height of the mixing layer in the lower atmosphere. The mixing layer of the lower atmosphere affects the diffusion of pollutants in the atmosphere, as well as the ozone concentration [13]. The mixing layer height is influenced by the wind speed changes which affect the dispersion of pollutants [14, 15]. Hence the proximity of Sahara desert and the dust winds over West Africa plays salient role in the visibility state of its member state. The additional anthropogenic emission over West Africa accounts for over 17.6% variance on the ozone [13]. The atmospheric particles and gas dictates the tropospheric ozone as shown over China in Figure 3. Recall that tropospheric ozone formation takes place when nitrogen oxides (from nitrates aerosols), carbon monoxide (from carbon based aerosols) and volatile organic compounds react in the atmosphere in the presence of sunlight. This is a further proof that air pollutant concentration dictates the changes in regional meteorology by the formation of tropospheric ozone via slow-moving, high-pressure weather systems [16].

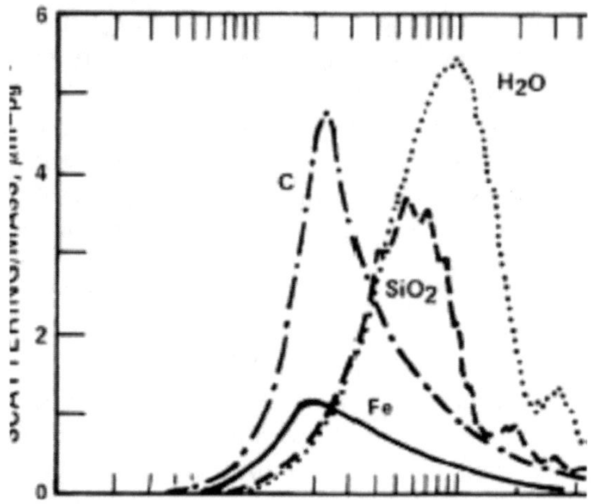

Figure 3. Particles and gas scattering potentials in the atmosphere [17]

4. Mitigating Air Disaster: The Theories Of Visibility Impairment

When there is visibility impairment, pilots rely on the run-way lightning and airport control tower to ensure successful landing. The airport control tower makes use of embedded mathematical software which obeys the Allard's law, Meteorological Optical Range and Koschmeiders's law to calculate the run-way visual range (RVR) amongst other to assist aviation tarmac operations. While the Allard's law predicts the visual range of self-luminous targets, meteorological optical range estimates the length of the path of the luminous flux in the atmosphere and Koschmeiders's law illustrate the connection per time between the inherent luminance contrast and the apparent luminance contrast of an object. The Allard's law and Meteorological Optical Range can be simplified in equations 1 & 2 respectively:

$$E = \frac{Ie^{-\sigma R}}{R^2} \quad (1)$$

where I is the intensity of the run-way lamp, σ is the extinction coefficient, and R is the distance from the lamp. σ is estimated by transmissometers, I is estimated by using the TMS photometrics.

$$MOR = -\frac{\log(0.05)}{\sigma} \approx \frac{3}{\sigma} \quad (2)$$

Though there are proven RVR algorithms used in the aviation sector, the reality of moving layers of aerosols [15] and the collective effect on visibility impairment cast upon the pilot and the operator in the control tower may be dangerous (Figure 4). Hence, scientist had suggested slant visual range (SVR). It is suggested that it is a better gauge of safe landings than VRV. However, measuring the SVR may be very difficult based on the relative motions of the aerosols layer and the path of sight. Hence, this is a major problem in the aviation industry.

5. Conclusion

In this review, two major problems were spotted to be responsible for air disaster in any region of the world. While some developed countries had almost resolved the challenge of visibility impairment by seeking relevant solutions, most developing countries do not have a recovery plan. Therefore, the resolution of this major challenge may be the leverage for most developing nations to draw out a recovery plan.

References

[1] Xu Xuezhe *et al* 2016 Optical properties of atmospheric fine particles near Beijing during the HOPE-J3A campaign *Atmos. Chem. Phys.* **16** 6421

[2] Emetere M E, Akinyemi M L and Akinwumi S A 2016 Aerosols loading trends and its environmental threats over Cotonou-Benin, *Nature Environment and Pollution Technology* **15**(3) 1111

[3] Emetere M E, Akinyemi M L and Uno U E 2015 Computational analysis of aerosol dispersion trends from cement factory *Proceedings 2015 International Conference on Space Science & Communication* pp 288-291

[4] Li R, Yunjie H, Ling L, Hongbo F and Jianmin C 2017 Real-time aerosol optical properties, morphology and mixing states under clear, haze and fog episodes in the summer of urban Beijing *Atmos. Chem. Phys. Discuss*. doi:10.5194/acp-2016-976

[5] Seinfeld JH and Pandis SN 1998 *Atmospheric chemistry and physics: from air pollution to climate change*, 1st edition (New York: J. Wiley) p 101

[6] Horvath H 1991 Spectral extinction coefficients of background aerosols in Europe, North and South America: a comparison *Atmos. Environ.* **25**A 725

[7] Horvath H 1995 Estimation of the average visibility in central Europe *Atmos. Environ.* **29** 241

[8] Lee D O 1990 The influence of wind direction, circulation type and air pollution emissions on summer visibility trends in southern England *Atmos. Environ.* **24**A 195

[9] Akinyemi M L, Emetere M E and Akinwumi S A 2016 Dynamics of wind strength and wind direction on air pollution dispersion *Asian J. Appl. Sci.* **4**(2) 422

[10] Tsai Y I. Lin Y H and Lee S Z 2003 Visibility variation with air qualities in the metropolitan area in southern Taiwan *Water, Air and Soil Pollution* **144** 19

[11] Lee C T, Cheng, J P 1996 The effects of aerosol species and meteorological factors on visibility in the Taipei metropolitan area *J. Chinese Inst. Environ. Eng.* **6** 21

[12] Doyle M and Dorling S 2002 Visibility trends in the UK 1950–1997 *Atmos. Environ.* **36** 3161

[13] Emetere M E, Akinyemi M L and Akinojo O 2016 Effects of band superposition on the satellite imagery of aerosol optical depth over west Africa *J. Eng. Appl. Sci.* **11**(1) 17

[14] Seibert P, Beyrich F, Gryning S.-E., Joffre S, Rasmussen A and Tercier P 2000 Reviewand intercomparison of operational methods for the determination of the mixing height *Atmospheric Environment* **34** 1001

[15] Emetere M E, Akinyemi M L and Akinojo O 2015 Parametric retrieval model for estimating aerosol size distribution via the AERONET, LAGOS station *Environ. Pollut.* **207**(C) 381

[16] Jacob D J and Winner D A 2009 Effect of climate change on air quality *Atmos. Environ.* **43**(1) 51

[17] EPA 1979 *Protecting Visibility: An EPA Report to Congress* United States Environmental Protection Agency, Research Triangle Park, NC

Atmospheric configurations of aerosols loading and retention over Bolgatanga-Ghana

M E Emetere,[1] S E Sanni[1] and P Tunji-Olayeni[1]

[1]Covenant University Canaan land, P.M.B 1023, Ota, Nigeria

E-mail : moses.emetere@covenantuniversity.edu.ng, adex07@yahoo.com, patience.olayeni@covenantuniveristy.edu.ng

Abstract. Bolgatanga is located on latitude of 10.78 °N and longitude 0.85 °W. This research is aimed to estimate the aerosols loading and retention over Bolgatanga-Ghana through a conceptual model that is made up of analytical, statistical and Matlab curve-fitting tool. The model's accuracy was established over the aerosol optical depth for a thirteen-year satellite data set from the Multi-angle imaging specto-reflectometer (MISR). The maximum aerosol retention was 31.73%. Its atmospheric constants, tunning constants and the phase difference over Bolgatanga was found to 0.67, 0.24 and $\pm\pi/4$ respectively. The phase difference expresses the different kinds of network topologies and beam forming methods for measuring devices that may be used in Bolgatanga. Therefore a good estimation of the aerosols loading and retention over Bolgatanga, we may be in the best position to controlling its effect on health, farming, rainfall pattern, cloud formation and regional climate.

1. Introduction

Studies on the significance of aerosols on climate change have received considerable attention in recent years. Several remote sensing techniques have been deployed in the past with the zest to gain insight or better understanding of the effects of aerosols on climate change. Such remote sensors read-off large volumes of data which suffices for the spatial and temporal coverage of the areas or vicinities under study. For scientific applications, it is pertinent that such data are qualitative. For example, data obtained from ground remote sensing have been previously used for understanding aerosol effects on atmospheric heat budget as reported in Cloud Screening of Sky Radio Meter Data Manual [1-4], validating satellite measurements [5] and model simulations [6].

Mathematical modeling of the removal of liquid aerosols from gases flowing through channels without any form of packing and channels filled with fine random packing was carried out by Anatoly et al [7]. Some conservation/governing equations for aerosols were established. For the separation efficiency of aerosols in the packed bed layer, a convective mass transfer equation was adopted. Some assumptions were made and a 3D equation describing steady state mass transfer of aerosols was obtained. A particle deposition velocity was established in relation to the particle relaxation time. The finite difference technique was employed in the numerical solution. Also, the work of Chan et al [8] presents an equilibrium kinetic model of secondary organic aerosols (SOA). They employed the model as a way of examining the effects of SOA formation of the gas phase oxidation of the products initially generated to more or less volatile species. From the findings, it is evident that, while different controlling/interacting mechanisms can lead to differing SOA growth behavior, it becomes generally impossible to deduce the actual mechanism for SOA formation with absolute dependence of the SOA growth data.

Content from this work may be used under the terms of the Creative Commons Attribution 3.0 licence. Any further distribution of this work must maintain attribution to the author(s) and the title of the work, journal citation and DOI.
Published under licence by IOP Publishing Ltd

In this paper we examine a conceptual model to estimate the aerosols loading and retention over Bolgatanga. It would greatly assist in projecting its negative effects on health, agriculture and communication in the future.

2. Source of Data and location

Bolgatanga is the fourth most crowded city in northern Ghana and it is situated on longitude 0.85°W and latitude 10.78°N in the Sahelian geographic locale south of the Sahara, henceforth, we expect a high effect of the north east winds from close by Sahara desert. Likewise, it is additionally affected by the neighborhood steppe atmosphere. Its metropolitan enclave is around 499 km^2. Bolgatanga has normal temperature and precipitation of 26.1 °C and 752 mm individually. The distance of Bolgatanga to the Sahara is around 2826 km. Before now, no aerosols ground data was accessible; consequently, the satellite is generally accepted. Fourteen years satellite observation was gotten from the Multi-angled Imaging SpectroRadiometery (MISR). The details of the MISR are given by Emetere [9].

3. Methodology

A statistical approach [9] was adopted. The raw MISR dataset was processed using Excel. The mean for every month was figured all as the year progressed. We tried the precision of the information by applying the airborne scattering model that was propounded by Emetere et al.[10]. An expansion of the scattering model utilized is given as

$$\psi(\lambda) = a_1{}^2\cos\left(\frac{n_1\pi\tau(\lambda)}{2} + \alpha\right)\cos\left(\frac{n_1\pi\tau(\lambda)}{2} + \beta\right) + \cdots \ldots a_n{}^2\cos\left(\frac{n_n\pi\tau(\lambda)}{2} + \alpha\right)\cos\left(\frac{n_n\pi\tau(\lambda)}{2} + \beta\right) \quad (1)$$

Here α and β are the stage contrasts or the phase difference, k is the diffusivity, τ is the AOD, ψ is the convergence of contaminant, λ is recognized as the wavelength, 'a' is the environmental or atmospheric constant and and 'n' is the tuning constants.

The rate of maintenance can be resolved from the coefficient of difference for every year. This was finished by considering the past and current years which are indicated separately as G_P and G_r respectively. Thus we propound that the aerosol retention between two years as:

$$A = \left|\frac{G_r - G_P}{G_r}\right|^2 \times 100\% \quad (2)$$

The aerosols retention can be figured from Tables 2-3 to acquire Tables 4-5. Any factual instrument could be utilized to get the aerosols retention. In this paper, the Matlab and Excel bundles were utilized to acquire the outcomes that are reported in the ensuing segment

4. Results and Implications

The Bolgatanga monthly AOD trend agreed perfectly with the proposed model (see Figures 1 to 2). The AOD pattern over Bolgatanga is a gamma distribution with the average maximum in March. From Figure 2, it can be inferred that the type of aerosols in the area are majorly dust particulates from the northeast wind. The AOD retrieval technique by MISR is perfect over Bolgatanga. A trivial explanation is that the area falls within the satellite orbit. Beyond the trivial reason, Bolgatanga has a more stable weather compared to other parts of the country.

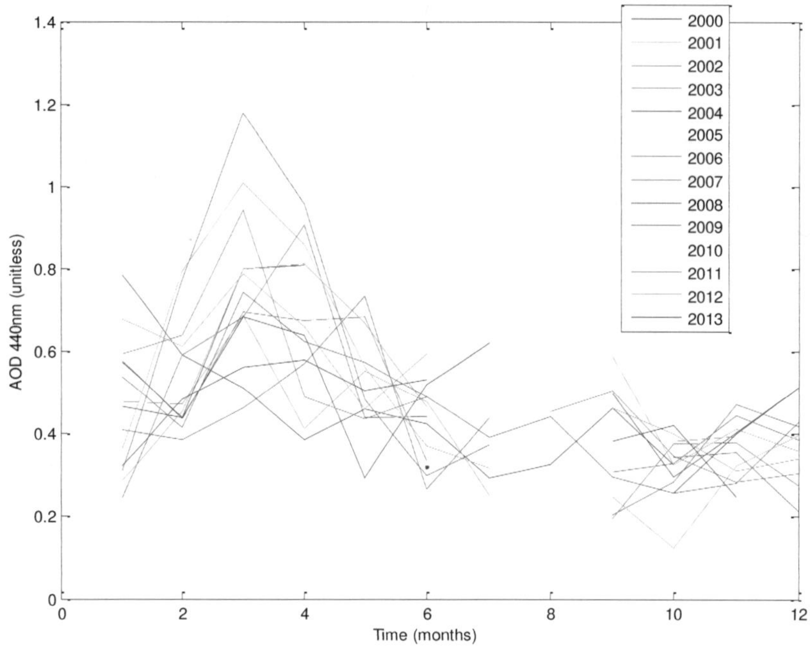

Figure 1. AOD pattern for Bolgatanga (2000 – 2013)

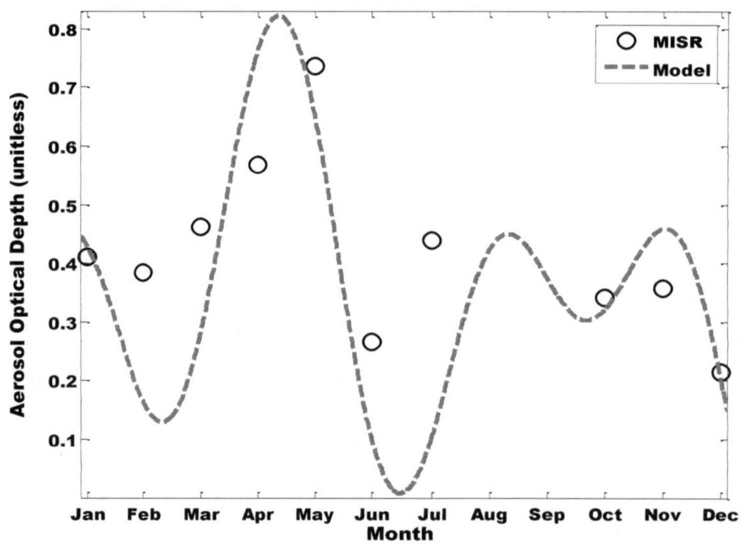

Figure 2. AOD for new model and MISR for the year 2002

Figures 2 give the model predictions and the data generated from satellite. In Figure 2, the aerosol optical depths as obtained from the satellite (MISR) image agrees with the AOD predictions of the model for the year 2000 for all other months except that slight deviations can be seen in the months of July, August, October and November. In Figure 4, the AOD estimated for the year 2002 from by the model seems to agree with those of the MISR for all other months except for the clear deviation in March and slight deviations in the months of May, June, October, November and December. From

Figure 1 to 2, the environmental constants, stage contrasts and tuning constants can be acquired from the Matlab and equation (2) as appeared in Table 1 underneath

Table 1. Atmospheric constants over Bolgatanga

Location	a_1	a_2	n_1	n_2	α	**β**
Bolgatanga	0.667	0.6403	0.242	0.1515	$\pm\frac{\pi}{4}$	$\pm\frac{\pi}{4}$

Note that this paper do not query the unwavering quality of the ITU model but suggests that aerosol retention should be incorporated into the ITU model as shown in equation (3).

$$N = \frac{77.6P}{T} + 3.73 \times 10^5 \frac{e}{T^2} = N_{dry} + N_{wet} \ (N - units) \quad (3)$$

where e represent the water vapor pressure, P is the atmospheric or barometric pressure (hPa) and T is the total temperature (K). The numerical connection between relative humidity and water vapor pressure is communicated in the accompanying equation:

$$e = \frac{RH}{100} a \ exp\left[\frac{bT}{T+c}\right] \quad (4)$$

Here, temperature T in the above equation is given in °C and the coefficients a, b and c is given as: a = 6.1121, b = 17.502 and c = 240.97. Leck and Svensson [11] gave the assurance that the coefficients a, b and c are impacted by the optical state over a topographical area. This paper hence proposes the ITU model requires the addition of the environmental constants to understand the signal budgeting over Bolgatanga. By this idea, we factually analyzed the AOD loading over Bolgatanga as appeared in Tables 2 and 3.

In 2005, the highest 99% confidence interval, standard deviation variance, AOD mean, 95% confidence interval and coefficient of variation were obtained. The highest skew and kurtosis can be found in 2010. The highest Kolmogorov-Smirnov stat can be found in 2004. The most noteworthy AOD mean, 95% certainty interim, 99% certainty interim, fluctuation, standard deviation and coefficient of variety were acquired in 2005. The most astounding skew and kurtosis can be found in 2010. The most noteworthy Kolmogorov-Smirnov detail can be found in 2004. This outcomes demonstrates that the lower air of Bolgatanga may not be as dynamic as those of urban communities in in West Africa [12]. Henceforth we analyzed the aerosol retention in Tables 5 and 6.

Table 2. Statistical AOD analysis 2000-2006

Statistical Tool	**2000**	**2001**	**2002**	**2003**	**2004**	**2005**	**2006**
Standard Error	0.047	0.047	0.047	0.036	0.094	0.048	0.045
95% confidence interval	0.105	0.107	0.105	0.08	0.209	0.108	0.101
99% confidence interval	0.151	0.154	0.15	0.114	0.298	0.155	0.146
Variance	0.022	0.022	0.025	0.014	0.097	0.023	0.02
Standard deviation	0.147	0.149	0.157	0.119	0.312	0.15	0.142
Coefficient of variation	0.323	0.357	0.326	0.271	0.595	0.275	0.293
Skew	1.258	0.945	0.236	0.468	1.234	-0.12	-0.231
Kurtosis	1.799	1.326	-1.588	0.818	0.526	-1.05	-0.845
Kolmogorov-Smirnov stat	0.185	0.184	0.168	0.196	0.279	0.203	0.132

Considering the data in Table 2, the estimated variances show that the highest deviation of the estimated monthly means from the 13-year mean was obtained for the year 2005 which reveals that it is farthest from the 13-year mean with that of 2001 showing the least deviation. From the estimated standard deviations, it is clear that the amount/degree of aerosol dispersion of the monthly means from the 13-year mean is lowest for the year 2001 with the highest seen in the year 2005; those of other years lie between these values.

Table 3. Statistical AOD analysis 2007-2013

Statistical Tool	**2007**	**2008**	**2009**	**2010**	**2011**	**2012**	**2013**
Standard Error	0.062	0.063	0.061	0.064	0.042	0.085	0.038
95% confidence interval	0.142	0.142	0.135	0.148	0.093	0.189	0.087
99% confidence interval	0.207	0.205	0.192	0.216	0.131	0.269	0.126
Variance	0.034	0.04	0.04	0.037	0.021	0.079	0.013
Standard deviation	0.185	0.199	0.201	0.193	0.146	0.281	0.113
Coefficient of variation	0.358	0.417	0.424	0.372	0.319	0.576	0.251
Skew	0.689	1.204	1.164	0.723	0.437	0.794	-0.646
Kurtosis	-0.56	1.282	2.211	-1.08	-0.23	-0.54	-0.624
Kolmogorov-Smirnov stat	0.172	0.266	0.193	0.271	0.125	0.262	0.176

In Tables 4 and 5, two years that have the highest atmospheric aerosols retention were 2004 and 2013. Hence, skew and kurtosis may be used as perfect indicators to determine atmospheric aerosols retention. The relevance of the atmospheric aerosols retention over a region has incredible impact on aeronautics plans [13], human wellbeing [14], measuring instruments, earth's energy budget and meteorology [15].

Table 4. Atmospheric aerosols retention over Bolgatanga 2001-2006

	2001	**2002**	**2003**	**2004**	**2005**	**2006**
Aerosol deposition	1.13	0.76	2.89	29.72	29.02	0.44

Table 5. Atmospheric aerosols retention over Bolgatanga 2007-2013

	2007	**2008**	**2009**	**2010**	**2011**	**2012**	**2013**
Aerosol deposition	4.89	2.75	0.03	1.51	2.01	64.27	31.73

5. Conclusion

The new model seems to be somewhat reliable based on the accurate predictions as compared with satellite data. Hence, the estimation of the aerosols loading and retention over Bolgatanga-Ghana through a conceptual model that is made up of analytical, statistical and Matlab curve-fitting tool was a huge success. However, slight modifications ought to be made in order to improve on the model's accuracy. The aerosol distribution for all other years was found to be platykurtic except for the year 2010 which shows a leptokurtic distribution. However, all other years had normal aerosol distribution. Also, all the years considered, show a random distribution of the aerosols. The maximum aerosol retention was 31.73%. Its atmospheric constants, tunning constants and the phase difference over

Bolgatanga was found to 0.67, 0.24 and $\pm\pi/4$respectively. The phase difference expresses the different kinds of network topologies and beam forming methods for measuring devices that may be used in Bolgatanga.

Appreciation
The authors recognize the sponsorship from Covenant University. The authors appreciate NASA for the satellite data set.

References

[1] Aoki K and Fujiyoshi Y 2003 Sky radiomter measurements of aerosol optical properties over Sapporo, Japan *J. Meteorol Soc.Jpn.* **81** 493
[2] Kim D H, Sohn B J, Nakajima T and Takamura T 2005. Aerosol Radiative Forcing Over East Asia Determined from Ground-based Solar Radiation Measurements *J. Geophys. Res.* **110** D10S22
[3] Uchiyama A, Yamazaki A, Togawa H and Asano J 2005 Characteristics of aeolian dust observed by sky-radiometer in the intensive observation period 1 (IOP1) *J. Meteorol Soc. Jpn.* **83** 291
[4] Takamura, T., Sugimoto, N., Shimizu, A., Uchiyama, A., Yamazaki, A., Aoki, k., Nakajima, T., Sohn, B. J and Takenaka, H. 2007. Aerosol radiative Characteristics at Gosan, Korea, during the atmospheric brown cloud east asian regional experiment 2005 *J. Geophys. Res.* **112** D22S36
[5] Higurashi A, Nakajima T, Holben B N, Smirnov A, Frouin R and Chatenet B 2000 A study of global aerosol optical climatology with two channel AVHRR remote sensing *J. Clim.* **13** 2011
[6] Takemura T, Okamoto H, Maruyama Y, Numaguti A, Higurashi A and Nakajima T 2000 Global three-dimensional simulation of aerosol optical thickness distribution of various origins *J. Geophys. Res.* **105**(17) 853
[7] Anatoly G L, Marat M B, Timur M F and Albert R I 2014 Calculating Efficiency of Separation of Aerosol Particles from Gases in Packed Apparatuses *Advances in Chemical Engineering and Science, Apparatuses* **2014** (4) 143
[8] Chan A, Kroll W H, Ng J H, N L and Seinfeld J H 2007 Kinetic Modeling of Secondary Organic Aerosol Formation: Effects of Particle- and Gas-phase Reactions of Semi-volatile Products *Atmos. Chem. Phys.***7** 4135
[9] Emetere, Moses Eterigho 2016 Statistical Examination Of The Aerosols Loading Over Mubi-Nigeria: The Satellite Observation Analysis *Geogr. Panonica* **20**(1) 42
[10] Emetere M E, Akinyemi M L and Akinojo O 2015 Parametric retrieval model for estimating aerosol size distribution via the AERONET, LAGOS station *Environ. Pollut.* **207**(C) 381
[11] Leck C and Svensson E 2015 Importance of aerosol composition and mixing state for cloud droplet activation over the Arctic pack ice in summer *Atmos. Chem. Phys.* **15** 2545
[12] Emetere M E, Akinyemi M L and Akin-Ojo O 2015. Aerosol optical depth trends over different regions of nigeria: thirteen years analysis *Modern Appl. Sci.* **9**(9) 267
[13] Gettelman A and Chen C 2013 The climate impact of aviation aerosols *Geophys. Res. Lett.* **40** 1
[14] Ronald E, Wyzga and Lawrence J F 1995 Health effects of acid aerosols *Water, Air, and Soil Pollution* **85**(1) 177
[15] Emetere M E and Akinyemi M L 2013 Modeling Of Generic Air Pollution Dispersion Analysis From Cement Factory *Analele Universitatii din Oradea–Seria Geografie* 231123-628 181

Altering rainfall patterns through aerosol dispersion

M E Emetere[1], M Bakeko,[1] L Onyechekwa[1] and W Ayara[1]

[1]Physics Department, Covenant University, P.M.B. 1023, Ota-Ogun State, Nigeria

E-mail : moses.emetere@covenantuniversity.edu.ng

Abstract. The possibility of recirculation mechanism on rainfall patterns is salient for sustenance of the human race through agricultural produce. The peculiarity of the lower atmosphere of south west region of Nigeria was explored using theoretical and experimental approach. In the theoretical approach, the reconstruction of 1D model as an extraction from the 3D aerosol dispersion model was used to examine the physics of the recirculation theory. The experimental approach which consists of obtaining dataset from ground instruments was used to provide on-site guide for developing the new recirculation theories. The data set was obtained from the Davis weather station, Nigeria Meteorological agency and Multi-angle Imaging Spectro-radiometer (MISR). We looked at the main drivers of recirculation and propounded that recirculation is a complex process which triggers a reordering of the mixing layer- a key factor for initiating the type of rainfall in this region.

1. Introduction

The role of aerosols in understanding the climatic system is enormous because it controls verse processes in the atmosphere like atmospheric circulation [1], rainfall pattern [2] cloud formation [3]. Its direct and indirect influence on the global climate can be very complicating when understudying the comparative role of aerosols to global energy balance [4]. Atmospheric aerosols have been reported to form heterogeneous or homogenous layers in the atmosphere [5]. Models are propounded to amplify the usefulness and exactness of the measuring instruments. Sadly, these model needs change in a few zones e.g. sufficient estimation of airborne scattering and statement, rates of new molecule arrangement and their reliance on controlling factors e.g. development and maturing. The aerosols dispersion and deposition technique requires a systemic evaluation of aerosols not only at ground level but also in the planetary boundary layer (PBL) to determine the aerosol transport mechanism-in relation to weather formation at the PBL [5, 6]. The planetary boundary layer contains a large variety of aerosols from anthropogenic and natural sources [1]. The planetary boundary layer (PBL) comprises of sub-layers e.g. mixed layer, surface layer, stable layer and residual layer. The surface layer is closest to the earth i.e. 10 percent of the height of the PBL. Turbulent stresses are relatively constant. The mixed layer or convective layer are directly above the surface layer and are characterized by convective motion and significant mixing of air. It makes up the entire layer of the PBL above the surface layer - during day and reduces drastically during the night. The stable layer is located above the mixed layer. It is the entry point where air from above the PBL entrains into the mixed layer. This stable layer restricts turbulence, prevents frictional influences from reaching above the PBL and prevents the continued upward motion of thermals. The residual layer is majorly residuals of the mixed layer and is not influenced by turbulent stresses. The PBL are influenced by environment forces like differential heating, energy budgets, moisture, diurnal variations, buoyancy, shear, and roughness length. The PBL takes out aerosol retention by means of regular components which works

Content from this work may be used under the terms of the Creative Commons Attribution 3.0 licence. Any further distribution of this work must maintain attribution to the author(s) and the title of the work, journal citation and DOI.
Published under licence by IOP Publishing Ltd

either as wet deposition, dry deposition or chemical reactions. These procedures can happen exclusively or partially. Layering of the PBL is subject to its temperature. All the more for all intents and purposes, the review site that is featured in Emetere et al. [6] transverses the beach front area (in the south of Nigeria) to the sub-Sahel locale (in the north of Nigeria). Significantly, the wind transport instrument starts turbulent weight on the PBL. For instance, amid harmattan, the wind transports vast amounts of particulates from the Sahara Desert in the north toward the south. Likewise, aerosols due to bush burning, automobile activities in the south increase the aerosols content over the study site [6]. Amid the non-harmattan, the West African Sub area is washed by the sticky SW Monsoon winds from the Atlantic Ocean [7].

The recirculation theory helps us to understand the anomalies in both the aerosols dispersion and deposition points of the atmospheric field settings. The mysteries of the recirculation zones in the mixing layer of the lower atmosphere are yet to be explained i.e. comparing the pre-circulation and post circulation. The turbulent boundary layers formed during recirculation are somewhat complex when explaining the minute changes that triggers continuous recirculation [8, 9]. In this paper, we understudied a 3D dispersion model which was developed from the aerosols dispersion dynamics around a cement factory. We looked at the main drivers of recirculation and propounded that recirculation is a complex process which triggers a reordering of the mixing layer- a key factor for initiating the type of rainfall in this region.

2. Methodology

The Davis weather station used for this study was installed at Covenant University, Ota Nigeria. Ota is on 6.21°N and 3.21°E. The weather station was used to obtain the wind properties i.e. peculiar to the region. The Meteorological data for rainfall was obtained from the Nigerian Meteorological Agency for five States in the south-west i.e. Osogbo-Osun, Ilorin-Kwara, Ibadan-Oyo, Abeokuta-Ogun, Ondo-Ondo and Ikeja-Lagos. The main objective of the theoretical approach is to derive recirculation equations from micro scale. The theoretical approach to this studies can be obtained from the aerosol dispersion model [5, 6]. The mathematical representation of type 3 is given as

$$\frac{\partial c}{\partial t} + V_x\frac{\partial c}{\partial x} - V_z\frac{\partial c}{\partial z} - V_y\frac{\partial c}{\partial y} = \frac{\partial}{\partial z}\left(K_z\frac{\partial c}{\partial z}\right) + \frac{\partial}{\partial y}\left(K_y\frac{\partial c}{\partial y}\right) + \frac{\partial}{\partial z}\left(K_{z2}\frac{\partial c}{\partial z}\right) + \frac{\partial}{\partial y}\left(K_{y2}\frac{\partial c}{\partial y}\right) - B + D \quad (1)$$

Here D represents the transport equations for recirculation; B is the pressure gradient force.

$$D = \frac{\partial(V_x\epsilon)}{\partial x} = \frac{\partial}{\partial x}\left(\frac{V_t}{\sigma_\epsilon}\frac{\partial\epsilon}{\partial x}\right) - k \quad (2)$$

$$B = dP.dA \quad (3)$$

Here V_x is the mean velocity in x direction, P is the pressure, A is the surface area, V_t is the kinematic viscosity, ϵ is the turbulence kinetic energy, k is the dissipation rate, σ_ϵ turbulent diffusivity coefficient.

$$\frac{\partial c}{\partial t} + V_x\frac{\partial c}{\partial x} = V_x\frac{\partial c}{\partial x} + \frac{\partial}{\partial x}\left(\frac{V_t}{\sigma_\epsilon}\frac{\partial\epsilon}{\partial x}\right) - k \quad (4)$$

Equation 7 has important implications for precipitation formation via the recirculation region. The recirculation region initiates the nucleation scavenging which grows the precipitation size. The ability of aerosol particles to act as a cloud condensation nuclei (CCN) depends on its size [10]. The solution of equation (4) is given as

$$\epsilon = \frac{\sigma_\epsilon}{V_t}(-V_x(kt + C - x) - PAx) \quad (5)$$

The solution i.e. equation (5) shows that the updraft term $xV_x\frac{\sigma_\epsilon}{V_t}$ was able to initiate the nucleation scavenging. Hence, large scale recirculation may influence precipitation over an area provided that other terms in equation (5) are quite low. This assumption was verified by the rainfall and aerosol optical depth over the research site.

3. Results and Discussion

In this section we tried verifying how intense is the relationship between the aerosol content of an area and the rainfall pattern. The data duration is thirteen years. First we examine the wind speed data (Figures 1(a) and 1(b)) from the Davis weather station to determine the various expected possibilities in the recirculation zones.

The wind data set for January, 2013 was used to estimate the recirculation zone in 1D where x=z. More specifically, the recirculation zone occurs more frequently along the SE, S, ESE, ENE, E, SSW and SSE as shown in Figure 1(b). If we assume that all the negative terms in equation 8 are minute, then the turbulence kinetic energy which defines the sustenance of the recirculation zone can be written as

$$\epsilon \approx x V_x \frac{\sigma_\epsilon}{V_t} \tag{6}$$

where

$$V_x(\theta) = \begin{cases} V\sin\theta & if \;\; 0 < x < 480, \quad 0 < \theta < pi/16 \\ V\cos\theta & if \;\; 480 \leq x < 960, \;\; pi/16 < \theta < pi/8 \\ V(\sin\theta + \cos\theta) & if \;\; 960 \leq x < 1440, \;\; pi/8 < \theta < pi/4 \\ V\sin^2\theta & if \;\; 1440 \leq x < 1920, \;\; pi/4 < \theta < pi/2 \\ V\cos^2\theta & if \;\; 1920 \leq x < 2400, \;\; pi/2 < \theta < pi \end{cases}$$

$\frac{\sigma_\epsilon}{V_t} = 0.01$, and V =7.8m/s i.e. the highest speed with an equivalently high frequency. Let x be the altitude of the lower atmosphere where altitude 1440 m to 2400 m represents the free troposphere. The recirculation activity was theoretically represented in Figure 2.

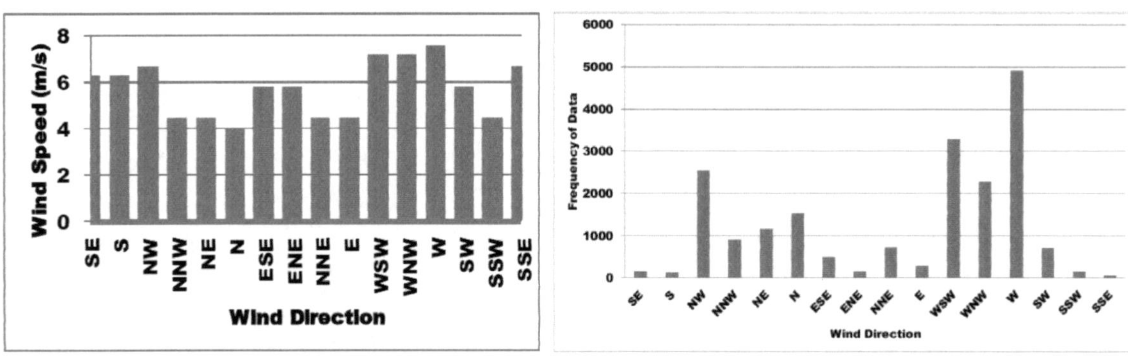

Figure 1: Magnitude of wind speed within different directions

Figure 1. (a) Magnitude of wind speed within different directions and (b) Frequency of wind directions

The 3D representation of the recirculation parameters illustrated in equation (9) where the maximum turbulence kinetic energy over the research area-which is about 261 J/Kg and occurs between 1000 m to 1500 m altitude (see Fig. 3). The divergence of recirculation features as seen by the red dotted lines is evidence that the recirculation zones are unpredictably dynamic. Hence, it possesses a verse ability to influence rainfall over an area. We understudy the rainfall pattern between the years 2000 to 2012 over the research area. A comparative analysis between the rainfall pattern and the aerosol optical depth was carried to further explain the dynamic nature of the recirculation zones. This study provides the hint into understanding the tropospheric anomaly noticed by Lee et al. [11]. Lee et al. [11] proposed that via statistical analysis that biomass burning aerosol (BBA) affects liquid clouds- leading to an increase in cloud droplet number concentrations and a decrease of droplet effective radii. In the long run, this process leads to geographical rearrangements of rainfall patterns. The theory of recirculation as explain in this paper sheds more light on the possibility of rainfall suppression and promotion [12, 13]. The recirculation activity can influence the normal deceleration (in form of normalization, accelerating or decreasing) of auto-conversion processes in the lower atmosphere [14].

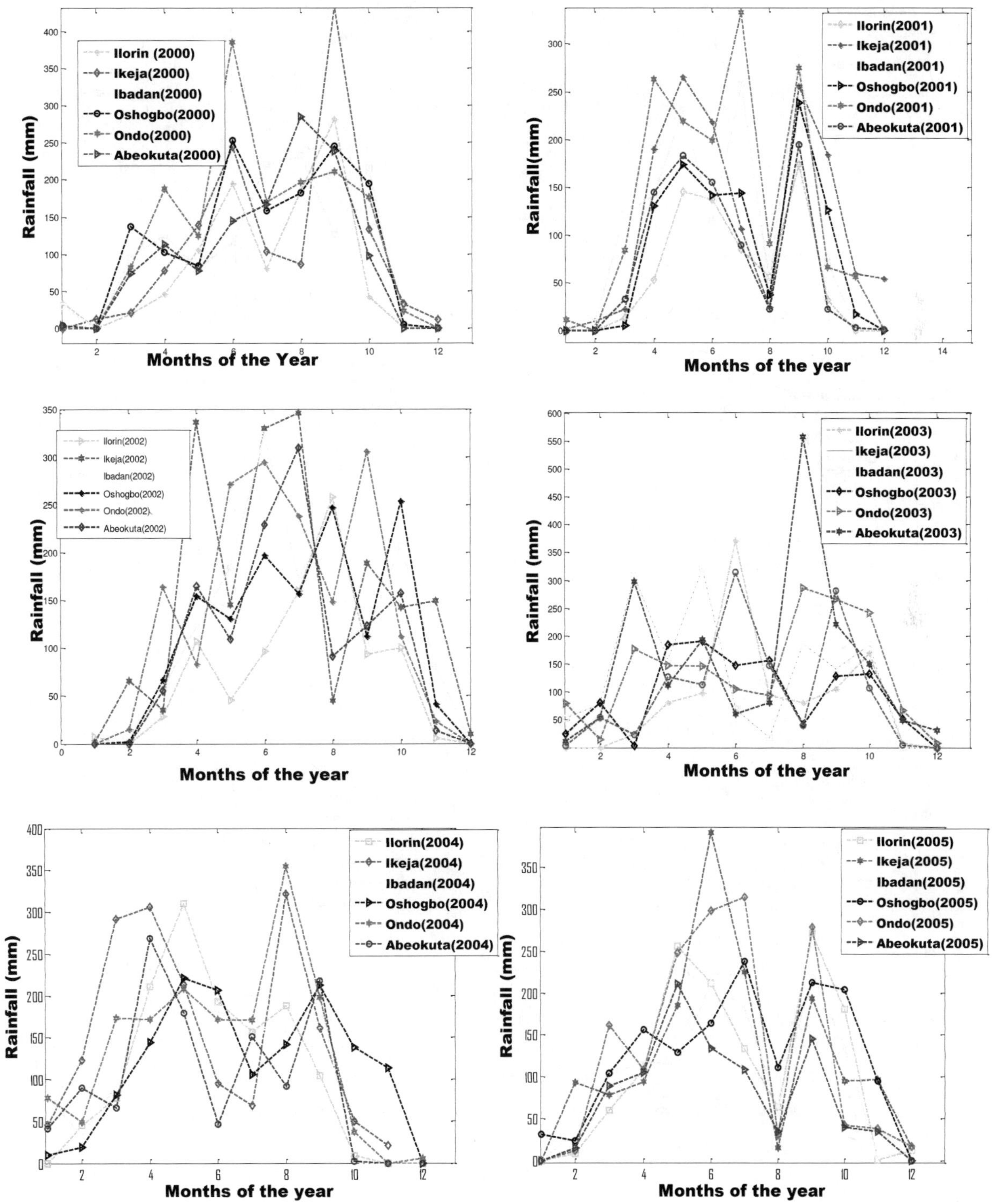

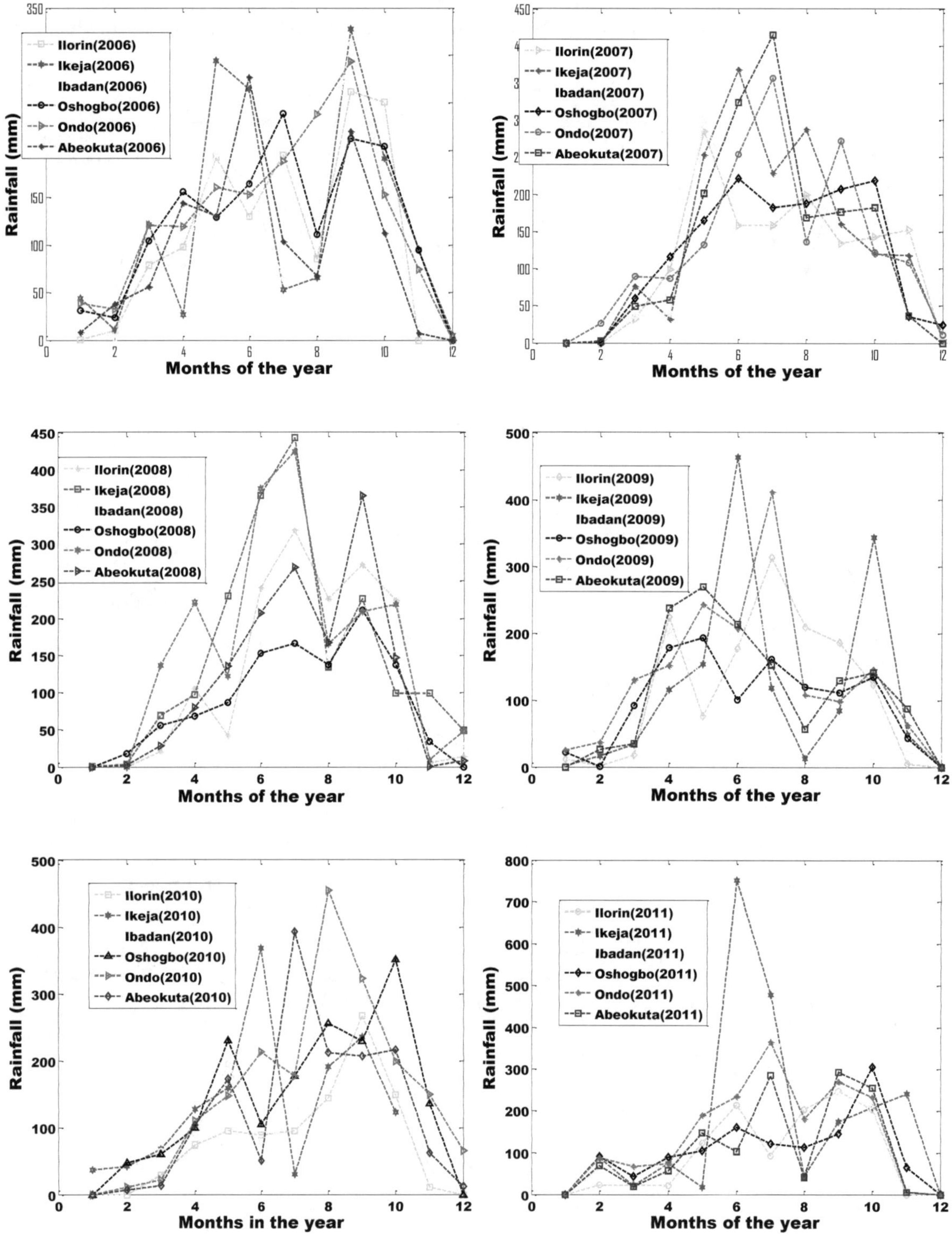
Ilorin(2006)
Ikeja(2006)
Ibadan(2006)
Oshogbo(2006)
Ondo(2006)
Abeokuta(2006)
Rainfall (mm)
Months of the year
Ilorin(2007)
Ikeja(2007)
Ibadan(2007)
Oshogbo(2007)
Ondo(2007)
Abeokuta(2007)
Rainfall (mm)
Months of the year
Ilorin(2008)
Ikeja(2008)
Ibadan(2008)
Oshogbo(2008)
Ondo(2008)
Abeokuta(2008)
Rainfall (mm)
Months of the year
Ilorin(2009)
Ikeja(2009)
Ibadan(2009)
Oshogbo(2009)
Ondo(2009)
Abeokuta(2009)
Rainfall (mm)
Months of the year
Ilorin(2010)
Ikeja(2010)
Ibadan(2010)
Oshogbo(2010)
Ondo(2010)
Abeokuta(2010)
Rainfall (mm)
Months in the year
Ilorin(2011)
Ikeja(2011)
Ibadan(2011)
Oshogbo(2011)
Ondo(2011)
Abeokuta(2011)
Rainfall (mm)
Months of the year

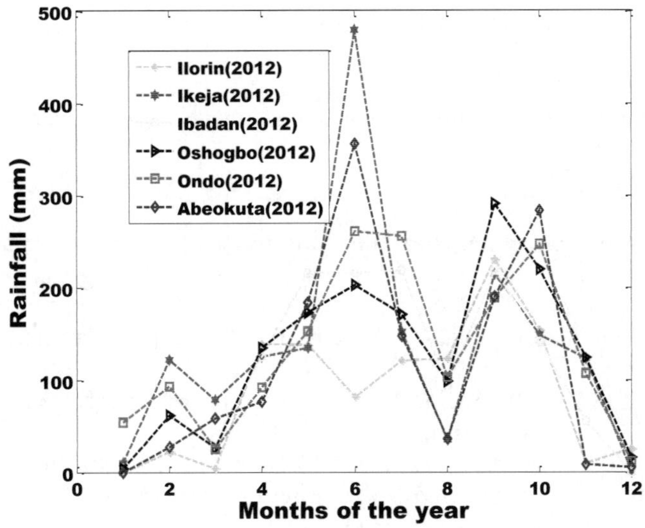

Figure 2. Rainfall pattern within the years 2000 to 2012

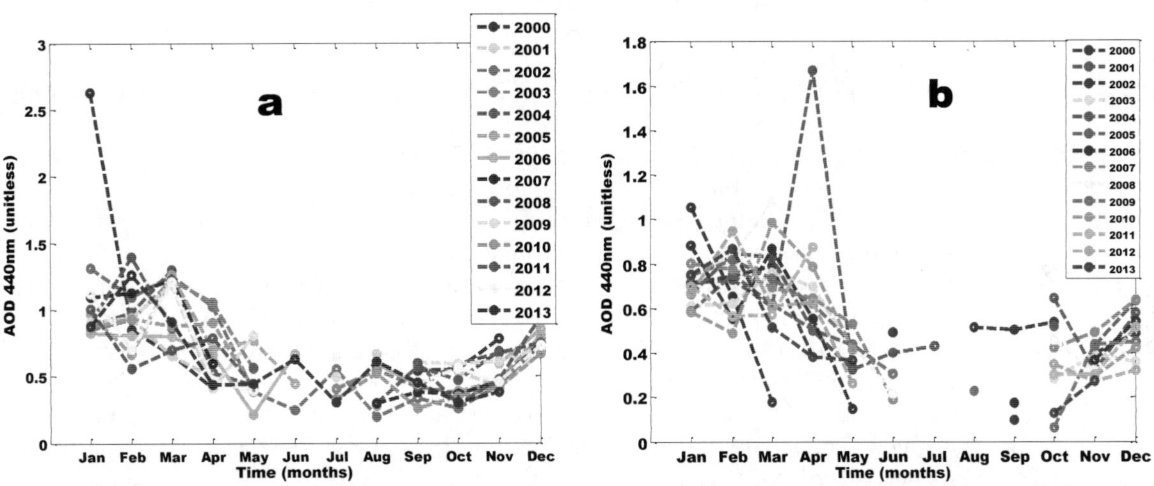

Figure 3. (a) AOD pattern within the years 2000 to 2012 for Lagos and (b) AOD pattern within the years 2000 to 2012 for Ilorin.

4. Conclusion

The recirculation zone is comprised the negative recirculation which supports the downdraft activities and positive recirculation which supports the updraft activities within the lower atmospheric setting. The negative recirculation has no influence on the rainfall pattern of an area-as seen in Osogbo-Osun. The positive recirculation has significant influence on the rainfall pattern of an area-as seen in Ikeja-Lagos. The maximum turbulence kinetic energy over the research area is about 261 J/kg and occurs between 1000 m to 1500m altitude. This altitude shows the role of recirculation in cloud formation and rainfall. Hence, the meteorological data set for the six locations shows that Ondo and Ikeja-Lagos has the highest value of turbulence kinetic energy. Also, we discover that the turbulence kinetic energy of negative recirculation is low i.e. compared with the turbulence kinetic energy of positive influence. The rainfall pattern of 2001 seems to be uniquely uniform with a clear evidence of uniform recirculation activity throughout the six locations. It was shown that recirculation activity influences the nature of clouds. The glaring results are the increase in cloud droplet number concentrations and a

IOP Conf. Series: Journal of Physics: Conf. Series **852** (2017) 012008 doi:10.1088/1742-6596/852/1/012008

decrease of droplet effective radii. This result is possible because nucleation scavenging triggers rainfall both at the long or short term as shown in the results.

Acknowledgement
The authors acknowledge the Nigeria Meteorological Agency, NASA and Covenant University for allowing us to use their data set.

References

[1] Andreas Bott 1999 A numerical model of the cloud-topped planetary boundary-layer: chemistry in marine stratus and the effects on aerosol particles *Atmospheric Environment* **33**(12)1921

[2] Rotstayn L D and Lohmann U 2002 Tropical rainfall trends and the indirect aerosol effect *J. Clim* **15**(15) 2103

[3] Allen R J, Sherwood S C, 2010 The aerosol-cloud semi-direct effect and land-sea temperature contrast in a GCM *Geophys. Res. Lett.* **37** L07702

[4] Mao K B, Ma Y, Xia L, Wendy Y C *et al.* 2014 Global aerosol change in the last decade: An analysis based on MODIS data *Atmospheric Environment* **94** 680

[5] Emetere M E, Akinyemi M L and Akin-Ojo O 2015 Aerosol Optical Depth Trends over Different Regions of Nigeria: Thirteen years Analysis *Modern Applied Science* **9**(9) 267

[6] Emetere M E , Akinyemi M L , Uno U E 2015 Computational analysis of aerosol dispersion trends from cement factory *Proceedings 2015 International Conference on Space Science & Communication* p 288

[7] Okpani P E 2009 Determination of aerosol mass loading in ilorin using remote sensing technique *The Pacific Journal of Science and Technology* **10**(2) 5

[8] Sadek R, Soulhac L, Brocheton F, Buisson E 2011 Evaluation of wind field and dispersion models in complex terrain, 14th HARMO conference pp 1-23

[9] Kim H G and Patel V C 200 Test of turbulence models for wind flow over terrain with separation and recirculation *Boundary-Layer Meteorology* **94** 5

[10] Westervelt D M, Pierce J R and Adams P J 2014 Analysis of feedbacks between nucleation rate, survival probability and cloud condensation nuclei formation *Atmos. Chem. Phys.* **14** 5577

[11] Lee D, Sud Y C, Oreopoulos L *et al.* 2014 Modeling the influences of aerosols on pre-monsoon circulation and rainfall over Southeast Asia *Atmos. Chem. Phys.* **14** 6853

[12] Iacobellis S F and Somerville R C J 2006 Evaluating parameterizations of the autoconversion process using a single-column model and atmospheric radiation measurement program measurements *J. Geophys. Res.* **111** D02203

[13] Diem J E 2013 Response to comments on Influences of the Bermuda High and atmospheric moistening on changes in summer rainfall in the Atlanta, Georgia region, the United States *International Journal of Climatology* **33** 3086

[14] Weber T and J Quaas 2012 Incorporating the subgrid-scale variability of clouds in the autoconversion parameterization using a PDF-scheme *J. Adv. Model. Earth Syst.* **4** M11003

IOP Conf. Series: Journal of Physics: Conf. Series **852** (2017) 012009 doi:10.1088/1742-6596/852/1/012009

Satellite observation analysis of aerosols loading effect over Monrovia-Liberia

M E Emetere,[1] F Esisio, [1] and F Oladapo [1]

[1]Department of Physics, Covenant University Canaan land, P.M.B 1023, Ota, Nigeria

E-mail : moses.emetere@covenantuniversity.edu.ng

Abstract. The effect of aerosols loading most often results in aerosols retention in the atmosphere. Aside the health hazards of aerosol retention, its effect on climate change are visible. In this research, it was proposed that the effect of aerosol retention also affects rain pattern. The Tropical Rainfall Measuring Mission (TRMM) layer 3 observations and the multi-imaging spectro-reflectometer (MISR) was used for the study. The aerosols loading over were investigated using sixteen years satellite observation in Monrovia-Liberia. Its effect on the rain rate over the region was documented. The results show that aerosol loading over the region is high and may have effect on farming in the nearest future. It was affirmed that the scanty AOD data was as a result of the rain rate that is higher within May and October.

1. Introduction

Several studies have been dedicated to modeling of aerosols and a good number of these models have established a proven degree of accuracy. For example, a multilayer analysis of a ground-based sun photometer of aerosol optical properties was carried out by some researcher [1] and compared with satellite observations over West Africa. Daily observations of satellite imaging of aerosol precipitations were made from 2005 – 2009 at three sites in West Africa and in an area within the Atlantic Ocean. Specific locations observed include Agoufou area of Mali, Banzoumbou area of Niger, Cape Verde in Tropical Atlantic Ocean and Ilorin Area of Nigeria. The results of the findings showed that the four stations had different annual precipitation, temperature and relative humidity. Spatial, seasonal and interannual variation of the aerosol loading over sahelian West Africa were detected by satellite (MODIS and TOMS) and ground-based AERONET Sunphotometer sensor between 2005-2009 as determined daily, monthly and annually showed that the Moderate resolution imaging spectro radiometer (MODIS) and TOMS retrieved as a ratio of aerosol optical depth to the aerosol interval AOD/AI were in good agreement with ground-based AERONET data.

Yang et al [2] carried out a study on the simulation of aerosol dynamics which includes a comparative review of algorithms used in Air Quality Models. The paper is focused on four areas including coagulation, condensational growth, nucleation, and gas particle mass transfer. The sectional and modal approaches were the analytical techniques adopted for analysis where they confirmed that the sectional approach has the capacity to make good predictions for coagulation and condensational growth phenomena. Based on the work, they recommended apt simulation of the rate of condensation with nucleation rates because of the inherent influence of environmental conditions. For mass transfer situations, they suggested an explicit treatment of the problem especially in situations where the

Content from this work may be used under the terms of the Creative Commons Attribution 3.0 licence. Any further distribution of this work must maintain attribution to the author(s) and the title of the work, journal citation and DOI.
Published under licence by IOP Publishing Ltd

volatile species undergo equilibrium reactions for components of different particle sizes where coarse particles are Present. Simulations were carried out for urban conditions in an area and over a period of 12 hours. The formulated algorithms for the four cases mentioned gave good results. The numerical solution adopted to describe the condensational growth rate of particles in the range of 0.001-10 µm has its limitation in the area of 3D simulation. However, the review confirms the reliability of the CIT, GATOL, Models-3, SAQM-AERO, UAM-AERO and UAM-AIM the following tools for handling coagulation and condensational growth of aerosol particulates. Based on the models mentioned, the nucleation rates predicted showed variations. They noted that all simulations were conducted in a standalone mode i.e. outside their 3-D host air quality models however, they suggested the need for further evaluation of some promising algorithms for aerosol dynamics and thermodynamics for a 3-D gridded setting where emissions, transport, dispersion, and deposition of aerosols are inherent.

Concentration profiles were generated based on calculations for an empty hollow tube and for another filled with rash rings. It was confirmed that a tube filled with packing has higher separation efficiency relative to an empty tube filled with a gaseous mixture (i.e. air and water) comprising aerosols. Separation efficiencies varied along the channel length as well as their flow regimes that were estimated from Reynolds Number. The determination of particle concentration profiles and the estimation of separation efficiencies using mathematical correlations/models can compare accurately with those obtained using contact devices can compare with model calculations. The obtained equations can be used for calculating the concentration distribution of aerosols and the efficiencies of industrial gas separators present at petrochemical and gas producing stations.

In this study, we intend to investigate the aerosol loading over Monrovia using established model [1] and estimate its effect on the rain rate over the location.

2. Methodology

In the past, no aerosols ground observation was available; hence, the satellite observation was adopted. Fourteen years satellite observation was obtained from the Multi-angle Imaging SpectroRadiometer (MISR). The MISR operates at various directions i.e. nine different angles (70.5°, 60°, 45.6°, 26.1°, 0°, 26.1°, 45.6°, 60°, 20.5°) and gathers data in four different spectral bands (blue, green, red, and near-infrared) of the solar spectrum. The blue band is at wavelength 443nm, the green band is at wavelength 555nm, the red band wavelength 670nm and the infrared band is at wavelength 865nm. MISR acquire images at two different levels of spatial resolution i.e. local and global mode. It gathers data at the local mode at 275 meter pixel size and 1.1 km at the global mode. Typically, the blue band is to analyze coastal and aerosol studies. The green band is to analyze Bathymetric mapping and estimating peak vegetation. The red band analysis the variable vegetation slopes and the infrared band analysis the biomass content and shorelines.

The raw MISR dataset was processed using the Excel package. The mean for each month were calculated for each year. We tested the accuracy of the data by applying the aerosol dispersion model that was propounded by Emetere et al. [1]. An extension of the dispersion model used is given as

$$\psi(\lambda) = {a_1}^2\cos\left(\frac{\mathrm{n}_1\pi\tau(\lambda)}{\mathrm{k_y}} + \alpha\right)\cos\left(\frac{\mathrm{n}_1\pi\tau(\lambda)}{\mathrm{k_z}} + \alpha\right) + {a_2}^2\cos\left(\frac{\mathrm{n}_2\pi\tau(\lambda)}{\mathrm{k_y}} + \beta\right)\cos\left(\frac{\mathrm{n}_2\pi\tau(\lambda)}{\mathrm{k_z}} + \beta\right) \quad (1)$$

Here α and β are the phase differences, k is the diffusivity, τ is the AOD, ψ is the concentration of contaminant, λ is the wavelength, a and n are atmospheric and tuning constants respectively.

3. Results and Discussion

The Monrovia has the highest cases of scanty data in the locations considered (Figure 1). There were little or no data from March to November. Since Monrovia falls within the tropical monsoonal climate, the scanty AOD data may be as a result of the moisture content [3], cloud scavenging [4], precipitable water content[5] and high rain drop rate [6]. The Tropical Rainfall Measuring Mission (TRMM) layer 3 observations in Figure 3 show the daily rain rate in Monrovia for the year 2012.

Figure 4 shows a monthly rain rate analysis. Hence, it can be affirmed that the scanty AOD data was as a result of the rain rate that is higher within May and October as shown in Figure 4. The AOD pattern over Monrovia agreed with proposed model (Figure 2). Monrovia is constantly under the oceanic wind influence from the Atlantic oceanic. The retrieved AOD may not capture salient events in the lower atmosphere; hence, a radiosonde station is essential in Monrovia to adequately capture the inadequacies in its lower atmosphere.

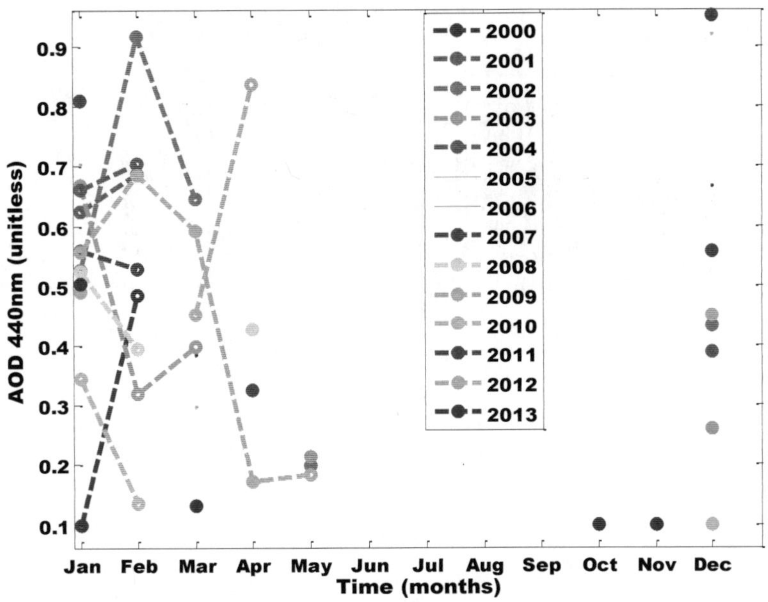

Figure 1. AOD pattern for Monrovia (2000 – 2013)

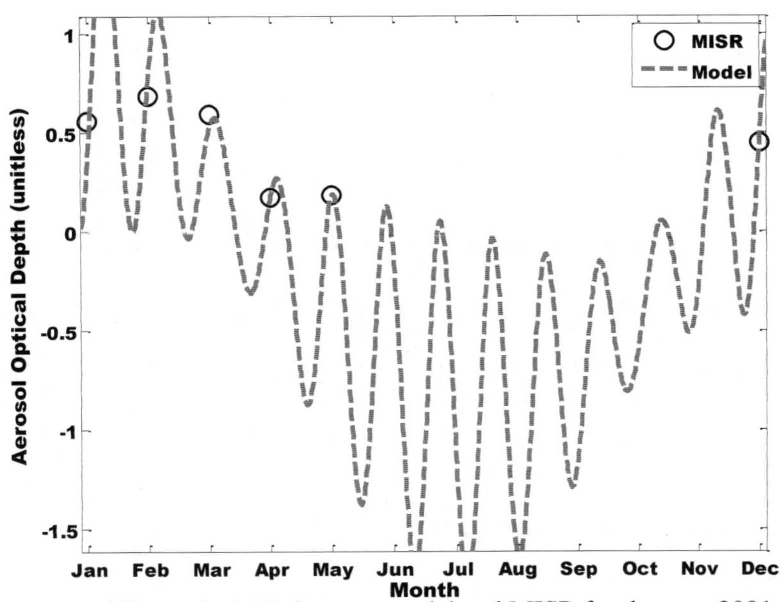

Figure 2. AOD for new model and MISR for the year 2001

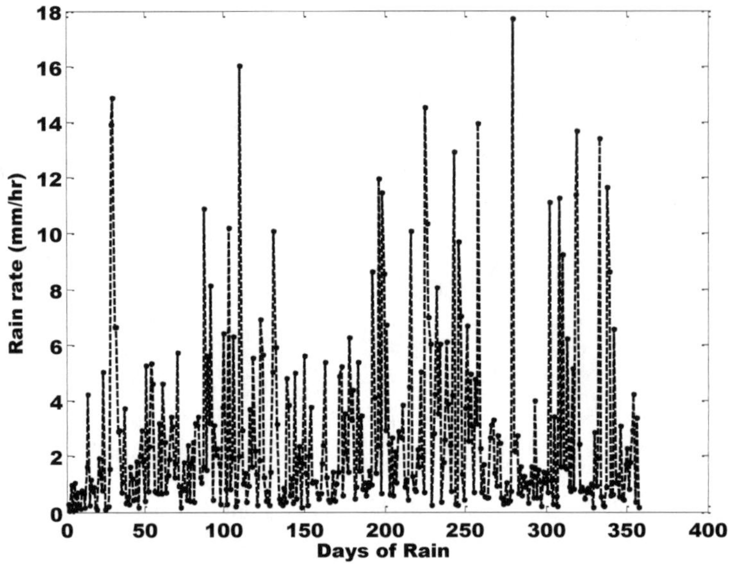

Figure 3. Daily Precipitation rate over Monrovia 2012

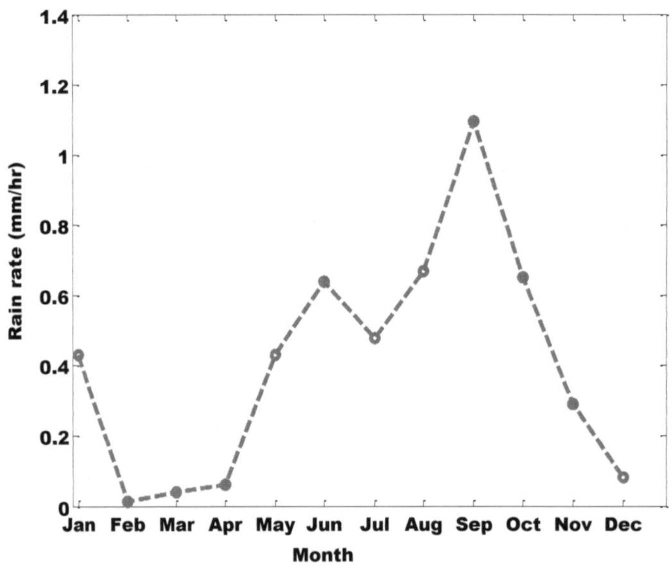

Figure 4. Monthly Precipitation rate over Monrovia 2012

The statistical results of the AOD are shown in Table 1 below. The highest frequency of data was found in 2012. The AOD for 2002, 2005 and 2011 was high with 0.92, 0.92 and 0.95 respectively. The highest mean was recorded in 2007. However, the result is not reliable becomes of the number of data for the year i.e. 2. Hence, the highest mean for 2011 was chosen because it had an average number of data i.e. 4. From the statistical analysis, the aerosols loading in 2011 is very significant. It shows that though the scanty measurement, the aerosol loading over Monrovia is far beyond the WHO limit [7].

Table 1. Statistical dataset for aerosol loading over Monrovia

Year	**2000**	**2001**	**2002**	**2003**	**2004**	**2005**	**2006**
Number of values	3	3	5	5	3	3	3
Maximum	0.48	0.69	0.92	0.67	0.7	0.92	0.67
Mean	0.23	0.47	0.54	0.37	0.58	0.64	0.52
Standard error	0.13	0.19	0.12	0.08	0.1	0.18	0.08
Standard deviation	0.22	0.32	0.27	0.18	0.17	0.32	0.14
Coefficient of variation	0.97	0.69	0.49	0.48	0.29	0.49	0.27

Year	**2007**	**2008**	**2009**	**2010**	**2011**	**2012**	**2013**
Number of values	2	4	4	3	4	6	3
Maximum	0.81	0.52	0.84	0.35	0.95	0.69	0.5
Mean	0.68	0.36	0.47	0.19	0.59	0.44	0.25
Standard error	0.13	0.09	0.15	0.08	0.13	0.09	0.13
Standard deviation	0.18	0.18	0.3	0.13	0.26	0.22	0.22
Coefficient of variation	0.26	0.51	0.64	0.69	0.44	0.49	0.92

In recent time, the focus of the West Africa atmospheric exploration by our team changed with specific emphasis on how to maximally control pollution by a unique awareness programme. Pilot studies on mobile apps were embarked upon. The initial successes were without its challenge. However, at this stage of the research we have designed an algorithm shown in Figure 5.

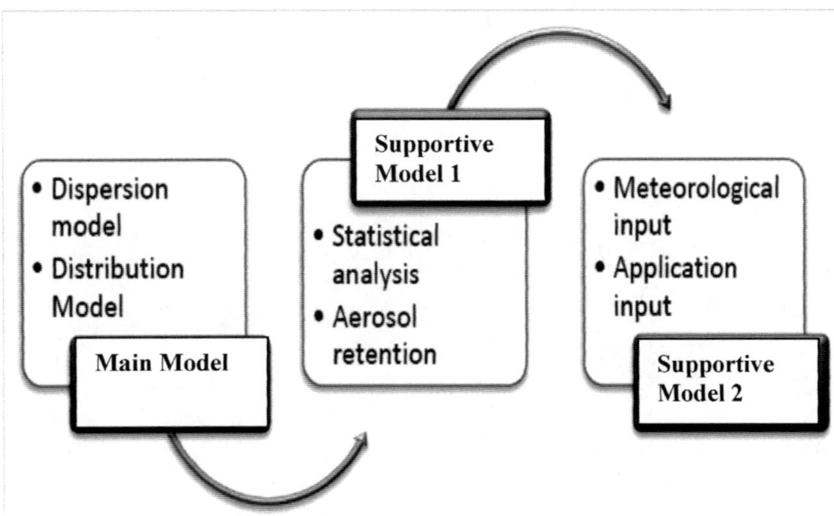

Figure 5. Proposed software description of environmental software

4. Conclusion

The aerosols loading over Monrovia is very high. The effect was observed on the overall rain rate over the region. Hence, if nothing is done to ameliorate the continuous aerosols loading, it may lead to partial famine in the nearest future.

IOP Conf. Series: Journal of Physics: Conf. Series **852** (2017) 012009 doi:10.1088/1742-6596/852/1/012009

Appreciation
The authors gratefully acknowledge Covenant University.

References

[1] Emetere Moses E., Akinyemi M.L. & Oladimeji T.E. 2016. Statistical Examination Of The Aerosols Loading Over Kano-Nigeria: The Satellite Observation Analysis, *Scientific Review Engineering and Environmental Sciences,* **72** 167-176

[2] Yang, Y., Liao, H. and Lou, S.-J., 2014. Simulated impacts of sulphate and nitrate aerosol formation on surface-layer ozone concentrations in China. *Atmospheric and Oceanic Science Letters*, **7(5)** 441-446.

[3] Adebiyi, A.A., Paquita, Z. and Steven, J.A., 2015, The convolution of dynamics and moisture with the presence of shortwave absorbing aerosols over the southeast atlantic. *Journal of Climate*, **28** 1997–2024.

[4] Dani, K.K., Maheskumar, R.S., and Devara, P.C.S., 2003. Study of total column atmospheric aerosol optical depth, ozone and precipitable water content over Bay of Bengal during BOBMEX-99. *Journal of Earth System Science*, **112 (2)** 205-221.

[5] Vijayakumar, K. and Devara, P.C.S., 2013. Study of aerosol optical depth, ozone, and precipitable water vapour content over Sinhagad, a high-altitude station in the Western Ghats. *International Journal of Remote Sensing*, **34 (2)** 613-630.

[6] Boucher, O. and Quaas, J., 2013. Water vapour affects both rain and aerosol optical depth, *Nature Geosciences*, **6** 4–5

[7] UN Report (2016). http://www.un.org/sustainabledevelopment/blog/2016/09/vast-majority-of-world-6-76-billion-people-living-with-excessive-air-pollution-un-report/

IOP Conf. Series: Journal of Physics: Conf. Series **852** (2017) 012010 doi:10.1088/1742-6596/852/1/012010

Effect of aerosols loading and retention on surface temperature in the DJF months

M E Emetere[1], L Onyechekwa[1] and P Tunji-Olayeni[1]

[1]Covenant University Canaan land, P.M.B 1023, Ota, Nigeria

E-mail : moses.emetere@covenantuniversity.edu.ng, onyechekwa@yahoo.com, patience.olayeni@covenantuniveristy.edu..ng

Abstract. The effect of aerosols loading most often results in aerosols retention in the atmosphere. Aside the health hazards of aerosol retention, its effect on climate change are visible. In this research, it was proposed that the effect of aerosol retention also affects the fluctuation of the surface temperature. The location of study is Enugu, Nigeria (6.4584° N, 7.5464° E). Twenty-nine years GISS Surface Temperature Analysis (GISTEMP) data set and sixteen years MISR aerosol optical data set were used. The fluctuations in the sixteen years aerosol optical depth (AOD) tallied with the surface temperature. The curve-fitting tool of Matlab was used to generate a polynomial for the surface temperature and used to project a five years prediction of the surface temperature.

1. Introduction

Aerosols comprise mainly of dust particulates and their dynamics in open space will require a systemic approach to the modelling of their dispersion, loading, precipitation, their molecular interactions and kinetics. Aerosols have benefits attached to their use. In agriculture, with large particle entrainment in the clouds, the air is polluted thus, poisoning the carbon dioxide required for green plants to photosynthesize; this may also result in abnormal growth. Aerosols are sprayed over farmlands as pesticides which help to eradicate plant pests but at high concentrations, the soil nutrients may be altered which makes the soil/farmland unfavourable for cultivation/planting. In addition to some of the already mentioned disadvantages, aerosols may be poisonous and injurious to health, plant and animal life if it escapes into the atmosphere. It is also believed that contagious diseases such as flu are trapped in the air by aerosols and are transported to their targets as they disperse. Aside the anthropogenic sources of aerosols loading of the atmosphere, the north-east dust winds are constant influence in all parts of West Africa.

The concept of aerosol retention shows that all aerosols dispersed into the atmosphere do not decay or fizzle-out quickly [1, 2]. These aerosols are retained in the atmosphere and stay for as long as their natural lifetime [3-5]. The estimation technique of the retained aerosol i.e. aerosols retention was propounded by Emetere [1]. Hence, the season of highest aerosols retention in the year is December, January and February [2]. This concept may be arguable looking at the AOD data for these months and the likely events of biomass burning due to farm activities. The statistical approach of Emetere [1] showed a clear argument of highest aerosols retention within the DJF.

The major properties of atmospheric aerosol have been substantiated in mathematical expressions in order to understand its predictability and applicability in meteorological terrains. For example, Hansen *et al.* [6] simplified the radiative forcing of climate.The shown in Figure 1 below.

Content from this work may be used under the terms of the Creative Commons Attribution 3.0 licence. Any further distribution of this work must maintain attribution to the author(s) and the title of the work, journal citation and DOI.
Published under licence by IOP Publishing Ltd

The largest radiative forcing (Figure 2.1) are the positive forcing initiated by greenhouse gases (GHG) while the negative forcing are initiated by aerosols. Also, it can be seen in Figure 1 that both the positive and negative forcing have significant effect on the global temperature [6]. Aerosols indirect radiative forcing (especially in warm clouds) is traceable to two sources: aerosols cloud condensation nuclei and change in precipitation efficiency.

Popular models used to estimate aerosols' indirect radiative forcing is the aerosol size-resolved model which has been used to show that the influential aerosols physical properties are important to estimate radiative forcing. For instance, the wind field activates the sea salt aerosols' size-segregated surface emission rates and its collective impact on the radiative forcing [8]. The objective of this research is to show the effect of aerosol loading and retention over surface temperature by considering a large data sets.

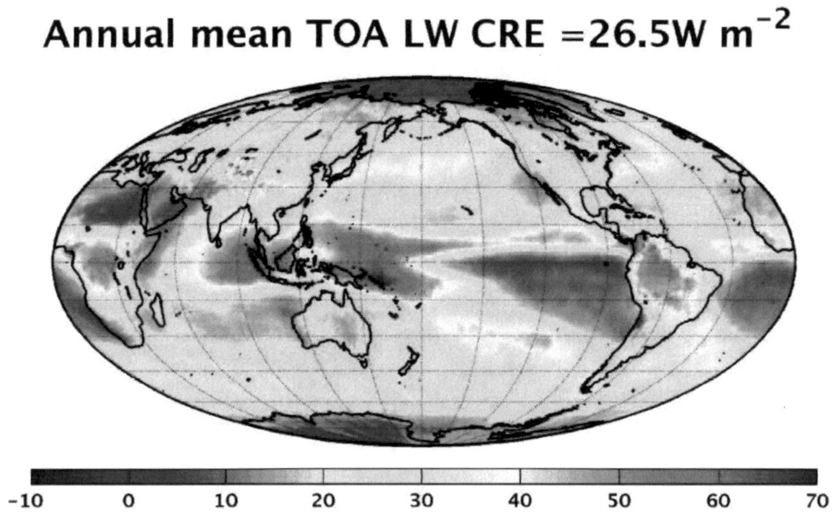

Figure 1. Global effect of atmospheric effect on radiative forcing [7]

2. Methodology

The location of study is Enugu-east Nigeria. The site is located on 6.4584° N and 7.5464° E. Twenty-nine years GISS Surface Temperature Analysis (GISTEMP) data set was obtained from the NASA site. Also, sixteen years MISR aerosol optical data sets were obtained from MISR website. The data processing was accomplished using the Microsoft Excel and Matlab programme. The curve-fitting tool on the Matlab was used to determine the polynomial expression for 29 years GISTEMP data sets. The polynomial expression was later used to project five years event.

3. Results and Discussion

The aerosol loading over the research location for sixteen years is presented in Figure 2 below. It can be observed that the highest and consistent aerosol loading record can be found in the DJF. Hence, the various fluctuations within the DJF should have impact on the climatic forcing. The scanty data sets between March and November is a clear indication of the active aerosols loading and retention in the DJF. Also, it has been reported that scanty over most parts of West Africa was due to atmospheric constant configurations in the MISR algorithm [1, 5, 6] We examined the twenty-nine years GISTEMP data sets for DJF. The GISTEMP for December is shown using the solid line in Figure 3. It can be observed that the scanty data of 2004 (Figure 2) can be corroborated with the drastic fall of the surface temperature in Figure 3. The other AOD fluctuations could also be observed in the surface temperature. Hence, there is a good relationship established between the AOD and surface temperature. We projected the expected surface temperature using mathematical expression from the polynomial curve-fitting for five years (Figure 4).

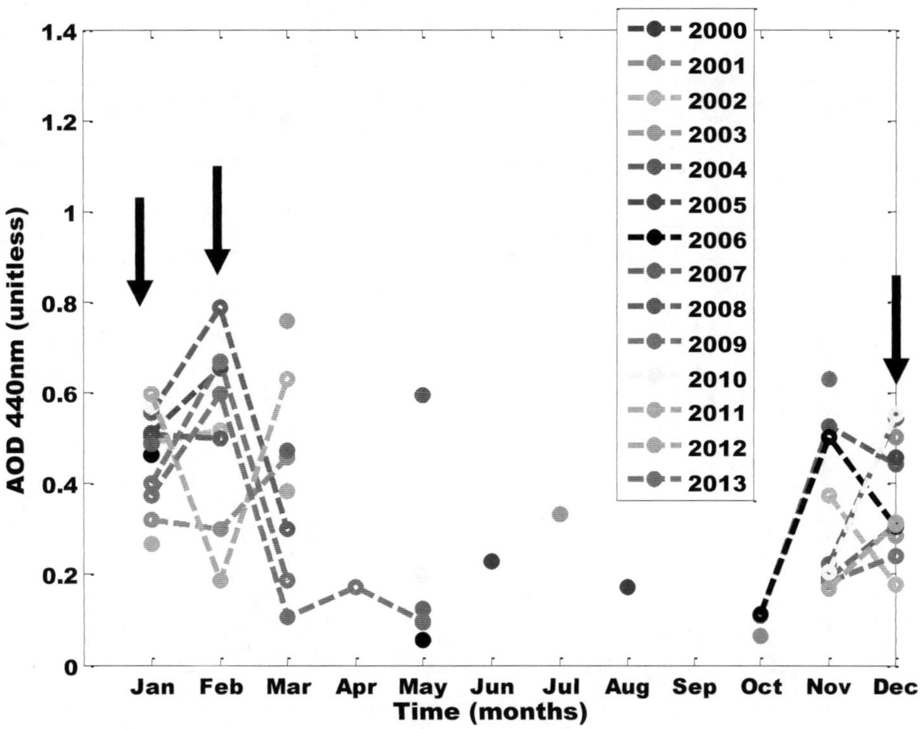

Figure 2. AOD pattern for Enugu (2000 – 2013)

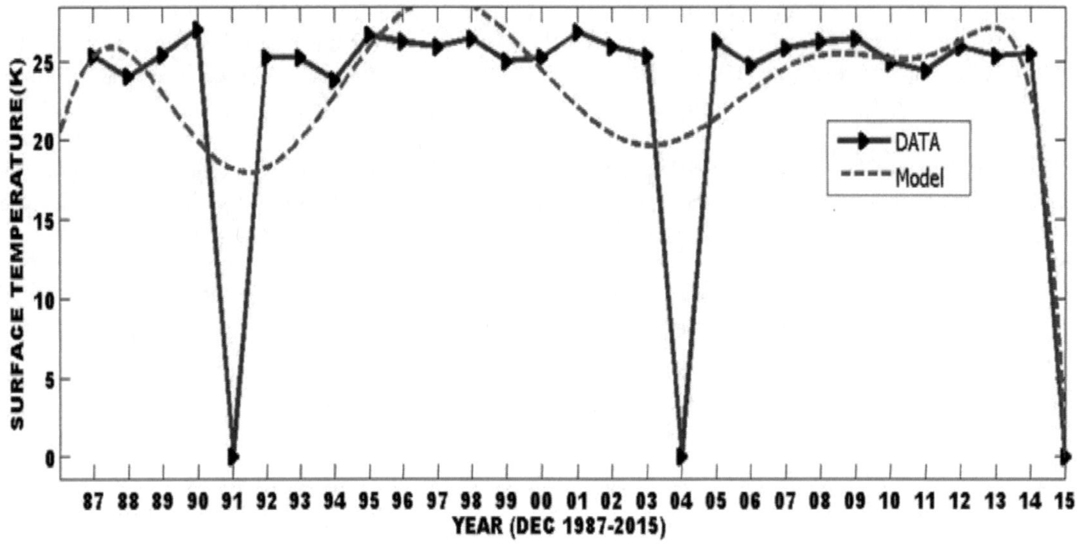

Figure 3. Surface temperature and the polynomial fit from 1987-2015 for December

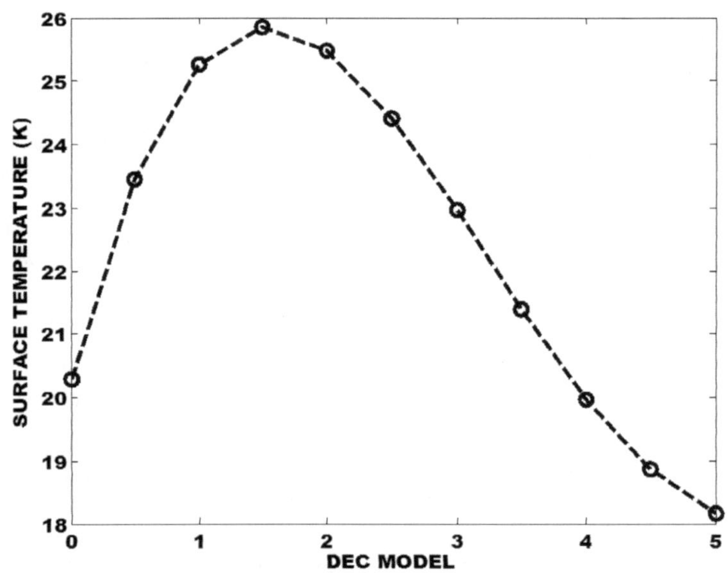

Figure 4. Five years surface temperature forecast for December

It was assumed that the AOD at the five years remains the over the location. It was projected that if the aerosol loading over remains constant for the next five years, the highest surface temperature over Enugu may be observed around mid-2017 while the lowest maybe observed in the year 2020.

The GISTEMP for January is shown using the solid line in Figure 5. It can be observed that the scanty data of 2006 (Figure 5) can also be corroborated with the drastic fall of the surface temperature in Figure 3. It was noted that this sharp decrease of surface temperature was visible in 1992, 1993 and 2015. Unlike the December dataset who had a drastic fall in 1991 and 2015 i.e. showing fair perturbations / fluctuations, the January data set had higher perturbations / fluctuations. We projected the expected surface temperature using mathematical expression from the polynomial curve-fitting for five years (Figure 6). The highest surface temperature over Enugu may be observed around late 2017 while the lowest maybe observed in the year 2020.

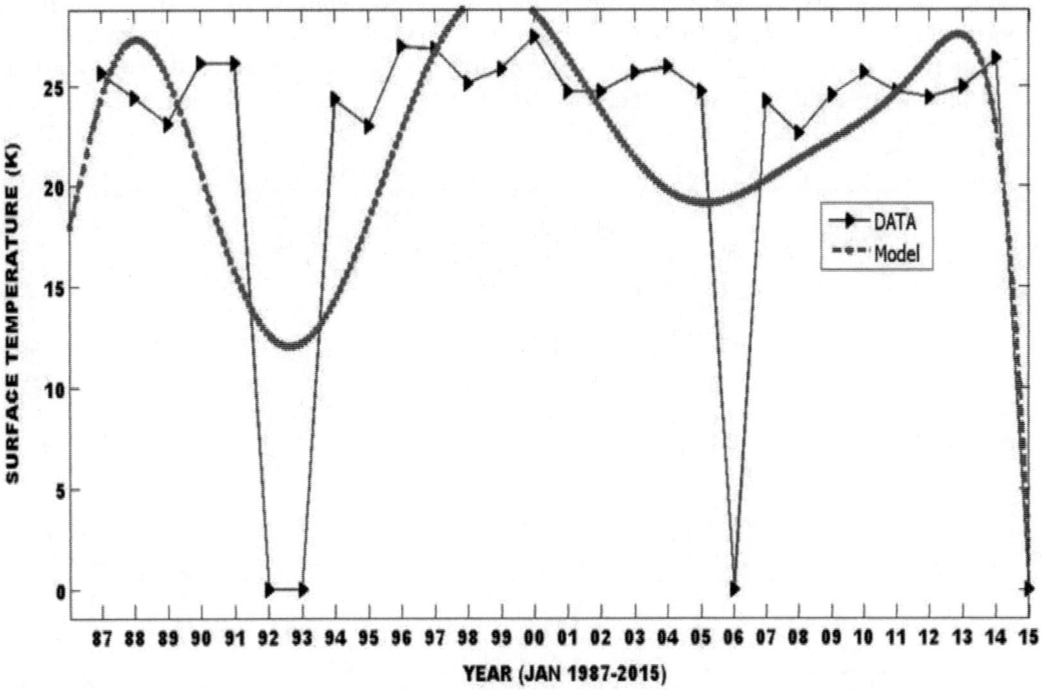

Figure 5. Surface temperature and the polynomial fit from 1987-2015 for January

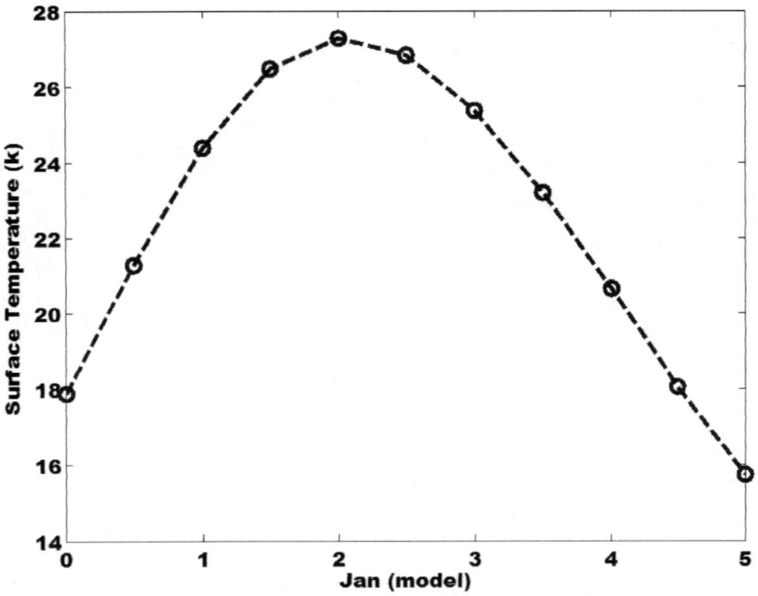

Figure 6. Five years surface temperature forecast for January

The GISTEMP for February is shown using the solid line in Figure 7. It can be observed that the scanty data of 2010 and 2013 (Figure 7) can also be corroborated with the drastic fall of the surface temperature in Figure 3. It was noted that this sharp decrease of surface temperature was visible also in 1993 and 1997. January and February had almost the same surface temperature fall in 1993. Unlike the December and January dataset, the February data set do not tend to fall by 2015. We projected the expected surface temperature using mathematical expression from the polynomial curve-fitting for five years (Figure 8). The highest surface temperature over Enugu may be observed around mid to late 2017 while the lowest maybe was in the year 2015.

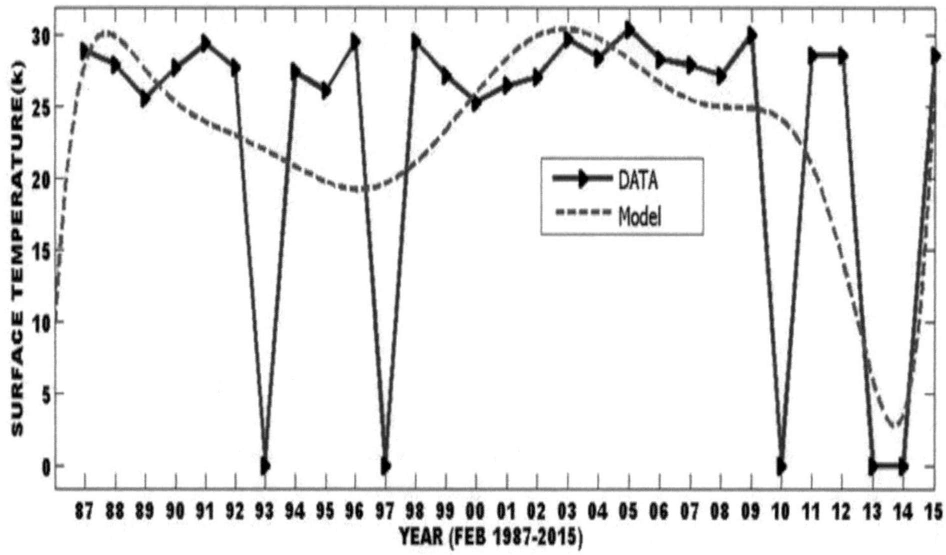

Figure 7. Surface temperature and the polynomial fit from 1987-2015 for January

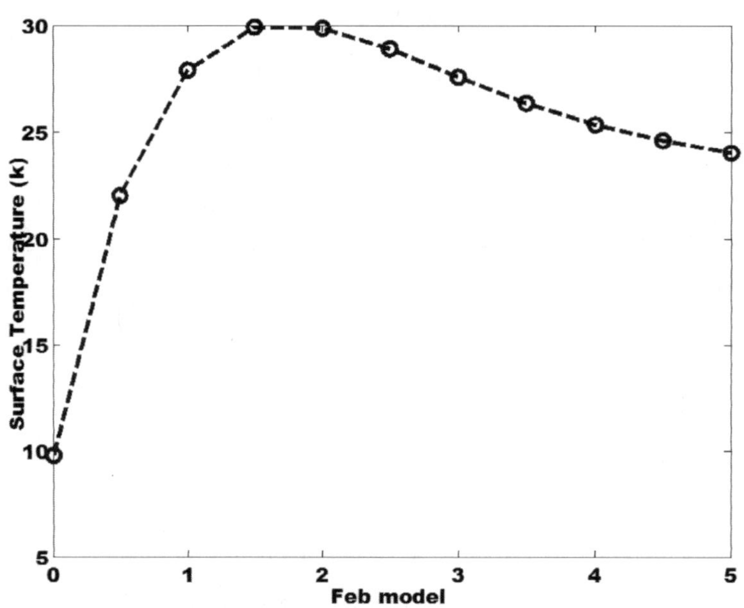

Figure 8. Five years surface temperature forecast for January

IOP Conf. Series: Journal of Physics: Conf. Series **852** (2017) 012010 doi:10.1088/1742-6596/852/1/012010

4. Conclusion

The impact of aerosol loading which is expected to lead to the increase of the surface is expected to have highest impact in 2017. The effect of tropospheric aerosol is therefore very visible in climate forcing. The five year projection of February shows that the highest aerosol loading and retention is obtained within this period.

Appreciation

The authors appreciate the partial sponsorship of Covenant University. The authors thanks NASA for the satellite imagery

References

[1] Emetere M E 2016 Statistical examination of the aerosols loading over Mubi-Nigeria: the satellite observation analysis *Geographica Panonica* **20**(1) 42

[2] Emetere M E, Akinyemi M L and Oladimeji T E 2016 Statistical examination of the aerosols loading over Kano-Nigeria: The Satellite Observation Analysis *Scientific Review Engineering and Environmental Sciences* **72** 167

[3] Emetere M E, Akinyemi M L and Oladimeji T E 2016 a conceptual mathematical examination of the aerosols loading over Abuja-Nigeria *Communications in Mathematics and Applications* **7**(1) 47

[4] Emetere M E, Akinyemi M L and Akinwumi S A 2016 Aerosols loading trends and its environmental threats over Cotonou-Benin *Nature Environment and Pollution Technology* **15** (3) 1111

[5] Emetere M E, Akinyemi M L and Akinojo 2016 Effects of band superposition on the satellite imagery of aerosol optical depth over west Africa *Journal of Engineering and Applied Sciences* **11**(1) 17

[6] Hansen J, Sato M and Ruedy R 1997 Radiative forcing and climate response *Journal of Geophysical Research* **102** 6831

[7] Kato S, Loeb N G, Rose F G *et al.* 2013 Surface irradiances consistent with CERES-derived top-of-atmosphere shortwave and longwave irradiances *J. Clim.*,**6** 2719

[8] Bjorn S 2015 Rethinking the lower bound on aerosol radiative forcing *J. Clim.* **28** 4794

IOP Conf. Series: Journal of Physics: Conf. Series **852** (2017) 012011 doi:10.1088/1742-6596/852/1/012011

Aerosols loading statistical dimensions over Serekunda-Gambia

M E Emetere[1], I Ebubechukwu[1], O Ekene[1] and E Charles[1]

[1]Department of Physics, Covenant University Canaan land, P.M.B 1023, Ota, Nigeria

E-mail : moses.emetere@covenantuniversity.edu.ng

Abstract. Aerosols loading and retention justifies the efficiency of the 'self-cleansing' mechanism of the geographical air space. The advantage of the proposed technique which is based on the coefficient of variation helps to estimate the aerosols retention over a geographical area without necessarily going through the complexities of formulating sectional equations to illustrate the physics of the atmosphere. The system of aerosol retention over Serekunda suggests that air pollution should be taken more seriously to avoid natural disasters in the nearest future. It was discovered that the 'self-cleansing' cycle over Serekunda is four years. The year with a fairly high aerosols retention and correlation was between 2006 & 2007 i.e. 36.82 and 0.61 respectively.

1. Introduction

West Africa has the highest biomass burning, dust interference from Sahara and anthropogenic pollution [1-3]. The technicality of aerosol retention over the air space of any geographical region cannot be over emphasized because of its importance in ascertaining the state of the natural 'self-cleansing' mechanism of the atmosphere. Therefore, the state of aerosols dispersion and retention is equally important for the adequate understanding of aerosols loading and transport. Within the tropospheric layer, the shape of the size distribution, number of aerosol particulates and the chemical composition of the average aerosol particles vary as a function of altitude.

Since, this scientific omission of aerosols retention extends to the radiosonde temperature biases noticed on a wide variation between the temperature of device sensor and its surrounding air [4], there is the need solve this challenge to aid sondes manufacturers. However, this is not the objective of this study. In this paper, we investigated the aerosols loading over Serekunda. Serekunda is the most populous city in Gambia and it is located on longitude 16.6667 ^{o}W and latitude 13.4333 ^{o}N in the Sahelian geographic region south of the Sahara (see Figure 1), hence, we expect a high impact of the north east winds alongside Sahara dust. Over the past forty years Serekunda as well as Gambia has experienced a decline in mean total annual rainfall [5]. The volume of automobiles and anthropogenic activities in Serekunda increases the carbon aerosols within its atmosphere. The anthropogenic activities is biomass burning from farm activities in the mangrove forest. Since there is poor ventilation in the evening, the carbon and Sahara dust aerosols would accumulate over few months.

In summary, we propose that the success of the any of the parameter for determining aerosols loading depends strictly on the aerosols retention over a geographical area. In this paper, we intend to statistical analyze the aerosol loading over Serekunda for further analysis.

Content from this work may be used under the terms of the Creative Commons Attribution 3.0 licence. Any further distribution of this work must maintain attribution to the author(s) and the title of the work, journal citation and DOI.
Published under licence by IOP Publishing Ltd

IOP Conf. Series: Journal of Physics: Conf. Series **852** (2017) 012011 doi:10.1088/1742-6596/852/1/012011

2. Methodology

We shall be adopting major Microsoft excel statistical tool. We also adopted already validated aerosol models. Aerosols retention between two consecutive years has been postulated by Emetere [6] as:

$$A = \left|\frac{G_r - G_P}{G_r}\right|^2 \times 100\% \quad (1)$$

Here, the previous and current years are denoted as G_P and G_r respectively. The beauty of this formulation is the inclusion of the possibility of obtaining CV that is above unity. However, the possibility of obtaining above 100% is inevitable. Hence the proposition of the second equation:

$$A = \left|\frac{G_P}{G_P - G_r}\right|^2 \times 100\% \quad (2)$$

Equation (2) is valid only if $G_P - G_r > G_P$. Upon this salient assumption, the aerosol retention can be controlled below unity or 100%.

Fourteen years satellite observation was obtained from the Multi-angle Imaging Spectro Radiometer (MISR). The raw MISR dataset was processed using the Excel package. The aerosol optical depth (AOD) mean for each month were calculated for each year.

3. Result and Discussion

Serekunda and its environs are prone to increased frequency and severity of drought events (Figure 1). The rainy season which runs from mid-June to October has little influence on the AOD, hence, the type of aerosols in abundance would be carbon-compound aerosols. The AOD trend agrees with the Emeteres' model which shows that the model can be used to forecast future events over Serekunda. In 2010, the AOD was highest between January and July while the previous year i.e. 2009 had almost uniform. After 2010 i.e. 2011 the AOD trend was the most turbulent as shown in Figure 1. Hence, the maximum AOD for 2009, 2010 and 2011 are 0.65, 1.83 and 0.72 respectively. After 2009-2011, it was observed that the value of AOD increased in 2012 and 2013 i.e. 0.8 and 0.82 respectively.

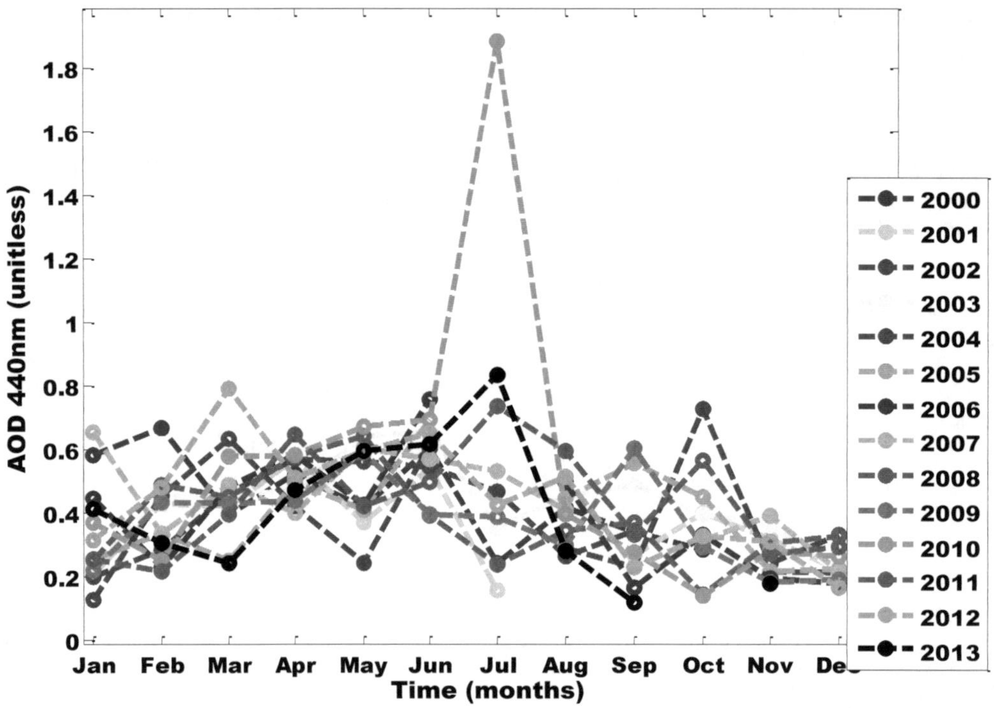

Figure 1. AOD pattern for Serekunda 2000 - 2013

The statistical analysis is shown in Tables 1 & 2. The highest AOD mean, 95% confidence interval, 99% confidence interval, variance, standard deviation and coefficient of variation was in 2010. The highest skew and kurtosis can be found in 2005 and 2003 respectively. The highest Kolmogorov-Smirnov stat can be found in 2002. This results shows that the lower atmosphere of Serekunda has high rate of turbulence. Hence we examine the atmospheric aerosol retention shown in Tables 3 & 4.

Table 1. Statistical AOD analysis 2000-2006

Statistics Tool	2000	2001	2002	2003	2004	2005	2006
Mean	0.448	0.356	0.377	0.432	0.393	0.395	0.391
Standard error	0.045	0.041	0.045	0.039	0.059	0.046	0.057
95% confidence interval	0.102	0.091	0.100	0.086	0.13	0.103	0.127
99% confidence interval	0.146	0.129	0.142	0.122	0.186	0.147	0.181
Variance	0.02	0.018	0.022	0.018	0.038	0.024	0.036
Standard deviation	0.142	0.135	0.149	0.136	0.194	0.154	0.190
Coefficient of variation	0.318	0.379	0.396	0.314	0.494	0.389	0.485
Skew	0.152	0.161	0.157	0.082	0.534	0.557	0.469
Kurtosis	-1.17	-0.926	-1.90	-0.043	-0.960	-0.375	-0.41
Kolmogorov-Smirnov stat	0.184	0.110	0.186	0.165	0.167	0.18	0.165

Table 2. Statistical AOD analysis 2007-2013

Statistical Tool	2007	2008	2009	2010	2011	2012	2013
Mean	0.448	0.379	0.374	0.533	0.436	0.44	0.407
Standard error	0.041	0.042	0.04	0.134	0.049	0.05	0.071
95% confidence interval	0.091	0.091	0.089	0.294	0.108	0.12	0.16
99% confidence interval	0.129	0.129	0.126	0.415	0.153	0.16	0.23
Variance	0.018	0.021	0.02	0.215	0.029	0.03	0.05
Standard deviation	0.135	0.144	0.14	0.463	0.171	0.18	0.224
Coefficient of variation	0.302	0.38	0.375	0.869	0.391	0.41	0.551
Skew	-0.03	0.372	0.253	2.567	0.396	0.32	0.651
Kurtosis	-0.83	-0.19	-0.744	7.571	-1.141	-0.4	-0.269
Kolmogorov-Smirnov stat	0.154	0.127	0.151	0.28	0.143	0.15	0.175

The highest correlation was found between 2001 & 2002 i.e. 0.74. The highest aerosols retention is within 2010 & 2011 i.e. 67.06. The year with a fairly high aerosols retention and correlation was between 2006 & 2007 i.e. 36.82 and 0.61 respectively. This means that 2007 was the trigger point that results in the events around 2010 and 2011.

Table 3. Atmospheric aerosols retention over Serekunda 2001-2006

	2001	2002	2003	2004	2005	2006
Aerosol retention	2.6	0.18	6.76	13.3	7.37	3.95
Correlation	0.59	0.74	0.29	0.29	0.53	0.59

Table 4. Atmospheric aerosols retention over Serekunda 2007-2013

	2007	2008	2009	2010	2011	2012	2013
Aerosol retention	36.82	4.18	0.014	32.31	67.06	0.26	6.32
Correlation	0.61	0.37	0.2	0.63	0.29	0.26	0.61

The year of highest atmospheric aerosols retention was found between 2010 and 2011. This explains the sudden increase in the AOD data for July, 2010 (Figure 2). Therefore, knowing the aerosols retention in the atmosphere is as important as knowing the aerosols dispersed into the atmosphere. The significance of the aerosols retention over Serekunda portrays that its high value in 2004, 2007, 2010 and 2011 are not mere coincidence but shows that the 'self-cleansing' cycle is four years. However, climate change influence is gradually altering the 'self-cleansing' cycle as shown in 2011. The significance of the atmospheric aerosols retention in a geographical region has great influence on aviation schedules [7], human health [8], measuring instruments, energy budget and meteorology [6]. It is therefore suggested that the recovery of the regional atmosphere can be done via accurate instrumentation configuration. Hence, the flow chart in Figure 2 below.

IOP Conf. Series: Journal of Physics: Conf. Series **852** (2017) 012011 doi:10.1088/1742-6596/852/1/012011

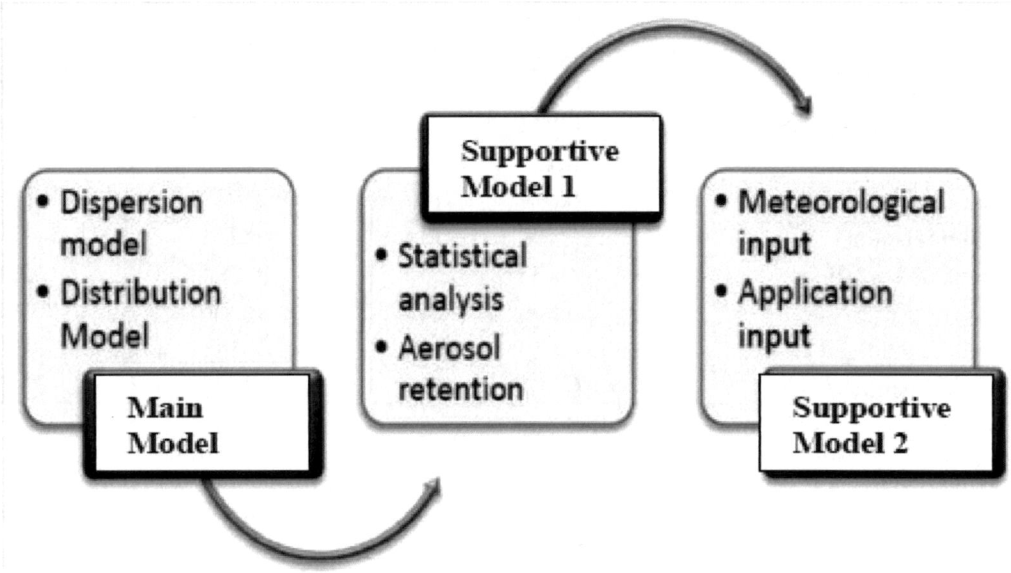

Figure 2. Architecture of aerosol retrieval Technique

4. Conclusion

The influences of aerosols retention over all aerosols dispersion or loading parameters have been proven. It was shown that there is a correspondence between aerosols loading and retention. The year with a fairly high aerosols retention and correlation was between 2006 & 2007 i.e. 36.82 and 0.61 respectively. Therefore, aerosols retention in the atmosphere is as important as knowing the aerosols dispersed into the atmosphere. Also it was discovered that the 'self-cleansing' mechanism in the atmosphere has a unique cycle. For Serekunda, the 'self-cleansing' cycle is four years i.e. 2007 to 2010.

Acknowledgements

The author appreciates NASA for using their satellite imagery. The author acknowledges the partial sponsorship of Covenant University. The author declares no conflict of interest.

References

[1]. Emetere M E, Akinyemi M L and Oladimeji TE 2016 Statistical examination of the aerosols loading over Kano-Nigeria: The Satellite Observation Analysis *Scientific Review Engineering and Environmental Sciences* **72** 167

[2]. Emetere M E, Akinyemi M L and Akinwumi S A 2016 Aerosols loading trends and its environmental threats over Cotonou-Benin *Nature Environment and Pollution Technology* **15**(3) 1111

[3]. Emetere M E, Akinyemi M L and Oladimeji T E 2016 A conceptual mathematical examination of the aerosols loading over Abuja-Nigeria *Communications in Mathematics and Applications* 7(1) 47

[4]. Sun B, Reale A, Schroeder S, Seidel D J and Ballish B 2013 Toward improved corrections for radiation induced biases in radiosonde temperature observations *J. Geophys. Res. Atmos.* **118** 4231
[5]. Njie M 2007 The gambia national adaptation programme of action (NAPA) on climate change. Banjul, The Government of The Gambia
[6]. Emetere M E 2016 statistical examination of the aerosols loading over mubi-nigeria: the satellite observation analysis *Geographica Panonica* **20**(1) 42
[7]. Gettelman A and Chen C 2013 The climate impact of aviation aerosols *Geophys. Res. Lett.* **40** 1
[8]. Wyzga R E and Folinsbee L J 1995 Health effects of acid aerosols *Water, Air, and Soil Pollution* **85** (1) 177

Impact of meteorological parameters over Covenant University, Ota, Nigeria

E O Falayi[1,2], M R Usikalu[2], T V Omotosho[2], O S Ojoniyi[1] and S A Akinwumi[2]

[1]Department of Physics, Tai Solarin University of Education, P.M.B. 2118, Ijebu Ode, Ogun State, Nigeria
[2]Department of Physics, Covenant University, P.M.B. 1023, Ota, Ogun State, Nigeria

Email: moji.usikalu@covenantuniversity.edu.ng

Abstract. This study utilized monthly daily values of solar radiation, humidity, ultraviolet dose, temperature, pressure and wind speed obtained at Department of Physics, Covenant University, Ota, Nigeria, for 2013. It was observed that correlation coefficients vary from 0.134 - 0.955 across the months. This was further demonstrated with values of coefficient of determination R^2 which vary from 0.018-0.911 for Ota stations. Based on the RMSE, MPE, MBE and p values, we obtained an Equation that produces the best correlation. Our result reveals the models that produce the smallest p values which are considered as the best model for estimating the solar radiation at Ota site with an acceptable error. Also, Wavelet spectrum based approach was employed to analyse the meteorological monthly series in a sequence of monthly scales of January to December 2013.

1. Introduction

Solar radiation is the key parameter in the design of the solar energy devices used by engineers, agriculturist, hydrologist and architects. Solar energy received on the Earth surface is not equally distributed [1, 2]. Nigeria is among the nations with abundant solar radiation due to its tropical location. Utilization of solar energy is essential in Nigeria where there is insufficient supply of electricity generation and distribution. The amount of the solar energy received varies significantly with climatic condition and weather patterns of locations.

The direct beam and diffuse solar energy are produced from the global solar energy incident on a horizontal surface. Pyranometers, Solarimeters, or Actinography are instrument used to measure diffuse solar energy, while the Pyrheliometer device measure direct beam solar radiation. The measuring devices can be employed to develop an empirical model for solar energy that describes the numerical relations between the solar energy and the meteorological variables such as maximum and minimum temperature, relative humidity, pressure and wind speed.

Different models developed by different researchers for the prediction of solar radiation using different meteorological data [3, 4, 5, 6, 7]. These parameters include sunshine hours, relative humidity, sunshine duration, maximum and minimum temperature, cloud cover, vapor pressure and pressure. Numerical linear and nonlinear functions, artificial neural network and fuzzy logic have been presented by literatures in the prediction of solar energy [8, 9]. The correctness of the developed model in global solar radiation is evaluated using statistical analysis such as the mean error (MPE), mean bias error (MBE) and root mean square error (RMSE) and t-test statistics [9, 10, 11, 12, 13].

True prediction of the solar radiation in a given location depends on the exact determination of the strength of solar radiation at that location. Determining the correct method for evaluating the daily solar radiation on horizontal surface using obtainable data has influenced this research study. In this article, we present the measurement of solar radiation for Ota from reliable model pyranometer. We developed equations that correlate monthly average daily solar radiation with meteorological

Content from this work may be used under the terms of the Creative Commons Attribution 3.0 licence. Any further distribution of this work must maintain attribution to the author(s) and the title of the work, journal citation and DOI.
Published under licence by IOP Publishing Ltd

parameters (solar radiation, humidity, ultraviolet dose, temperature, pressure and wind speed) for Ota, Nigeria. The Wavelet spectrum based approach was also employed to analyze the meteorological monthly series.

2. Data Analysis

Davis weather station was used for the experimental measurement. It incorporates Very Small Aperture Terminal (VSAT) satellite dish antenna, installed at the top-roof of College of Science and Technology of Covenant University, Ota. One minute data interval extracted for daily, monthly, and yearly for 2013. It is used for the experimental measurement which incorporates Very Small Aperture Terminal (VSAT) satellite dish antenna, installed at the top-roof of College of Science and Technology of Covenant University, Ota. The measurement is on the latitude 6.7° N and longitude 3.23° E while the elevation angle of the receiver antenna is 59.9°.

2.1 Correlation

Multiple linear regressions of five parameters (relative humidity RH, ultraviolet dose UV, temperature T, pressure P and wind speed WS were employed to estimate the solar radiation for Ota. One variable correlation is presented as

$$SR = 355.3 - 2.256RH \quad (\mathrm{r} = 0.255;\ \mathrm{R}^2 = 0.065) \tag{1}$$

$$SR = 231.17 + 14.72T \quad (\mathrm{r} = 0.782;\ \mathrm{R}^2 = 0.612) \tag{2}$$

$$SR = 136.11 + 8.195W \quad (\mathrm{r} = 0.134;\ \mathrm{R}^2 = 0.018) \tag{3}$$

$$SR = 11397 - 11.126P \quad (\mathrm{r} = 0.745;\ \mathrm{R}^2 = 0.555) \tag{4}$$

$$SR = 128.10 + 3674.04UV \quad (\mathrm{r} = 0.426;\ \mathrm{R}^2 = 0.181) \tag{5}$$

Eq. 1 to Eq. 5 show the relationship between solar radiation and other variables (temperature T, pressure P and wind speed WS). It was observed that solar radiation correlated well with temperature. The correlation of 0.782 exists between the solar radiation and temperature, also coefficient determination is 0.612 implies that 61.2% of solar radiation can be accounted for using temperature. Two variables correlations:

$$SR = -314.79 + 16.010T + 19.41WS \ (\mathrm{r} = 0.841;\ \mathrm{R}^2 = 0.708) \tag{6}$$

$$SR = -752.83 + 4.175RH + 20.61T \ (\mathrm{r} = 0.858;\ \mathrm{R}^2 = 0.737) \tag{7}$$

$$SR = 11999 - 11.755P + 4253.9UV \ (\mathrm{r} = 0.893;\ \mathrm{R}^2 = 0.797) \tag{8}$$

$$SR = 12295 + 18.456W - 12.062P \ (\mathrm{r} = 0.801;\ \mathrm{R}^2 = 0.642) \tag{9}$$

$$SR = 109.43 + 7.40W + 3641UV \ (\mathrm{r} = 0.443;\ \mathrm{R}^2 = 0.196) \tag{10}$$

$$SR = -260.68 + 1473T + 3682UV \ (\mathrm{r} = 0.891;\ \mathrm{R}^2 = 0.795) \tag{11}$$

$$SR = 571.144 + 13.805T - 0.7702P \ (\mathrm{r} = 0.783;\ \mathrm{R}^2 = 0.613) \tag{12}$$

$$SR = 16382 + 4.533RH - 16.45P \ (\mathrm{r} = 0.831;\ \mathrm{R}^2 = 0.691) \tag{13}$$

$$SR = 361.5 - 2.6978.5RH - 12.68W \text{ (r = 0.3248; R}^2 = 0.105) \quad (14)$$

$$SR = -341.8 + 16.010W + 19.412T \text{ (r = 0.842; R}^2 = 0.708) \quad (15)$$

Two variable correlations were exhibited between Eq. 6 and Eq.15, we observed that solar radiation correlated well with pressure and ultraviolet dose in Eq. 8. The correlation of 0.893 exists between the solar radiation pressure and ultraviolet dose, the coefficient of determination is 0.797implies that 79.9 % of solar radiation can be accounted using pressure and ultraviolet dose. Three variables correlations

$$SR = -340.15 + 15.97T + 18.59W + 3602.8UV \text{ (r = 0.842; R}^2 = 0.708) \quad (16)$$

$$SR = 5258.36 + 4.600RH - 5.82P + 14.28T \text{ (r = 0.867; R}^2 = 0.751) \quad (17)$$

Three variable correlations were exhibited between Eq. 16 and Eq.17. We observed better correlation in Eq. 17 between solar radiations with relative humidity, pressure and temperature. 0.867 is the correlation that exists between the solar radiation with relative humidity, pressure and temperature. 0.751 is the coefficient of determination; this implies that 75.1 % of solar radiation can be accounted for using relative humidity, pressure and temperature. Four variables correlations

$$SR = -482.1 + 17.60T + 17.86W + 3084.8UV + 1.19RH \text{ (r = 0.942; R}^2 = 0.889) \quad (18)$$

$$SR = 4890.9 + 4.160RH - 5.48P + 15.15T + 16.57W \text{ (r = 0.905; R}^2 = 0.820) \quad (19)$$

Four variable correlations were exhibited between Eq. 18 and Eq.19. We observed better correlation in Eq. 18 between solar radiations with temperature, wind speed, ultraviolet dose and relative humidity. 0.942 is the correlation that exists between the solar radiation with temperature, wind speed, ultraviolet dose and relative humidity. 0.889 is the coefficient of determination; this implies that 88.9 % of solar radiation can be accounted for using temperature, wind speed, ultraviolet dose and relative humidity. Five variables correlations:

$$SR = 6904.4 + 9.62T - 7.136P + 17.68W + 3268.5UV + 1.57RH \quad \text{(r = 0.955; R}^2 = 0.911) \quad (20)$$

Five variables in Eq. 20 exhibit relationships between solar radiation with temperature, pressure, wind speed, ultraviolet dose and relative humidity, 0.955 is the correlation that exists between the solar radiation with temperature, wind speed, ultraviolet dose and relative humidity. 0.911 is the coefficient of determination; this implies that 91.1 % of solar radiation can be accounted for using temperature, pressure, wind speed, ultraviolet dose and relative humidity.

The RMSE, MBE, MPE and p-values were computed to show the correctness of the estimated values in Eq. 21- Eq.24 (see Table 1).

$$RMSE = \left\{ \left[\frac{1}{n} \sum \left(SR_{pred} - SR_{obs} \right)^2 \right] \right\}^{\frac{1}{2}} \quad (21)$$

$$MBE = \frac{1}{n} \left[\sum \left(SR_{pred} - SR_{obs} \right) \right] \quad (22)$$

$$MPE = \frac{\left\lfloor \sum \left(SR_{obs} - SR_{pred} \right) \times 100 \right\rfloor}{n} \tag{23}$$

$$p = r\sqrt{\frac{(n-2)}{(1-r^2)}} \tag{24}$$

where SR_{pred}, SR_{obs}, n and r are the predicted, observed, no of samples and correlation coefficient respectively.

Table 1. Correlation and Statistical Indicators

Equations No	r	R^2	RMSE	MBE	MPE	P-VALUES
1	0.255	0.065	22.905	3.671E-14	-2.049	0.424
2	0.782	0.612	14.750	-1.776E-14	-0.839	0.003
3	0.134	0.018	23.475	-3.434E-14	-2.197	0.678
4	0.745	0.555	15.796	-7.970E-14	-0.929	0.005
5	0.426	0.18	21.432	-8.290E-15	-1.845	0.167
6	0.841	0.708	12.793	-2.487E-14	-1.062	0.001
7	0.858	0.736	12.151	4.026E-14	-0.593	0.000
8	0.893	0.797	10.677	5.921E-14	-0.429	0.000
9	0.802	0.643	13.745	1.906E-13	-0.783	0.003
10	0.443	0.196	18.680	-2.013E-14	-1.833	0.191
11	0.891	0.795	10.737	1.421E-14	-0.468	0.000
12	0.783	0.613	14.746	-7.105E-14	-0.836	0.003
13	0.831	0.691	13.177	8.491E-13	-0.644	0.001
14	0.326	0.106	19.342	-1.303E-14	-1.991	0.539
15	0.842	0.708	12.793	2.487E-14	-0.679	0.001
16	0.939	0.882	8.116	2.250E-14	-0.497	0.000
17	0.867	0.752	11.803	7.460E-14	-0.015	0.000
18	0.943	0.889	7.891	-4.832E-14	-0.311	0.000
19	0.906	0.820	10.039	-8.657E-13	-0.435	0.000
20	0.955	0.911	7.065	-8.704E-13	-0.242	0.000

2.2 Wavelet Power Spectrum

Wavelet transformation is an important mathematical device that generates a time-frequency representation in the time domain signals [14, 15, 16, 17]. These mathematical devices can be used in meteorology and climatology parameters investigation. It is now a day to day device for examining variation of power within a time series. Braking down time series into time-frequency space, it is an easy ways to establish both the variability of dominant modes and to examine its variation dominate modes with time. This section, we examined the wavelet analyses of wind speed, temperature and relative humidity data obtained from department of physics, Covenant University, Ota. The wind speed, temperature and relative humidity were examined using wavelet power spectrum analysis (see Figure 1(a-c)).

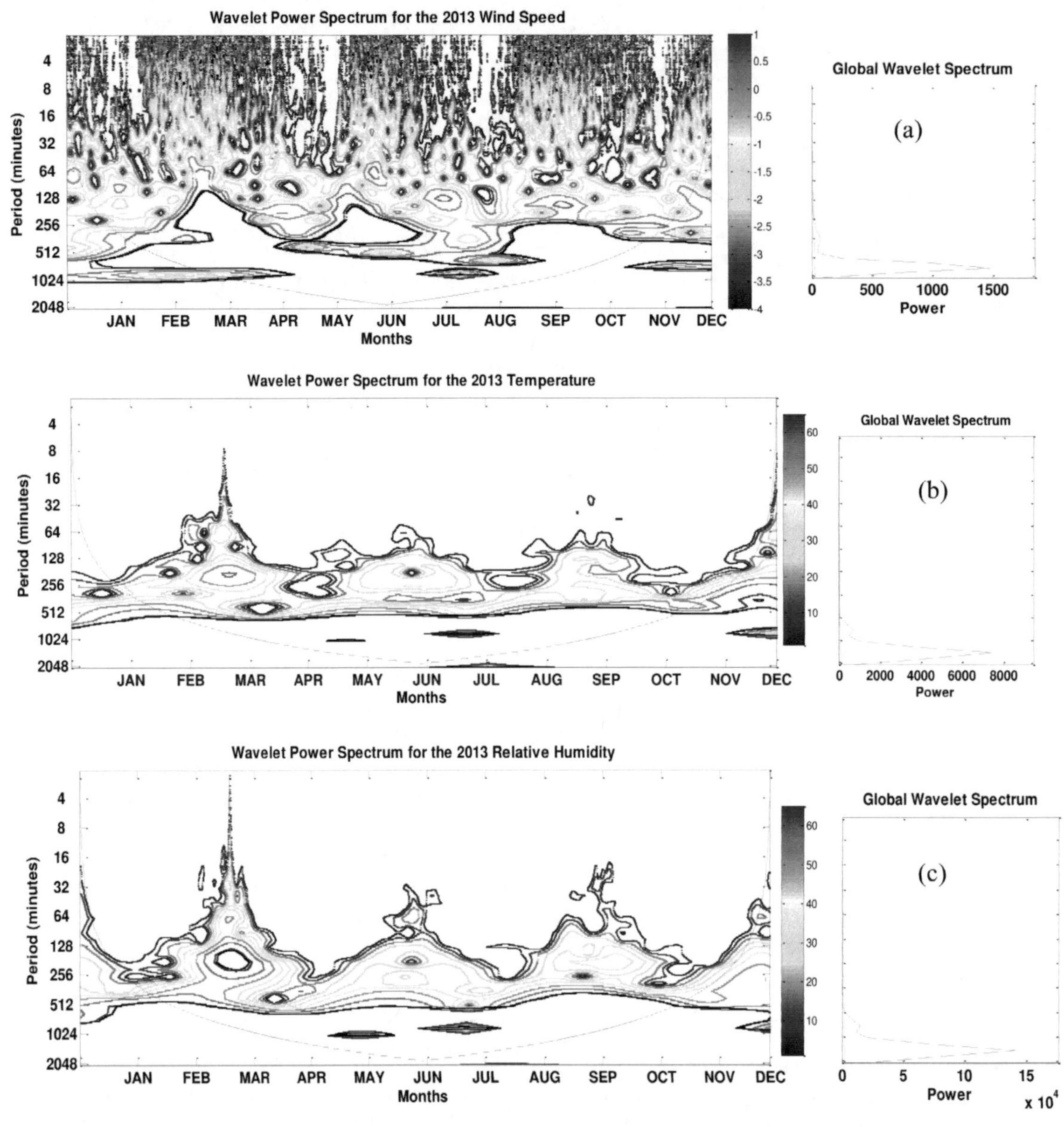

Figure 1. (a-c) Wavelet Powers Spectrum (WPS) and Global Wavelet Spectrum (GWS) of wind speed, temperature and relative humidity.

3. Discussion of the results

The monthly global solar radiation was estimated using relative humidity, maximum temperature, Ultraviolet radiation, pressure and wind speed to develop regression coefficient. Table 1 presented regression statistics that is correlation coefficient r, coefficient of determination R^2 obtained from Eq. (1) - Eq. (20). The MBE, RMSE, MPE and p values are computed from Eq. (21) – Eq. (24). We obtained positive and negative values of MBE across the models. This depicts that the models vary from over to under approximation of solar radiation. The RMSE is often used to evaluate the variation between values predicted by a model. MPE provides long term performance of the examined regression equations. A low value of MPE is sensible. The correlation coefficient is significant if the calculated t is higher than the critical t (P = 0.05). In summary, correlation coefficients vary from

0.134 - 0.955 across the months. This implies that, there are statistically significant relationships between the solar radiation, the relative humidity, maximum daily temperature, ultraviolet and wind speed, this is further demonstrated with values of coefficient of determination R^2 vary from 0.018-0.911 for Ota stations. The best model is obtained from Equation 20 based on the statistical computation of RMSE, MBE, MPE and p values and with high correlation coefficient as shown in Tables 1. From our result, the models that produce the smallest values of the p values is considered as the best model for estimating the solar radiation at Ota site with moderate error. This means that the models of equations (20) and (18) are good estimate for the solar radiation in Ota location during the period time in the present study.

Figure 1(a-c) shows the WPS (wavelet power spectrum) and GWS (global wavelet spectrum). The colours in the WPS denote signal energy, which is referred to as magnitude of wavelet transforms. The frequency giving maximum to the energy of the signal over the period is denoted with red colour. We observed that that periodic cycle bands of 512-1024 minutes bands have the highest variability (strongest intensity) during January - April, July – August and November- December months. This implies that these time series have a strong annual signal. The large empty space signifies the missing data gaps. The GWS power decreases with increase in wind speed, pressure, temperature and relative humidity which suggest decrease in variability (see Figure 1a). As for the periods of variations of monthly average of temperature (Fig. 1b), the periodicity is characterized by 12months period. It is noted that at 256-540 period show more concentration of power of high temperature from November to March. Figure 1c depicts periodicity properties of monthly average of relative humidity of Ota area. The 1-year period is significant and dominant in the May-July with low surface temperature, high relative humidity within the periodicity of relative humidity of Ota. The relative humidity increases when the air temperature dropped due to presence of moist air. When the temperature decreases, the relative humidity was at peaks, meaning that the amount of rainfall depended on relative humidity, but independent of maximum temperature.

4. Conclusion

Different variables were engaged in the study solar radiation, humidity, ultraviolet dose, temperature, pressure and wind speed to obtained Eq. (1) - Eq. (20). Based on the RMSE, MPE, MBE, r, R^2 and p values, Equation (20) produces the best regression correlation, while Equation (3) give the weakest regression correlation with high values of RMSE, MPE, p-value and low values of r, and R^2 . The equation could be employed to estimate solar radiation of Ota location in the South- west, Nigeria.

Acknowledgements

The authors acknowledge the Management of Department of Physics, College of Sciences and Technology, Covenant University, Ota, Ogun State, Nigeria, for provision of solar radiation and other meteorological data.

References

[1] Ahmad F and Ulfat I 2004 Empirical models for the correlation of monthly average daily global solar radiation with hours of sunshine on a horizontal surface at Karachi Pakistan *Turkish J. Physics* **28** 301

[2] Falayi E O and Rabiu A B 2005 Modeling global solar radiation using sunshine duration data *J. Physics* **17S** 181

[3] Angstrom A S 1924 Solar and terrestrial radiation. Report to the international commission for solar research on actinometric investigations of solar and atmospheric radiation *Meteorological Society* **50** 121

[4] Prescott J A 1940 Evaporation from water surface in relation to solar radiation *Trans. Roy. Soc. S. Austr.* **64** 114

[5] Awachie I R N and Okeke C E 1990 New empirical model and its use in predicting global solar irradiation *Nig. J. Solar Energy* **9** 143

[6] Akpabio L E and Etuk S E 2002 Relationship between global solar radiation and sunshine duration for Onne, Nigeria *Turkish J. Phys.* **27** 161

[7] Falayi E O, Adepitan J O and Rabiu A B 2008 Empirical models for the correlation of global solar radiation with meteorological data for Iseyin, Nigeria *Int. J. Phys. Sci.* **3**(9) 210

[8] Reddy K S 2003 Solar Resource Estimation Using artificial neural networks and comparison with other correlation models *Energy Conversion and Management* **44**(15) 2519

[9] Yadav A K and Chandel S S 2012 Artificial neural network based prediction of solar radiation for Indian stations *Int. J. Comp. App.* **50** 1

[10] Duffie J A and Beekman W A 1994 Solar engineering of thermal processes (*New York*: *John Wiley)*

[11] Gueymard G 2000 Prediction and performance assessment of mean hourly global radiation *Solar Energy* **68** (3) 285

[12] Burari F W and Sambo A S 2001 *Nig. J. of Renewable Energy* **91**30

[13] Udo S O 2002 Contribution to the relationship between solar radiation and sunshine duration in the tropics: a case study of experimental data at Ilorin, Nigeria *Turkish Journal of Physics* **26** 229

[14] Torrence C and Compo G P 1998 A Practical guide to wavelet analysis *Bull. Amer. Meteor. Soc.* **79**(1) 61

[15] Celso A G SANTOS, Carlos de O GALVÃO, Koichi SUZUKI and Ricardo M. TRIGO 2001 *Annual Journal of Hydraulic Engineering* **45** 1

[16] Rehmana S and Mohandes M 2008 Artificial neural network estimation of global solar radiation using air temperature and relative humidity *Energy Policy* **36**(2) 571

[17] Avdakovic S, Lukac A, Nuhanovic A and Music M 2011 Wind Speed Data Analysis using Wavelet Transform *International Journal of Geological and Geophysical EngineerinG* **5** 1

Year to year variation of rainfall rate and rainfall regime in Ota, southwest Nigeria for the year 2012 to 2015

T V Omotosho[1], O O Ometan[1,2], S A Akinwumi[1], O M Adewusi[1,2], A O Boyo[2] and M S J Singh[3]

[1]Department of Physics, Covenant University, PMB 1023, Ota, Ogun State, Nigeria
[2]Department of Physics, Lagos State University, Ojo, Lagos State, Nigeria
[3]Institute of Space Science, Universiti Kebangsaan Malaysia 43600 UKM Bangi, Selangor Darul Ehsan, Malaysia

Email : omotosho@covenantuniversity.edu.ng[1], luwafunmilayo.ometan@lasu.edu.ng[1,2], oluwasayo.akinwumi@covenantuniversity.edu.ng[1], madewusi@gmail.com[1,2], adenikeboyo@gmail.com[2]

Abstract. The tropics is characterized to have convective type of rainfall which has high occurrence of rainfall compared to the temperate regions of the world. In this paper, the accumulation of rainfall in Ota, Southwest, Nigeria (6° 42 N, 3° 14 E) has been analysed to present the one-minute rainfall rate and the predominant type of rainfall. Four years' data used for this study was taken using the Davis Wireless vantage Pro2 weather station at Covenant University, Ota, Ogun State. The data collected were used to analyse the one-minute rainfall rate and different types of rainfall predominant in this region. For the prediction and modelling of rain attenuation at microwave frequencies for a region like the Nigeria at various percentage of time, one-minute rainfall rate is required. Nigeria falls into the P zone of 114 mm/hr. as per International Telecommunication Union – Recommendation (ITU-R). The analysis carried out indicated that the measured yearly averaged maximum one-minute rainfall rate for 2012, 2013, 2014 and 2015 are 157.7 mm/h, 148.0 mm/h, 241.2 mm/h and 157.3 mm/h respectively. It also indicated that the drizzle type of rainfall is predominant in contrast to established fact that thunderstorm occurs more in the tropics.

1. Introduction

One of the features that is most readily identified with the tropical atmosphere is the Intertropical Convergence Zone (ITCZ) [1]. The ITCZ lies in the equatorial trough, a permanent low-pressure feature that marks the meteorological Equator where surface trade winds, composed of heat and moisture from surface evaporation and sensible heating, converge to form a zone of increased mean convection, cloudiness, and precipitation [1]. Due to this, the tropical region of the world is known for their high rainfall rate. Amongst other atmospheric phenomenon, rainfall has major attenuation effect on propagation of communication signals at frequencies above 10 GHz [2]. Rain is formed when large cloud droplets from water vapour coalesce until the cloud can no longer hold together. Rain can occur as stratiform rain or convective rain. In high latitudes, stratiform rain is more dominant while in the tropics, the convective rain is dominant [3]. Rain, amongst the most common form of precipitation is the easiest to measure.

The cumulative rainfall is recorded using the recording rain gauge type which also gives the duration of rainfall events. The tipping bucket type is used for measuring the amount of accumulated rainfall and the rainfall rate [4]. Obiyemi et al. [5] estimated the point rain rate for Akure for 0.01% to be 120 mm/h. The 0.01% rain rate is about 120mm/h over locations in Malaysia as estimated by

Content from this work may be used under the terms of the Creative Commons Attribution 3.0 licence. Any further distribution of this work must maintain attribution to the author(s) and the title of the work, journal citation and DOI.
Published under licence by IOP Publishing Ltd

Abdulrahman et al. [6]. The point rainfall rates estimated by Ong and Zhu in Singapore were 104, 118, 115, 111mm/h observed over 2 years using a network of rain gauges' installations at Nanyang Technological University (NTU), Shunfu, Paya Lebar and Tampines respectively [7]. These results are also very similar with the result obtained by Pontes et al. for 0.01% percentage exceedance of rainfall rate in Brazil for different locations which were 118, 123 and 126 mm/h respectively for Rio de Janeiro, Belem and Manaus [8].

Also, precipitating clouds in the tropics have been reported to be entirely convection-generated by cumulonimbus clouds which forms either convective or stratiform rainfall type [9]. Stratiform type of rainfall is presumed to be of major occurrence in the tropics due to inadequate data, but in temperate regions ice particles are formed by nimbostratus cloud layer which are stable. The upward movement of air causes water vapour to be deposited on the ice particles. The ice particles fall to the earth as rainfall as they melt and fall from upper level [9]. Sometimes, the particles may never melt but reach the surface as snow rather than rain under certain cold conditions [9].

The Inter Tropical Convergence Zone (ITCZ) at the equator is the major phenomenon that causes changes in rainfall in the tropics [9]. According to Houze (1997), the Global Atmospheric Research Programme Atlantic Tropical Experiment (GATE), radar studies have confirmed, many times over, that a large component of tropical precipitation exhibits stratiform radar-echo structure [9]. The ratio between stratiform and convective precipitation may have significant impacts on other aspects of the simulated climate because of their different heating profiles and other characteristics. [9].

There are mainly four types of rainfall: drizzle, widespread, shower and thunderstorm. Drizzles are made up of small droplets of water that usually last longer. Its diameter ranges from 0.1 and 0.5 mm/hr. Showers are heavy, large drops of rain and usually only last for a short period [9]. Omotosho *et al.*, observed from the measurement obtained from the first year that there were elevated values of rain accumulation and rain rates compared to predictions of precipitation for the tropics by ITU-R. [10]. Rainfall measurement continued in the location of interest as this research was done to show the rainfall rate and the predominant rainfall type in Ota, Southwest Nigeria which is a predominant factor to determine the level of attenuation of communication signals for the tropics.

2. Methods and Data

The four years' rainfall data used for this research work were obtained from the Davis Wireless vantage Pro2 weather station installed at Covenant University, Ota, Nigeria with geographical coordinates 6.7°N, 3.23°E. It measures meteorological data such as temperature, pressure, wind speed, humidity, water vapour, rainfall rate etc. The receiver antenna is pointed towards the ASTRA satellite located at 36°E in the geostationary orbit. The satellite transmitted power is 50.8 dBW and the stability of the power over time is 98%. The beacon signal frequency is 12.245 GHz and it is vertically polarized. The tipping bucket type rain gauge is used which is connected to the Weather station to obtain rainfall data, all in one package called Integrated Sensor Suite. A sensor interface module (SIM) collects outside data from the ISS and transmits the data to a vantage Pro console via a low power radio at a frequency of 868.0 – 868.6 MHz with a maximum line-of-sight range of 300 m. The beacon measurement was taken through a signal transmitted from the operational transponder. The output signal of the Low Noise Block Converter (LNB), at the satellite dish, was connected to a data logger via personnel computer (PC). The rain collector type is tipping bucket with 0.01 inches per tips (0.2 mm with metric rain adapter). The temperature sensor is PN junction silicon with a range of -40° to +65° and an accuracy of ±0.5°C above -7°C, ±1°C under -7°C. The pressure sensor has a range of 540 to 1100 hPa / mb and a sensor accuracy of ±1.0hPa, the humidity sensor has a range of 1 to 100% and an accuracy of ±3% (0 to 90% RH), ±4% (90 to 100% Rh). The data logger was programmed to stored data every second continuously which was recorded as averaged data every 1-minute. 1440 data is recorded per day with 44640 for 31 days per month. The one minute rain rate experimental data obtained were used to analyze rain rate variation from April 2012 to December 2015.

3. Results and Discussion

Table 1 presents yearly rainfall accumulation from 2012 to 2015 showing that year 2014 has the highest total rainfall accumulation of 1348.7 mm with the lowest in 2012, 965.4 mm. The peak of the maximum rainfall rate is 241.2 mm/hr in 2014 while the minimum is 148.0 mm/hr. Tables 2 and 3 presents the yearly different rainfall regimes from 2012 to 2015.

Table 1. Yearly Rainfall accumulation for Ota from 2012 to 2015

Year	2012	2013	2014	2015	Average
Total Yearly Rainfall (mm)	956.4	1057.9	1348.7	954.2	1079.3
Total no of rainy days	116.0	127.0	159.0	114.0	129.0
Max. Rain rate (mm/hr.)	157.7	148.0	24.0	157.3	176.1

Table 2. The yearly different rainfall regime for 2012 and 2013

Year	2012				2013			
Rain Regime	Numb er of events	Total rain intensity (mm/hr.)	Rain (Min utes)	Percentage (%) Rain Minutes	Number of events	Total rain intensity (mm/hr.)	Rain (Min utes)	Percenta ge (%) Rain Minutes
Drizzle $0 < R \leq 5$	5895	11116.4	5895	72.1	5966	11348.2	5966	71.1
Widespread $5 > R \leq 10$	929	6623	929	11.4	796	5696.2	796	9.5
Showers $10 > R \leq 40$	840	17420.6	840	10.3	1049	22056.4	1049	12.5
Thunderstorm $40 > R$	514	40058.6	514	6.3	575	45240.6	575	6.9
Total	8178	75218.6			8386	84341.4		

Table 3. The yearly different rainfall regime for 2014 and 2015

Year	2014				2015			
Rain Regime	Numb er of event s	Total rain intensity (mm/hr.)	Rain (Min utes)	Percentage (%) Rain Minutes	Number of events	Total rain intensity (mm/hr.)	Rain (Min utes)	Percenta ge (%) Rain Minutes
Drizzle $0 < R \leq 5$	7963	15027.8	7963	72.5	5241	10112.6	5241	70.8
Widespread $5 > R \leq 10$	1086	7592	1086	9.9	697	4951.6	697	9.4
Showers $10 > R \leq 40$	1253	25806.9	1253	11.4	983	20795	983	13.3
Thunderstorm $40 > R$	677	53163	677	6.2	477	35650	477	6.4
Total	10979	101589.7			7398	71509.2		

Investigations from Table 1 shows that the drizzle type of rainfall has a greater percentage of occurrence of 72.1 %, 71.1 %, 72.5 % and 70.8 % for 2012, 2013, 2014 and 2015 respectively. Figures 1 to 4 shows the variations in the occurrence of the different rainfall regimes. In 2012, there were two peaks in June and October with December recording the lowest. In 2013, there were three peaks: May, July and September with July recording the highest. Also, December shows the lowest in 2013. There is a variation in 2014 compared to 2012 and 2013. There were two peaks: June and October with February recording the lowest. In 2015, two peaks were also seen in June and October respectively with December being the lowest.

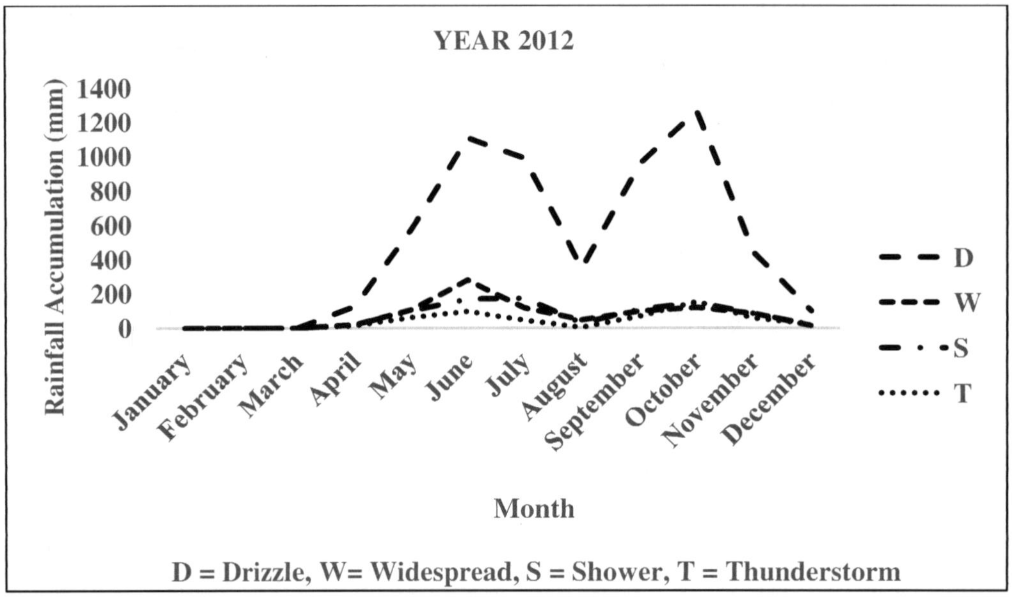

Figure 1. The variation in rainfall regime for Ota in 2012

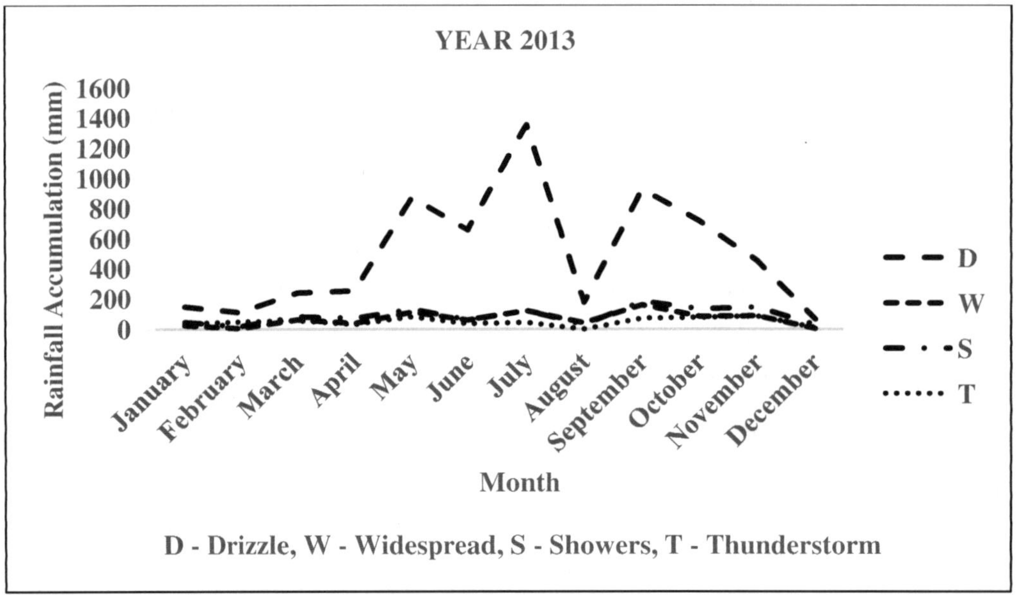

Figure 2. The variation in rainfall regime for Ota in 2013.

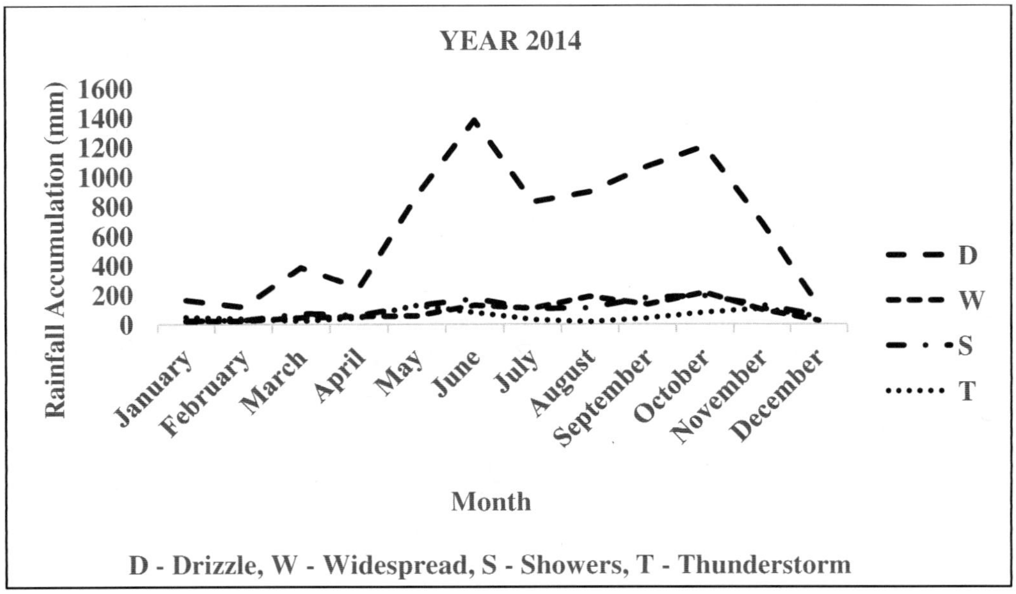

Figure 3. The variation in rainfall regime for Ota in 2014.

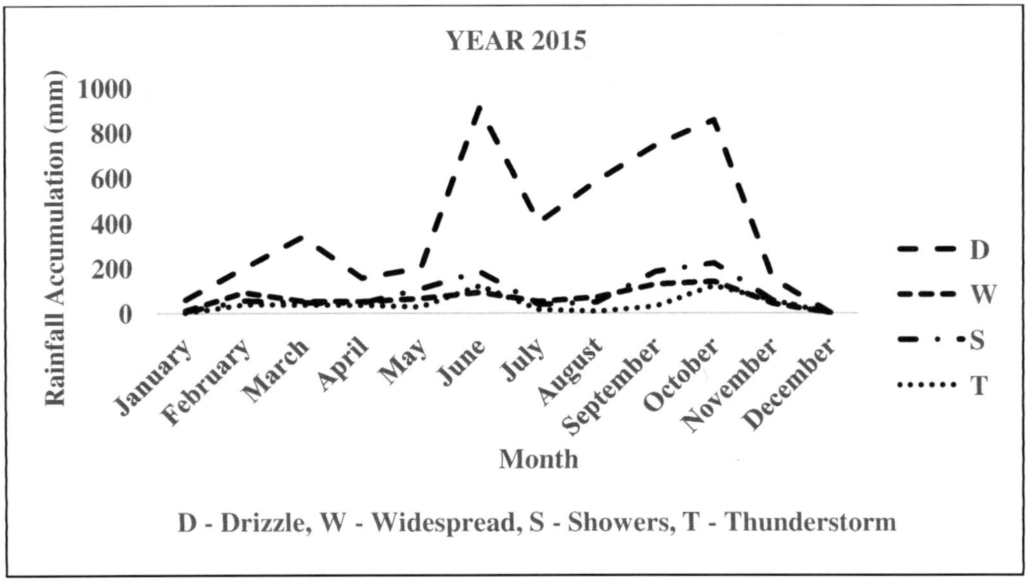

Figure 4. The variation in rainfall regime for Ota in 2015.

4. Conclusion

The result reported in this study is used for the analysis of the rainfall regimes at Ota in terms of rainfall regime, number of event and total rainfall intensity which shows the weather dynamics of this region. The drizzle type of rainfall is evidently predominant in Ota on the average of 71.6 %. The total numbers of rainy days from 2012 to 2015 are 116, 127, 159 and 114 respectively. The four years of our experiment results indicated that the measured $R_{0.01}$ rainfall rate at Ota is ~ 200 mm/h. The result obtained also indicated that year 2014 had the maximum number of rainy days, maximum total rainfall and maximum rainfall rate which are 159.0 mm, 1348.7 mm and 241.2 mm/h respectively. The results were compared with ITU-R recommendation on the characteristics of precipitation for propagation modelling and it was discovered that the ITU-R [11] model underestimated the rainfall rate. Therefore, the P region recommendation by ITU-R model is under estimated for Ota, the Q-region of ITU-R model would be more suitable. This result will provide adequate information to the communication engineers about the possible attenuation or link reliability that is likely to occur in Ota. Therefore, new model of rainfall rate should be developed using the rainfall rate measurement from the place of interest.

Acknowledgement
The authors would like to thank Covenant University for full sponsorship pf this research work.

References

[1] Waliser D E 2002 Intertropical Convergence Zone doi:10.1006/rwas 04.

[2] Zhou Z X, Li L W, Yeo T S and Leong M S 2000 Cumulative distributions of rainfall rate and microwave attenuation in Singapore's tropical region *Radio Science* P **35** 751–756.

[3] Werner A and Christian M Synthetic aperture radar marine user's Manual *NOAA/NESDIS office of Research and applications* 355-371

[4] World Meteorological Organization

[5] Obiyemi O O, Ojo J S and Ibiyemi T S 2014 Performance analysis of rain rate models for microwave propagation designs over tropical climate *Progress in Electromagnetics Research M* **39** 115–122.

[6] Abdulrahman A *et al.* 2013 An improved slant path attenuation prediction method in tropical climates *Radio engineering* **22** 1083

[7] Ong, J T and Zhu C N 1997 Rain rate measurements by a rain gauge network in Singapore *Electronics Letters* **33** 240–242

[8] Ponte M S, Migliora, C G S and Da Silva Mello L A R 1989 Preliminary results of 12 GHz radiometric measurements in Brazil *Electronics Letters* **25**

[9] Houze R A, Jr 1997 Stratiform precipitation in regions of convection: a meteorological paradox? *Bulletin of the American Meteorological Society* **78**

[10] Omotosho T V, Willoughby A A, Akinyemi M L, M S J Singh and Abdullah M 2013 One year results of one minute rainfall rate measurement at Covenant University, Southwest Nigeria *Proceeding of the IEEE 2013 International Conference on Space Science and Communication (IconSpace)* p 98-101

[11] International Telecommunication Union 2003 Characteristics of precipitation for propagation modelling, *Recommendation ITU-R P* 837–4 *Geneva, Switzerland*

Investigation of zonal velocity of equatorial plasma bubbles (EPBs) by using GPS data

I Sarudin[1], N S A Hamid[1*], M Abdullah[2,3], and S M Buhari[4]**

[1]School of Applied Physics, Faculty of Science and Technology,
[2]Space Science Centre (ANGKASA), Institute of Climate Change,
[3]Department of Electrical, Electronics and Systems Engineering, Universiti Kebangsaan Malaysia, 43600 UKM, Bangi, Selangor, Malaysia.
[4]Department of Physics, Faculty of Science, Universiti Teknologi Malaysia, 81310 UTM, Skudai, Johor, Malaysia.

[*]Email: shazana.ukm@gmail.com, [**]Email: mardina@ukm.edu.my

Abstract – The zonal velocity of equatorial plasma bubbles (EPBs) have been studied using various techniques in the past few years. However, the derivation of the zonal drift of EPBs using GPS ROTI have not been studied before. This study aims to investigate the zonal drifts of EPBs using GPS ROTI keogram. The Malaysia Real–Time Kinematic GNSS Network (MyRTKnet) which consists of 78 GPS receivers was used to study the occurrence of EPBs along 96°E - 120°E longitude. The EPBs are detected from daily ROTI keogram that derived from east-west cross section of two dimension of ROTI maps at 5°N for every 5 minutes. On the night of 29 March 2011, 10 consecutive EPBs with periodic spacing between 50 km to 100 km were recorded by MyRTKnet. In this study, we obtained that highest drift velocity is about 277.7 m s-1 at 1400 UT which denoted by EPB 7 whereas the lowest drift velocity is EPB 4 with 27.8 m s-1 at 1600 UT. Besides, the EPBs are propagated towards the east from 200 km to 2800 km with average velocities is about 55.6 m s-1 to 138.9 m s-1.

1. Introduction

Equatorial plasma bubble (EPBs) is a 3D structure of depleted plasma density that first appear at bottom side of F layer from a few centimeters to thousands of kilometers ranged and may take a few hours. The development and movement of the EPBs are primarily controlled by the zonal electric fields. In the E and F region, the electric field is created by neutral winds moving across the magnetic field lines (dynamo action) [1]. EPBs have been widely studied for decades using different techniques including all sky imager and in situ satellite measurement. The observation of the EPBs have been carried out using ground and space based instruments. Among early studied using ground based data are from Martinis et al. [2] presented studies of plasma drift using OI 630.0 nm measurements from South American Sector. The range of eastward velocity of EPBs is about 140-160 m s-1 [3]. These drifts are largest at about 21:00-22:00 local time (LT) and they decrease towards sunrise. In spite of this ground based techniques, the manipulation of space based data such as satellite and GPS data is very limited. The previous studied made through these space based instrument is the zonal plasma drift speeds of EPBs observed using the imager aboard high apogee IMAGE satellite during March-May 2002 had a strong longitudinal dependence and are maximum over the Indian sector [4]. In this paper we present the zonal drift velocity of EPBs calculated from high-density GPS receivers in Southeast Asia (SEA) on night of 29 March 2011.

Content from this work may be used under the terms of the Creative Commons Attribution 3.0 licence. Any further distribution of this work must maintain attribution to the author(s) and the title of the work, journal citation and DOI.
Published under licence by IOP Publishing Ltd

These drift velocities are calculated from longitudinal change at significant time that can be seen from plotted ROTI keogram.

2. Method and Analysis

2.1 Data collection
GPS data in the format of Receiver Independent Exchange (RINEX) were collected from the Department of Survey and Mapping Malaysia (JUPEM). Since 2003, about 78 receiver stations over Malaysia called Malaysia Real-Time Kinematics GNSS Network (MyRTKnet) has installed by JUPEM.

Integrated electron density along line of sight of the receiver and satellite (1TECU=10^{16} electrons/m^2) is known as the total electron content (TEC). Therefore, TEC is derived from the time delay of the two signals transmitted by GPS satellites with both frequencies f_1=1575.42 MHz and f_2=1227.60 MHz. Carrier phase data that provides high precision of TEC measurement is used to calculate the TEC. The small scale fluctuations inside EPBs still can be recognized from the high precision of TEC measurement, although the carrier phase data suffered from the opacity between 0 to 2π.

The EPBs structure was extended in north-south directions and drifted towards the east with time. In order to show the longitudinal and temporal variations in the EPBs structure, the keogram was created by taking an east-west cross-section of ROTI at 5°N for every 5 minutes. The development of ROTI can be perceived after 2000 LT which is linked with the tens of kilometer-scale irregularities inside the EPBs. The first EPBs structure initiate to appear at 120°E and followed by the succeeding structures to the west due to the passage of the solar terminator [5,6]. The striations respectively represent a single structure of EPBs that moves to the east direction with time [7,8]. In this keogram, we can be perceived that all the EPBs structure drifted eastward and vanished approximately after 1800 UT.

2.2 Data Analysis
The present study is generally considered only one day ROTI keogram derived from the MyRTKnet was examined to identify the occurrence of EPBs in Malaysia area. The occurrence of EPBs was verified if ROTI was larger than 0.1 TECU/min happened at anyplace in the keogram. High threshold value (0.06 TECU/ min) was taken into the attention to confirm the EPBs are truly present in the observational area. Then, the zonal drift velocity of EPBs can be calculated from its spatial displacement divided by time [9]. Zonal drift velocity was calculated for each EPBs from the onset highest ROTI values to the final highest ROTI values as can be seen from the keogram. For instance, we estimate the velocity drift of EPB 7 from 1220 UT to 1550 UT. The time range chosen for all EPBs are stated in Table 1.

3. Results and Discussions

By using the methods described earlier, we have calculated plasma drift velocities for one night on 29 March 2011. By using GPS TEC measurement in SEA, we can be observe the spatial variations of EPBs.

Figure 1 shows a keogram generated from the two-dimensional maps of ROTI and their longitudinal variations. The figure as a clear overview of the characteristics of the EPBs on night 29 March 2011, which is a cross section of ROTI by choosing the horizontal profiles of the ROTI at 5°N latitude with several times and longitudes. There is a red line located on the left side of the keogram is actually represents the sunset terminator. Figure 1 shows 10 continuous generation of striations along the 550 km sunset line. For giving a clear understanding, every single striations is symbolized as EPBs number as shown in Figure 1.

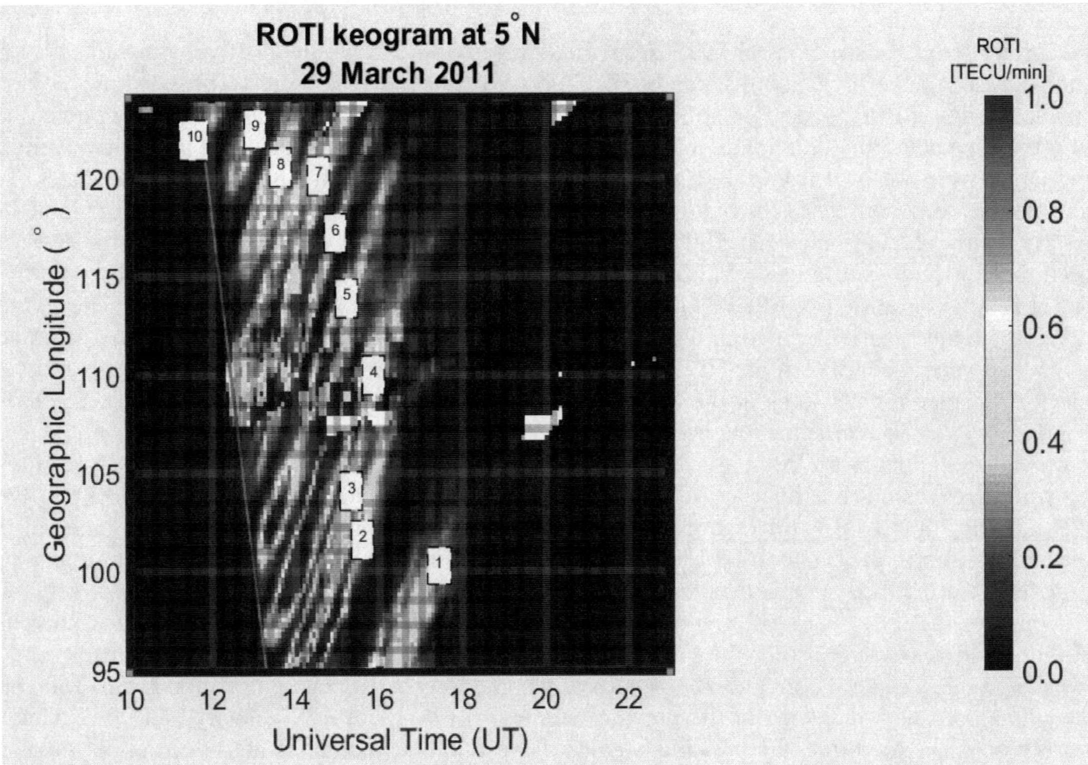

Figure 1. ROTI keogram at 5°N latitude obtained from GPS networks in SEA from 1000 UT to 2230 UT on 29 March 2011. The keogram shown that 10 striations associated with EPBs were generated continuously at approximately the 550 km altitude sunset time.

Table 1. Onset Times, Final Time, Initial Longitude, Final Longitude, Maximum Zonal Width, Lifetime, Distance and Average Velocity of the 10 EPBs Observed in the ROTI Keogram Generated at 5°N Latitude.

EPB	ONSET TIME (UT)	FINAL TIME (UT)	INITIAL LONGITUDE (KM)	FINAL LONGITUDE (KM)	MAX WIDTHS (±50KM)	LIFETIME (±5MIN)	DISTANCE (±0.5°)	AVERAGE VELOCITY (m **s-1**)
1	-	18:00	0	600	350	-	-	-
2	-	16:00	0	1300	200	-	15	-
3	-	16:00	0	1300	200	-	15	-
4	13:15	17:00	200	1700	250	3h 45min	21	104.7
5	13:22	17:00	600	2400	200	3h 38min	20	125.0
6	13:18	16:00	1000	2800	50	2h 42min	9	138.9
7	12:20	15:50	1400	2600	400	2h 30min	18	111.1
8	13:25	15:00	2000	2600	250	1h 35min	10	83.3
9	13:20	14:00	2400	2800	200	30min	6	55.6
10	-	-	-	-	100	-	-	-

We further select six striations of EPBs from the keogram to calculate the drift velocities. These EPB are choose based on their striation that can be seen fully from the ROTI keogram. Table 1 shows some of the characteristics of EPBs generated at 5°N latitude observed from ROTI keogram. The onset time, final time, initial longitude, final longitude, maximum zonal width, lifetime, distance and average velocity of each striation were obtained from the ROTI keogram at 5°N were identify and listed as in Table 1. The separations between consecutive striations estimated from distance between the onsets longitudes of successive EPBs obtained between 100 and 550 km, which is listed in Table 1. The time intervals between the consecutive striations were between 4 to 50 min. Another interesting feature observed from the ROTI keogram is a lifetime of EPBs that varied from 30 min to 3 h 45 min respectively. The shortest and longest lifetime can be seen from the Table 1, which is EPB 9 and EPB 4. Besides, the maximum zonal widths were about 50 km for EPB 6 and 400 km for EPB 7. Higher ROTI values were always represent the larger zonal widths at the beginning of each striation (as in Figure 1), and as the ROTI keogram are weaker with time the width will be smaller.

Next, we select only five EPBs to further analyze in this section. The selection are based on EPB with a longer lifetime which is one hour above. The EPBs that have been selected are denoted by EPB 4, EPB 5, EPB 6, EPB 7 and EPB 8 (in Figure 2). Figure 2 shows the variations in the drift velocities of the selected EPBs from 1300 UT to 1600 UT. The comparison of the zonal drift velocities as a function of time for five selected EPBs that is denoted as EPB 4, EPB 5, EPB 6, EPB 7 and EPB 8 respectively with clear signatures of EPBs. The drift velocities are calculated for each one or half an hour to shows the variation of drift velocities between the EPBs.

From Figure 2, we can see roughly the pattern of the graph were decrease from 1300 UT to 1600 UT. Our results agreed with the some of the previous studies by [9] where drift velocity of EPBs gradually decrease with time. Although, the previous studied using different technique and location but obtained same results as well.

The zonal drift of the EPBs shows significant different during 1300 UT to 1500 UT. The highest drift velocity is about 277.7 m s-1 at 1400 UT which denoted by EPB 7 while the lowest drift velocity is EPB 4 and EPB 8 at 1500 UT is about 55.6 m s-1. As can be seen from ROTI keogram plotted in Figure 1, EPB 7 is the biggest EPB and give the highest drift. Meanwhile, the drift velocities at 1500 UT to 1600 UT for EPB 4 is 27.8 m s-1 whereas for EPB 5, EPB 6, EPB 7 and EPB 8 are 55.6 m s-1 respectively.

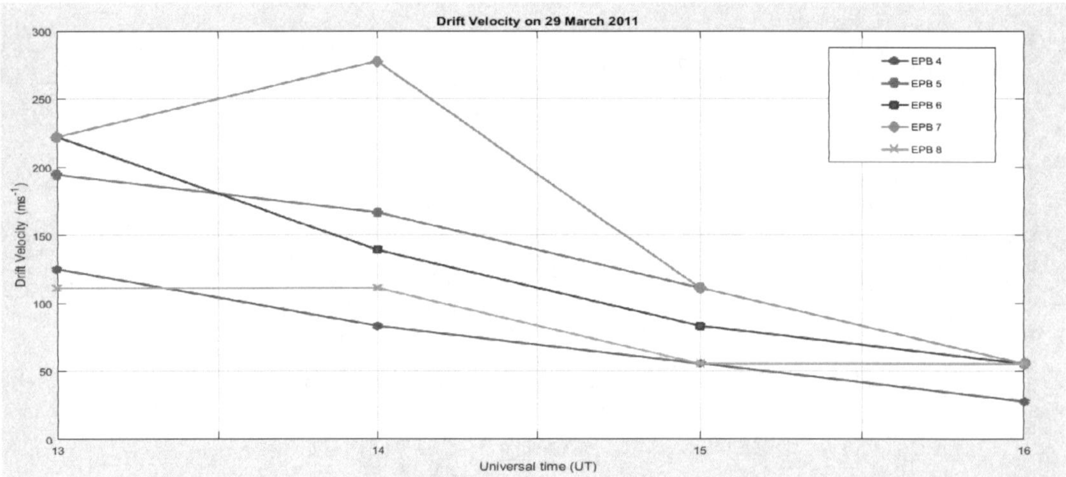

Figure 2. The zonal velocities calculated of different EPBs on 29 March 2011. The figure plotted with five selected EPB denoted as EPB 4, EPB 5, EPB 6, EPB 7 and EPB 8 based on ROTI keogram.

4. Conclusion

In this work we have presented zonal drift of 10 consecutive EPBs from high-density GPS receivers in SEA on 29 March 2011. In general, the EPBs propagated towards the east from 200 km to 2800 km with average velocities is about 55.6 m s-1 to 138.9 m s-1. Our results agreed with previous study that shows the drift velocity of EPB is gradually decrease with time. Besides, we found that the highest drift velocity is about 277.7 m s-1 at 1400 UT which denoted by EPB 7 whereas the lowest drift velocity is EPB 4 with 27.8 m s-1 at 1600 UT.

Acknowledgements

The GPS data are available from MyRTKnet by downloading through http://www.rtknet.gov.my/ and also from Department of Survey and Mapping Malaysia (JUPEM) data policy. This work was supported by the FRGS/1/2016/WAB08/UKM/01/1 from the Ministry of Education Malaysia and GUP-2016-016 from Universiti Kebangsaan Malaysia.

References

[1] Rishbeth H, Bauer P and Hanson W B 1972 Molecular ions in the F2 layer *Journal of Planet and Space Science* **20** 1287

[2] Martinis C, Eccles J V, Baumgardner J, Manzano J and Mendillo M 2003 Latitude dependence of zonal plasma drifts obtained from dual-site airglow observation *Journal of Geophysical Research* **108** 1129 doi: 10.1029/2002JA009462

[3] de Paula E R, Kantor I J, Sobral J H A, Takahashi H, Santana D C, Gobbi D, Medeiros A F, Limiro L A T, Kil H, Kintner P M and Taylor M J 2002 Ionospheric irregularity zonal velocities over Cachoeira Paulista *Journal of Atmospheric Solar Terrestrial Physics* **64** 1511

[4] Immel T J, Frey H U, Mende S B and Sagawa E 2004 Global observations of the zonal drift speed of equatorial ionospheric plasma bubbles *Annales Geophysicae* **22** 3099 doi: 10.5194/angeo-22-3099-2004

[5] Buhari S M, Abdullah M, Hasbi A M, Otsuka Y, Tokoyama T, Nishioka M and Tsugawa T 2014 Continuous generation and two-dimensional structure of equatorial plasma bubbles observed by high-density GPS receivers in Southeast Asia *Journal of Geophysical Research* **119** doi: 10.1002/2014JA020433

[6] Yokoyama T and Fukao S 2006 Upwelling backscatter plumes in growth phase of equatorial spread F observed with the Equatorial Atmosphere Radar *Journal of Geophysical Research* **33**

[7] Paulino I, de Medeiros A F, Buriti F A, Takahashi H, Sobral J H A and Gobbi D 2011 Plasma bubble zonal drift characteristics observed by airglow images over Brazilian tropical region *Brazilian Journal of Geophysics* **29(2)** 239

[8] Chapagain N P, Michael J Taylor Nielsen K and Jarvis M J 2011 Airglow observations and modelling of F region depletion zonal velocities over Christmas Island *Journal of Geophysical Research* **116** A02301 doi:10.1029/2010JA015958

[9] Nade D P, Sharma A K, Nikte S S, Patil P T, Ghodpage R N, Rokade M V Gurubaran S, Taori A and Sahai Y 2013 Zonal velocity of the equatorial plasma bubbles over Kolhapur, India *Annales Geophysicae* **31** doi: 10.5194/angeo-31-2077-2013

IOP Conf. Series: Journal of Physics: Conf. Series **852** (2017) 012015 doi:10.1088/1742-6596/852/1/012015

Peak time of equatorial electrojet from different longitude sectors during fall solar minimum

N S A Hamid[1*], H Liu[2,3], T Uozumi[3], A Yoshikawa[2,3] and N M N Annadurai[1]

[1]School of Applied Physics, Faculty of Science and Technology, Universiti Kebangsaan Malaysia, 43600 UKM Bangi, Selangor, Malaysia.
[2]Department of Earth and Planetary Sciences, Faculty of Sciences, 33 Kyushu University, 6-10-1 Hakozaki, Higashi-ku, Fukuoka 812-8581.
[3]International Center for Space Weather Science and Education, Kyushu University 53, 6-10-1 Hakozaki, Higashi-ku, Fukuoka 812-8581.

[*]Email: shazana.ukm@gmail.com

Abstract. Equatorial electrojet is an eastward flowing current at about ±3° dip equator. This current intensity is always higher during noontime as it is greatly influenced by the ionization of Earth ionosphere. Apart from that, previous study had shown that EEJ current varies with longitude and latitude as well as solar cycle. The aim of this study is to investigate the peak time of EEJ current at different longitude sectors using simultaneous data in 2009. By eliminating Sq current contribution and normalizing ground-based data from MAGDAS/CPMN, IIG and WDC network, we managed to reach our purpose and gain reliable output. We found out that EEJ is strongest in South American sector. Our results show that the peak EEJ during fall solar minimum is observed at 11 LT for South American, Indian and Southeast Asian sector but it is 1 hr earlier in African sector.

1. Introduction

The equatorial electrojet current or EEJ has always been conceived as a phenomenon confined to narrow area around ±3° of dip equator. It is affected by solar cycle [1, 2, 3, 4], seasons [5, 6, 7, 8] as well as varies with longitude [4, 5, 6, 8] and latitude [2, 9, 10]. At the dip-equator, this current overlaps with a part of global Sq current which consequently affect the reading of geomagnetic component measured at station near the dip equator. Thus, most of researchers will subtract the Sq contribution while studying the EEJ variability using geomagnetic data.

EEJ intensity is known to be greater during noontime due the highest rate of ionization at this period. Based on this, previous study of EEJ tends to choose the highest value of geomagnetic northward during 10 to 14 LT to represent this current component [11]. However, study by [1] had shown that the EEJ strength reached its maximum around 11 LT during solar minimum and around 12 LT during solar maximum. They conclude that the time for the peak EEJ is around local noon during solar maximum and shifts to earlier local times during solar minimum. This output is obtained using ground based data at three longitude sector which are east of South American, east of African, and Indian without considering the latitudinal variation of the observation stations. As EEJ is greatly known to vary with latitude, their result might contain uncertainties when a direct comparison is made using data in those three longitude sectors.

Apart from that, study by [8] has shown that the peak time of EEJ can be also affected by the existence of counter electrojet (CEJ) that happened during noontime. In their study, the peak off EEJ was observed 1 hr earlier which is at 10 LT time compare to the days without CEJ. This important feature of EEJ has recently been raised by [10] as the factor that causes the difference in EEJ peak time during two phases of solar cycle is still a remaining question.

2. Data and Analysis

This study utilizes two stations located along the same longitude, with one located near dip equator around ±3° dip latitude while the other one is outside this region. The stations are grouped into six pairs, one in the South American sector, two in the South African sector (east and west), one in Indian sector and two in Southeast Asian sector (east and west). The data were obtained from various networks which are Magnetic Data Acquisition System/ Circum-pan Pacific Magnetometer Network

Content from this work may be used under the terms of the Creative Commons Attribution 3.0 licence. Any further distribution of this work must maintain attribution to the author(s) and the title of the work, journal citation and DOI.
Published under licence by IOP Publishing Ltd

(MAGDAS/CPMN), Indian Institute of Geomagnetism (IIG) and International Real-time Magnetic Observatory Network (INTERMAGNET). Apart from that, we also requested data from one stand-alone station in South American sector. Table 1 shows the geomagnetic and geographic coordinate of the stations while and Figure 1 illustrates the geographic distribution.

Analysis is carried out using the *EUEL* index from September 16 to 30, 2009. In order to compare the result from all longitude sectors, we normalize the *EUEL* index observed at these stations to the dip equator with the aid of the CM4 model [12]. The method is introduced by [9] to overcome the latitudinal variation effect of EEJ current. In this technique, the latitudinal profile of Sq is obtained by applying fitting of polynomial of degree 2 to the CM4 model. This is done after the region of 6° across dip equator was masked in order to avoid the influence of the EEJ in the determination of Sq profile. Using this latitudinal profile, we normalize the observed *EUEL* at off-dip equator station (θ° dip latitude) to dip equator (0°dip latitude) to obtain normalized Sq at dip equator, $EUEL_{Sq(0°)}$. On the other hand, *EUEL* observed at station near dip equator (θ° dip latitude) is normalized to dip equator (0° dip latitude) directly using CM4 model profile to give normalized total current at dip equator, $EUEL_{total(0°)}$. We further calculate EEJ current by subtracting $EUEL_{Sq(0°)}$ from $EUEL_{total(0°)}$. The details procedure of this technique can be found in [9]. We further apply the same steps to all stations pairs.

Table 1. Geomagnetic and geographic coordinates of stations used in this study

Region	Station		Geographic		Geomagnetic	
	Name	Code	Lat.(°)	Lon.(°)	Lat.(°)	Lon.(°)
South America	Ancon	ANC	-11.77	-77.15	0.77	354.33
	Fuquene	FUQ	5.40	-73.73	15.72	357.99
African	Ilorin	ILR	8.50	4.68	-1.82	76.80
	Tamanrasset	TAM	22.8	5.5	25.4	80.6
	Adis Ababa	AAB	9.04	38.77	0.18	110.47
	Nairobi	NAB	-1.16	36.48	-10.65	108.18
India	Tirunelveli	TIR	8.70	77.80	0.21	149.30
	Alibag	ABG	18.62	72.87	10.36	146.54
Southeast Asia	Langkawi	LKW	6.30	99.78	-2.32	171.29
	Kototabang	KTB	-0.20	100.32	-10.63	171.93
	Davao	DAV	7.00	125.40	-1.02	196.54
	Muntinlupa	MUT	14.37	121.02	6.79	192.25

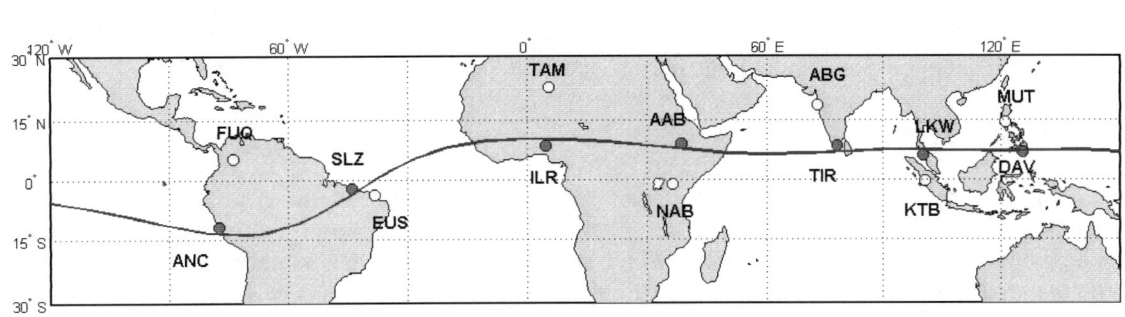

Figure 1. Map of stations used in this study

Figure 2(a) illustrates the original and normalized data of ILR-TAM pair in the west of African sector. The black thick lines indicate the mean of EUEL with the highest is at ILR station (near dip equator) while the lower is at TAM station (off-dip equator). The normalized EUEL from ILR which is the total current is represented by the thick blue dashed line while the thick green dashed line is the Sq current obtained by the normalization of EUEL at TAM station.

We further calculate the EEJ current from the difference between the total current and Sq obtained in Figure 2(a). The output is presented in Figure 2(b). At this stage, we only consider the reading between 09 LT to 15 LT. The EEJ current within this period is shown by the thick red line. The blue and green lines here are the same as the one in Figure 2(a).

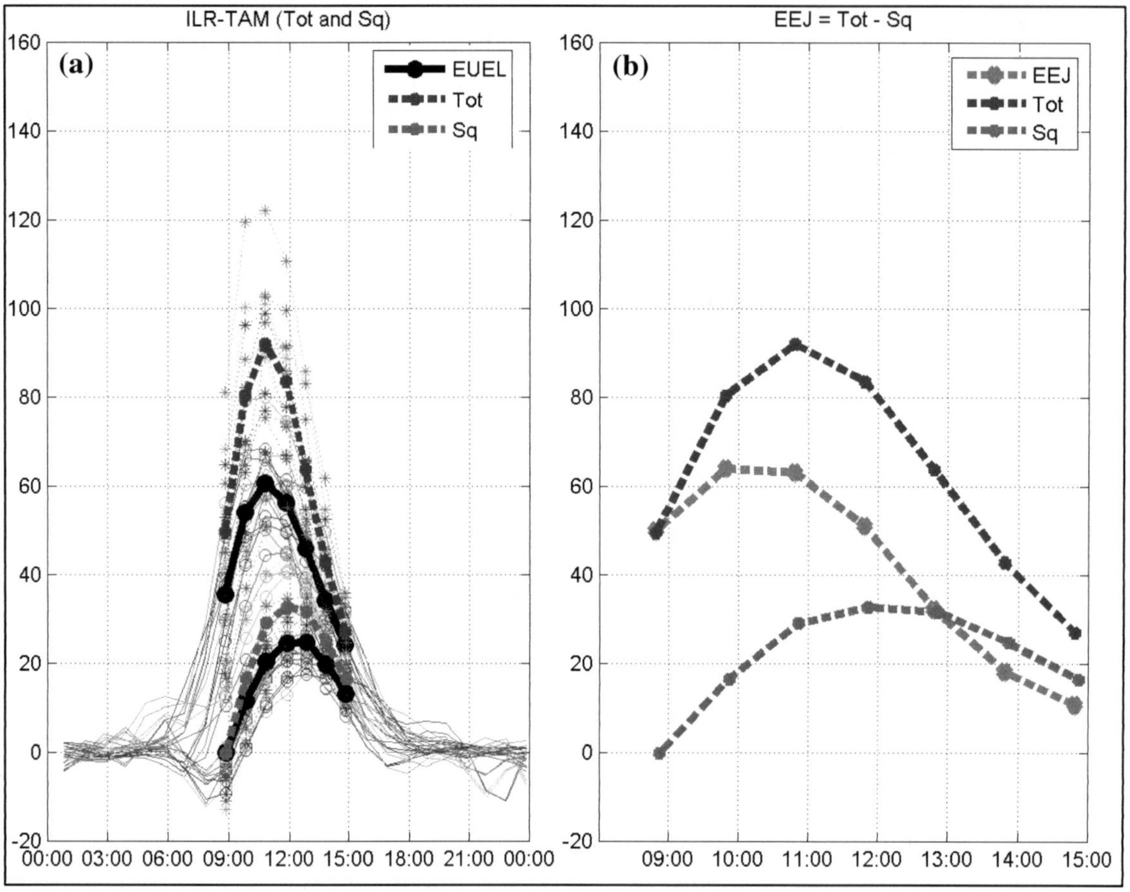

Figure 2 a) EUEL from dip equator station (ILR) and off-dip equator station (TAM) are indicate by thick black lines and their normalized values are indicate by thick blue line and thick green line respectively **b)** EEJ current (thick red line) is calculated from total current (normalized EUEL at ILR, thick blue line) minus Sq current (normalized EUEL at TAM, thick green line)

3. Results and Discussion

Simultaneous observations from September 16 to 20, 2009 were recorded from 12 magnetometer stations located in six different longitude sectors. Figure 3 (a) shows the magnitude of total current obtained from CM4 model while Figure 3 (b) shows the one observed from ground based magnetometer data for the period between 09 LT and 15 LT. Their values are presented in Table 2. It can be seen that CM4 model result for peak EEJ current is at 11LT or 12LT. Nevertheless, analysis of total current from observations demonstrate every station pairs have the same result for peak EEJ current which is at 11 LT. These results indicate that the agreement between CM4 model and observation is only can be observed at east of African and west of Southeast Asian sector which both shows the peak time of EEJ is at 11 LT. The discrepancies between the model and observation data might be caused by the controlling parameter used in the model. In this model, only F10.7 and Dst were considered whereas the total current (i.e. EEJ and Sq) is known to be influenced by several other factors such as the local wind and gravity waves. This difference however will not be discussed in this current paper.

We further analyze the magnitude of EEJ which is shown in Figure 4. Here we can compare the mean of EEJ from all station pair for each noontime from 09 LT to 15LT. We found that the magnitude of EEJ is strongest in South American sector regardless of local time and weakest in Indian sector during 09 and 10 LT but shifted to African sector during 11 to 14 LT.

In addition, we also examine the peak time of EEJ current as indicate by the circle symbols with the values are listed in Table 3. The results obtained show that the EEJ peak time is observed at 11 LT for South American, Indian and Southeast Asian sector while at African sector, it is observed at 10 LT. The peak time of 11b LT is in agreement with previous study by [1] for east of South American and Indian sector. Our EEJ peak time at African sector is however 1 hr early than the one obtained in their study. This discrepancy might cause by the latitudinal variation of EEJ which is neglected in their study.

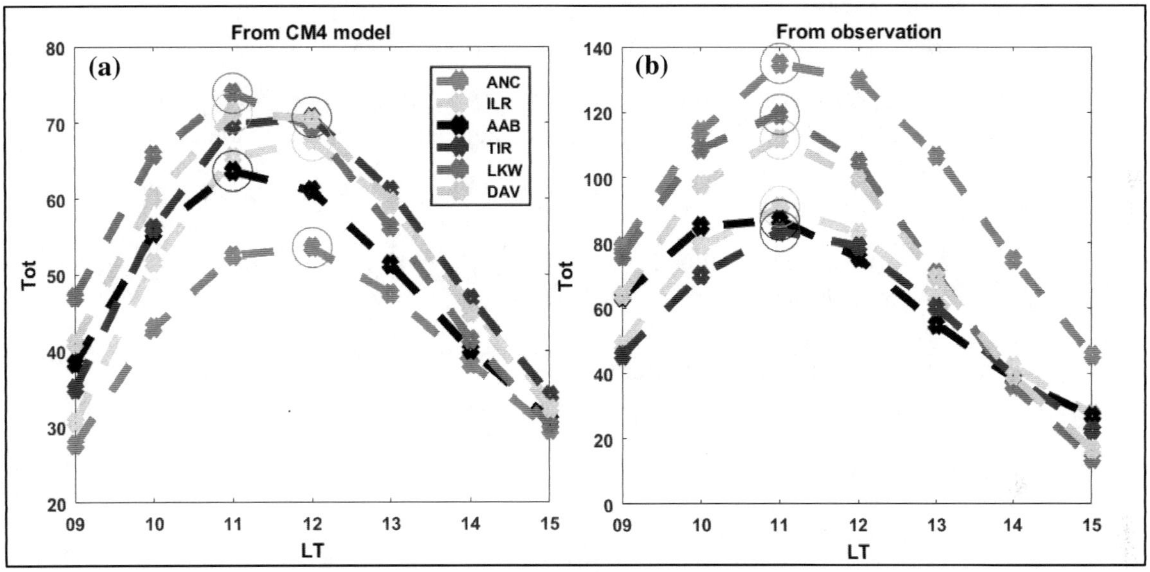

Figure 3. Magnitude of total current from a) CM4 model and b) observation between 09LT and 1500 LT from six longitude sectors (stations)

Table 2. Peak time of total current from CM4 model and observation

Station	CM4 Model	Observation
ANC	12	11
ILR	12	11
AAB	11	11
TIR	12	11
LKW	11	11
DAV	11	11

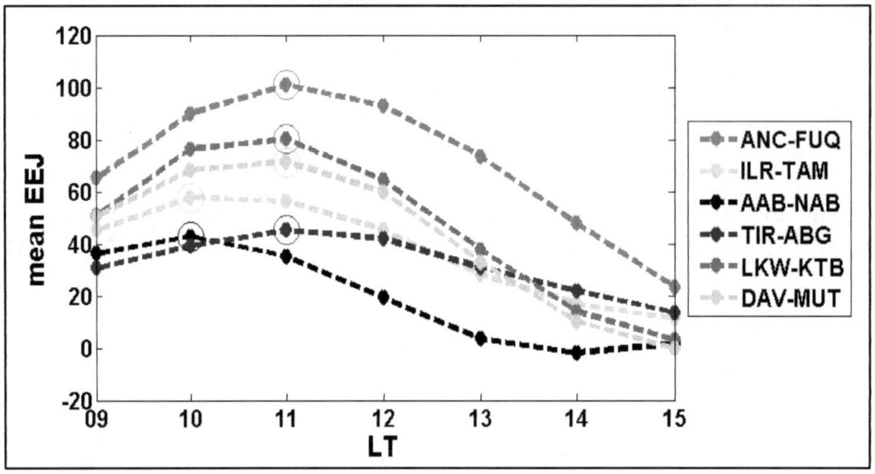

Figure 4. Mean EEJ from 09 LT to 15 LT calculated from all station pairs. Circle symbol indicate the maximum mean value which represents the peak time of EEJ

Table 3. Maximum local time (LT) where mean EEJ amplitude is the highest (see Figure 4)

Pair(region)	Local time (LT)
ANC-FUQ(South America)	11
ILR-TAM(west African)	10
AAB-NAB(east African)	10
TIR-ABG(India)	11
LKW-KTB(west Southeast Asia)	11
DAV-MUT(east Southeast Asia)	**11**

5. Conclusions

In this study, the total current from CM4 model and observation are found to be different where our analysis shows that the peak time of total current measured from ground based magnetometer is observed at 11 LT for all six longitude sectors. Other than that, we found out that EEJ peak time at most longitude sector is at 11 LT except for African sector which EEJ there is observed to be maximum at 10 LT in the solar minimum phase. Further research on this phenomenon should be conducted using solar maximum data in order to understand the factor that causes the longitudinal variation of EEJ peak time.

Acknowledgements
We thank all the members of the MAGDAS project for their cooperation and contribution to this study. Financial support was provided by the Universiti Kebangsaan Malaysia and Ministry of Education, Malaysia, using grants GGPM-2015-020 and FRGS/1/2015/ST02/UKM/02/1. A. Yoshikawa was supported in part by JSPS Core-to-Core Program (B. Asia-Africa Science Platforms), Formation of Preliminary Center for Capacity Building for Space Weather Research, and JSPS KAKENHI grants 15H05815. We acknowledge the National Oceanic and Atmospheric Administration (NOAA) for providing Kp index data, Goddard Space Flight Center/Space Physics Data Facility (GSFC/SPDF) OMNIWeb at http://omniweb.gsfc.nasa.gov for providing F10.7 data, and the National Geophysical Data Center (NGDC) for the estimated values of the magnetic inclination component.

References

[1] Rastogi R G and Iyer K N 1976 Quiet day variation of geomagnetic H-field at low latitudes *Journal of Geomagnetism and Geoelectricity* **28** 461

[2] Onwumechili C A and Ezema P O 1992 Latitudinal and vertical parameters of the equatorial electrojet from an autonomous data set *Journal of Atmospheric and Terrestrial Physics* **54** 1535

[3] Hamid N S A, Liu H, Uozumi T and Yumoto K 2013 Equatorial electrojet dependence on solar activity in the Southeast Asia sector *Antarctic Record* **57** 329

[4] Hamid N S A, Ismail, W N I and Yoshikawa A 2016 Longitudinal profile of EEJ current and its dependence on solar activity *Advanced Science Letter* **23** 1357 doi:10.1166/asl.2017.8372

[5] Le Mouel J L, Shebalin P and Chulliat A 2006 The field of the equatorial electrojet from CHAMP data *Annales Geophysicae* **24** 515

[6] Alken P and Maus S 2007 Spatio- temporal characterization of the equatorial electrojet from CHAMP, Ørsted, and SAC- C satellite magnetic measurements *Journal of Geophysical Research* **112** A09305

[7] Shume E B, Denardini C M, Paula E R and Trivedi N B 2010 Variabilities of the equatorial electrojet in Brazil and Perú *Journal of Geophysical Research* **115** A06306

[8] Chandrasekhar N P, Arora K and Nagarajan N 2014 Characterization of seasonal and longitudinal variability of EEJ in the Indian region *Journal of Geophysical Research* **115** A06306

[9] Hamid N S A, Liu H, Uozumi T, Yumoto K, Veenadhari B, Yoshikawa A, Sanchez J A 2014 Relationship between the equatorial electrojet and global Sq currents at the dip equator region *Earth, Planets and Space* **66** 146 doi: 10.1186/s40623-014-0146-2

[10] Yamazaki Y and Maute A 2016 Sq and EEJ-A Review on the Daily Variation of the Geomagnetic Field Caused by Ionospheric Dynamo Currents *Space Science Review* 55-56 doi:10.1007/s11214-016-0282-z

[11] Hamid N S A, Liu H, Uozumi T and Yoshikawa A 2015 Empirical model of equatorial electrojet based on ground-based magnetometer data during solar minimum in fall *Earth, Planets and Space* **67** 205 doi: 10.1186/s40623-015-0373-1

[12] Sabaka TJ, Olsen N and Purucker M E 2004 Extending comprehensive models of the Earth's magnetic field with Ørsted and CHAMP data *Geophysical Journal International* **159** 521

IOP Conf. Series: Journal of Physics: Conf. Series **852** (2017) 012016 doi:10.1088/1742-6596/852/1/012016

Comparison of ionospheric profile parameters with IRI-2012 model over Jicamarca

S A Bello[1, 2], M Abdullah[2, 3], N S A Hamid[4,*] and B W Reinisch[5,6]

[1]Department of physics, Faculty of physical sciences, university of Ilorin, Nigeria
[2]Space Science Centre (ANGKASA), Institute of Climate Change, Universiti Kebangsaan Malaysia, Malaysia
[3]Department of Electrical, Electronic and Systems Engineering, Faculty of Engineering and Built Environment, Universiti Kebangsaan Malaysia
[4]School of Applied Physics, Faculty of Science and Technology, Universiti Kebangsaan Malaysia, 43600 UKM Bangi, Selangor, Malaysia
[5]Center for Atmospheric Research, University of Massachusetts, Lowell, MA, USA
[6]Lowell Digisonde International, Lowell, MA, USA

*shazana.ukm@gmail.com

Abstract. We used the hourly ionogram data obtained from Jicamarca station (12° S, 76.9° W, dip latitude: 1.0° N) an equatorial region to study the variation of the electron density profile parameters: maximum height of F2-layer (*hm*F2), bottomside thickness (*B0*) and shape (*B1*) parameter of F-layer. The period of study is for the year 2010 (solar minimum period).The diurnal monthly averages of these parameters are compared with the updated IRI-2012 model. The results show that *hm*F2 is highest during the daytime than nighttime. The variation in hmF2 was observed to modulate the thickness of the bottomside F2-layer. The observed *hm*F2 and *B0* post-sunset peak is as result of the upward drift velocity of ionospheric plasma. We found a close agreement between IRI-CCIR *hm*F2 model and observed *hm*F2 during 0000-0700 LT while outside this period the model predictions deviate significantly with the observational values. Significant discrepancies are observed between the IRI model options for *B0* and the observed B0 values. Specifically, the modeled values do not show *B0* post-sunset peak. A fairly good agreement was observed between the observed *B1* and IRI model options (ABT-2009 and Bill 2000) for *B1*.

1. Introduction

The bottomside ionospheric electron density profile below the F- layer are described by the maximum electron density (*Nm*F2), the maximum height of F2 layer (*hm*F2), the thickness (*B0*) and the shape parameter (*B1*). These parameters are important for understanding the morphology of the ionosphere and for ionospheric modeling, e.g. the International Reference Ionosphere (IRI).The illustration of these profile parameters is given in figure 1. The altitude distributions of these profile parameters exhibit diurnal, seasonal and solar activity variations, which depend on the geographical location of the observational station. The *B0* and *B1* parameter were introduced by the IRI model to describe the ionospheric electron density thickness and shape below the F2 region. The *B0* parameter is obtained from the ionospheric profile *Ne (h)* and calculated as the height difference between the peak height of F2-layer (*hm*F2) and the height where the electron density equals to 0.24*NmF2 in the absence of Fl-

Content from this work may be used under the terms of the Creative Commons Attribution 3.0 licence. Any further distribution of this work must maintain attribution to the author(s) and the title of the work, journal citation and DOI.
Published under licence by IOP Publishing Ltd

layer or the F1 peak height (*hm*F1) if the latter occur [1]. *B1* determines the shape of the profile between the two heights from which *B0* parameter is estimated [2].

In the previous works, the comparative study between the observational profile parameters and IRI-2001 model at Jicamarca station (12° S, 76.9° W, dip latitude: 1.0° N) has been earlier studied during solar maximum [3] and during solar minimum conditions [4]. A comparison between the observational B-parameter (*B0* and *B1*) and the IRI-2012 model was recently carried out in the African sector [5], where they observe a fair agreement between the model predictions and *B1* parameter. However, they observe that the model prediction agrees with the observational *B0* values during the nighttime period and obtained some degree of differences during the pre-sunrise towards the midday period. A similar result was obtained at the Dibrugarh (27.5° N, 95° E, 43° dip), a low mid-latitude station where they have attributed the discrepancies between the IRI-2012 model option as due to the limitation of the data composition (location) used in building the ABT-2009 model option and the auto-scaling of F1 layer in the ionogram inversion software[6]. IRI model is a data-driven model which has been undergoing periodic updates in order to improve the model prediction. For this reason, the present study, re-examine the ionospheric parameters obtained from digisonde measurements at Jicamarca station, located in the Southern American sector and make a comparison with the updated version of the IRI-2012 model during the period (the year 2010) following the deep solar minimum.

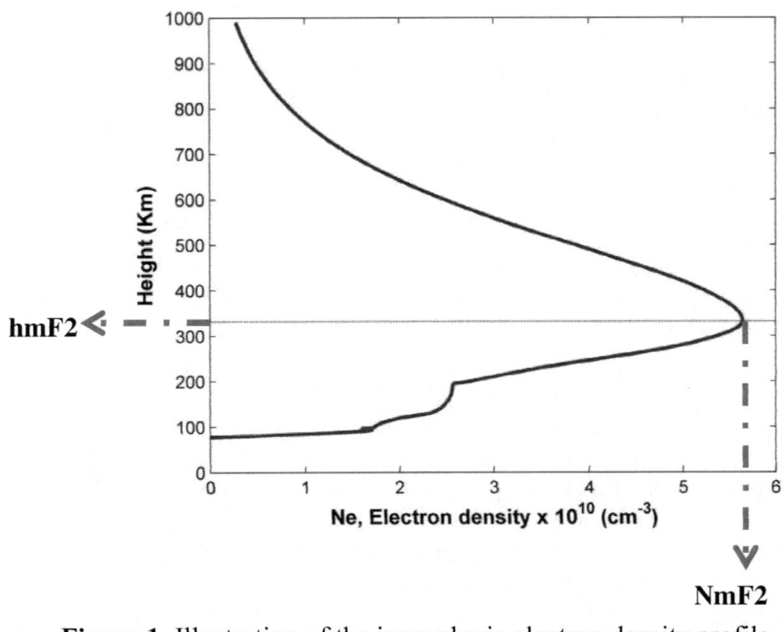

Figure 1. Illustration of the ionospheric electron density profile.

2. Data Analysis

The DPS-4 digisonde measurement used in this study is located at Jicamarca (12° S, 76.9°W, dip latitude: 1.0° N), an equatorial location in the Southern American sector. We have used hourly ionograms during the year 2010, a year of low solar activity with 27-day averaged solar radio flux, F10.7 = 80 sfu. The ionogram traces were inverted into "true" height electron density profiles Ne(h) by the True Height Profile Inversion program (NHPC), which is run in the Digisonde Ionogram Data Visualization/ Editing Tool (SAO-X) software. The SAO explorer program allows us to estimate the profile parameters; *hm*F2, *B0,* and *B1* from the raw ionogram data. The routine techniques for calculating *B0* and *B1* is by best fitting using the least–square fitting approach, the experimental profiles with the formula used in IRI model in equation (1)

$$Ne\ (h) = Nm\mathrm{F2}\frac{exp^{-X^{B1}}}{Cosh(X)} \qquad where\ \ X = \frac{hm\mathrm{F2} - h}{B0} \tag{1}$$

Monthly averages of these parameters are calculated from the selected ten (10) days quietest period. The quietest period is days with no record of geomagnetic disturbances. All the selected quiet period have $\Sigma Kp \leq 24$. Largely due to the paucity of data, only the months of January to July and November to December data are analyzed for this study. The months are then grouped into seasons: equinoctial months (March and April), summer months (November to February) and winter months (May and June), respectively.

The observational results are compared with the IRI-2012 model. The IRI- 2012 model is accessible via http://irimodel.org. In the IRI-2012 model, the IRI-CCIR option was used to compared with observational *hm*F2 because CCIR-maps of *hm*F2 was obtained through its close correlation with propagation factor M(3000)F2 [3]. For the *B0* parameter, the IRI-2012 offers three choices: ABT-2009 model option [1], Bill-2000 model option [2] and Gul-1987 model option [7]. For the *B1* parameter, Bil-2000 and Gul-1987 model options used the same *B1* model and thus give same values of *B1* prediction. The performance of the IRI model prediction relative to the observational results was estimated by finding the deviation of the model options from the experimental (observational) values.

$$\Delta_{IRI\ prediction} = \alpha_{obervational} - \varphi_{IRI\ option} \tag{2}$$

Where the term $\Delta_{IRI\ prediction}$ is the deviation of the model prediction from the observational values ($\alpha_{obervational}$) and $\varphi_{IRI\ option}$ indicates the model option use for various profile parameters.

3. Results and Discussion

3.1. Diurnal variations of hmF2, B0, and B1

For the equinoctial months, the diurnal monthly mean values of *hm*F2, *B0,* and *B1* are given in the left panel in figure 2(a, d and g). In figure 2(a), *hm*F2 generally increases during the sunrise hour to ~1000 LT. The observed increase in *hm*F2 can be attributed to the plasma drift during the daytime and this occurrence can be explained as due to the cross product of electric field (E) and magnetic field (B) at the equatorial latitude. Consequently, the ExB drift raises electron density to a higher altitude where the ratio of gyrofrequency to the ion collision frequency is large. This process is generally known as Equatorial Anomaly (EIA). At midday, the increase in *hm*F2 ceased and the magnitude remains fairly constant. The restriction of the growth of plasma height is known as ionospheric ceiling [8]. The ionospheric ceiling is due to the isoelectron density delineations which are aligned with the magnetic field lines near the dip equator [8]. These mechanisms affect the morphology of *B0* parameter as well (see figure 2d). Observe that *B0* increases during sunrise towards the midday. At post sunset, around 1800-1900 LT, *hm*F2 increased gradually and reached a peak value at about 2000 LT for the months of March and April. The observed *hm*F2 post sunset peak is as a result of the upward velocity of the pre-reversal enhancement (PRE). The PRE resulting from eastward polarized electric field causing the rise of plasma to drift upward (ExB drift velocity) because of F-region dynamo mechanism when the conductivity of E-layer already decayed at post-sunset. The post-sunset peak is observable for *B0* as well. The increase in the magnitude of *B0* around this time indicates that PRE upward velocity not only modulates the height of F2-layer but also increases the bottomside thickness (*B0*) below the peak height of F2-layer. During the nighttime period towards the pre-noon around 2100 to 0700 LT, the magnitude of *hm*F2 decreases and this also affect the values of *B0* parameter. In general, the *B0* values during the daytime period are higher than the nighttime period. We observe that *B1* do not respond to any of the mechanism (earlier mentioned) that causes redistribution of electron density at the bottomside F2 region. The *B1* values are centered on ~2 during the periods from 0800-1800 LT and outside this time; its peak value is ≥ 2.6. The *B1* values generally exhibit morning and nighttime peaks.

The diurnal monthly average values of the ionospheric profile parameters of *hm*F2 (figure 2b), *B0* (figure 2e) and *B1* (figure 2h) for the summer months are given in figure 2 (middle panel). Generally, the morphologies of these parameters are similar to that in the equinoctial months. We observe that post sunset peak values of *hm*F2 and *B0* are smaller as compared to the equinoctial months. The late reversal time of PRE values during the summer month at about 2100-2200 LT is attributed to the longer persistent of the upward velocity of the lifted ionospheric plasma to higher altitude [3].

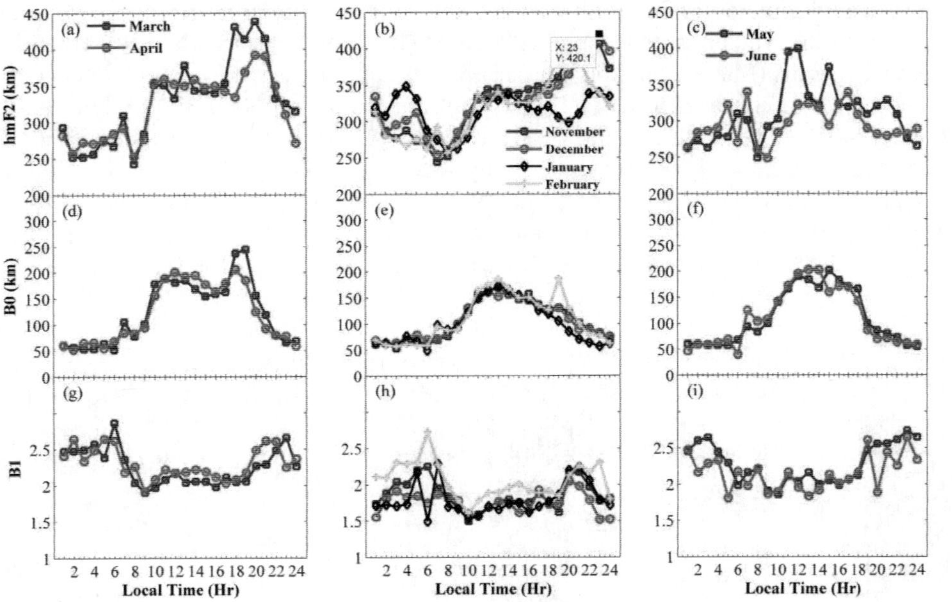

Figure 2. The diurnal seasonal average values of the ionospheric profile parameters of *hm*F2, *B0* and *B1* at Jicamarca station under quiet time condition for the year 2010. Each panel indicates different seasons: the left panel gives the equinoctial months (2a, 2d, and 2g), the middle panel indicates summer months (2b, 2e, and h) and the right panel shows the winter months (2c, 2f, and 2i).

The diurnal monthly averages of the ionospheric profile parameters *hm*F2 (figure 2c), *B0* (figure 2f) and *B1* (figure 2i) for the winter months are given in the right panel of figure 2. We observe that the amplitude of *hm*F2 is lowest in the winter than in the other seasons. The post-sunset peak of *B0* parameter diminishes and the midday values are lower when compared with other seasons. Since B0 is measured from the peak height of F2 layer to the peak height of F1 layer or to the h0.24 (height of 0.24**Nm*F2) when F1 does not exist, this explains why the variations in the magnitude of either *hm*F2 or *Nm*F2 affect the observed values of *B0*. A close observation in the variation of B1parameter revealed a similar trend across the three seasons. Observe in figure 2, that the midday minimum values of *B1* coincide with the period in which the thickness parameter B0 reaches its maximum value. This observation has been reported in similar findings at Cyprus (35° N, 33° E) by Panda and Haralambous [9].

3.2. Comparison between observational results (hmF2, B0, and B1) and IRI model

The contour plot of the diurnal monthly values of observational *hm*F2 and the IRI-CCIR *hm*F2 model option in the IRI-2012 model are given in figure 3a-b. The estimated model deviation (see equation 2) from the observational value is given in figure 3c. The IRI-CCIR gave a fairly representation of *hm*F2 during the pre-sunrise period between (0000 -0700 LT). During the day, the model overestimates *hm*F2 between 0800-1500 LT for all the months presented. The existence of a post-sunset peak in the observational results around the period of about 1800-1900 LT was not modeled by the IRI-CCIR option. From figure 3c, a small deviation can be found between the model values and the observational results during the period of 0000-0800 LT and 14000-1800 LT.

The observational values of *B0* obtained from digisonde measurement (figure 4a) and the model options for *B0* in the IRI-2012 model: ABT-2009 option (figure 4b), Bill-2000 option (figure 4d), and Gul-1987 option (figure f) are presented as a contour plot in figure 4. The location of the annual maxima differs between the IRI-2012 model options predictions and the observational results. For the Gul-1987 model option, a small deviation between the modeled values and the observational results can be observed during 0000-7000 LT for all the months. In term of the season, the greatest discrepancies are observed during the summer months (January, February, November, and December) and equinoctial months (March and April). For Bill-2000 option, a small deviation between the model prediction and the observational results is found between the period 0000-3000 LT for all the months. During 0400-1000LT a negative deviation was observed. Our result is consistent with a study carried out by Lee et al. [4] and with similar findings by Kalita and Bhuyan [6]. For the ABT-2009 model option, a small deviation between the modeled values and the observational results is observed during 1400 LT and 2200-2400 LT for all the months. A negative difference between the observational results and ABT-2009 model prediction is observed during the daytime between 1000-1300 LT in all the months.

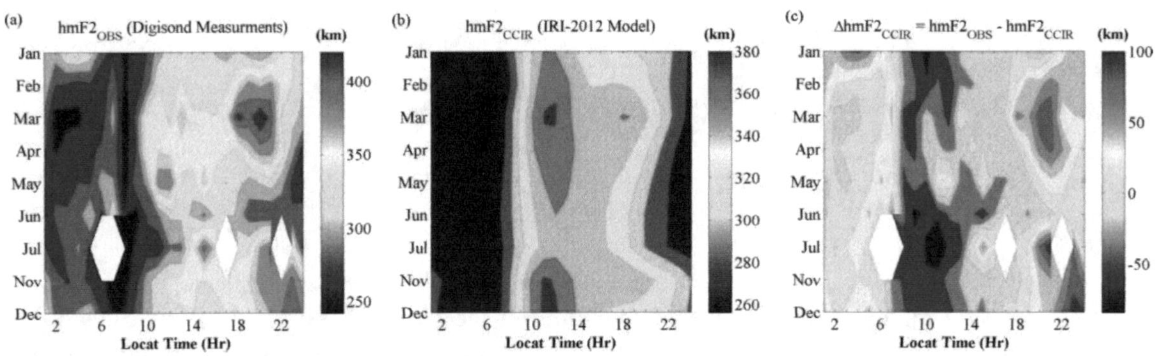

Figure 3. (**a**) Diurnal variation of *hm*F2 obtained from digisonde measurement, (b) IRI-CCIR *hm*F2 model option in the IRI-2012 model (c) and the model deviation.

The observational values of *B1* obtained from digisonde measurement (figure 5a) and the model options for *B1* in the IRI-2012 model; ABT-2009 option (figure 5b), and Bill-2000 option (figure 5d) are shown in figure 5, respectively. The observational values of *B1* during the day are smaller than during the morning and nighttime periods for all the months. Observe that the model options of *B1* in the IRI-2012 model are similar with that of the observational values. In the IRI model, the Bill-2000 and Gul-1987 use the same *B1* model and hence give a similar prediction. As a consequence, only the contour plot of Bill-2000 is presented in figure 5. The deviation of the model prediction and the observational results given in figure 5c and 5e shows that IRI-2012 fairly represent the observational *B1* in all the months.

4. Conclusion

The electron density profile parameters obtained from the digisonde sounder installed at Jicamarca station during the year 2010 were used to study the variation of *hm*F2, *B0,* and *B1* during solar minimum. We observed that the magnitude of both *hm*F2 and *B0* are highest during the equinoctial months (April and March) than the rest of the months. We observe the effect of upward PRE velocity to be responsible for the observable *hm*F2 and *B0* post-sunset peak. The prediction made by the

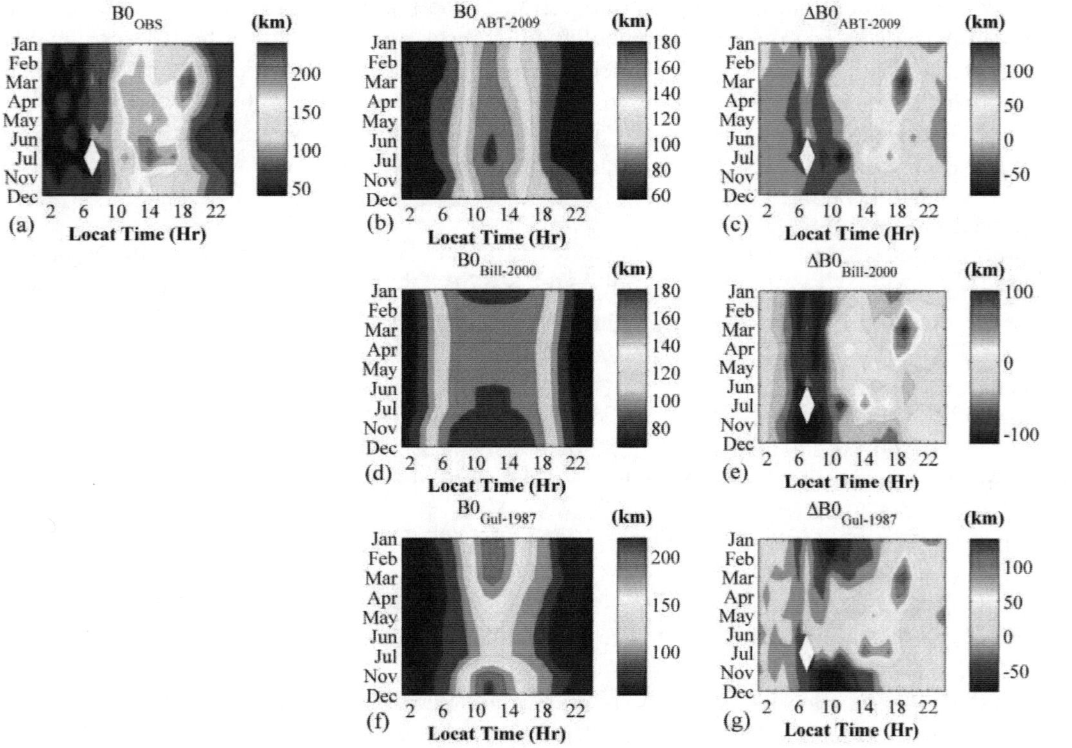

Figure 4. (a) The observational values of *B0* obtained from digisonde measurement and the model options for *B0* in the IRI-2012 model: (b) ABT-2009 option, (d) Bill-2000 option and (f) Gul-1987 option. The IRI-2012 model deviations from the observational values (details in the text) are: (c) $\Delta B0_{ABT\text{-}2009}$, (e) $\Delta B0_{Bill\text{-}2000}$ (g), and $\Delta B0_{Gul\text{-}1987}$

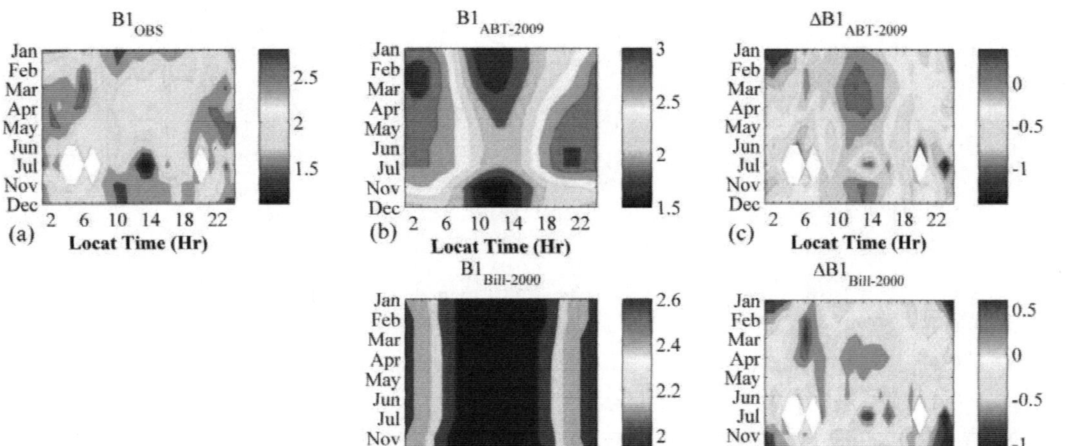

Figure 5. (a) The observational values of *B1* obtained from digisonde measurement and the model options for *B1* in the IRI-2012 model: (b) ABT-2009 option and (d) Bill-2000 option. The IRI-2012 model deviations from the observational values (details in the text) are: (c) $\Delta B0_{ABT\text{-}2009}$ and (e) $\Delta B0_{Bill\text{-}2000}$

IRI-CCIR hmF2 model during the pre-noon period is close to the measured hmF2 values. However, significant disagreements between the model and the observational values of hmF2 were found during 0700-1500 LT at all months. For the observed B0 compared with IRI model, all the model options give a positive and negative deviation from the observational values. The B1 parameter does not exhibit a significant seasonal variability and has small value during the daytime than during the morning and nighttime periods. The IRI model options gave a closer value with the observed B1result.

Acknowledgements
The authors are grateful to Global Ionospheric Radio Observatory (GIRO) and to IRI working group members for making data accessible online for public use. This work is supported by the grants GGPM-2015-020 and GUP-2016-016 of Universiti Kebangsaan Malaysia.

References

[1] Altadill D, Torta J M and Blanch E 2009 *Advance Space Research* **43**(11)1825 doi: 10.1016/j.asr.2008.08.014.
[2] Bilitza D, Radicella S M, Reinisch B W, Adeniyi J O, Mosert Gonzalez M E, Zhang S R and Obrou O 2000 *Advance Space Research* **25** 89
[3] Lee C -C and Reinisch B W 2006 *Journal Atmopherc. Solar Terrestrial Physic* **68**, 2138-2146.
[4] Lee C -C, Reinisch B W, Su S -Y, Chen W S 2008 *Journal Atmopheric and Solar Terrestrial Physic* **70**184
[5] Bello S A, Abdullah M, Hamid N S A, Yoshikawa A, and Olawepo A O 2017 *Advance Space Research* doi: 10.1016/j.asr.2017.02.003
[6] Kalita B R and Bhuyan P K 2016 *Advance Space Research,* doi: 10.1016/j.asr.2016.06.041
[7] Gulyaeva T L 1987 *Adv. Space Res.* **7**(6) 39
[8] Maruyama T, Uemoto J, Ishii M, Tsugawa T, Supnithi P, and Komolmis T 2014 J. *Geophysical Research Space Physics* **119** 10595 doi:10.1002/2014JA020215
[9] Panda S K and Haralambous H 2016 *Advance Space Research* doi:10.1016/j.asr.2016.08.025

IOP Conf. Series: Journal of Physics: Conf. Series **852** (2017) 012017 doi:10.1088/1742-6596/852/1/012017

Irregularities of ionospheric VTEC during lightning activity over Antarctic Peninsula

W Suparta[1] and W N A Wan Mohd Nor[1]

[1] Space Science Centre (ANGKASA), Institute of Climate Change, Universiti Kebangsaan Malaysia, 43600, Bangi, Selangor Darul Ehsan, Malaysia.

wayan@ukm.edu.my

Abstract. This paper investigates the irregularities of vertical total electron content (VTEC) during lightning activity and geomagnetic quiet days over Antarctic Peninsula in year 2014. During the lightning event, the ionosphere may be disturbed which may cause disruption in the radio signal. Thus, it is important to understand the influence of lightning on VTEC in the study of upper-lower interaction. The lightning data is obtained from World Wide Lightning Location Network (WWLLN) and the VTEC data has analyzed from Global Positioning System (GPS) for O'Higgins (OHI3), Palmer (PALV), and Rothera (ROTH). The results demonstrate the VTEC variation of ~0.2 TECU during low lightning activity which could be caused by energy dissipation through lightning discharges from troposphere into the thermosphere.

1. Introduction

The performance of precise navigation system is essential to ensure smooth communication between satellite transmitter and ground based station receiver. However, variation in the ionosphere can lead to disruption of the signal and significant error through solar events [1], geomagnetic activity [2], atmospheric gravity waves (AGW) [3, 4], lightning [5, 6], and plasma bubble [7]. Other authors have studied the effect of lightning activity on ionospheric irregularities in America [8], Europe [5] and South Africa [6] for a mid-latitude region. This paper explores the vertical total electron content (VTEC) disturbance with lightning activity in Antarctic Peninsula. The objective of the study is to investigate the VTEC irregularities during lightning activity with a low geomagnetic storm. Whether with the low geomagnetic activity, the lightning will somehow affect radio signals through the VTEC variation.

Lightning is an electrical discharge between positive and negative regions of a thunderstorm. The cloud interacts with ice particle and begins separating smaller particles. The separation of net positive and negatively charged cloud forms huge electrical potential between cloud and cloud (CC) and between cloud and ground (CG) [9]. As the lightning is discharged, it couples energy through quasi electrostatic (QE) and electromagnetic pulsed (EMP) field and travels to mesosphere and lower ionosphere [6]. The EMP caused an increased ionization by heating the atmosphere which changes the atmospheric conductivity. It may have an indirect impact on the lower atmosphere or even interact with an electron that can diffuse from mesosphere to upper atmosphere.

Content from this work may be used under the terms of the Creative Commons Attribution 3.0 licence. Any further distribution of this work must maintain attribution to the author(s) and the title of the work, journal citation and DOI.
Published under licence by IOP Publishing Ltd

The ionosphere can be measured using Incoherent Scatter Radar (ISR), ground and space based ionosonde, Low Earth Orbit (LEO) satellite and Global Navigation Satellite System (GNSS). In this case, Global Positioning System (GPS) is used as it offers global availability of signal and accurate performance. GPS is used to measure vertical TEC (VTEC) enhancement which shows scintillation in F region.

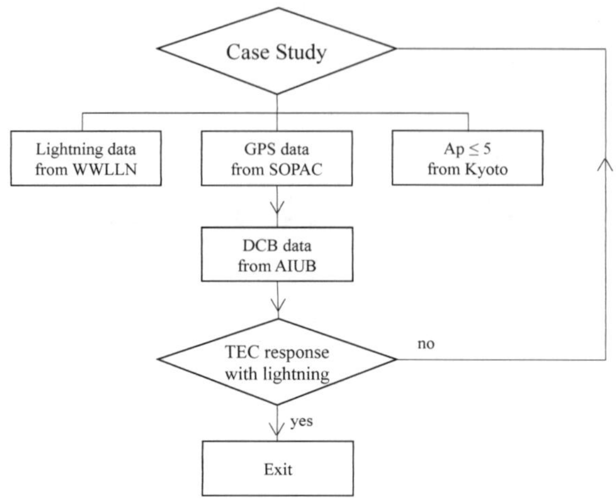

Figure 1. Method of analysis

2. Methodology

The lightning data is collected for Antarctic Peninsula throughout the year 2014 which is obtained from World Wide Lightning Location Network (WWLLN). The data contain information on lightning energy in Joules (J), latitude and longitude in Fractional Degrees, date and time of lightning occurrence in UTC and number of station participate in the observation. Root Mean Square (rms) is the uncertainty energy in Joules. The sensor installed in Antarctic Peninsula is located in Rothera (ROTH) which is 360 km away from Palmer station (PALV) and 666 km away from O'Higgins station (OHI3).

The GPS data is collected from Scripps Orbit Permanent Array Centre (SOPAC). Three GPS station is selected which are O'Higgins (OHI3: 63.32°S, 57.90°W) at UTC – 4 hour, Palmer (PALV: 64.77°S, 64.05°W) at UTC – 4 hour and Rothera (ROTH: 67.57°S, 68.12°W) at UTC – 3 hour. The GPS data is processed using Translation Editing Quality Check (TEQC), a toolkit provided by UNAVCO to translate binary format data and extract variables in GPS data. The VTEC is calculated based on Suparta et al. [10] with a 30 seconds interval. The data is then smoothed and averaged at 150 seconds to remove noise. Additionally, Astronomisches Institut University of Bern (AIUB) provides Differential Code Bias (DCB) to rectify the instrumental bias caused by ionospheric pierce point.

Several studies [11, 12] have shown that high geomagnetic index has significant impact on VTEC. In order to investigate the impact of VTEC on lightning event, it is important to ensure that the study is isolated from geomagnetic storm. There is several methods such as using Kp index however, we have used *Ap* indices from Geophysical Institute Gottingen University to provide a daily average level of geomagnetic activity. *Ap* less than or equal to 5 nT is selected to identify low geomagnetic activity in order to focus on the lightning event and investigate the VTEC irregularities. Figure 1 shows the method of analysis for this study.

The data is plotted in 3 consecutive days together with VTEC irregularities, lightning energy and geomagnetic indices. The data is plotted at Universal Time (UT). The duration of lightning is taken between 0 to 75 minutes before and after the lightning occurrence. The distance between the GPS base station and the location of lightning data is measured using Haversine formula which is determined based on their latitude and longitude. The delta VTEC (ΔVTEC) is calculated to determine the enhancement of VTEC. The formula for slant TEC (STEC) and ΔVTEC are as in equation (1) and (2).

$$VTEC = STEC(\cos z)^{-1} \tag{1}$$

$$\Delta VTEC = |VTEC_i - VTEC_{i\pm n}| \tag{2}$$

where STEC is the slant TEC in TECU and z is the angle between STEC and VTEC in radian [13], i is the time of VTEC during lightning occurrence in minute and n is the time of enhancement of VTEC before or after lightning occurrence in minute.

3. Results and Discussion

3.1. Ionospheric VTEC irregularities and lightning activity during a low geomagnetic storm

From a total of 72 numbers of lightning occurrence detected in Antarctic Peninsula in 2014, 29 numbers of lightning activity are found to occur during the low geomagnetic event with *Ap* index less than or equal to 5 nT. However, only 22 numbers of lightning shows a sufficient result to analyze the VTEC irregularities before or after the lightning occurrence as the remainder has missing VTEC data.

As the data is plotted in three consecutive days, the data is combined to produce 14 cases. Figure 2 demonstrates the VTEC response and lightning activity during geomagnetic quiet days. Typical VTEC response can be observed as it progressively increases from 20:00 to 10:00 hour UT as the sun is rising and decreases gradually from 10:00 to 20:00 hour UT as the sun is setting. As the sun is rising, the atmosphere is heated up which causes VTEC to gradually increase from 10 TECU to peak of 40 TECU at Palmer station. The number of lightning appeared in a day is one only in Case 1 to 10, 13 and 14, 2 numbers in Case 12, and the highest is 6 numbers in Case 11.

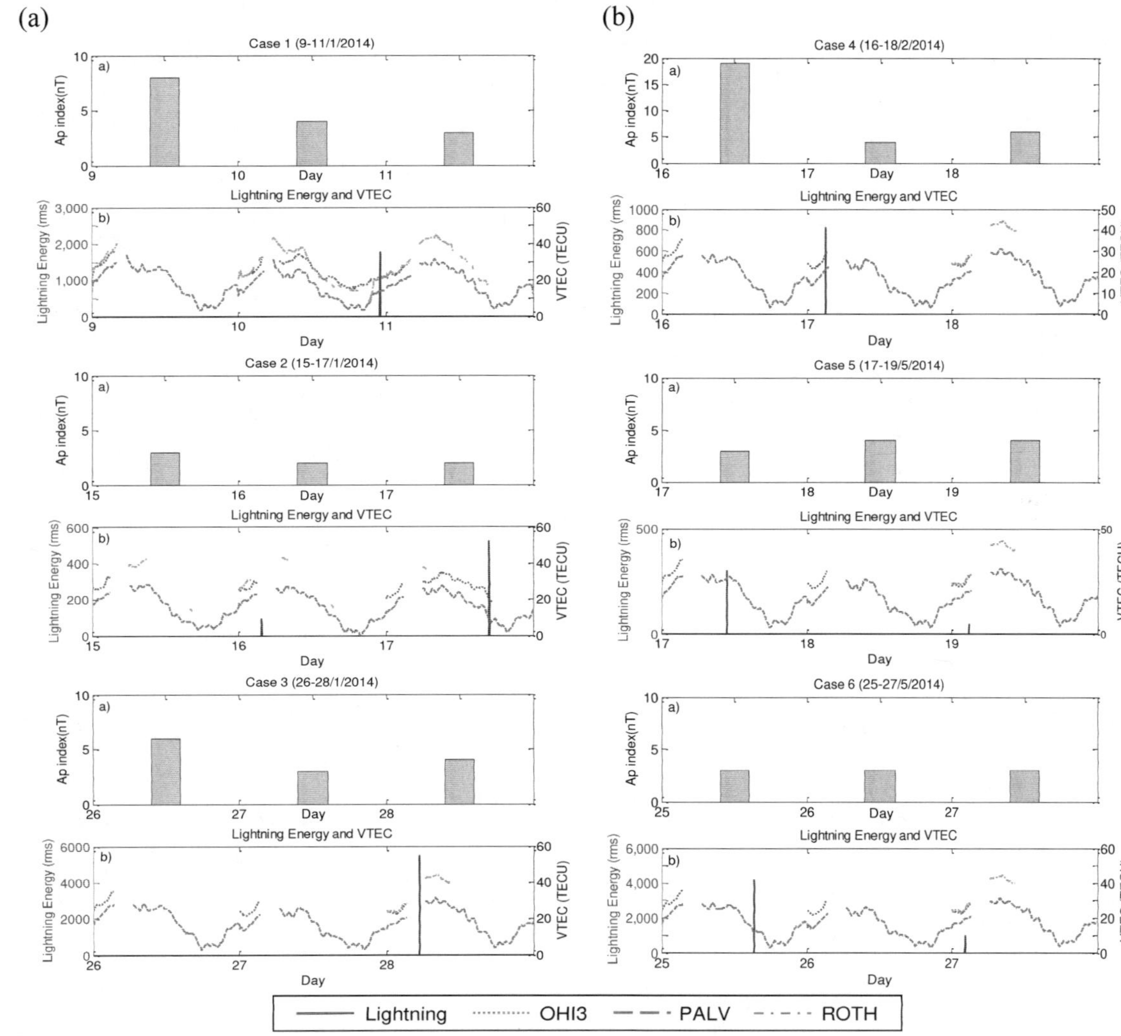

Figure 2. VTEC irregularities and lightning energy in line graph and Ap indices in bar graph. From top to bottom is **a)** Case 1 to 3 and **b)** Case 4 to 6

The relationship between *Ap* index and VTEC can be clearly observed in Figure 2(a) in Case 1. It shows the lightning event that occurs in 10th January 2014 with *Ap* equal to 4 nT. The impact of geomagnetic activity can be observed on the 9th January at *Ap* equal to 8 nT which shows VTEC peak for PALV at 34 TECU as compared with a peak of 31 TECU on the 10th January 2014. A similar observation is also demonstrated for Case 4 in figure 2(b) and Case 14 in figure 3(b). The VTEC irregularities are further examined by concentrating on the specific duration of lightning discharge.

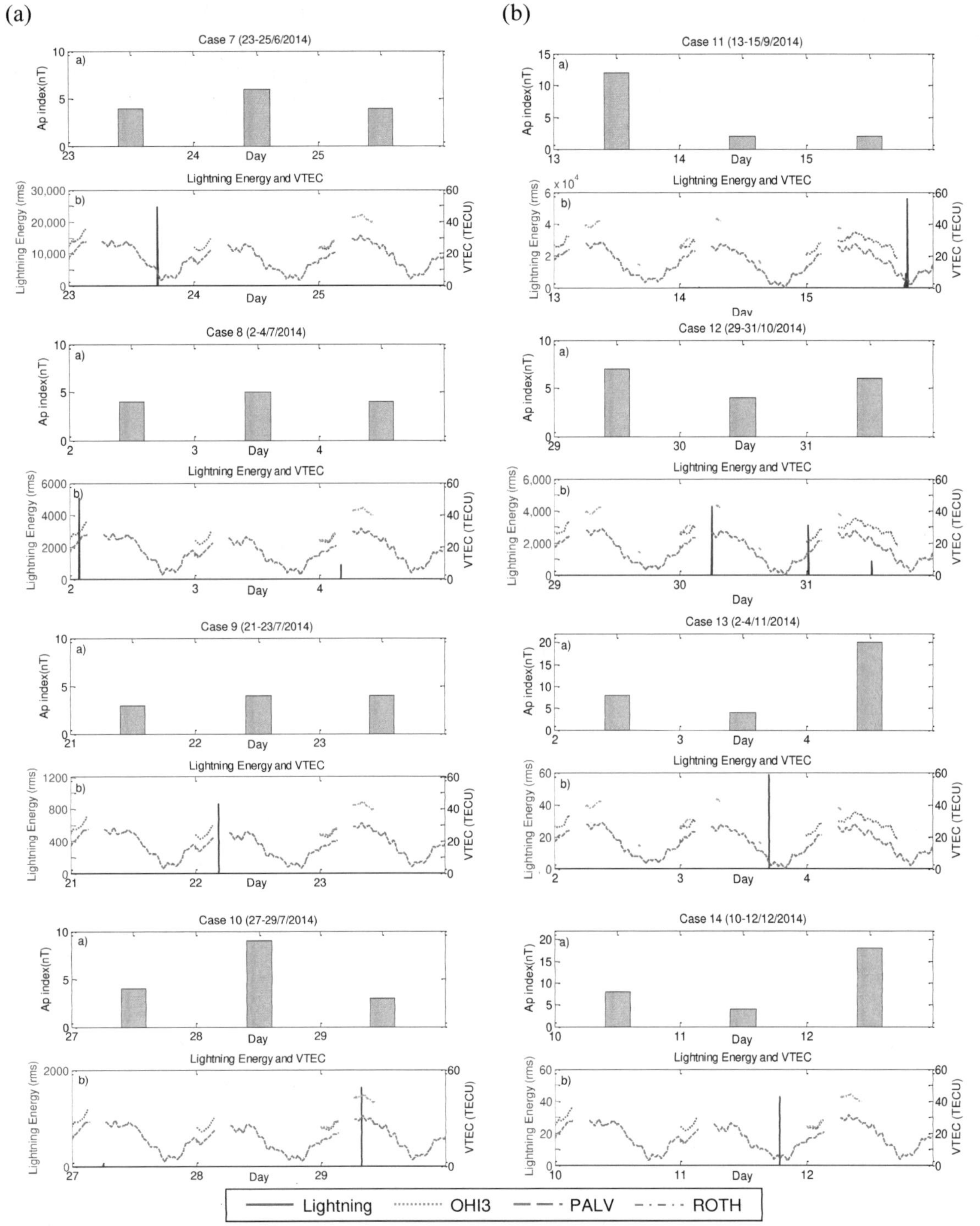

Figure 3. VTEC irregularities and lightning energy in line graph and Ap indices in bar graph. From top to bottom is **a)** Case 7 to 10 and **b)** Case 11 to 14

3.1.1. Ionospheric VTEC irregularities and lightning event. The VTEC result is further analyzed to identify its response before and after the lightning events. The lightning discharge is plotted in 0 minute (Min) while the VTEC is plotted 75 minutes before and after the event. Figure 4 and 5 show the compilation of VTEC irregularities and lightning energy. There are two typical trends which are a positive and a negative gradient of VTEC. The positive gradient occurs as the sun is rising and vice versa for the negative gradient. The positive gradient is shown in figure 4 and 5 for lightning numbers 1, 4, 6, 8, 10, 12, 13, 17, 25 and 26. Meanwhile, the negative gradient of VTEC is depicted for lightning numbers 3, 7, 9, 11, 19 to 24 and 27 to 29. Other lightning numbers such as 2, 5, 14 to 16 and 18 show a missing VTEC result.

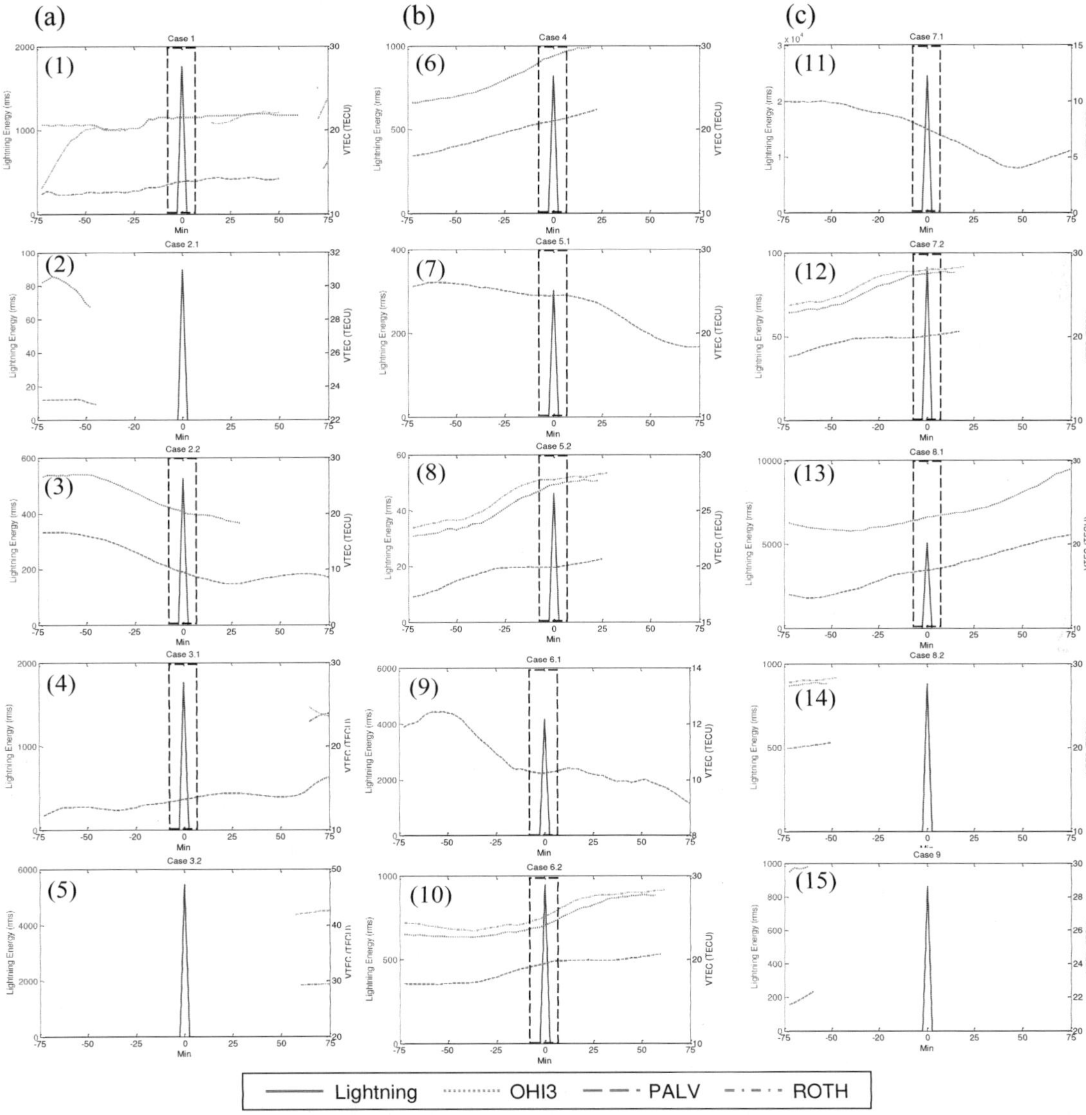

Figure 4. VTEC irregularities and lightning energy. From top to bottom shows lightning **a)** No. 1 to 5, **b)** No. 6 to 10, and **c)** No. 11 to 15

Figure 4(a)(1) shows the positive gradient of VTEC for Case 1 which is highlighted in the rectangular dashed line that overlapped with lightning energy and VTEC of OHI3, PALV and ROTH. During lightning strike, PALV shows VTEC of 14 TECU at 22:59 UTC. Similarly in figure 4(b)(6), there is a positive gradient of VTEC as shown by OHI3 and PALV which gradually increasing from 28 to 29 TECU and 20 to 21 TECU respectively. However, figure 4(c)(11) shows a negative gradient of VTEC. During lightning event, the VTEC is at 7 TECU and progressively decrease to 6 TECU within 15 minutes.

It can be observed that 35% of the results show lightning strikes with positive gradient and 41% of the events occur with negative gradient. Typically, only one lightning is recorded within 150 minutes of VTEC as depicted in figure 5. However, figure 5(b)(19 to 24) shows both positive and negative gradient of VTEC with six (6) lightning discharges within 39 minutes ranging from 1.6 kJ to 56.2 kJ. The sudden increase in lightning number could be due to intense negatively charge cloud or due to seasonal transition from spring to summer. The VTEC enhancement that occurs after lightning strikes may be due to ionosphere is heated up.

Figure 5(c)(28) shows VTEC enhancement before lightning strikes. This is because a tremendous amount of charge has to build up first. Other factors that can contribute to lightning development are ice, hail, semi frozen water drops or graupel [14]. Precipitation could also potentially be responsible for lightning development. Data collected from www.wunderground.com shows a positive precipitation in lightning event number 12, 28 and 29. This means that 10% of the total lightning events may be caused by precipitation.

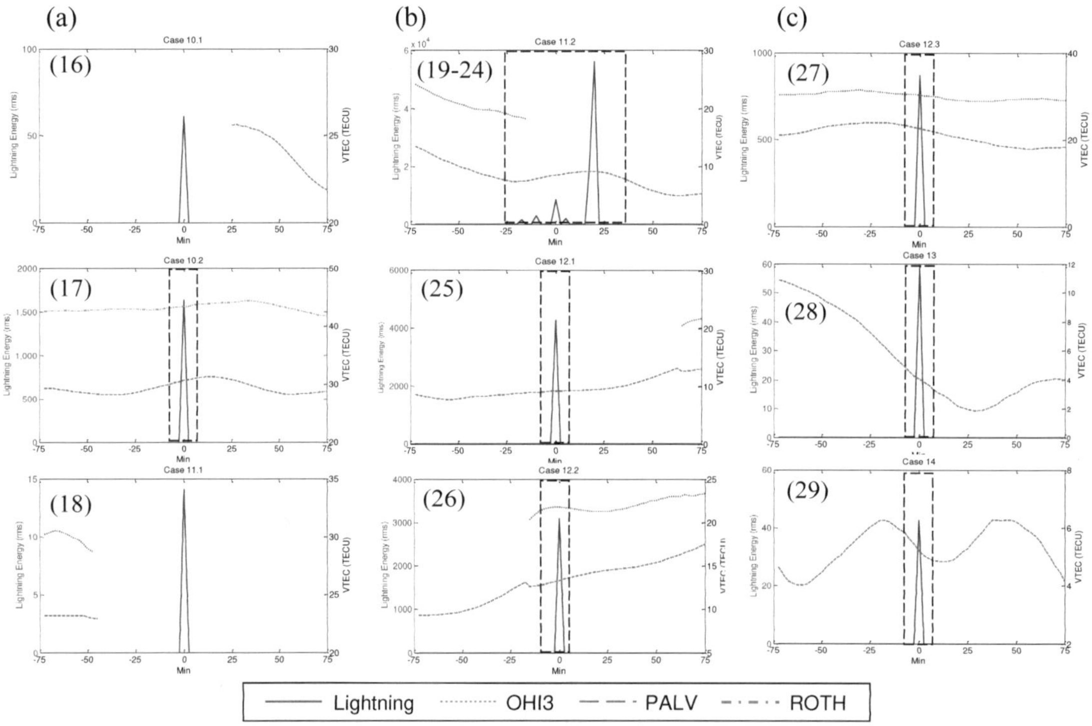

Figure 5. VTEC irregularities and lightning energy. From top to bottom shows lightning **a)** No. 16 to 18, **b)** No. 19 to 26, and **c)** No. 27 to 29

Table I summarized the result of ΔVTEC before and after the lightning occurrence of the twenty-nine (29) number of lightning detected during a low geomagnetic event. The VTEC differential value is calculated and the result is tabulated. The average distance is calculated for the three GPS stations is between 296 km and 1396 km. Approximately 83% of the total lightning event is located within 1000 km average distance from GPS station. It can be noted that there are some unavailable VTEC data from ROTH.

The average ΔVTEC before the lightning event is 0.32, 0.28 and 0.29 TECU for OHI3, PALV and ROTH respectively. Meanwhile, after a lightning strike the average ΔVTEC is 0.17, 0.15 and 0.16 TECU respectively. It is clear that the average irregularities of VTEC during lightning activity are 0.23 TECU in Antarctic Peninsula.

Table I. The ΔVTEC irregularities for lightning event during low geomagnetic activity in 2014

No	*Case*	*Date*	*Average distance (km)*	*ΔVTEC (TECU) before lightning event*			*ΔVTEC (TECU) after lightning event*			*Positive (P) or Negative (N) gradient*
				OHI3	*PALV*	*ROTH*	*OHI3*	*PALV*	*ROTH*	
1	1	9-11 Jan	1051	0.20	0.32	-	0.13	0.26	-	P
2	2	15-17 Jan	677	-	-	-	-	-	-	-
3			537	0.38	0.30	-	0.25	0.27	-	N
4	3	26-28 Jan	906	-	0.35	-	-	0.31	-	P
5			327	-	-	-	-	-	-	-
6	4	16-18 Feb	418	0.40	0.21	-	0.42	0.25	-	P
7	5	17-19 May	1033	-	0.02	-	-	0.07	-	-
8			1221	0.41	0.01	0.01	0.15	0.11	0.13	P
9	6	25-27 May	1025	-	0.04	-	-	0.05	-	N
10			853	0.54	0.45	0.69	0.26	0.10	0.32	P
11	7	23-25 Jun	817	-	0.45	-	-	0.37	-	N
12			1396	0.22	0.24	0.21	0.00	0.04	0.01	P
13	8	2-4 Jul	936	0.42	0.24	-	0.19	0.08	-	P
14			1283	-	-	-	-	-	-	-
15	9	21-23 Jul	498	-	-	-	-	-	-	-
16	10	27-29 Jul	607	-	-	-	-	-	-	-
17			707	-	0.57	0.24	-	0.17	0.19	P
18	11	13-15 Sep	764	-	-	-	-	-	-	-
19			794	-	0.07	-	-	0.08	-	N
20			787	-	0.19	-	-	0.13	-	N
21			764	-	0.17	-	-	0.07	-	N
22			780	-	0.15	-	-	0.15	-	N
23			741	-	0.03	-	-	0.01	-	N
24			731	-	0.01	-	-	0.02	-	N
25	12	29-31 Oct	319	-	0.15	-	-	0.05	-	P
26			439	0.01	0.45	-	0.07	0.15	-	P
27			581	0.34	0.71	-	0.04	0.27	-	N
28	13	2-4 Nov	296	-	0.75	-	-	0.22	-	N
29	14	10-12 Dec	317	-	0.61	-	-	0.10	-	N
	Average VTEC (TECU)			*0.32*	*0.28*	*0.29*	*0.17*	*0.15*	*0.16*	

Figure 6 reveals the scatter plot of ΔVTEC results obtained from Table I. The ΔVTEC distribution is mainly observed at ~0.2 TECU. The maximum ΔVTEC is 0.75 TECU and the minimum is 0.01 TECU. Other authors have found that the VTEC enhancement during intense lightning activity is between 1.2 and 1.4 TECU [6] in South Africa that has high number of the lightning activity. These results could possibly be due to low number of lightning detected and the distance between location of lightning and GPS station is more than 300 km. Hence, the dissipation of energy through lightning discharge caused a very low energy infiltration in the ionosphere.

Figure 7 also shows boxplot of VTEC on each GPS station which corresponds to the median VTEC of less than ~0.2 TECU after lightning strikes. The maximum shown by PALV before lightning strikes could indicate the ΔVTEC irregularities can reach up to 0.75 TECU which is significantly higher due to the distance of lightning strikes is nearer to PALV than OHI3 or ROTH.

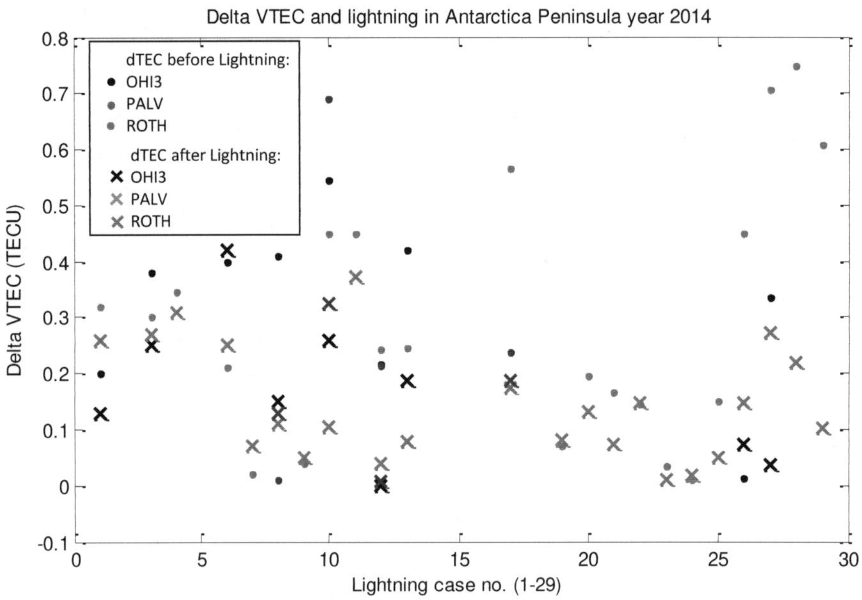

Figure 6. ΔVTEC irregularities distribution against lightning number from 1 to 29. The dot and 'x' marks ΔVTEC before after lightning strikes respectively for OHI3, PALV and ROTH.

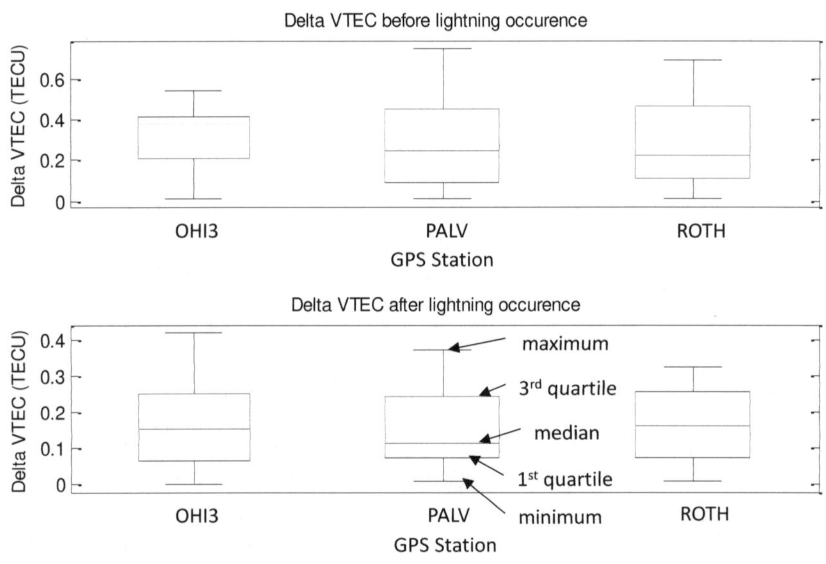

Figure 7. Boxplot of ΔVTEC irregularities before and after lightning strikes

3.2. Lightning distance from GPS station

Several authors [5, 6] have investigated the lightning activity and used 300 km as a distance limitation. In this case, using similar approach would only restrict the study to only one case which is insufficient for statistical analysis. Figure 8 reveals the majority location of lightning strike occurs over the ocean during autumn, spring and summer. The red dot shows the GPS station of OHI3, PALV and ROTH, the 'x' marks the lightning location and the number represents lightning numbering sequence. The distance between PALV and dashed small circle is 500 km and big circle is 1000 km. The lightning location outside the dashed circle is less than 1500 km.

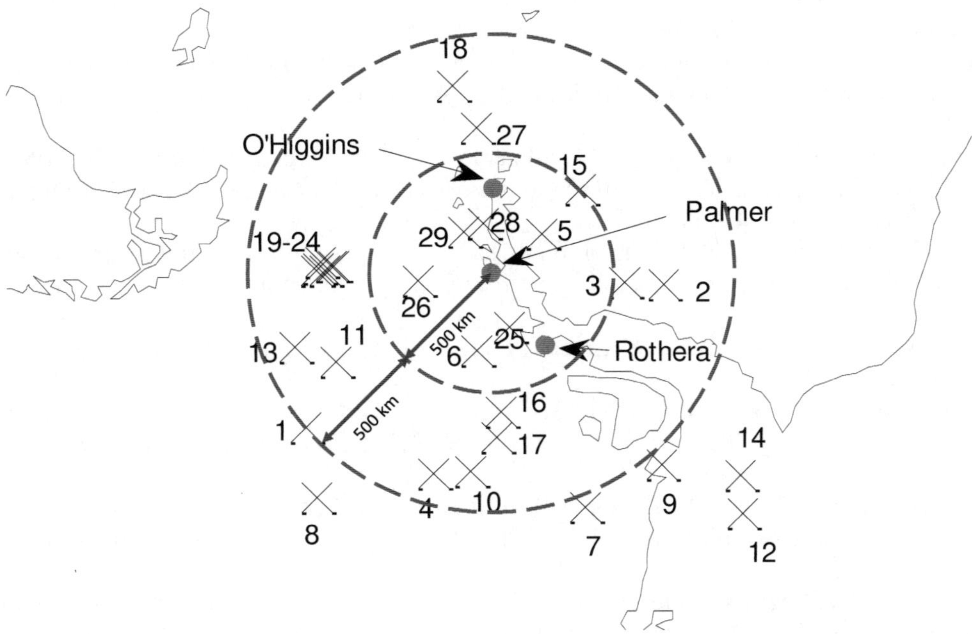

Figure 8. Lightning map in Antarctica Peninsula

The ionosphere over land when GPS measurement is taken could be disturbed due to the infiltration of lightning energy into the upper atmosphere during winter which increases the ionospheric irregularities occurrence [6]. Additionally, the absolute ΔVTEC decreases with distance which indicates that VTEC enhancement reduces the further away lightning occurrence is from GPS station. This is because gravity waves launched by lightning have very small vertical component and would need to propagate hundreds of kilometres horizontally to reach ionospheric altitude.

4. Conclusion
It can be concluded that the VTEC response conducted on quiet days of geomagnetic activity during lightning strikes can be observed approximately 0.2 TECU for overall stations. The distance between 300 and 1000 km of the location of a lightning strike from GPS station could also give an impact on the VTEC enhancement as the lightning detected within that range is found have higher VTEC variation. Nonetheless, causes of VTEC irregularities isolated from lightning activity needs much investigation as it could be due to solar activity, atmospheric gravity waves, plasma bubble, lower atmosphere condition and/or any of the combination of the above. A statistical analysis needs to be performed in future work based on long-term data to clarify whether during a low geomagnetic it will produce the lightning strike.

Acknowledgements
We would like to extend our gratitude to Ministry of Science, Technology and Innovation Malaysia (MOSTI) that funded the research through the Flagship Program under ZF-2014-016 grant. The authors would like to thank Scripps Orbit Permanent Array Centre (SOPAC) for providing important GPS data at O'Higgins, Palmer and Rothera station. We also appreciate the essential CODE data given

by Astronomisches Institut University of Bern (AIUB) Data Centre of University Switzerland and lightning data from World Wide Lightning Location Network (WWLLN). We also sincerely acknowledge the World Data Center-C2, Kyoto, Japan, for archiving geomagnetic data.

References

[1] Bilitza D, McKinnel L-A, Reinisch B and Fuller-Rowell T 2010 International reference ionosphere today and in the future *J. Geod.* **85** 909
[2] Hajj G A, Wilson B D, Wang C, Pi X and Rosen I G 2004 Data assimilation of the ground GPS total electron content into a physics-based ionospheric model by use of the Kalman filter *Radio Sci.* **39** RS1S05
[3] Bishop R L, Aponte N, Earle G D, Sulzer M, Larsen M F and Peng G S 2006 Arecibo observations of ionospheric perturbations associated with passage of Tropical Storm Odette *J. Geophys. Res.* **111** A11320
[4] Vadas S L and Liu H-I 2009 Generation of large scale gravity waves and neutral winds in the thermosphere from the dissipation of convectively generated gravity waves *J. Geophys. Res.* **114** A10310
[5] Johnson C G and Davis C J 2005 Lightning induced intensification the ionospheric sporadic-E layer *Nature* **435** 799
[6] Amin M M, Inggs M and Cilliers P J 2014 *Influence of lightning on total electron content in the ionosphere using WWLLN lighting data and GPS data Proc. of SA Institute of Phys.* 411-16
[7] Li G, Ning B, Liu L, Wan W and Liu J Y 2009 Effect of magnetic activity on plasma bubbles on equatorial and low-latitude region in East Asia *Ann. Geophys.* **27** 303
[8] Lay E H, Shao X-M and Carano C S 2013 Variation in total electron content above large thunderstorms *Geophys. Res. Lett.* **40** 1945
[9] Uman M A 1987 *The Lightning Discharge* (San Diego: Courier Corporation) chapter 1pp 10-7
[10] Suparta W, Rashid Z A A, Mohd Ali M A, Yatim B and Fraser G J 2008 Observations of Antarctic precipitable water vapor and its response to the solar activity based on GPS sensing *J. Atmos. Terr. Phys.* **17** 1419
[11] Meza A, Zele M A V, Brunini C and Cabassi I R 2005 Vertical electron content and geomagnetic perturbations at mid- and sub-auroral southern latitudes during geomagnetic storms *J. Atmos. Terr. Phys.* **67** 315
[12] Trivedi R, Jain A, Jain S and Gwal A K 2011 Study of TEC changes during geomagnetic storms occurred near the crest of the equatorial ionospheric ionization anomaly in the Indian sector *Adv. Space Res.* **48** 1617
[13] Ya'acob N, Abdullah M and Ismail M 2010 GPS total electron content (TEC) prediction at ionosphere layer over the equatorial region *Intech Open Access Publisher* **23** 493
[14] Dugan P R 2013 Climates and attitude change naturally (United States: AuthorHouse) p 19

Derivation of GPS TEC and receiver bias for Langkawi station in Malaysia

W L Teh[1,*], **W S Chen**[2] **and M Abdullah**[1]

[1]Space Science Centre, Institute of Climate Change, Universiti Kebangsaan Malaysia, Malaysia.
[2]General Education Center, Chien Hsin University of Science and Technology, Taiwan.

[*]E-mail: wteh@ukm.edu.my

Abstract. This paper presents the polynomial-type TEC model to derive total electron content (TEC) and receiver bias for Langkawi (LGKW) station in Malaysia at geographic latitude of 6.32° and longitude of 99.85°. The model uses a polynomial function of coordinates of the ionospheric piercing point to describe the TEC distribution in space. In the model, six polynomial coefficients and a receiver bias are unknown which can be solved by the least squares method. A reasonable agreement is achieved for the derivation of TEC and receiver bias for IENG station in Italy, as compared with that derived by the IGS analysis center, CODE. We process one year of LGKW data in 2010 and show the monthly receiver bias and the seasonal TEC variation. The monthly receiver bias varies between –48 and –24 TECu (10^{16} electrons/m^2), with the mean value at –37 TECu. Large variations happen in the monthly receiver biases due to the low data coverage of high satellite elevation angle ($60^{\circ} < \alpha \leq 90^{\circ}$). Post-processing TEC approach is implemented which can resolve the wavy pattern of the monthly TEC baseline resulted from the large variation of the receiver bias. The seasonal TEC variation at LGKW exhibits a semi-annual variation, where the peak occurs during equinoctial months, and the trough during summer and winter months.

1. Introduction

The Earth's ionosphere is a thin atmospheric layer consisting of charged particles (ions and electrons) and neutral particles. The charged particles are resulted from the photoionization process by solar EUV radiations. It is found that the electron density plays a crucial role in manipulating the ionospheric conditions. Since the Global Positioning System (GPS) satellites have been launched for positioning and velocity determination, the dual-frequency GPS signals have then been implemented for estimating the total electron content (TEC) of the ionosphere. Here TEC is denoted as a line integral of electron density along the signal path between GPS satellite and ground receiver. Over the past decades, the TEC parameter had been widely used for studying the ionospheric phenomena such as equatorial ionization anomaly, plasma bubbles, equatorial spread F, and ionospheric storms.

The TEC calculation is based on the group delay and phase advance of the dual-frequency GPS signals, which are caused by the dispersive characteristic of the ionosphere. The accuracy of TEC estimation is affected by systematic biases and random noises. The systematic biases can be attributed to satellite bias (e.g., orbital errors and clock bias) and receiver bias (e.g., clock bias, multipath, and antenna phase center variation). The Center for Orbit Determination in Europe (CODE), one of the International GPS Service (IGS) analysis centers, has provided the monthly differential code biases (DCBs) for all active GPS satellites and for all the IGS stations processed at CODE [1]. However, the GPS stations in Malaysia are not included in the list of the IGS stations.

For a single GPS station, there are three approaches in the literatures to estimate the TEC and the DCB for the receiver, namely, polynomial method [2], minimum variance method [3], and IONOLAB-bias method [4]. The first approach is using polynomial function of coordinates to model the TEC distribution, where the receiver bias and the polynomial coefficients are unknown which can be solved by the least squares method. The second approach is to minimize the standard deviation of

Content from this work may be used under the terms of the Creative Commons Attribution 3.0 licence. Any further distribution of this work must maintain attribution to the author(s) and the title of the work, journal citation and DOI.
Published under licence by IOP Publishing Ltd

vertical TEC for different satellites over a certain elevation angle by use of the trial receiver biases, for which the optimal receiver bias will yield the minimum deviation. The third approach is using the global ionosphere maps (GIMs) and the satellite DCBs from CODE to estimate the receiver bias. We note that the polynomial method is able to estimate the TEC along the non-measured lines of sight as compared to the second approach.

In this paper we apply the polynomial-type TEC model to derive the receiver bias and TEC for Langkawi (LGKW) station in Malaysia. We use one year of LGKW data for demonstration and show the variation of the monthly receiver bias and the seasonal TEC variation. The paper is organized as follows. In section 2, the polynomial-type TEC model is presented. In section 3, we apply the polynomial-type TEC model to the GPS data from IENG station in Italy and compare our results of TEC and receiver bias with that derived from the CODE. The results of receiver bias and TEC are presented for LGKW station in section 4, where the post-processing TEC approach are proposed for resolving the monthly TEC baseline issue. Finally, summary and discussion are given in section 5.

2. Polynomial-type TEC model

Two L-band frequencies (f_1 = 1575.42 MHz and f_2 = 1227.60 MHz) are used in the GPS transmitted signals. The GPS observables are code and phase pseudoranges which are deduced from measured time and phase differences between received signals and receiver generated signals, respectively. The code (phase) pseudoranges are denoted as P_1 and P_2 (L_1 and L_2) for the two different frequencies. The slant TEC (S) along the ray path between the satellite and the ground receiver can be calculated from P_2–P_1 or from $\lambda_1 L_1 - \lambda_2 L_2$, as shown in equations (1) and (2), where λ_1 and λ_2 are the wavelength at f_1 and f_2, respectively.

$$S_P = \frac{1}{40.3} \frac{(f_1 f_2)^2}{(f_1^2 - f_2^2)} (P_2 - P_1) \tag{1}$$

$$S_L = \frac{1}{40.3} \frac{(f_1 f_2)^2}{(f_1^2 - f_2^2)} (\lambda_1 L_1 - \lambda_2 L_2) \tag{2}$$

Due to the ambiguity of phase measurement, S_L is a relative value in equation (2), but has higher precision than S_P. To retain the higher precision for slant TEC, a baseline value, $B_r = \langle S_P \rangle - \langle S_L \rangle$, is computed for S_L. Here $\langle\ \rangle$ denotes the mean value. Therefore, a slant TEC is calculated as $S = S_L + B_r$. Note that the satellite biases from the CODE are used in the slant TEC calculation and the cycle slips in L_1 and L_2 phase measurements are processed. To convert slant TEC to vertical TEC, the ionosphere is assumed as a thin shell at a given height. By considering the receiver bias, the vertical TEC (VTEC) is then calculated as

$$\mathrm{VTEC} = (S - b)m \tag{3}$$

$$m = \cos(\sin^{-1}(\frac{R_E \cos\alpha}{R_E + h})) \tag{4}$$

Here b is the receiver bias, m is mapping function, α is the elevation angle of the satellite, R_E is the Earth's radius, and h = 400 km is the height of the ionosphere.

In this paper, the VTEC can be expressed as a polynomial function of coordinates of the ionospheric pierce point (IPP), which is the intersection point of the ray path from satellite to receiver through the thin shell of the ionosphere. The polynomial function is defined as

$$E_V(\phi,\theta) = \mathrm{VTEC} = c_1\phi^2 + c_2\theta^2 + c_3\phi + c_4\theta + c_5\phi\theta + c_6 \tag{5}$$

where ϕ and θ are the longitude and latitude of IPP, and c_1–c_6 are constant. One can then compute the variance as

$$\sigma^2 = \frac{1}{N}\sum_{i=1}^{N} w^i[(S^i - b)m^i - E_V^i]^2 \tag{6}$$

Here N is the number of measurements and $w^i = \sin^2\alpha_i$ is the weighting function. By minimizing σ^2 with respect to b and the coefficients c_1–c_6, one can then obtain equations which can be solved for the receiver bias and the polynomial coefficients. In the present study, we use all measurements from different satellites over one hour period to perform the minimization.

3. Comparison results for IENG station

Figure 1a shows the monthly receiver bias derived from the CODE and the polynomial model for IENG station in Italy for the year 2008. Figure 1b shows the difference of receiver bias (ΔDCB) between the polynomial model and the CODE. The mean deviation is ~1.7 TECu (10^{16} electrons/m^2), as shown by the dashed line in Figure 1b, while the maximum deviation is ~3.0 TECu. The overall result indicates that the receiver bias derived from the polynomial model agrees reasonably with the CODE. Figure 2 shows the daily TEC from the two different approaches for IENG station during February (low ΔDCB) and October (high ΔDCB). It is seen that these two approaches agree well in the TEC calculation, albeit with a bias shift in Figure 2b. This bias shift is attributed to the deviation of the receiver bias from the CODE.

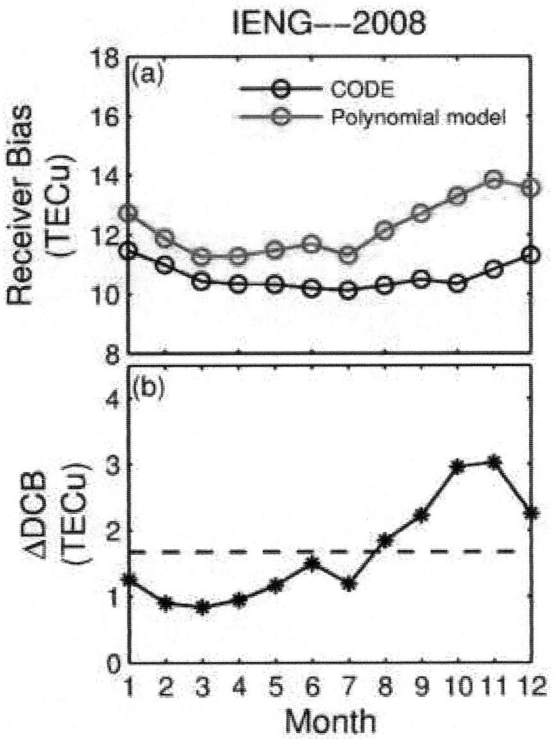

Figure 1. (a) The monthly receiver bias derived from the CODE and the polynomial model for IENG station in Italy. (b) The difference of receiver bias between the CODE and the polynomial model. The dashed line denotes the mean value of the deviations.

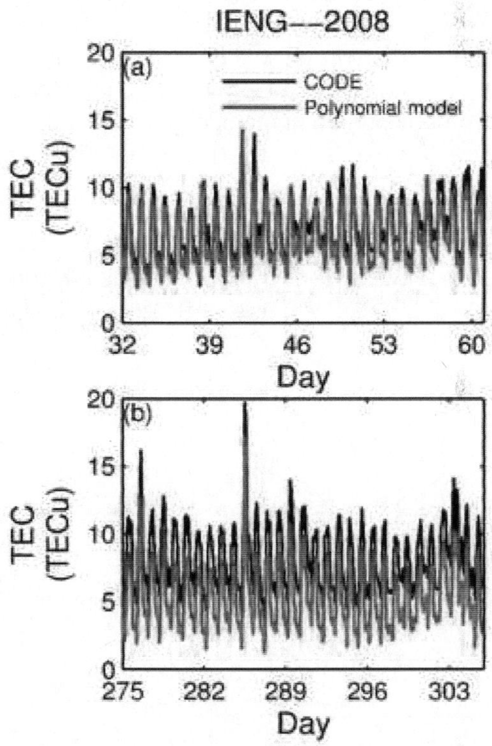

Figure 2. Plots of the daily TEC computed from the CODE and the polynomial model for IENG station during (a) February and (b) October. Note that the ΔDCB is low in February while it is high in October.

4. Receiver bias and TEC for Langkawi station

The GPS data from LGKW station is obtained from the Malaysia Real Kinematics GNSS Network under the Department of Survey and Mapping Malaysia (JUPEM). In this study, we use the data in the year 2010 for demonstration. Note that the LGKW data is using C1 signal, which is less accurate than P1 signal in the TEC calculation from the code pseudoranges. Figure 3 shows the monthly receiver bias derived from the polynomial-type TEC model for LGKW station. The monthly receiver bias varies between −48 and −24 TECu, with a mean value at −37 TECu (as shown by dotted line in Figure 3). Large variations happen in the monthly receiver biases (e.g., August and September) which is likely due to the low data coverage of high satellite elevation angle ($60^{o} < \alpha \leq 90^{o}$).

Figure 4a shows the daily TEC data over an one year period for LGKW station. It is seen that the monthly TEC baseline is wavy, which is due to the large variation of the receiver bias. To tackle this issue, we implement five post-processing steps as follows:

(1) Finding the daily peak (P) and trough (T) TEC over an one year period.
(2) Calculating the mean value κ of P and T, i.e., $\kappa = (P + T)/2$.
(3) Calculating the median value of κ, κ_0.
(4) Repeating step 1 to step 3 for monthly data to get κ_M.
(5) Calculating the new TEC, $TEC_R = TEC_P - (\kappa_M - \kappa_0)$. Here TEC_P is the TEC obtained from the polynomial model for M month.

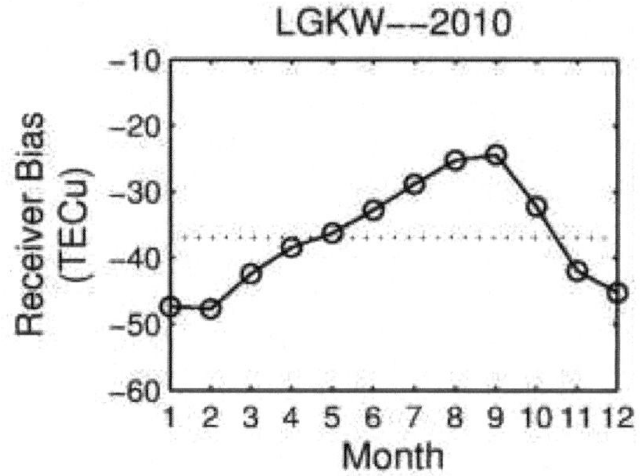

Figure 3. The variation of the monthly receiver bias derived from the polynomial-type TEC model for LGKW station.

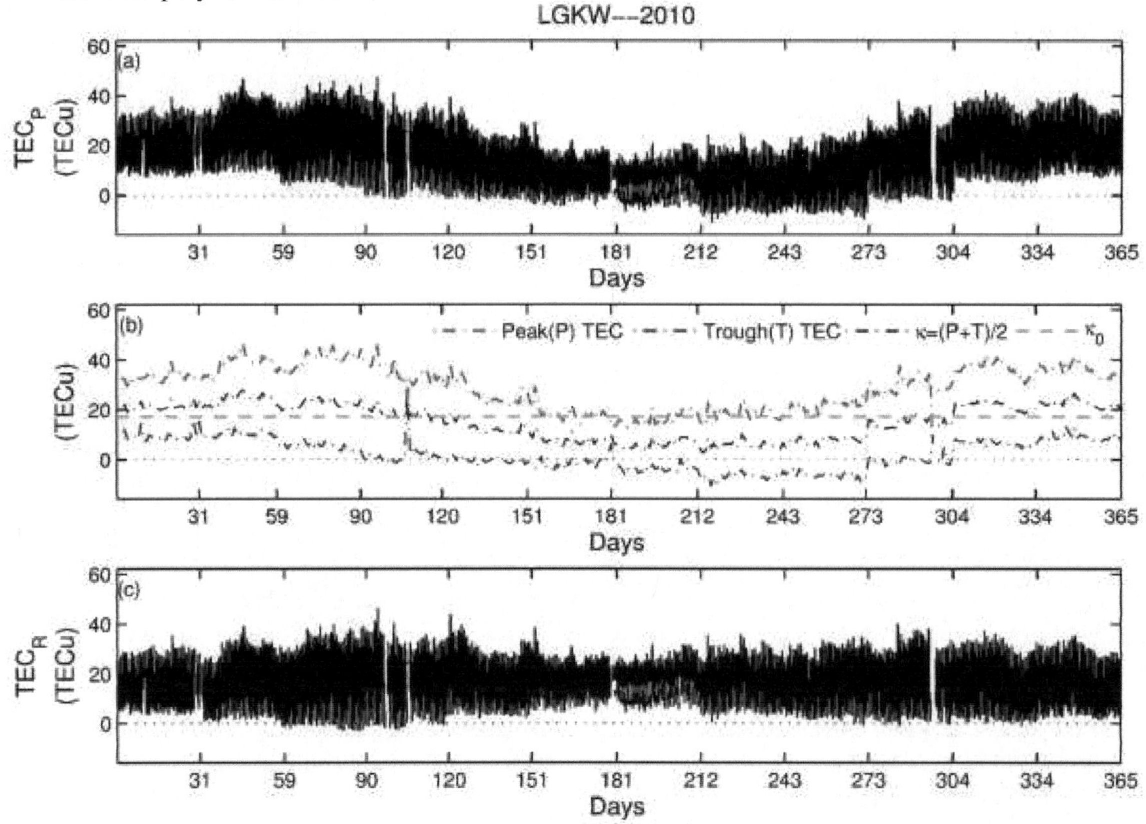

Figure 4. (a) Daily TEC obtained from polynomial-type model for the whole year 2010. (b) Variations of the peak TEC, trough TEC, and κ. The green dashed line is the median value of κ. (c) Corrected TEC by the post-processing procedures.

Figure 4b shows the variation of the peak TEC, the trough TEC, and κ over the whole year period. As compared to κ_0, the κ has large deviations at June, July, August, and September. The post-processing TEC is shown in Figure 4c. Evidently, the monthly TEC baseline has been improved significantly, indicating that the post-processing approach can resolve the monthly TEC baseline problem resulted from the large variation of the receiver bias.

Figure 5 shows the seasonal variation of the peak TEC at LGKW station. It is seen that the seasonal TEC variation exhibits a semi-annual variation, where the peak occurs during equinoctial months and the trough during summer and winter months. This pattern is consistent with the previous studies for the equatorial regions [5].

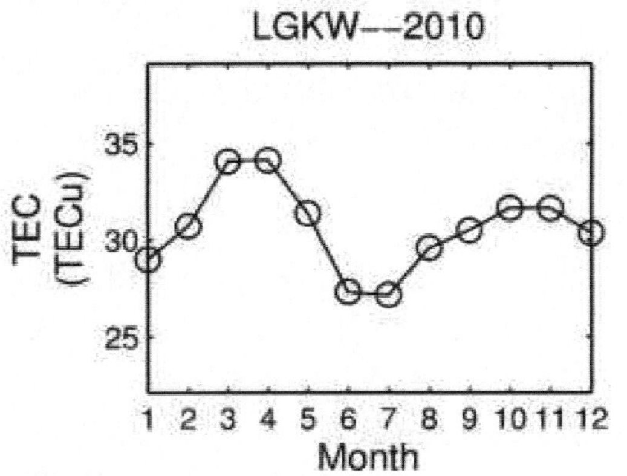

Figure 5. Seasonal variation of the peak TEC at LGKW station for the year 2010.

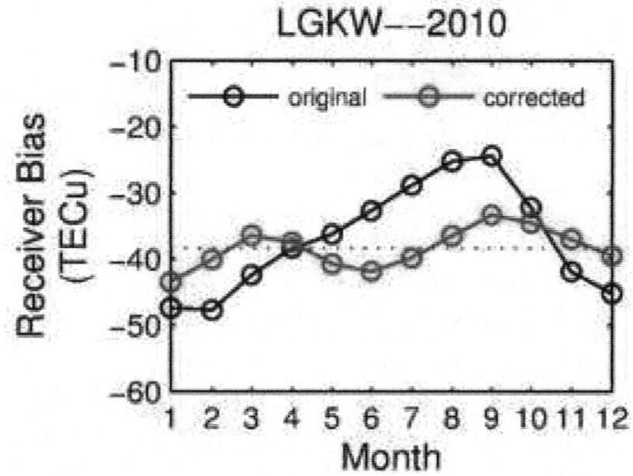

Figure 6. Comparison of the original receiver bias with the corrected one.

5. Summary and discussion

We have examined the polynomial-type TEC model to derive the receive bias and TEC for Langkawi station in Malaysia. In the model, six polynomial coefficients and a receiver bias are unknown, which can be solved by the least squares method. We have tested the performance of the polynomial model using GPS data from the IENG station in Italy. The comparison result indicates that the polynomial model is capable of deriving the receive bias and TEC, although there are some deviations between the model and the CODE. We have therefore applied the polynomial-type TEC model to the LGKW data in the year 2010. The monthly receiver bias of LGKW varies between −48 and −24 TECu, with the mean value at −37 TECu. Large variations in the monthly receiver biases are present due to the low data coverage of high satellite elevation angle ($60^o < \alpha \leq 90^o$). As a result, the TEC values obtained from the polynomial model are biased by the wavy pattern of the monthly TEC baseline. To tackle this issue, we implement the post-processing TEC procedures. Significant improvement is achieved for which the monthly TEC baseline is rectified over the whole year period. The seasonal variation of the peak TEC at LGKW exhibits a semi-annual variation, where the peak occurs during equinoctial months and the trough during summer and winter months.

It is found that the polynomial-type TEC model cannot efficiently extract the wavy pattern of the monthly TEC baseline from the TEC calculation. In doing the post-processing TEC, the $\kappa_M - \kappa_0$ is used for TEC correction. By considering that this corrected amount is treated as the receiver bias, one can recalculate the receiver bias by adding it into the original one. Figure 6 shows the comparison of the original receiver bias with the corrected one. Evidently, the corrected receiver biases become more stable as compared to the original one. The mean value of the corrected receiver biases is about −38 TECu, consistent with that obtained from the original one.

In Figure 5, one can find that the overall TEC is higher in the winter months than in the summer months. This result is closely related with the thermospheric neutral composition ratio O/N_2. During solstice, the global thermospheric circulation is generated as a result of the asymmetric heating of the globe that induces a prevailing summer-to-winter wind with upwelling at low latitudes in the summer hemisphere and downwelling in the winter hemisphere [6]. The upwelling in the summer hemisphere enhances the concentration of molecular species and thus decreases the O/N_2 ratio, while the downwelling in the winter hemisphere increases the concentration of atomic oxygen and thus enhances the O/N_2 ratio. Rishbeth et al. [7] have shown that NmF2 is closely related to the O/N_2 ratio, a large ratio of O/N_2 resulting in a large NmF2 at low latitudes. As a result, the TEC is on average higher in winter than in summer.

In conclusion, we have successfully applied the polynomial-type TEC model to derive the receiver bias and TEC for LGKW station, with the post-processing TEC procedures. The seasonal TEC variation at LGKW is consistent with the previous studies for the equatorial regions.

Acknowledgments

This work was supported by the grant of Universiti Kebangsaan Malaysia (GGPM-2015-027). WSC was supported by the grant of Ministry of Science and Technology, Taiwan MOST 105-2811-M-231-001.

References

[1] Schaer S 1999 Mapping and predicting the Earth's ionosphere using the Global Positioning System *Ph.D. thesis* Uni. of Bern Switzerland
[2] Mitch R H, Psiaki M L, and Tong D M 2013 *Radio Sci.* **48** 671
[3] Ma G and Maruyama 2003 *Ann. Geophys.* **36** 356
[4] Arikan F, Nayir H, Sezen U, and Arikan O 2008 *Radio Sci.* **43** RS4004
[5] Bagiya M S, Joshi H P, Iyer K N, Aggarwal M, Ravindran S, and Pathan B M 2009 *Ann. Geophys.* **27** 1047
[6] Fuller-Rowell T J 1998 *J. Geophys. Res.* **103** 3951
[7] Rishbeth H, Muller-Wodarg I C F, Zou L, Fuller-Rowell T J, Millward G H, Moffett R J, Idenden D W, and Aylward A D 2000 *Ann. Geophys.* **18** 945

IOP Conf. Series: Journal of Physics: Conf. Series **852** (2017) 012019 doi:10.1088/1742-6596/852/1/012019

Longitudinal Variation of EEJ Current during Different Phases of Solar Cycle

W N I Ismail[1], **N S A Hamid**[1*], **M Abdullah**[2,3], **A Yoshikawa**[4,5] **and T Uozumi**[4,5]

[1]School of Applied Physics, Faculty of Science and Technology, Universiti Kebangsaan Malaysia, 43600 UKM Bangi, Selangor, Malaysia
[2] Space Science Centre (ANGKASA), Institute of Climate Change, Universiti Kebangsaan Malaysia, 43600 UKM Bangi, Selangor, Malaysia
[3]Department of Electrical, Electronic and Systems Engineering, Faculty of Engineering and Built Environment, Universiti Kebangsaan Malaysia, 43600 UKM Bangi, Selangor, Malaysia
[4]Department of Earth and Planetary Sciences, Faculty of Sciences, 33 Kyushu University, 6-10-1 Hakozaki, Higashi-ku,Fukuoka 812-8581,Japan.
[5] International Center for Space Weather Science and Education (ICSWSE), Kyushu University 53, 6-10-1 Hakozaki, Higashi-ku,Fukuoka 812-8581,Japan.

Email: shazana.ukm@gmail.com

Abstract - The equatorial electrojet (EEJ) is an eastward current flow around the dip equator in E-region of the ionosphere within the range of 90 to 120 km altitude. The longitudinal variation of EEJ was studied using the ground-based measurement from six different longitudinal sectors. The ground magnetometer data were provided by magnetometer networks such as those from Magnetic Data Acquisition System (MAGDAS) /Circum-pan Pacific Magnetometer Network (CPMN), Indian Institute of Geomagnetism (IIG) and International Real-time Magnetic Observatory Network (INTERMAGNET). The EUEL index used in this study was calculated from geomagnetic northward H component. The H-component was normalized to the dip equator using the CM4 model. This study present results of the longitudinal variation of EEJ during the three phases of solar cycle 24: solar minimum (2008), inclination phase (2011) and solar maximum (2014). Results show that EEJ current is higher in the American sector and lowest between African and Indian sector in 2008 and 2011. On the other hand, during the year of solar maximum, this current component is comparable in American and Southeast Asian sector.

1. Introduction

Ionospheric current at E layer at the altitude of 90-120km consist of two main current which is the Equatorial Electrojet (EEJ) and Solar-quiet (Sq) currents. The former is driven by the eastward electric field that generated by local winds [1] within the latitude of ±3° at dip equator. On the other hand, Sq current is a global current system that flows counter-clockwise in Northern hemisphere and clockwise in the Southern hemisphere [2]. Both current overlap at the dip equator which give the total current that affects the ground-based magnetometer measurement. Previous studies carried out using satellite measurement found out that both EEJ and Sq vary with longitude [3,4]. Another approached by previous study using ground-based magnetometer by [5] where the EEJ current is stronger in South America at longitude 77°W and weakest in India between longitude 70°E and 90°E. However, their study does not use the normalization technique and their results might contain some uncertainty in order to get the longitudinal profile. A study by [6] found EEJ to be higher at about longitude 93° than longitude 77°. They use normalization of the ratio of the range of Sq at off-dip equator station for 10 quiet days. Their longitudinal profile is limited to 10 months data and focused on Indian region only. The longitudinal variation of EEJ was also studied by [7] using the simultaneous data during solar minimum. The present study adopts the technique of [8] to investigate the longitudinal variability of EEJ during different phases of solar cycle 24.

Content from this work may be used under the terms of the Creative Commons Attribution 3.0 licence. Any further distribution of this work must maintain attribution to the author(s) and the title of the work, journal citation and DOI.
Published under licence by IOP Publishing Ltd

2. Method and analysis

We are using the EUEL index that was derived from the northward component. In order to calculate the EUEL index, the median value of the northward component H is subtracted from the original data to obtain Er. The equatorial disturbance in storm time (EDst) is subtracted from the Er to get the EUEL index. The EDst index represents the magnetic disturbance at the equatorial region particularly from ring current and sudden storm commencement (SSC) [8,9]. In order to avoid the latitudinal variation effect at off-dip equator stations, we subtracting $EDst^{*}cos(\varphi)$ where φ is the geomagnetic latitude of station [10]. This technique is to ensures that the EUEL index are free from any part of magnetosphere disturbances.

The magnetometer data was taken from the six longitude region that located at South American sector, Indian sector and Southeast Asian sector. The station pairs are including the station located at dip equator and off-dip equator. The station pair includes ANC-FUQ at South America sector, TAM-ILR and AAB-NAB located at Africa, TIR-ABG located at Indian sector, while DAV-MUT and LKW-KTB located at Southeast Asian sector. Station details are presented in Table 1 and location of the observatory stations used are illustrated in Figure 1. We analysis data period to cover the three phases of solar cycle 24: 2008 (solar minimum), 2011 (inclination phase) and 2014 (solar maximum). Figure 2 is the P parameter that shows the solar activity level from the year 2005 until 2015. In order to avoid counter electrojet effect (CEJ) we consider data at noontime for selected days with Kp<3 are taken. The same normalization technique from [7,8] is applied. The difference between EUEL of total current and EUEL Sq current will get the EEJ current, as shown in this formula:

$$EEJ = EUEL_{total\ current} - EUEL_{Sq\ current} \quad (1)$$

The longitudinal profile of EEJ was obtained using the spline interpolation and linear interpolation fittings.

Table 1. Geographic and Geomagnetic coordinates of the observatories used in the study

Stations	Code	Geographic		Geomagnetic	
		Lat. (°)	Lon.(°)	Lat. (°)	Lon.(°)
Ancon	ANC	-11.77	-77.15	0.77	354.33
Fuquene	FUQ	5.40	-73.73	15.72	357.99
Iiorin	ILR	8.50	4.68	-1.82	76.80
Tamanrasset	TAM	22.8	5.5	25.4	80.6
Adis Adaba	AAB	9.04	38.77	0.18	110.47
Nairobi	NAB	-1.16	36.48	-10.65	108.18
Tirunelvelvi	TIR	8.70	77.80	0.21	149.30
Alibag	ABG	18.62	72.87	10.36	146.54
Langkawi	LKW	6.30	99.78	-2.321	171.29
Kototabang	KTB	-0.20	100.32	-10.63	171.93
Davao	DAV	7.00	125.40	-1.02	196.54
Muntinlupa	MUT	14.37	121.02	6.79	192.24

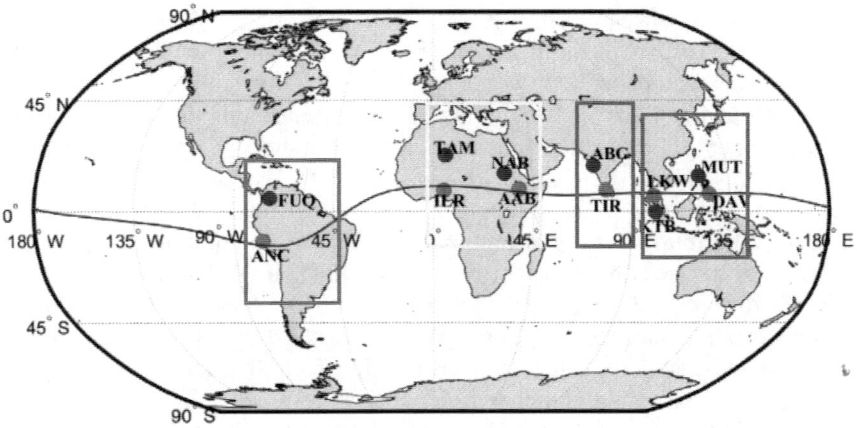

Figure 1. The location of the observatory station. Red dots show the stations at dip-equator while the blue dots show off-dip equator stations. The red box, yellow box, blue box and green box represents South American, African, Indian and Southeast Asian sector respectively.

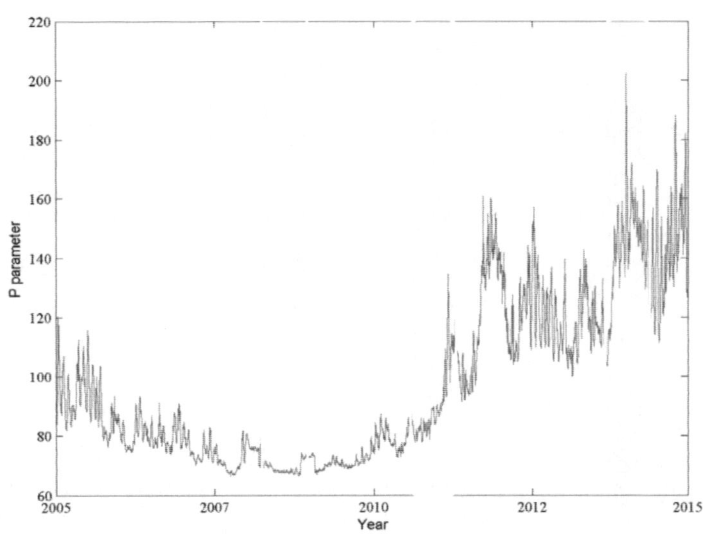

Figure 2. P parameter that indicates the level of solar activity from the year 2005 to 2015. Yellow box represents 2008, 2011 and 2014.

3. Results And Discussions.

3.1 Longitudinal Variation

The EEJ data at each longitude sector is calculated using average data for year 2008, 2011 and 2014.The spline interpolation is used to obtain the longitudinal variation and comparison is made with linear fitting as shown in Figure 3. Results show that the magnitude of EEJ is higher at ANC-FUQ stations pair which is in the South American sector in 2008 and 2011 and lowest between AAB-NAB and TIR-ABG which is in African and Indian sector respectively. Table 2 shows the average magnitude of EEJ. During 2008, EEJ value between LKW station and DAV station does not show any significant difference. The similar results obtained at station TIR and AAB in 2008 and station ILR and AAB in 2011. However, LKW and DAV show the significant result in 2011 and 2014. Apparently, our results also specify that the value of EEJ in 2014 between ANC and LKW are comparable. Possible factors that can be related to the differences on longitudinal dependence are atmospheric tides, gravity and planetary wave [6, 12]. Since it is the first time we discover this behaviour, further study is needed to explain the parameter that contribute to the difference of EEJ intensity at these longitude sector.

The results obtained agreed with the previous study by [5] where the EEJ is maximum in South America even though they are using a different approach. In their study, they only take data at dip-equator station or inside the EEJ band and they do not make any normalization technique. Our approach is better because by using the normalization technique we can make a comparison. The longitudinal variation of EEJ in this study also agrees with the study by [11,12] in which they separated the EEJ and Sq current at dip equator. By separating the EEJ and Sq it will give different results because of the contribution of Sq current that overlaps at dip-equator.

We continue the analysis by comparing the results in 2008 and 2011 with 2014 at Southeast Asian. This is due to the big difference longitudinal pattern at Southeast Asian sector in 2008 and 2011. The comparison is made between stations pair ILR-TAM, AAB-NAB and TIR-ABG. Unfortunately, data are not available in 2014 at stations ILR and NAB and it affects the normalization results.Therefore, future works using complete data with other station pair is recommended.

Apart from EEJ, we also perform spline interpolation to Sq data at all six sectors. The result is presented in Figure 4. Our result confirms that the Sq current does not show any longitudinal variation.

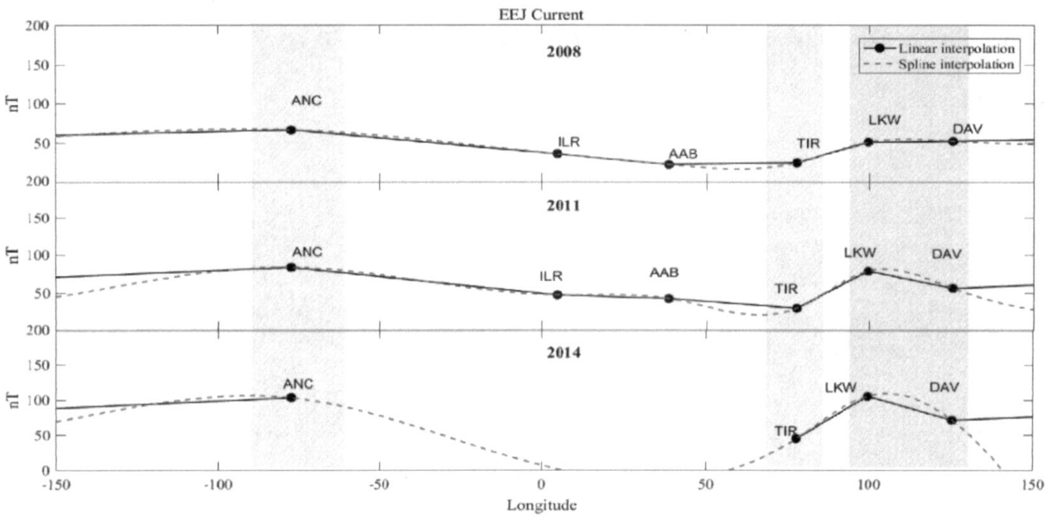

Figure 3. Longitudinal profile of EEJ from six different sectors for three different years (2008, 2011 and 2014). The black line indicates the linear interpolations while the dashed red line represents the spline interpolation.

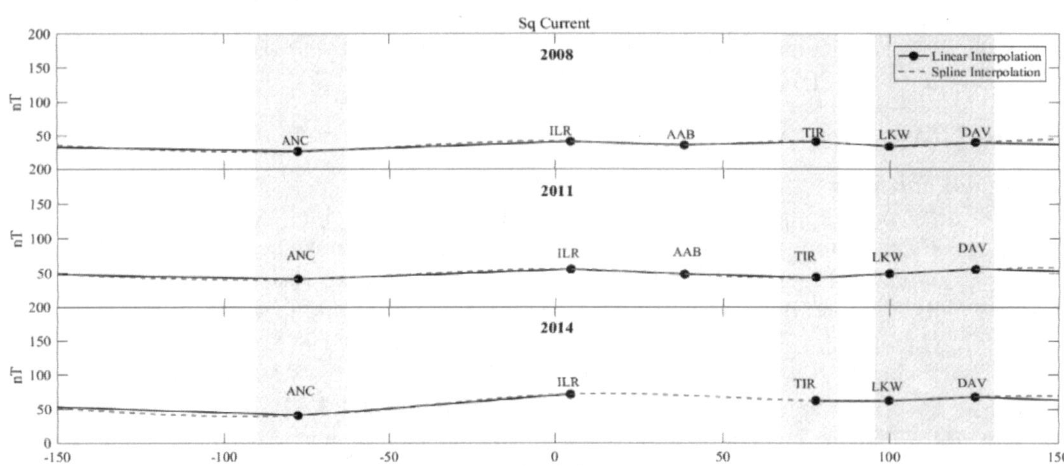

Figure 4. Longitudinal profile of Sq current from six different sectors for three different years (2008, 2011 and 2014)

Table 2. Average for the magnitude of EEJ current for 2008 (solar minimum), 2011 (inclination phase) and 2014 (solar maximum).

Station	Average value of EEJ current (nT)		
	2008	2011	2014
ANC	66.43	83.86	102.55
ILR	35.49	46.61	NaN
AAB	22.40	40.99	NaN
TIR	24.63	28.72	44.55
LKW	50.64	78.66	104.81
DAV	51.91	52.25	70.79

4. Conclusion

The longitudinal profile of EEJ has been constructed using yearly average data for 2008, 2011 and 2014 where six sectors have been considered in this study. Results show that the magnitude of EEJ current is higher at South America sector and lowest between African and Indian sector in 2008 and 2011. However, we found that in 2014 the value of EEJ is comparable between ANC and LKW. Due to the limitation of data in 2014, we could not compare the Southeast Asian sector with the 2008 and 2011 results. Magnitude of EEJ is higher during solar maximum due to increasing ionization in the ionosphere. Further study will be done in near future by changing the stations pair of AAB-NAB and TAM-ILR to other station and comparing ground-based magnetometer data with satellite data in order to find the discrepancy of longitudinal variations of EEJ.

Acknowledgements

The authors thank all the member of the MAGDAS project for their cooperation and contribution to this study. We thank the national institutes that support them and INTERMAGNET for promoting high standards of magnetic observatory practices (www.intermagnet.org). Financial support was provided by Universiti Kebangsaan Malaysia and Ministry of Education, Malaysia, using grants GGPM-2015-020 and FRGS/1/2015/ST02/UKM/02/1A. Yoshikawa were supported in part by JSPS Core-to-Core Program (B. Asia-Africa Science Platforms), Formation of Preliminary Center for Capacity Building for Space Weather Research, and JSPS KAKENHI grants 15H05815. We also acknowledge the National Oceanic and Atmospheric Administration (NOAA) for providing Kp index data, Goddard Space Flight Center/Space Physics Data Facility (GSFC/SPDF) OMNIWeb at http://omniweb.gsfc.nasa.gov for providing F10.7 data, and the National Geophysical Data Center (NGDC) for the estimated values of the magnetic inclination component.

References

[1] Stening R J 1995 What drives the equatorial electrojet? *Journal of Atmospheric and Terrestrial Physics* **57** 10 1117
[2] Yumoto K and MAGDAS group 2007 Space Weather Activities at SERC for IHY: MAGDAS B*ull. Astr. Soc. India* **35** 511
[3] Yamazaki Y, Richmond A D, Maute A, Wu Q, Ortland D A,Yoshikawa A, Adimula I A, Rabiu B, Kunitake M and Tsugawa T 2014 Ground magnetism effects of the equatorial electrojet simulated by the TIE-GCM driven by TIMED satellite data *Journal of Geophysical Research* **119** 3150 doi:10.1002/2013JA019487
[4] Pedatella N M, Forbes J M, and Richmond AD 2012 Seasonal and longitudinal variations of the solar quiet (Sq) current system during solar minimum determined by CHAMP satellite magnetic field observations *Journal of Geophysical Research* **116** doi:10.1029/2010JA016289.
[5] Ibrahim Khashaba A and Ghamry E 2015 Longitudinal dependence and seasonal effect in equatorial electrojet using MAGDAS data *Journal of Geology and Geophysics* **5** doi:10.4172/2381-8719,1000235

[6] Chandrasekhar N P, Arora K and Nagarajan N 2014 Characterization of seasonal and longitudinal variability of EEJ in the Indian region *Journal of Geophysical Research* **119** doi:10.1002/2014JA020183.

[7] Hamid N S A, Liu H, Uozumi T and Yoshikawa A 2015 2015 Empirical model of equatorial electrojet based on ground-based magnetometer data during solar minimum in fal *Earth, Planets and Space* **67** 205 doi: 10.1186/s40623-015-0373-1

[8] Hamid N S A, Ismail, W N I and Yoshikawa A 2016 Longitudinal profile of EEJ current and its dependence on solar activity *Adv. Sci Lett* **23** 1357

[9] Uozumi T, Yumoto K, Kitamura K, Abe S, Kakinami Y, Shinohara M , Yoshikawa A , Kawano H, Ueno T, Tokunaga T , McNamara D, Ishituka J K, Dutra S L G, Damtie B, Vafi D, Obrou O, Rabiu A B, Adimula I A, Othman M, Fairos M, Otadoy R E S and MAGDAS Group 2007 A New Index To Monitor Temporal and Long-Term Variations of The Equatorial *Earth Planets Space* **58** 1

[10] Hamid N S A, Liu H, Uozumi T, Yumoto K, Veenadhari B, Yoshikawa A, Sanchez J A 2014 Relationship between the equatorial electrojet and global Sq currents at the dip equator region *Earth, Planets and Space* **66** 146 doi: 10.1186/s40623-014-0146-2

[11] Hamid N S A, Liu H, Uozumi T and Yumoto K 2013 Brief study of equatorial electrojet and global Sq currents at Southeast Asia region *International Conferences On Space Science and Communication (ICONSPACE)* doi:10.1109/IconSpace.2013.6599463

[12] Shume E B, Denardini C M, Paula de E R, Trivedi N B 2010 Variabilities of the equatorial electrojet in Brazil and Peru *Journal of Geophysical Research* **115** A06306 doi:10.1002/2013JA019487

IOP Conf. Series: Journal of Physics: Conf. Series **852** (2017) 012020 doi:10.1088/1742-6596/852/1/012020

Response of lightning energy and total electron content with sprites over Antarctic Peninsula

W Suparta[1] and N Yusop[1]

[1]Space Science Centre (ANGKASA), Institute of Climate Change, Universiti Kebangsaan Malaysia, 43600 Bangi, Selangor, Malaysia

E-mail: wayan@ukm.edu.my

Abstract. This paper investigates the response of the lightning energy with the total electron content (TEC) derived from GPS over Antarctic Peninsula during St Patrick's geomagnetic storm. During this event, sprite as one of the mesospheric transient luminous events (TLEs) associated with positive cloud-to-ground (+CG) lightning discharges can be generated. In this work, GPS and lightning data for the period from 14 to 20 March 2015 is analyzed. Geomagnetic activity and electric field data are also processed to relate the geomagnetic storm and lightning. Results show that during St Patrick's geomagnetic storm, the lighting energy was produced up to ~257 kJ. The ionospheric TEC was obtained 60 TECU, 38 TECU and 78 TECU between 18:00 and 21:00 UT for OHI3, PALV and ROTH stations, respectively. The peak of lightning energy was observed 14 hours after peaked of TEC. Sprite possibly generated through the electrical coupling process between the top cloud, middle and upper atmosphere with the DC electric field found to be ~10 mVm^{-1} which leading to the sprite generation after the return strokes on 18 March 2015.

1. Introduction

In the electromagnetic coupling, an electromagnetic energy source and how it was transmitted from one region to the other region must be fully understood. This transmission depends on the electrical behaviour of the medium which is regulated by the existence of charged particles and its dynamics through electrical conductivity. The lightning discharges are the major source of electromagnetic energy in the lower atmosphere which radiating electromagnetic waves in a frequency range. Normally, convective clouds accumulate a negative charge in the lower and a positive charge at the top regions. The electric breakdown happens when the electric field strength passes 400 kVm^{-1}. At this time, it will occur as lightning. Lightning reacts as a source of electromagnetic radiation which stimulates the alternating current in the global circuit [1].

Some researchers indicate that the return current from the ground does not end in the clouds during lightning discharge [2]. It keeps moving upwards and ends in the lower ionosphere. This transient current is related with optical emissions such as sprites, elves, blue jets and blue starters in the space between the top of the cloud and the lower ionosphere [2]. Sprite is the members of a family of transient luminous events (TLEs) that illuminate the stratosphere and mesosphere above thunderstorms between 20 and 95 kilometres in altitude. Sprites are the results of the excitation and

Content from this work may be used under the terms of the Creative Commons Attribution 3.0 licence. Any further distribution of this work must maintain attribution to the author(s) and the title of the work, journal citation and DOI.
Published under licence by IOP Publishing Ltd

ionization of air molecules due to collisions with electrons accelerated by quasi-electrostatic (QE) field in the upper atmosphere established by cloud to ground (CG) and its current probably going to persist. Sprite propagating streamer upward may appear up to 90 km, downward initiated at 70–85 km and ends at about 40–50 km altitude [3].

Sprites and other mesospheric TLEs related with CG lightning discharges has been studied by many numbers of researchers. Worldwide sprites and halos over the thunderstorms in the mesosphere are due to large CG flashes almost exclusively with positive polarity. Roughly 5 to 10% of worldwide lightning activity is referable to the positive (+CG) lightning, which transfers the positive charge to ground mentioned by [4]. Sprite drove by the sustained mesospheric quasi-static electric field generated by lightning. Referring to the Maxwell equations; the potential energy is relative to the charged moment change in the lightning discharge that generates the sprite caused of the linearity [5]. A parameter related with lightning intensity is the charge moment which the product of a charge took away from the thundercloud and the length of a lightning channel.

In this work, we studied the response of lightning discharges associated sprite during the St. Patrick day storm on 17 March 2015. St. Patrick day storm was the first storm of solar cycle 24 which reached a level of ''Severe'' on the NOAA geomagnetic storm scale [6]. Two different types of data have been analyzed to ascertain the influence of lightning on the ionosphere over the Antarctic Peninsula through the observation of total electron content (TEC). TEC is an indicator of ionospheric variability derived from GPS signals. While lightning data from worldwide lightning location network (WWLLN) and was used as a proxy to identify regions of intense thunderstorms since GPS signals cannot be directly used to track the actual occurrence of lightning events due to different in frequency domains.

2. Data and methods

2.1. Data

WWLLN data was used to study the variation of lightning energy during the geomagnetic storms which locates lightning using time of group arrival (TOGA) at VLF: 3-30 kHz from lightning stroke. WWLLN data was collected from 14 March to 20 March 2015 for this study. The WWLLN data files are in ASCII which including the date and time in UTC, Latitude and Longitude in Fractional Degrees and Energy in Joules of the lightning stroke detected, and residual fit error always $\leq$ 30 microseconds and are considered as a good data. Nsta is the number of WWLLN stations which observed the stroke in the range from 1000 to 8000 km distant. The WWLLN Network confirms that lightning has occurred when 5 or more stations have recorded to produce a good observation. RMS is the energy uncertainty in Joules. This network evenly capable of detecting CG and intra cloud (IC) flashes likewise detection efficiency (DE) [7]. In the area of Antarctic Peninsula, only one WWLLN sensor was installed in Rothera. The location of WWLLN sensors to the GPS stations is about 666.9 km from OHI3 to ROTH and 360.2 km from PALV to ROTH, respectively.

The geomagnetic index called planetary 3-hour-range index (Kp) and disturbance storm time (Dst) index was used to indicate the level of the geomagnetic disturbance. Kp is an excellent indicator of disturbances in the Earth's magnetic field and Dst is a measure of geomagnetic activity used to assess the severity of magnetic storms. The Kp index is obtained from the Space Weather Prediction Center (SWPC) at http://www.sec.noaa.gov and Dst index from the Kyoto data service (http://swdcwww.kugi.kyoto-u.ac.jp). Magnetic field data represented by interplanetary magnetic field B in z-component (IMF B_z) was obtained from NASA (http://omniweb.gsfc.nasa.gov/). GPS data used in this study were taken from Scripps Orbit Permanent Array Center (SOPAC). Data from three GPS stations of O'Higgins (OHI3: 63.32°S, 57.90°W), Palmer (PALV: 64.78°S, 64.05°W) and Rothera

(ROTH: 67.57°S, 68.13°W) was taken to observe the TEC fluctuation over the Antarctic Peninsula. The variation of TEC was discussed before and after the major geomagnetic storm on 17 March 2015.

2.2. Data processing

In this study, lightning energy and variation of GPS TEC were analyzed to investigate the existence of sprites during the geomagnetic storm event. In order to process the GPS data, code in MATLAB program was used to compute TEC. Prior the processing, it need to be cleaned first using Translate Edit Quality Check (TEQC) routine prepared by UNAVCO (http://www.unavco.org). The data average were performed in 30 seconds interval in order to shorten the processing time but was later changed into three hours interval. The GPS TEC algorithm calculates TEC with an elevation angle < 30° and applies satellite and receiver bias corrections. The software takes RINEX observation files utilize the Hatanaka (d-file compression) with at least C1 (course/acquisition code transmitted on the F1), L1, L2, P2 observables as inputs. The differential code bias (DCB), satellite (P1-C1) biases and the (P1-P2) DCB biases files provided by the AIUB Data Centre of University Switzerland are used to correct the instrumental bias. The vertical TEC (VTEC) is calculated based on the MATLAB code developed by Suparta et al. [8].

3. Result and discussion

3.1. The variation of lightning energy and TEC during the geomagnetic storm

The St. Patrick's Day storm of the low solar cycle 24 was occurred due to a magnetic filament erupted in concert with a slow C9-class solar flare from sunspot AR2297 [9]. During this event, the speed of solar wind had increment from 400 to 650 kms^{-1} on 17 March 2015, and follow by the next days, the speed of solar wind was fluctuated within 500–650 kms^{-1}. This storm is categorized as severe geomagnetic storm arises from two consecutive moderate storms forced by two consecutive southward IMF structures [10].

Fig. 1 shows the indication of severe geomagnetic storm represented by variations of Kp, Dst index and interplanetary magnetic field in *z*-component (IMF *Bz*) between 14 March and 20 March 2015 including the 3 days before and after the geomagnetic storm event. The gray background shows the starting and ending of the geomagnetic storms. During this event, Kp reached maximum value of 8 and it was started with a positive Dst around 06:00 UT and a minimum Dst equal to -223 nT at 23:00 UT. After the shock arrival, the northward IMF *Bz* component was strongly reached the value of ~+20 nT at 05:00 UT due to the coronal mass ejection (CME) hits the earth on 17 March before turning southward to the first minimum peak value of about ~-16 nT at 08:00 UT. Then it aggressively turned northwards and varied drastically between south and north around 11:00 UT. After 14:00 UT, the IMF *Bz* turned southward again with the second lowest value of ~-18 nT and move northward before it keep fluctuated violently between north and south directions at 05:00 UT on the next day until the end of 21 March.

IOP Conf. Series: Journal of Physics: Conf. Series **852** (2017) 012020 doi:10.1088/1742-6596/852/1/012020

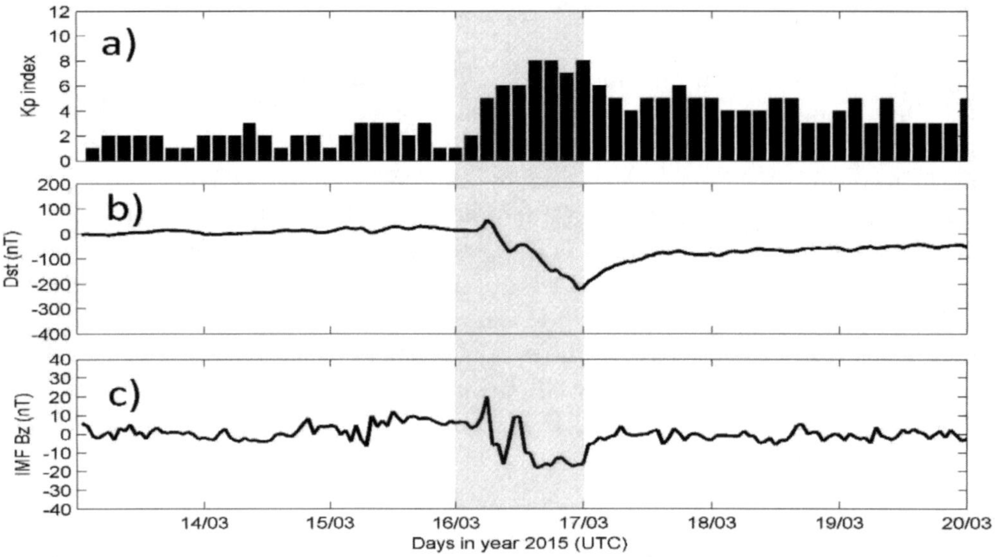

Figure 1. The variation of (a) Kp from NOAA/SEC, (b) Dst from World Data Center-C2 for geomagnetism, Kyoto and (c) IMF *Bz* from OMNIWeb between 14 and 20 March 2015

To relate the occurrence of geomagnetic storm with lighting, we analyzed the WWLLN data from 14 March to 20 March 2015 and found a total of 7 times of lightning occur in this study. The occurrences of lightning over Antarctic Peninsula are tabulated in Table 1. As show in the table, two days having twice lightning was detected occur on 14 & 18 March 2015. Lightning energy was in the range from ~13 J up to ~257 kJ throughout the observation. The highest energy recorded is ~257 kJ on 18 March 2015 at 11:19 UT. There was zero energy produced for a few days during the lightning activity. The gap time between first and second occurrences of lightning is 11 minutes on 18 March 2015 as shown in Fig. 2.

Table 1. The occurrences of lightning in March 2015

Date	Time (UTC)	*Latitude (Deg S)*	Longitude *(Deg W)*	*Residual Error*	Nsta	*Energy (J)*
2015/03/14	00:44:13.6598	-65.5968	-62.422	27.9	5	0
2015/03/14	10:28:06.9148	-74.3485	-53.1605	10.1	5	0
2015/03/15	15:16:34.3455	-64.7031	-70.6191	25.2	6	398.40
2015/03/17	11:12:22.2773	-63.5902	-70.755	24.6	5	13.44
2015/03/18	11:08:01.8062	-59.025	-72.6091	25.9	13	116074.01
2015/03/18	11:19:55.4806	-59.0645	-72.5936	17.3	7	257396.03
2015/03/20	14:01:21.1022	-61.1733	-79.3165	20.7	5	0

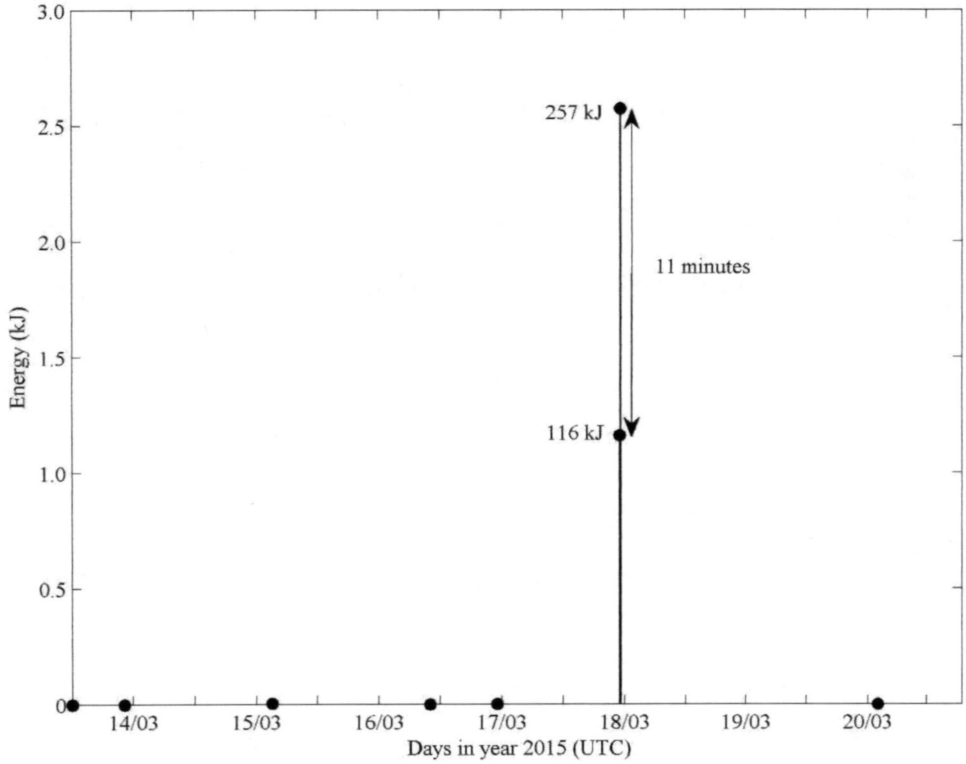

Figure 2. Variation of lightning energy based on WWLLN data over Antarctic Peninsula

To look the response of lower to upper atmosphere, the variations of vertical TEC derived from GPS data and electric field for selected station in Antarctic Peninsula between 14 and 20 March 2015 are demonstrated as in Fig. 3. The gray background refers to the response between TEC fluctuation and electric field during the severe geomagnetic storm. The TEC demonstrate almost similar patterns and the intensity of the fluctuations increase dramatically for three stations during the storm. It was noticed that highest TEC value was obtained for three stations is 60 TECU, 38 TECU and 78 TECU between 18:00 and 21:00 UT for OHI3, PALV, and ROTH, respectively. The magnetic storm event demonstrates the rate of TEC was maximal on most disturbed days of 17 March 2015. The electric field was observed ~10 mVm^{-1} during the lightning discharge, which is similar to the TEC response. The typically enhancements of TEC occur some distance from the region where the lightning activity is the largest.

As shown in Fig. 3, the night on 17 March have significant relation between lightning and geomagnetic storm that corresponds well in time and space to localized regions of enhanced TEC fluctuation at OHI3, PALV and ROTH. The structure of the phase fluctuations is increased caused by the presence of irregularities in the ionosphere including the coupling process of electric discharge between upper-lower altitudes. The most important that how possible sprites are generated from the response between TEC and lightning, which would be discussed in the next section.

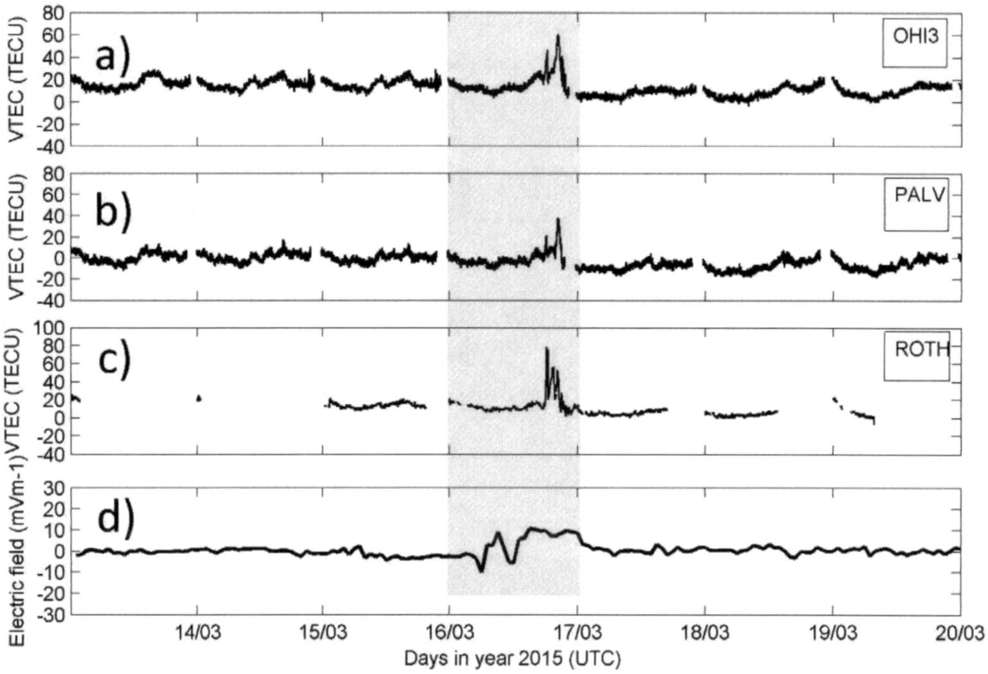

Figure 3. TEC and electric field variations for selected three stations in Antarctic Peninsula

3.2. Possibility sprites generated by the geomagnetic storm

Referring to Fig. 2, high-energy particles penetrating between upper-lower altitudes and increases the electrical conductivity of the upper-lower atmosphere during the lightning discharges. We suggest that the maximum energy produced is proportional to the charge moment changes (CMC), which possibly generate sprites. As reported by Takahashi et al. [11], absolute optical energies emitted from sprites were estimated for 14 streamer-type sprites for the first time and the averages of the time integrated optical energies are 176 kJ and 119 kJ for the N2 1PG and N2 2PG bands, respectively. On the other hand, sprites may occur during this time with the energy obtained reached up to 257 kJ which is much highest than the streamer sprites obtained by [11]. The important of sprite mechanism during geomagnetic storms will improve our understanding of all processes in the upper-lower atmosphere that lightning will also take part in this process.

From the perspective of the upper atmosphere, the good response between the TEC fluctuation and the electric field probably generates sprites after the events on 18 March. This assumption following Fullekrug and Rycroft which calculated the DC electric field from single sprite with a CMC of 500 C.km and transferring a charge of 20 C over a vertical distance of 25 km to be 4.4 mVm^{-1} [12]. This support that the sprites may be produced since the electric field in this event was observed to be ~10 mVm^{-1} and phase of TEC was maximum on 17 March 2015. The electric field is necessary for the production of sprites which is sensitive to the time scale of the thundercloud field due to the significant reduction of electrons. On the other hand, the schematic diagram of electrical processes between the cloud top, middle, and upper atmosphere is illustrated in Fig. 4. The figure shows the significant relation with sprites during lightning discharge and geomagnetic storm event. In the upper atmosphere, the GPS signals through TEC fluctuation will perturb the electric environment and in different ways coupled to the atmosphere. In addition, the upper atmosphere producing the ionosphere with high electrical conductivity and energetic charged particles from the Sun through the ionization process from solar X-ray and UV radiation. Furthermore, thunderstorm also plays a crucial role for transporting energy and constituting electrical coupling between atmosphere and ionosphere through the global electric circuit (GEC) [13]. Redistribution of charges in the thunderstorm resulted upward propagating QE field which is initiated from the upper charger-layer of a thunderstorm. The current in a lightning stroke can radiate the electromagnetic field of ~20 GW peak power [14] which affect the ionospheric disturbances due to the heating and acceleration of electrons, production of ionization and optical emissions.

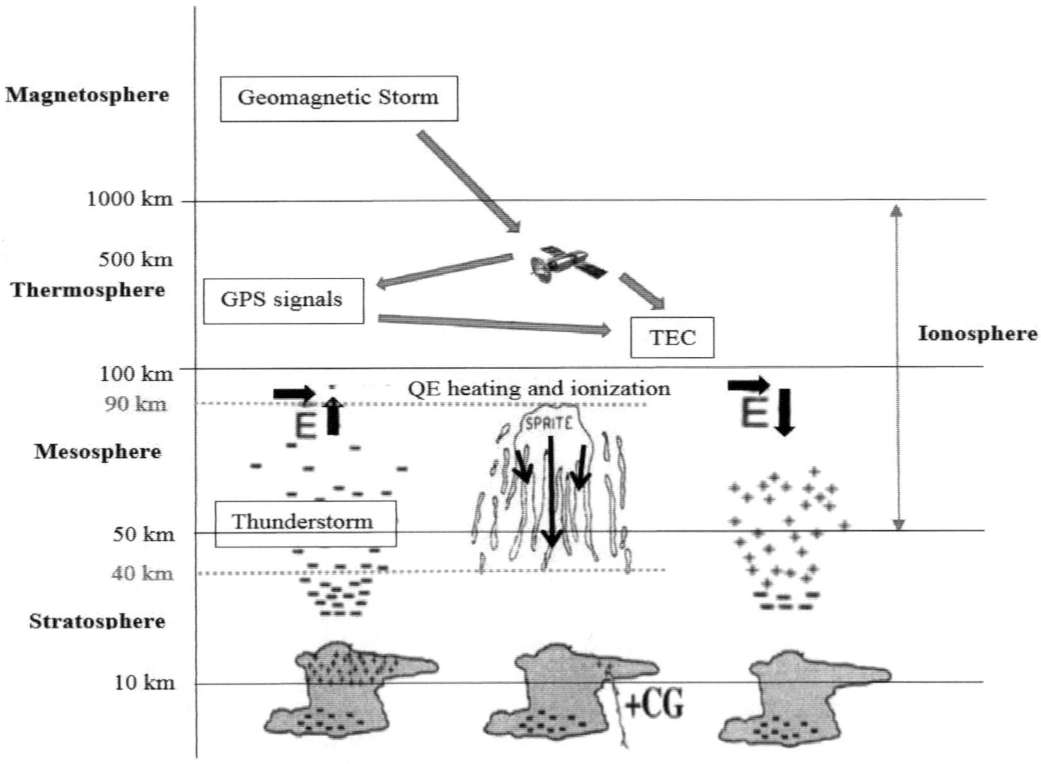

Figure 4. Schematic diagram in emergence of sprites between lower and upper atmosphere

Below the upper ionosphere layer (see Fig. 4), the interactions between electric fields and particles cause electrical breakdown at 80 km and 70 km altitude with 35 kA and 350 C km and 45 kA and 360 C km respectively [15] which leading in the production of sprites. Sprite is a type of optical emission occurs in the mesosphere layer. Their dynamics are governed by streamer discharges. Sprites are typically initiated at 70–85 km altitudes with downward propagating streamers which come to the end at about 40–50 km altitudes [4]. Upward propagating streamers may appear later and can reach about 90 km altitude as shown in Fig. 4. In other part, sprites are related with stratiform regions of mesoscale convective systems (MCS) adjoining to the stronger convective cores. They tend to occur during the mature to decomposing stages of thunderstorm development. Triggered by the sudden establishment of a strong QE field above thunderclouds due to intense +CGs lightning strokes which lower positive charges from a cloud to the ground.

4. Conclusion

Throughout the analysis, the response of lightning discharges and TEC associated sprite during the St. Patrick day storm on 17 March 2015 was overlooked. We observed the good response with almost similar patterns between TEC fluctuation and interplanetary electric field which the TEC value was highest for all the selected stations of OHI3, PALV, and ROTH in Antarctic Peninsula. TEC and geomagnetic activity are strongly correlated with the production of sprites. We also observed that high lightning energy was produced ~257 kJ with the DC electric field to be ~10 mVm^{-1}. With these values, ionospheric storm leading to the sprite generation after the return strokes on 18 March 2015. However, studies of sprites generated during high lightning energy are still in their infancy. Since sprites could change electrical properties of the atmosphere and influence the processes related to weather and climate, intense research activity in this area is required to more accurately predict the coupling between upper and the lower of the atmosphere.

Acknowledgments

This research is funded by the Ministry of Science, Technology and Innovation Malaysia (MOSTI) through the Flagship Program under ZF-2014-016 grant. The second author is PhD candidate supported by Universiti Teknikal Malaysia Melaka and the Ministry of Higher Education. The authors would like to thank the Space Weather Prediction Center (SWPC) for daily of Kp data; the World Data Center-C2, Kyoto, Japan, for hourly final of Dst data; NASA for providing the magnetic field and GPS data for Rothera station. We would also like to thank the Scripps Orbit Permanent Array Center (SOPAC) for archiving GPS data at O'Higgins and Palmer stations; the Astronomisches Institut University of Bern (AIUB) Data Centre of University Switzerland for CODE data.

References

[1] Aplin K L, Harrison R G and Rycroft M J 2008 Investigating Earth's Atmospheric Electricity: a Role Model for Planetary Studies *Space Sci Rev* **137** 11

[2] Lyons W A 1996 Sprites observations above the US high plains in relation to their parent thunderstorms systems *J Geophys Res* **101** 29641

[3] Stenbaek-Nielsen H C, Kanmae T, McHarg M G and Haaland R 2013 High-speed observations of sprite streamers *Surv. Geophys.* **34** 769

Rakov V A, Uman M A 2003 Lightning: Physics and Effects (New York: Cambridge University Press) pp 687

[4] Hu W, Cummer S A, Lyons W A and Nelson T E 2002 Lightning charge moment changes for the initiation of sprites *Geophys. Res. Let.t* **29(**8) 1279

[5] Jacobsen K S and Andalsvik Y L 2016 Overview of the 2015 St. Patrick's day storm and its consequences for RTK and PPP positioning in Norway *J. Space Weather Space Clim.* **6** A9

[6] Dowden R L, Brundell J B and Rodger C J 2002 VLF lightning location by time of group arrival (TOGA) at multiple sites *J. Atmos. Sol-Terr. Phys.* **64** 817

[7] Suparta W, Abdul Rashid Z A, Mohd Ali M A, Yatim B and Fraser G J 2008 Observations of Antarctic precipitable water vapor and its response to the solar activity based on GPS sensing *J. Atmos Sol-Terr. Phys.* **70** 1419

[8] B. Naval, Rodríguez-Zuluaga J, Alazo-Cuartas K, Kashcheyev A, Migoya-Orué Y, Radicella S M, Amory-Mazaudier C and Fleury R 2016 Middle- and low-latitude ionosphere response to 2015 St. Patrick's Day geomagnetic storm *J. Geophys. Res.: Space Phys.* **121** 3421

[9] Kamide Y and Kusano K 2015 No major solar flares but the largest geomagnetic storm in the

present solar cycle *Space Weather* **13** 365

[10] Takahashi Y, Yoshida A, Sato M, Adachi T, Kondo S, Hsu R R, Su H T, Chen A B, Mende S B, Frey H U and Lee L C 2010 Absolute optical energy of sprites and its relationship to charge moment of parent lightning discharge based on measurement by ISUAL/AP *J. Geophys. Res.* **115** A00E55

[11] Fullekrug M and Rycroft M J 2006 The contribution of sprites to the global atmospheric electric circuit *Earth Planets Space* **58** 1193

[12] Singh A K, Devendraa Siingh, Singh R P and Sandhya Mishra 2011 electrodynamical coupling of earth's atmosphere and ionosphere: an overview *Int. J. Geophys.* **971302**

[13] Neubert T, Rycroft M, Farges T, Blanc E, Chanrion O, Arnone E, Odzimek A, Arnold N, Enell C F, Turunen E, Bosinger T, Mika A, Haldoupis C, Steiner R J, Vander Velde O, Soula S, Berg P, Boberg F, Thejll P, Christiansen B, Ignaccolo M, Fullekrug M, Verronen P T, Montanya J and Crosby N 2008 Recent results from studies of electric discharges in the mesosphere *Surv. Geophys.* **29** 71

[14] Rycroft M J and Odzimek A 2010 Effects of lightning and sprites on the ionospheric potential and threshold effects on sprite initiation, obtained using an analog model of the global atmospheric electric circuit *J. Geophys. Res.* **115** A00E37

IOP Conf. Series: Journal of Physics: Conf. Series **852** (2017) 012021 doi:10.1088/1742-6596/852/1/012021

Distribution ozone concentration in Klang Valley using GIS approaches

A Sulaiman[1], A A Ab Rahman[1], K N Abdul Maulud[1,2,*], M T Latif[3,4], F Ahmad[5], M A Abdul Wahid[1], M A Ibrahim[1] and N D Abdul Halim[3].

[1]Earth Observation Centre, Institute of Climate Change (IPI), Universiti Kebangsaan Malaysia, 43600 UKM, Bangi, Selangor Malaysia
[2]Department of Civil & Structural Engineering, Faculty of Engineering &Built Environment, Universiti Kebangsaan Malaysia, 43600 UKM, Bangi, Selangor Malaysia
[3]Institute for Environment and Development (LESTARI), Universiti Kebangsaan Malaysia, 43600 UKM, Bangi, Selangor Malaysia
[4]School of Environmental and Natural Resources Sciences, Faculty of Science and Technology, Universiti Kebangsaan Malaysia 43600 UKM, Bangi, Selangor Malaysia
[5]Centre for Tropical Climate Change System, Institute of Climate Change, Universiti Kebangsaan Malaysia 43600 UKM, Bangi, Selangor Malaysia

Email: knam@ukm.edu.my

Abstract. Today, ozone has become one of the main air pollutants in Malaysia. The high ozone precursor concentrations have been encouraging the ozone production. The development of the Klang Valley, Malaysia has many types of physical activities such as urban commercial, industrial area, settlement area and others, which has increased the risk of atmospheric pollution. The purpose of this paper is to determine the spatial distribution between types of land use and ozone concentration that are occurred in the year 2014. The study areas for this paper include Shah Alam, Kajang, Petaling Jaya and Port Klang. Distribution of ozone concentration will be showed via spatial analysis tools in Geographic Information Systems (GIS) approached and the types of land use will be extracted using Remote Sensing technique. The result showed 97 ppb (parts-per-billion, 10^{-9}) and 161 ppb recorded at Port Klang and Shah Alam respectively that are mainly represented by the settlement area. Therefore, the physical land use need to be monitor and controlled by the government in order to make sure the ozone production for daily per hour will not exceed the regulation allowed.

1. Introduction

Nowadays, growth in population and rapid development were absolutely giving the negative impact on the air quality status around the world. Ozone concentration becomes one of the most significant air pollutants due to increasing sources of ozone precursors [1]. Ozone is created in the atmosphere when the sun's rays split oxygen molecules into single atoms. These atoms combine with nearby oxygen to form a three-oxygen molecule. It is known to be a secondary pollutant, photochemical oxidant and the main component of smog [2]. Besides that, the highest ozone formation can be harmful to human health and also air pollution can cause the negative impact on ecosystems, materials, buildings, works of art, vegetation, and visibility [4,5,6].

Based on the previous study that discussed the ozone effects on vegetation area in Europe, ozone concentrations are greater than 40 ppb may be harmful to the crop yield, biomass production, vitality

Content from this work may be used under the terms of the Creative Commons Attribution 3.0 licence. Any further distribution of this work must maintain attribution to the author(s) and the title of the work, journal citation and DOI.
Published under licence by IOP Publishing Ltd

and stress tolerance of forest trees [3]. In others study, the ozone concentration in the troposphere is highly variable are ranging from 10 ppb over the tropical oceans to 100 ppb over to land and can even exceed in the polluted urban area [5,7]. This variability to the increasing f the ozone concentration is might be dependent on available of solar radiation, temperature, winds, season and altitude or others factors [5,8]. Thus, air pollutants formation, dispersion, transport, and dilution are important roles that are highlighted in an aspect of the meteorology [9,10,11,12]. Based on another study also, average surface ozone level is rapidly increased in developing countries and slowly decreases in developed countries [13,14,17].

Land used are defined as a category land that are covered the physical activities for one place. Based on the previous study, the results on land use and the level of compliance mentioned that some problems in air quality status in the Malaysian peninsular only exist in highly spot on the urbanization areas [15]. According to the Malaysian department of environment and others study stated that motor vehicles getting 82%, power stations 9%, industrial fuel burning 5%, industrial production processes 3% domestic and commercial 0.2% and open burning at solid waste disposal site 0.8% to the total air emission load in the peninsular of Malaysian [15,16].

This paper will highlight the pattern of ozone concentration in Klang valley area. There are 4 areas that are chosen as a potential of high ozone concentration in Klang valley which are Port Klang, Petaling Jaya, Kajang and Shah Alam. The types of land use are defined to illustrate the category of land use. Then, the daily maximums are defined to shows the concentration of formation ozone and analysis high potential of ozone formation also types of category land use that is covered on the higher ozone concentration.

The details of the station are showed in Table 1 and Figure 1 is illustrated the location in 4 monitoring station of ozone data. Based on the Environmental Quality Report released by the Department of Environment in 2011, the status of air quality in Malaysia, especially in some major cities in Klang Valley, namely the Federal Territory of Kuala Lumpur, Shah Alam and Kajang is unhealthy as a whole for that year [18]. This evident can be proved with multiple episodes of air quality decreases and haze phenomenon that become more serious in this area in recent years. This condition can be detected through the Air Pollution Index, which has been recorded by the Department of Environment for a few major urban areas in Klang Valley, which are namely Shah Alam, Petaling Jaya, Gombak, Kajang, Klang and Kuala Lumpur. The status of air quality that is unhealthy level also hit in the Klang Valley area.

The air quality becomes decrease occurs due to the high of concentration levels of pollutants PM_{10} and O_3 become higher than in other pollutant materials [18,19]. Klang Valley had the air quality unhealthy in 2004 where Shah Alam one of the highest recorded air quality to unhealthy, which is 88 days in a year than the Federal Territory of Kuala Lumpur that are recorded air quality unhealthy for 63 days in the same period [18,19]. The air quality in the Klang Valley is influenced by several factors including the vehicle motorized, open burning, industrial and transboundary pollution [20,18]. Urban air pollution that hit several major cities in Klang Valley up resulting in the decreased of air quality, it is also due to the rapid of urbanization and liveliness activities transport in the region.

Table 1. Details of monitoring station ozone data

Station	Coordinates(N)	Coordinates(E)
Pelabuhan Kelang	N 03° 00' 37.02"	E 101° 24' 29"
Petaling Jaya	N 03° 06' 34.16"	E 101° 38' 20"
Kajang	N 02° 59' 38.07"	E 101° 44' 25"
Shah Alam	N 03° 06' 17.22"	E 101° 33' 22"

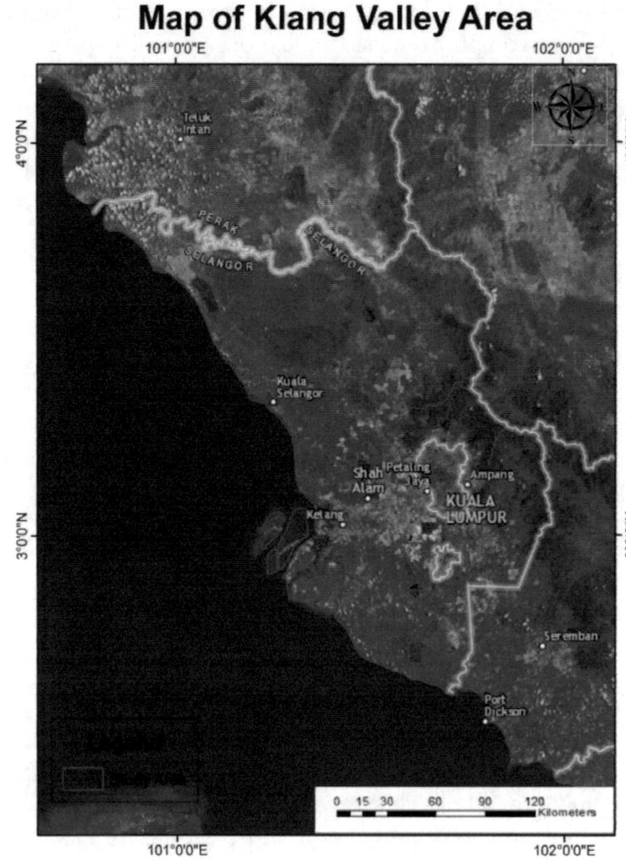

Figure 1. Study area

This study uses two types of data which are land use data and surface ozone concentration. The duration of data is provided for one year period (January 2014 to December 2014). Land use data are provided from the Institute of Climate Changes at University National of Malaysia and hourly surface ozone concentration data are provided from the Malaysian Department of Environment (DOE). The ozone data are given in hourly data for a one year.

2. Methodology

This research focus on the small area in Peninsular Malaysia which is the Klang Valley area. Starting with identifying types of land use for Klang valley area. Then, filtering ozone concentration data obtained from the Department of Environment Malaysia to determine maximum daily and annual maximum daily concentrations for every station in Klang Valley area. Analysis of the spatial distribution of ozone concentration is applying using interpolation techniques which are known as Topo to Raster contained in a GIS. Lastly, analysis using topo to raster is applied on that data to show the distribution of ozone concentration and types of land use that happen in the year 2014.

Arc Map software in version 10.2 is being chosed as the best options and appropriate software for the purposes of the analysis. GIS is a system to store, view and keep a type of information system that can be applied to geographical or spatial data. Other than that, the systems can be combine operating capacity to collect, acquire, store, manage and analyze the data to produce information during the display of spatial data. [21]. While, topo to raster in ArcGIS software that are in spatial interpolation tools or techniques are specially designed for the creation of hydrological correct digital elevation models (DEM). Other than that, these methods are interpolates a hydrologically correct raster surface from the point, line and polygon data [22]. The best result will be obtained if all input data is stored in the same planar coordinates system and has the same Z units. If more than one input points fall within an output cell, topo to raster will be used the average value for the interpolations (only the first 100 points that fall within a cell will be considered and the rest will be ignored) [22]. In the topo to raster method there have many types of feature data input such as point elevations, stream, lake, contour, sink, boundary, cliff, exclusion and coast.

Based on the other study [18] stated that analysis of the ozone concentration in Klang Valley using GIS techniques appearing the level of the ozone concentration followed by this Table 2 below.

Table 2. Level of patterns spatial distribution in the Klang Valley [20]

Level	Ozone Concentration (O_3)
1	0-50 ppb
2	50-100 ppb
3	100-150 ppb
4	150-200 ppb
5	200-250 ppb

3. Results and Discussion

Based on the analysis, by using satellite images and GIS techniques, the pattern distribution ozone concentration will be shown in a map. The types of land use categories will be shown in Table 3. The total areas for the land use classification on Klang Valley are 433479 hectares. The types of classification are covered by 7 types which are croplands, forestlands, grassland, mangrove, settlement area, wetlands and others. The types of land use are covered on that range are 48.26% of cropland area, 25.08% for forestland, 0.75% for grassland, 0.98% for mangrove, 20.24% for settlement area, 2.19% wetlands and 1.70% for others. According to Intergovernmental Panel on Climate Change (IPCC), 2006 [23] others lands category include bare soil, rock, ice, and all land areas that do not fall into any of the other five categories like cropland, forestland, mangrove, settlement, and wetlands. Figure 2 shows the types of land use in Klang Valley area.

Table 3. Types of land use in Klang Valley

Land Use Types	Area (ha)	Percentage (%)
Cropland	209191	48.26
Forestland	112192	25.08
Grassland	3251	0.75
Mangrove	4257	0.98
Settlement	87735	20.24
Others	7381	1.70
Wetlands	9471	2.19
Total	**433479**	**100**

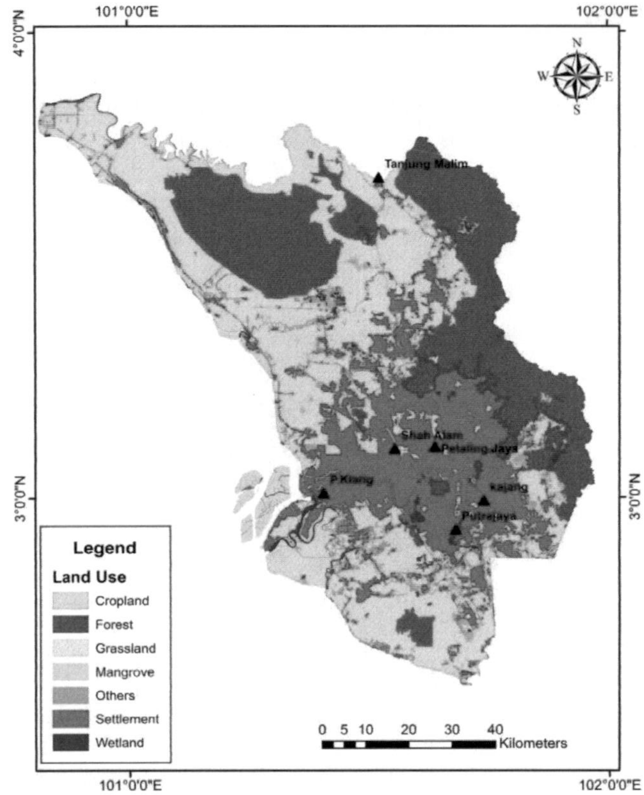

Figure 2. Types Of Land Use in Klang Valley

Figure 3 represent the graph of monthly maximum values for 4 selected areas that are in Klang Valley. Distribution of ozone concentration is different for every month from January until December 2014. The concentrations of ozone for the all sampling station are in the range between 40 ppb until 161 ppb. The red line shows the level of ambient ozone concentration of 100 ppb that is suggested by Malaysia Air Quality Guidelines (Department of Environment Malaysia) which is also discussed in other studies [24].

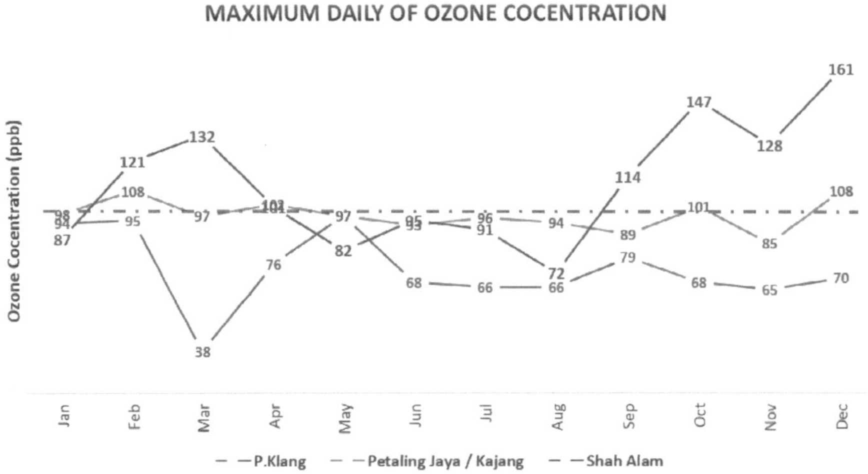

Figure 3. Monthly maximum daily of ozone concentrations at Klang Valley

The highest annual maximum value of surface ozone concentration at the 4 stations which are Port Klang, Petaling Jaya, Kajang, and Shah Alam. The annual maximum value per year at Port Klang is the maximum value is 97 ppb, Petaling Jaya and Kajang getting 108 ppb and Shah Alam recording highest valueof 161 ppb among the stations. This is because Shah Alam has many industrial areas and development where activities that are created by human-like industrial, lodging and others impact to the formation of the ozone. The level of ozone concentration illustrated by using GIS techniques as shown in Figure 4. The distribution annual maximum for the year 2014 illustrated using topo to raster method. The level of ozone concentration is shown by color. Shah Alam represents the level of highest ozone concentration in the Klang Valley starting from September until December in the year 2014. Other than that, Shah Alam is the place that has many physical activities like industrial, development and other man made that can be polluted air pollution.

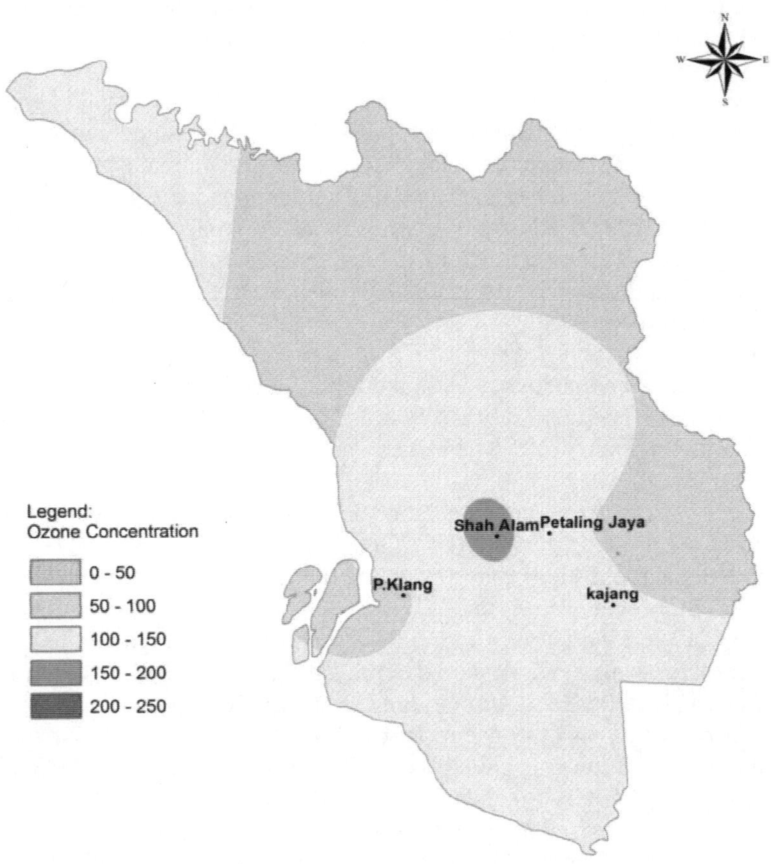

Figure 4. Distribution of annual maximum ozone concentration in the year 2014 at Klang Valley

4. Conclusion

In conclusion, ozone concentration recorded in Klang Valley area affected because of seasonal variations and physical activity that are happening in that area. Other than that, the effect from the types of land use also gives impact to the concentration of ozone. Klang Valley is the areas that have a many settlements and development land. The development is increasing, and hence it can be the effect population and physical activity in that area. This is indicated by analysis spatial distribution of ozone pattern by applying interpolation techniques topo to raster. The results also show Shah Alam area has exceeded the maximum allowable concentration of New Malaysian Air Quality Standard The concentration of ozone and others pollutants and also types of land use need to be controlled and planned. This paper attempts to understand the distribution of ozone concentration in Klang Valley and further research is required to find the relationship between ozone concentration and types of land use changes in Klang Valley area. This study also can be carried out by the interested parties by introducing green technologies to prevent the air pollution are archived to the dangerous level. Lastly, the government can do an adaptation to control strategies of future planning for physical activities and development that are happening for future in Klang Valley area as a proactive mitigation measures to reduce the level concentrations and emissions in that areas.

Acknowledgements
The authors would like to thank Universiti Kebangsaan Malaysia for the Research Grant (AP-2015-010). Special thanks to Institute of Climate Changes that are providing the land use data and Malaysian Department of the Environment (DOE) for providing all of the necessary air quality data for the process of conducting our research.

References

[1] Norrimi Rosaida Awang, Nor Azam Ramli, Nurul Izma Mohammed and Ahmad Shukri Yahya 2013 Time series evaluation of ozone concentration in malaysia based on location of monitoring stations *International Journal of Engineering and Technology (IJET)* **3**(3) 390
[2] Finlayson-Pitts B J and Pitts J N 1999 Chemistry of the upper and lower atmosphere: theory, experiments and applications (California: Academic Press)
[3] Fuhrer J, Skarby L and Ashmore M R 1997 Critical level for ozone effects on vegetation in Europe *Environ. Pollut.* **97** 91
[4] Ilyas S Z, Khattak Al, Nasir S M, Qurashi T, Durrani R 2009 Air pollution asessment in urban areas and its impact on human heatlh in the city of Quetta, Pakistan *Clean Technol. Environ. Policy* **1** doi: 10.1007/S10098-009-0209-4
[5] Shilenje Z W and Ongoma V 2014 Observed surface ozone trend in year 2012 over Nairobi, Kenya *Atmosfera* **27**(4) 377
[6] Kalabokas P D, Viras L G, Bartzis J and Repapis C C 2000 Mediterranean rural ozone characteristicsaround the urban area of Athens *Atmos. Environ.* **34** 5199
[7] Denman K L, Brasseur G, Chidthaisong A *et al.* 2007 Couplings between changes in the climate system and biogeochemistry. In: *Climate change 2007: Thephysical science basis.* Contribution of Working Group I to the Fourth Assessment Report of the Intergovernmental Panel on Climate Change (S Solomon, D Qin, M Manning, Z Chen, M Marquis, K B Averyt, M Tignor and H L Miller Eds.) Cambridge University Press p 499
[8] IPCC 2007 *Synthesis report.*Contribution of working group I, II and III to the Fourth Assessment Report of the Intergovernmental Panel on Climate Change (R K Pachauri and A Resinger Eds.) IPCC Geneva Switzerland, p 104
[9] Dueñas C, Fernández M C, Cañete S, Carretero J and Liger E 2002 Assessment of ozone variations and meteorological effects in an urban area in the Mediterranean coast *Science of the Total Environment* **299** 97
[10] Elminir H K 2005 Dependence of urban air pollutants on meteorology *Science of the Total Environment* **350** 225
[11] Satsangi G S, Lakhani A, Kulshrestha P R and Taneja A 2004 Seasonal and diurnal variation of surface ozone and a preliminary analysis of exceedance of its critical levels at a semi-arid site in India *Journal of Atmospheric Chemistry* **47** 271
[12] Jun Tu, Zong-Guo Xia, Hesheng Wang and Wenqing Li 2007 Temporal variations in surface ozone and its precursors and meteorological effects at an urban site in china *Atmos. Res.* **89** 310
[13] Wang T, Cheung V, Li Y and Anson M 2001 Ozone and related gaseous pollutants in the boundary layer of eastern china: overview of the recent measurements at a rural site

Geophys.Res. Lett. **28**, 2373-2376
[14] U S Environmental Protection Agency (EPA) 2008b Our nation's air e status and trends through 2008
[15] Siti Zawiyah Azmi, Mohd Talib Latif, Aida Shafawati Ismail, Liew Juneng and Abdul Aziz Jermai 2010 Trend and status air quality at three different monitoring stations in the Klang Valley *Malaysia. Air quality, atmosphere & heatlh* **3**(1) 53 doi : 10.1007/s11869-009-0051-1
[16] DOE 1996 1995 Environment Quality Report, Department of Environment, Ministry of Science, Technology and Environment Malaysia Kuala Lumpur
[17] Hang Lei, Donald J Wuebbles, Xin Zhong Liang 2012 Projected risk of high ozone episode in 2050 *Atmos. Environ.* **59**567
[18] Siti Haslina Mohd Shafie and Mastura Mahmud 2015 Analisis pola taburan reruang PM_{10} dan O_3 di Lembah Klang dengan mengaplikasikan teknik geographic information system (GIS) *Malaysian Journal of Society and Space* **11**(3) 61.
[19] Jabatan Alam Sekitar 2010 Laporan tahunan 2010
[20] Jabatan Alam Sekitar 2011 Laporan tahunan 2011
[21] Mohd Fariz Dziauddin and Ruslan Rainis 2003 *Pengenalan Kepada Kartografi* Pearson Malaysia Sdn Bhd Kuala Lumpur
[22] ArcGIS Help Library Manual ArcGIS 10.2 Help .Esri. Copyright© 1995-2013
[23] IPCC, 2006.*chapter 3*.consistent representation of lands of the IPCC(Kathryn bichel;gary richard; micheal kohl and ricardo leonado vianna rodrigues) . IPCC,Usa,Australia,Germany, Brazil p 42
[24] Mohd Talib Latif, Mohamed Rozali Othman and Zaharizam Johnny 2006 Kajian Kualiti Udara Di Bandar Kajang, Selangor *The Malaysian Journal of Analytical Sciences* **10**(2) 275

Alteration mineral mapping in inaccessible regions using target detection algorithms to ASTER data

A B Pour[1,2], M Hashim[1] and Y Park[2]

[1]Geoscience and Digital Earth Centre (Geo-DEC), Research Institute for Sustainability and Environment (RISE), Universiti Teknologi Malaysia (UTM)
[2]Korea Polar Research Institute (KOPRI), Songdomirae-ro,Yeonsu-gu, Incheon 21990, Republic of Korea

Email: beiranvand.amin80@gmail.com, mazlanhashim@utm.my

Abstract. In this study, the applications of target detection algorithms such as Constrained Energy Minimization (CEM), Orthogonal Subspace Projection (OSP) and Adaptive Coherence Estimator (ACE) to shortwave infrared bands of the Advanced Spaceborne Thermal Emission and Reflection Radiometer (ASTER) data was investigated to extract geological information for alteration mineral mapping in poorly exposed lithologies in inaccessible domains. The Oscar II coast area north-eastern Graham Land, Antarctic Peninsula (AP) was selected in this study to conduct a satellite-based remote sensing mapping technique. It is an inaccessible region due to the remoteness of many rock exposures and the necessity to travel over sever mountainous and glacier-cover terrains for geological field mapping and sample collection. Fractional abundance of alteration minerals such as muscovite, kaolinite, illite, montmorillonite, epidote, chlorite and biotite were identified in alteration zones using CEM, OSP and ACE algorithms in poorly mapped and unmapped zones at district scale for the Oscar II coast area. The results of this investigation demonstrated the applicability of ASTER shortwave infrared spectral data for lithological and alteration mineral mapping in poorly exposed lithologies and inaccessible regions, particularly using the image processing algorithms that are capable to detect sub-pixel targets in the remotely sensed images, where no prior information is available.

1. Introduction

Remote sensing satellite data are commonly used for reconnaissance mapping of inaccessible regions, lithological mapping, structural analysis and mineral exploration in arid and semi-arid and tropical regions around the world [1,2,3].

The Advanced Spaceborne Thermal Emission and Reflection Radiometer (ASTER) is a high spatial, spectral and radiometric resolution multispectral remote sensing sensor. It has three bands (0.52 and 0.86 μm) with a spatial resolution of up to 15 m in visible and near infrared (VNIR), six bands (1.6 to 2.43 μm) at a spatial resolution of 30 m in shortwave infrared (SWIR) and five bands (8.125 to 11.65 μm) with spatial resolution of 90 m. Each ASTER scene cover 60*60 km^2 area, which is suitable for geological regional mapping [4]. The possibility of applying several image processing algorithms to unique integral bands of ASTER is one the most important application of the data for hydrothermal alteration mapping [4]. SWIR bands of ASTER contain great ability for detecting and mapping hydrothermal alteration zones [2,3].

In this investigation, the applicability of target detection algorithms, including Constrained Energy Minimization (CEM), Orthogonal Subspace Projection (OSP) and Adaptive Coherence Estimator

Content from this work may be used under the terms of the Creative Commons Attribution 3.0 licence. Any further distribution of this work must maintain attribution to the author(s) and the title of the work, journal citation and DOI.
Published under licence by IOP Publishing Ltd

IOP Conf. Series: Journal of Physics: Conf. Series **852** (2017) 012022 doi:10.1088/1742-6596/852/1/012022

(ACE) algorithms were tested to ASTER SWIR bands for mapping hydrothermal alteration zones in the Oscar II coast area in eastern Graham Land, Antarctic Peninsula.

2. Materials and Methods

2.1 Geology of the study area

The Oscar II coast area is located in north eastern Graham Land, Antarctic Peninsula (AP). It is an inaccessible region due to the remoteness of many rock exposures and travel over sever mountainous and glacier-cover terrains for geological field mapping and sample collection. The lithological units in Oscar II coast area consisting of (i) volcanic and volcaniclastic rocks (Mapple Formation), (ii) intrusive rocks, (iii) sedimentary rocks (Botany Bay and Trinity Peninsula Groups). Figure 1 shows geological map of the Oscar II coast area [5].

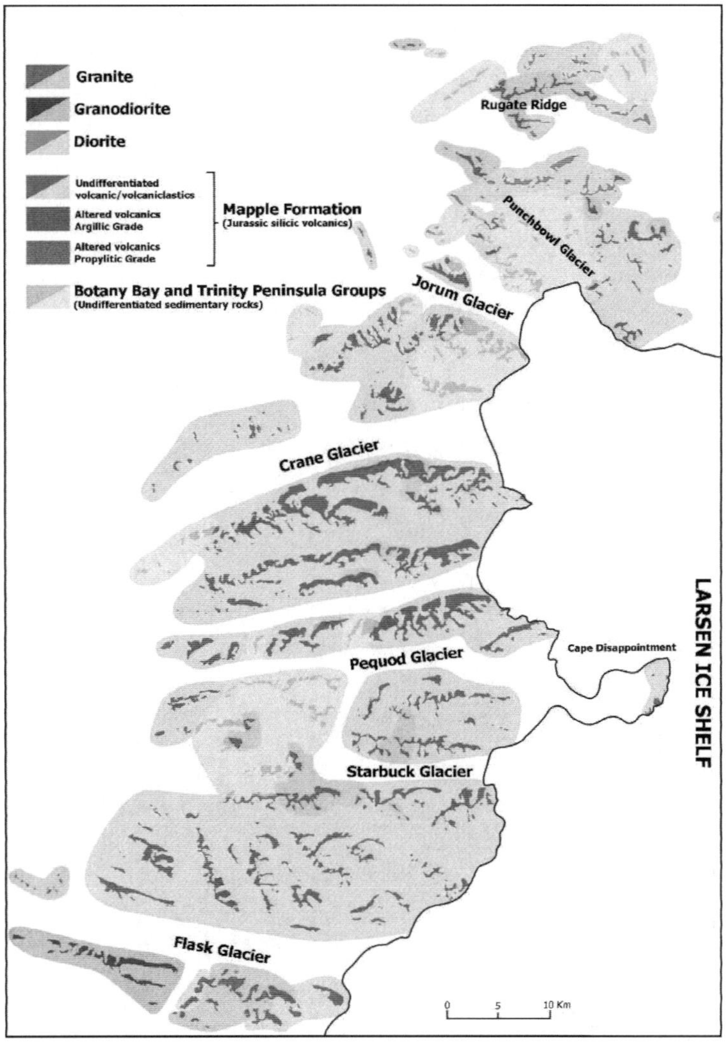

Figure 1. Geological map of the Oscar II Coast area [5]

2.2 ASTER remote sensing data

The ASTER data used in this investigation were obtained from U.S. Geological EROS (http://glovis.usgs.gov/). An ASTER level 1 T (Precision Terrain Corrected Registered At-Sensor Radiance) scenes covering the Oscar II coast region and surrounding areas

(AST_L1T_00311222001131924) (Path/Row 218/106), north eastern part of Graham Land, Antarctic Peninsula (AP) was acquired on November 22, 2001 for the remote sensing analysis.

2.3 Data processing

Atmospheric and crosstalk corrections were applied to shortwave infrared bands of ASTER data before application of target detection algorithms. CEM is a target signature-constrained approach, which constrains the desired target signature with a specific gain while minimizing effects caused by other unknown signatures [6]. OSP algorithm is an unconstrained linear unmixing method that developed by [7]. ACE algorithm is a uniformly most powerful invariant detection statistic that derived from the Generalized Likelihood Ratio (GLR) approach [7]. In this investigation, the reference spectra of selected end-member minerals for implementing target detection algorithms were extracted from the ASTER spectral library version 2.0 and were convolved to the spectral response function of ASTER's SWIR bands. Selected end-member spectra (target alteration minerals) in this study were muscovite, illite, montmorillonite, epidote, chlorite and biotite. The results of CEM, OSP and ACE appear as a series of gray scale (rule) images, one for each selected endmember. The value in resultant image represents the sub-pixel abundance of the target alteration mineral in each pixel. Rule image classifier tool was used for post classification of the resultant images. They were classified by maximum value option, and 0.050 was selected as appropriate threshold value for all resultant rule images [8,9].

3. Results and discussion

Figure 2 displays the reference spectra of selected end-member minerals for implementing CEM, OSP and ACE algorithms, which was convolved to the spectral response function of SWIR bands of ASTER. Muscovite, kaolinite, illite, montmorillonite, epidote, chlorite and biotite were target alteration minerals in the study area (Fig. 2). CEM, OSP and ACE algorithms were applied to a selected spatial subset scene the zone between Crane Glacier and Pequod Glacier of the Oscar II coast area (see Fig. 1).

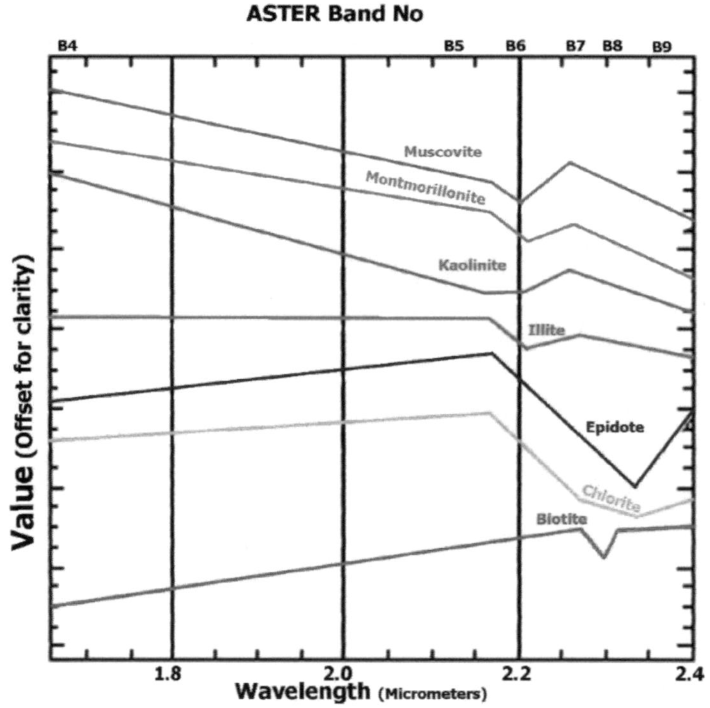

Figure 2. Reference spectra of selected end-member minerals from ASTER spectral library that convolved to ASTER's SWIR bands

Figure 3 shows the resultant image map derived from CEM algorithm for the zone between Crane Glacier and Pequod Glacier (Oscar II coast region). The distribution of sub-pixel abundance of alteration minerals was mapped. The CEM algorithm was capable to detect detailed surface distribution of fractional abundance of alteration target minerals in the selected zone. In fact, previous alteration classes that simply assigned to chlorite-muscovite, muscovite-chlorite and muscovite contain variety of mineral assemblages (see zoom parts in Figure 3). Montmorillonite is one the high abundance alteration mineral in muscovite-chlorite class especially associated with exposed rocks near Crane Glacier. Biotite was detected in this study; previously it has been classified in many zones as low albedo area in the geology map. In addition, detailed distribution of kaolinite and epidote in the study zone was mapped (see zoom parts in Figure 3).

OSP was applied to the selected spatial subset sector between Crane Glacier and Pequod Glacier (Oscar II coast region) for distinguishing subpixel distribution of end-member minerals. The resultant rule images were classified by 0.050 thresholding the images. But, the resultant image map was not informative and classified. However, the gray scale (rule) images show high DN (digital number) values of the sub-pixel abundance of the target end-member mineral in the image. Despite, testing different threshold values, classified image was not produced using rule image classifier tool. Therefore, RGB color composite was used to visualize the subtle differences in the rule images. This was done because the resultant image map contains information about diagnostic features of selected end-member minerals.

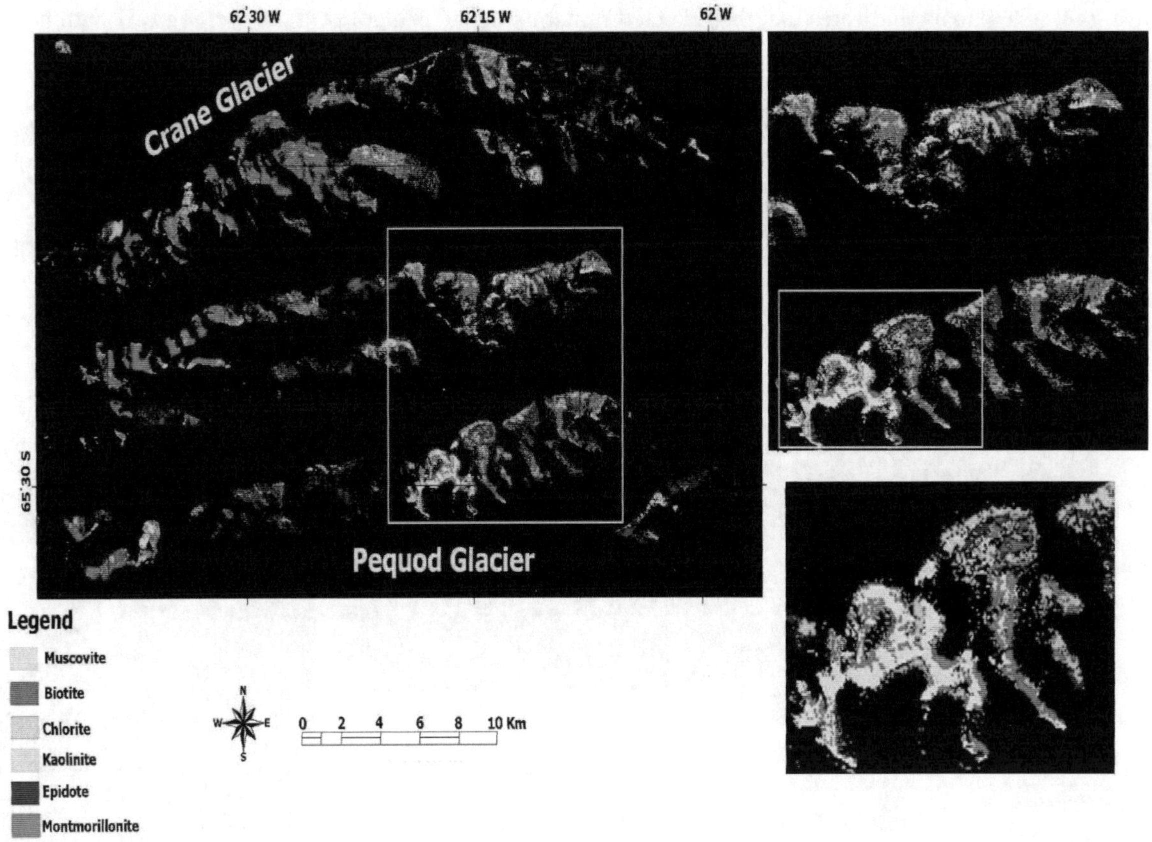

Figure 3. Resultant image map derived from CEM algorithm

Figure 4 displays RGB image map of kaolinite rule image (red), montmorillonite rule image (green) and chlorite (blue). Variety of colors is manifested in the pixel background of exposed lithology, including magenta, purple, pink, red, green, blue, yellow, cyan and white. The detected pixel classes are consistent with CEM resultant map (Fig. 3). High sub-pixel abundance value for kaolinite is coded in red (Fig. 4); therefore, zones contain high surface distribution of kaolinite appear in shades of red. Montmorillonite high abundance zones depict in shades of green in the rock exposure background. Blue color shade zones contain high sub-pixel abundance value of chlorite (Fig. 4). It is evident that in RGB image map the mixture pixel colors such as magenta, purple, pink, yellow, cyan and white represent combination of alteration mineral assemblages. For instance, yellow color represents a mixture of kaolinite and montmorillonite; purple color exhibits a mixture of kaolinite and chlorite; magenta, cyan, pink and white colors characterize different combinations of sub-pixel abundance of kaolinite, montmorillonite and chlorite mineral assemblages (Fig. 4).

ACE algorithm was also implemented to the selected spatial subset zone (between Crane Glacier and Pequod Glacier; Oscar II coast region) of the study area. Resultant gray scale (rule) images depict high DN values of the sub-pixel abundance of the target mineral in the study zone. Rule image classifier tool was used to produce a classified image map. But, the image map contains very noisy background, and classified mineral zones were not discernable from the background. Accordingly, RGB color composite was applied to produce an image map for the selected zone using the rule images comprise high DN values and low noisy background. Muscovite rule image was selected and assigned to red (R), chlorite rule image was assigned to green (G), and biotite rule image was assigned to blue (B), respectively. Figure 5 shows the resultant image map for the study zone. Magenta and red pixels show high fractional abundance of muscovite, and green pixels are chlorite high abundance area. Biotite papers as purple pixels in rock exposure region. Comparison with CEM and OSP resultant image maps indicates that the detected minerals by ACE algorithm are well agreed with high abundance zones of muscovite, chlorite and biotite.

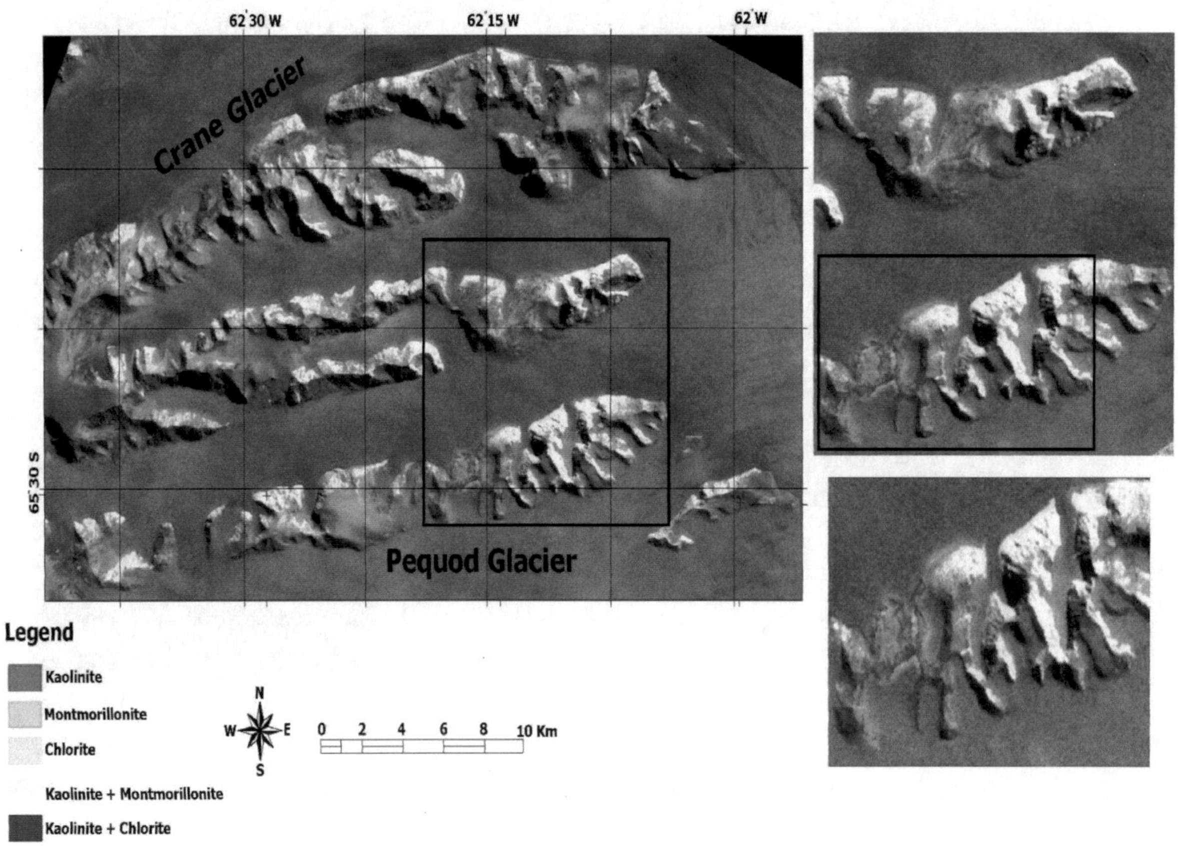

Figure 4. RGB image map of kaolinite (red), montmorillonite (green) and chlorite (blue) derived from OSP algorithm

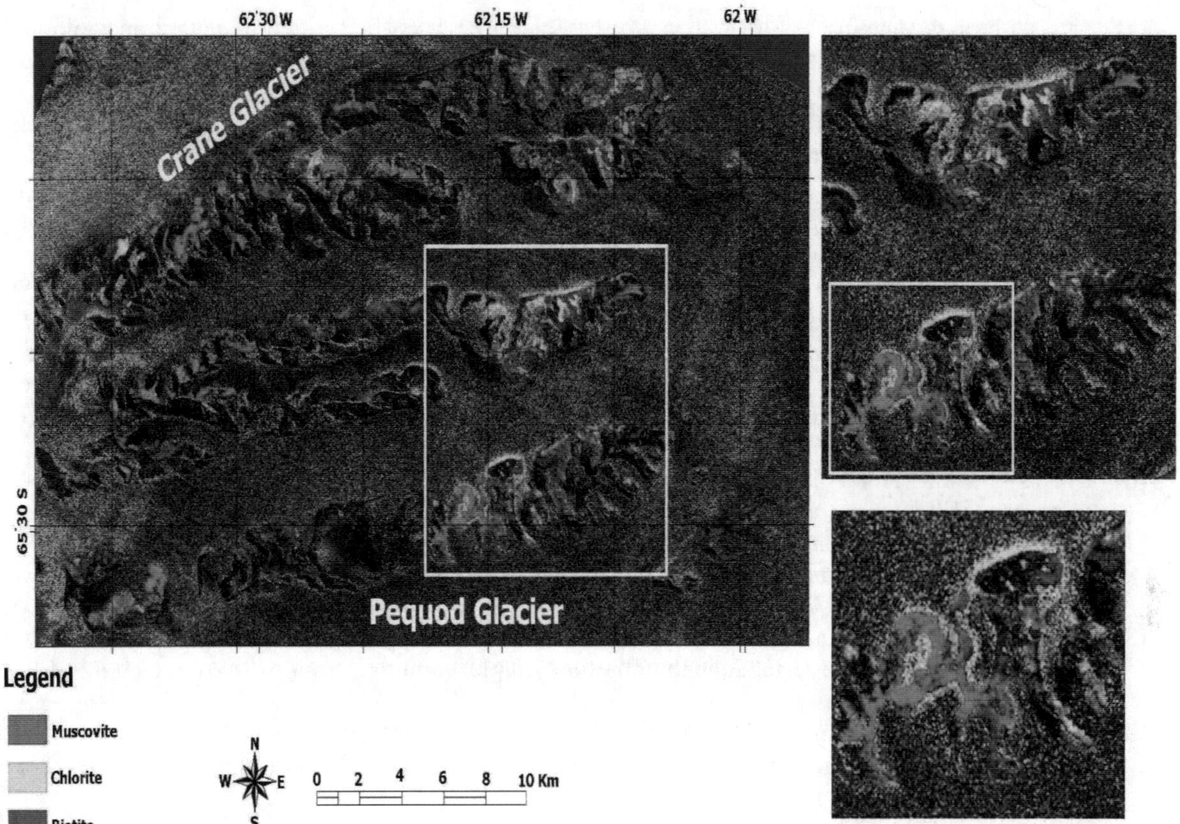

Figure 5. RGB image map of muscovite (red), chlorite (green) and biotite (blue) derived from ACE algorithm

4. Conclusion

The remote sensing satellite-based approach used in this investigation to map poorly exposed lithologies has been shown to perform extremely well in the Antarctic environments. It could be comprehensively used for alteration mapping and hydrothermal ore mineral exploration in other inaccessible regions. The capability to distinguish fractional abundance of alteration minerals using CEM, OSP and ACE algorithms provided invaluable information for discrimination lithological units and alteration mineral assemblages in remote and inaccessible terrains at district scale. Sub-pixel abundance value of mafic minerals such as epidote, chlorite and biotite (Fe, Mg-OH absorption), K-feldspar alteration products (kaolinite, illite, montmorillonite) and muscovite (Al-OH absorption) was detected using ASTER SWIR bands in poorly mapped and unmapped regions in the northern and southern sectors of Oscar II coast area, Graham Land.

Acknowledgments

This study was conducted as a part of Yayasan Penyelidikan Antartika Sultan Mizan (YPASM) research grant (Vote no: R.J130000.7309.4B221), Sultan Mizan Antarctic Research Foundation, Malaysia. We also would like to express our great appreciation to KOPRI (Korea Polar Research Institute) Asian Polar Science Fellowship Program 2016 for their great support during this research. This study was partially supported by KOPRI grant PE17050 and KOPRI's Asian Polar Science Fellowship Program. We are thankful to the Universiti Teknologi Malaysia for providing the facilities for this investigation.

References

[1] Pour A B and Hashim M 2011 The Earth Observing-1 (EO-1) satellite data for geological mapping, southeastern segment of the Central Iranian Volcanic Belt, Iran *International Journal of the Physical Sciences* **6**(33) 7638

[2] Pour A B and Hashim M 2014 ASTER, ALI and Hyperion sensors data for lithological mapping and ore mineral exploration *Springerplus* **3**(130) 1

[3] Pour A B and Hashim M 2015 Integrating PALSAR and ASTER data for mineral deposits exploration in tropical environments: a case study from Central Belt, Peninsular Malaysia *International Journal of Image and Data Fusion* **6**(2) 170-188

[4] Yamaguchi Y I, Fujisada H, Kahle A B, Tsu H, Kato M, Watanabe H, Sato I and Kudoh M 2001 ASTER instrument performance, operation status, and application to Earth sciences *IEEE Transactions of Geosciences and Remote Sensing* 1215

[5] Haselwimmer C E, Riley T R and Liu J G 2013 Lithologic mapping in the Oscar II Coast area, Graham Land, Antarctic Peninsula using ASTER data *International Journal of Remote Sensing* **32**(7)

[6] Chang C I and Heinz D C 2000 Constrained subpixel target detection for remotely sensed imagery *IEEE Transactions on Geoscience and Remote Sensing* **38**(3) 1144

[7] Harsanyi J C and Chang C I 1994 Hyperspectral image classification and dimensionality reduction: an orthogonal subspace projection approach *IEEE Transactions on Geoscience and Remote Sensing* **32**(4) 779

[8] Pournamdary M, Hashim M and Pour BA 2014a Application of ASTER and Landsat TM data for geological mapping of Esfandagheh ophiolite complex, southern Iran *Resource Geology* **64**(3) 233

[9] Pournamdary M, Hashim M and Pour B A 2014b Spectral transformation of ASTER and Landsat TM bands for lithological mapping of Soghan ophiolite complex, south Iran *Advances in Space Research* **54**(4) 694

Mapping land slide occurrence zones using Remote Sensing and GIS techniques in Kelantan state, Malaysia

M Hashim [1], A B Pour[1,2] and S Misbari[1]

[1]Geoscience and Digital Earth Centre (Geo-DEC), Research Institute for Sustainability and Environment (RISE). Universiti Teknologi Malaysia (UTM)
[2]Korea Polar Research Institute (KOPRI), Songdomirae-ro,Yeonsu-gu, Incheon 21990, Republic of Korea

Email: beiranvand.amin80@gmail.com, mazlanhashim@utm.my

Abstract. Integration of satellite remote sensing data and Geographic Information System (GIS) techniques is one of the most applicable approach for landslide mapping and identification of high potential risk and susceptible zones in tropical environments. Yearly, several landslides occur during heavy monsoon rainfall in Kelantan river basin, Peninsular Malaysia. In this investigation, Landsat-8 and Phased Array type L-band Synthetic Aperture Radar-2 (PALSAR-2) remote sensing data sets were integrated with GIS analysis for detect, map and characterize landslide occurrences during December 2014 flooding period in the Kelantan river basin. Landslides were determined by tracking changes in vegetation pixel data using Landsat-8 images that acquired before and after December 2014 flooding for the study area. The PALSAR-2 data were used for mapping of major geological structures and detailed characterizations of lineaments in the state of Kelantan. Analytical Hierarchy Process (AHP) approach was used for landslide susceptibility mapping. Several factors such as slope, aspect, soil, lithology, Normalized Difference Vegetation Index (NDVI), land cover, distance to drainage, precipitation, distance to fault, and distance to road were extracted from remote sensing satellite data and fieldwork to apply AHP approach. Two main outputs of this study were landslide inventory occurrences map during 2014 flooding episode and landslide susceptibility map for entire the Kelantan state. Modelled/predicted landslides with susceptible map generated prior and post flood episode, confirmed that intense rainfall in the Kelantan have contributed to weightage of numerous landslides with various sizes. It is concluded that precipitation is the most influential factor that bare to landslide event.

1. Introduction

Remote sensing imagery can play an important role for production of landslide inventory map and generation of thematic maps related to landslide occurrences in a region at both regional and district scales. Integration of GIS techniques with remote sensing imagery is a very effective method for potential landslide mapping and geological structural analysis especially in tropical environments [1,2,3].

Landslide and flooding are the most common geohazards leading to economic losses and death in Southeast Asia during the last few decades [4]. Natural force is one of the main reasons of their occurrences, followed by human interruption to the environment such as 'cut-and-fill' the hillside, mineral reserve exploitation and deforestation. Landslide hazard assessment and risk reduction can be accomplished by providing the accurate information using satellite remote sensing data and GIS analysis for risk management, which is urgent need to access easily and continuously on the landslide occurrence zones [5]. Thus, landslide susceptibility mapping after natural disaster such as flood episode, using recently launched satellite remote sensing data can provide important information for the majority of users in public and private sectors, government agencies and the scientific communities at the local and international levels.

Content from this work may be used under the terms of the Creative Commons Attribution 3.0 licence. Any further distribution of this work must maintain attribution to the author(s) and the title of the work, journal citation and DOI.
Published under licence by IOP Publishing Ltd

In this investigation, multitemporal Landsat-8, the Phased Array type L-band Synthetic Aperture Radar-2 (PALSAR-2) data and GIS techniques were integrated to detect and map landslide occurrence zones in the Kelantan state following by massive rainfall during December 2014.

2. Materials and Methods

2.1 Geology of the study area

The state of Kelantan is located in the north-eastern corner of Peninsular Malaysia (Fig. 1). A wide variety of rocks consisting of igneous, sedimentary and metamorphic rocks are distributed in a north-south trend in the Kelantan state. Typically, four types of rocks are classified in the region, including granitic rocks, sedimentary/metasedimentary rocks, extrusive rocks (volcanic rocks) and unconsolidated sediments [1,2]. Two distinct wet seasons from September to December and February to May are reported by Malaysian Meteorological Department (MMD) in Peninsular Malaysia [6].

2.2 Remote sensing data and GIS techniques

Two level 1T (terrain-corrected) Landsat-8 OLI images were obtained through the US Geological Survey Earth Resources Observation and Science Center (http://earthexplorer.usgs.gov). The image data (Path/Row 127/56) were acquired on December 5, 2014 (before the flooding event) and 11 March 2015 (after the flooding event) with low cloud cover. The image data cover four river basins in Kelantan, including Sungai Golok (103,918.37 hectares), Sungai Kelantan (1,308,690.95 hectares), Sungai Kemasin (30,927.82 hectares) and Sungai Semerak (59,076.48 hectares) (Fig. 1).

A ScanSAR mode dual polarization (level 3.1; acquired on 6 February 2015) and two Fine mode dual polarization (level 3.1; acquired on 6 February 2015) PALSAR-2 scenes were obtained from ALOS-2 data distribution consortium online system Remote Sensing Technology Center of Japan (RESTEC).

In this investigation, Analytical Hierarchy Process (AHP) approach was used for Landslide Susceptibility Mapping (LSM). The AHP is a multi-objective, multi-criteria decision-making approach that enables the user to arrive at a scale of preferences drawn from a set of alternatives [7]. In this investigation for applying AHP approach, 10 factors such as slope, aspect, soil, lithology, NDVI, land cover, distance to drainage, precipitation, distance to fault, and distance to road were extracted from Landsat-8 images, PALSAR data and fieldwork as shown in Figure 2. Figure 3 shows methodology flowchart to produce LSM by using AHP GIS-based statistical models in this study.

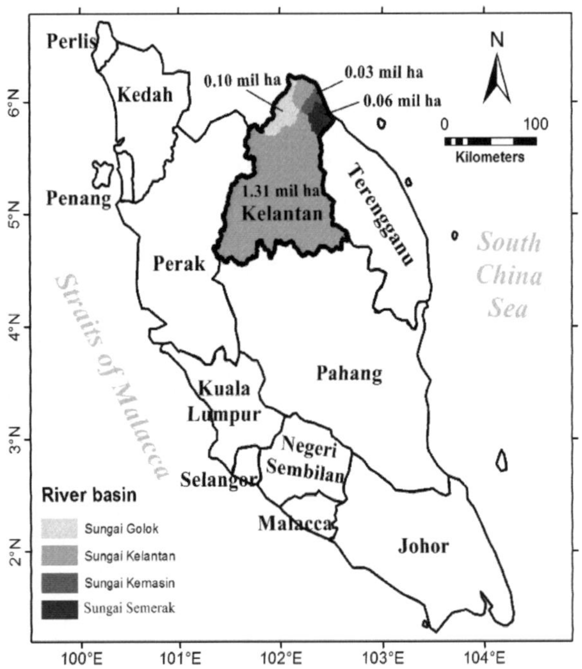

Figure 1. Location of the Kelantan state in Peninsular Malaysia. Major river basins were delineated in the image

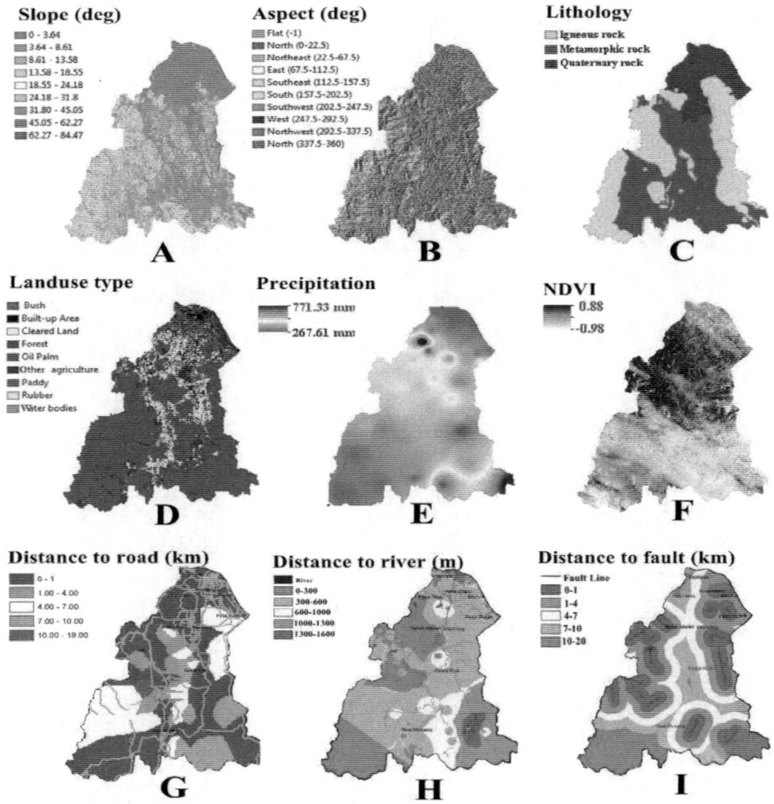

Figure 2. List of extracted factors from from Landsat-8, PALSAR, and fieldwork. A: Slope (degree), B: Aspect (degree), C: Lithology types, D:Landuse type, E:Precipitation (mm), F: NDVI, G: Distance to main road, H: Distance to main river (m), and I: Distance to fault line (km)

3. Results and discussion

A total of 36 landslides exceed 900 m^2 in size were identified and verified in the field. All 36 points has experienced the changes of NDVI from value of NDVI range for vegetation (> 0.50-1.0) to the NDVI range for bare soil (> 0.10 to 0.35). Statistic of landslide occurred on different types of land use shows that large landslides is most common in forestry areas (26 out of 36 identified), especially in the central part of the Kelantan and riverbank of the Sungai Kelantan. Active illegal logging may cause an increase of landslide occurrence during the rainy season. A total of 416 small and medium sizes of landslides that less than 900m^2 had been recorded through field observation. GPS surveying was used to record all locations of landslides as well as a rough observation in the surrounding area to assemble other supporting information. Based on the identified landslide distribution after December 2014 flood episode in the Kelantan, it indicates that the Gua Musang has the highest number of small landslides, mostly occurred in semi-sloping hill and along the roadside and the riverside. The similar observation was made in the Kuala Krai.

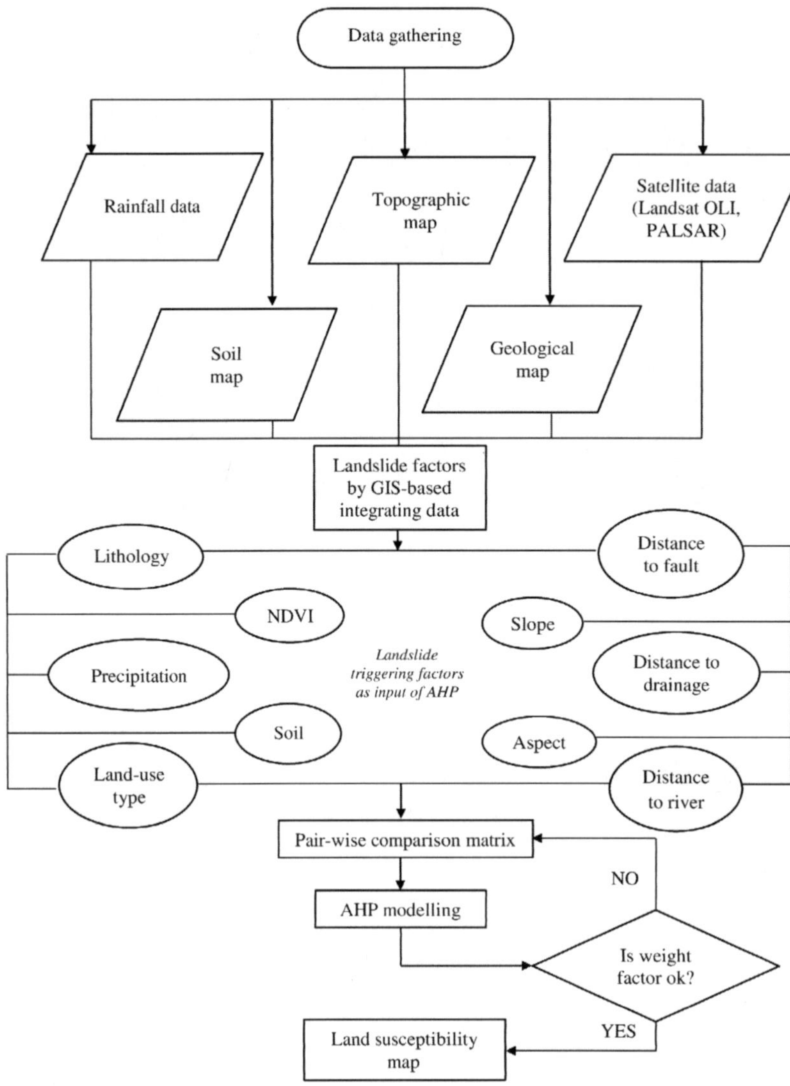

Figure 3. Methodology flowchart to produce landslide susceptibility map using AHP approach

Figure 4 shows the spatial distribution of landslide susceptibility map of each class extracted from the AHP approach during and before December 2014 flood episode in the Kelantan. Five classes category was used to indicate susceptibility of the area prone to landslide occurrences, including low risk (LR), fair risk (FR), moderate risk (MR), high risk (HR) and very high risk (VHR). Matrix of AHP category shifting before and after December 2014 flood episode is shown in Table 1. Landslide Susceptibility Maps (LSM) for before and during December 2014 flood episode was produced using related factors contributed to landslides event in Kelantan state (Fig. 5).

In overall, the number of landslides consisting of large and small size is 452. About 36.7% occurred in the area of rubber plantation area, while 34.3% occurred in forest areas. Location of landslides occurrences at different land use/cover type in the Kelantan is shown in Figure 6. Forest is the dominant land use/cover in the Kelantan besides rubber and oil palm. Young rubber trees that typically planted on a hillside are proved to be a triggering factor for increment of the risk of landslides occurrences. The study found that mature rubber areas have less risk of landslides compared to young rubber. Even the number of lower debris avalanche occurred in the forest than in the area of rubber cultivation, but the size of the landslide in the rubber is smaller than in the forest. Observing the scene showed that cleared the forest in a secluded place mostly experienced large landslide for horizontal and vertical dimensions.

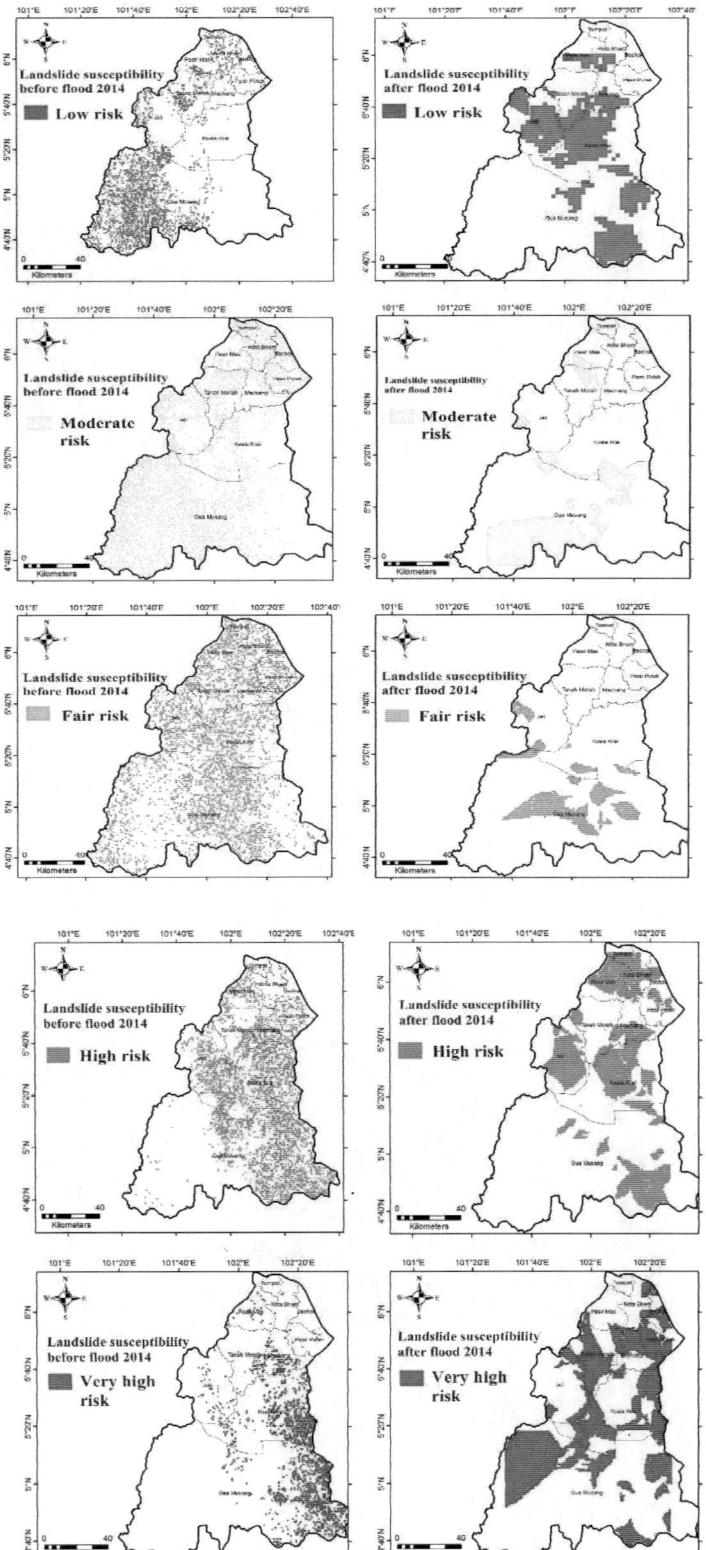

Figure 4. Spatial distribution of landslide susceptibility map of each class before and after flood episode in Kelantan 2014

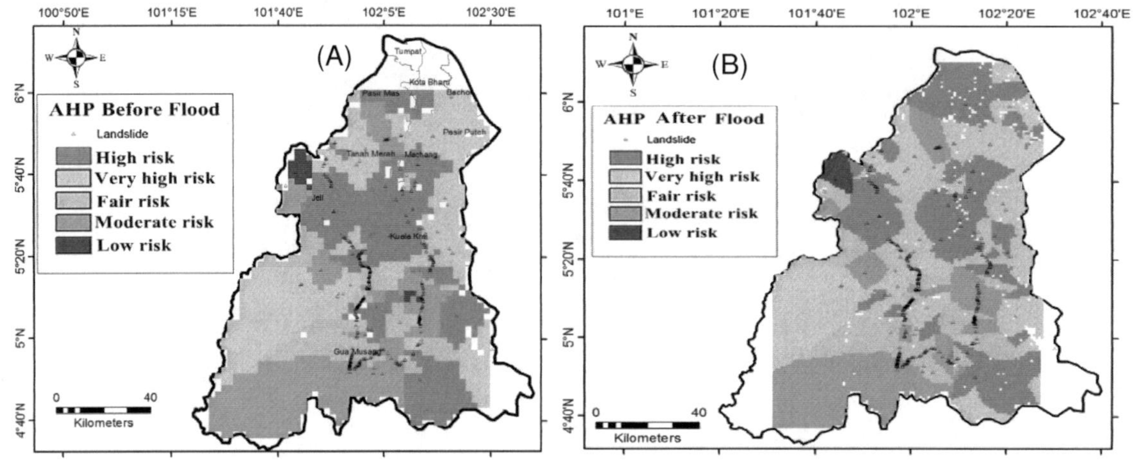

Figure 5. LSM before (A) and after (B) 2014 flooding episode in Kelantan state

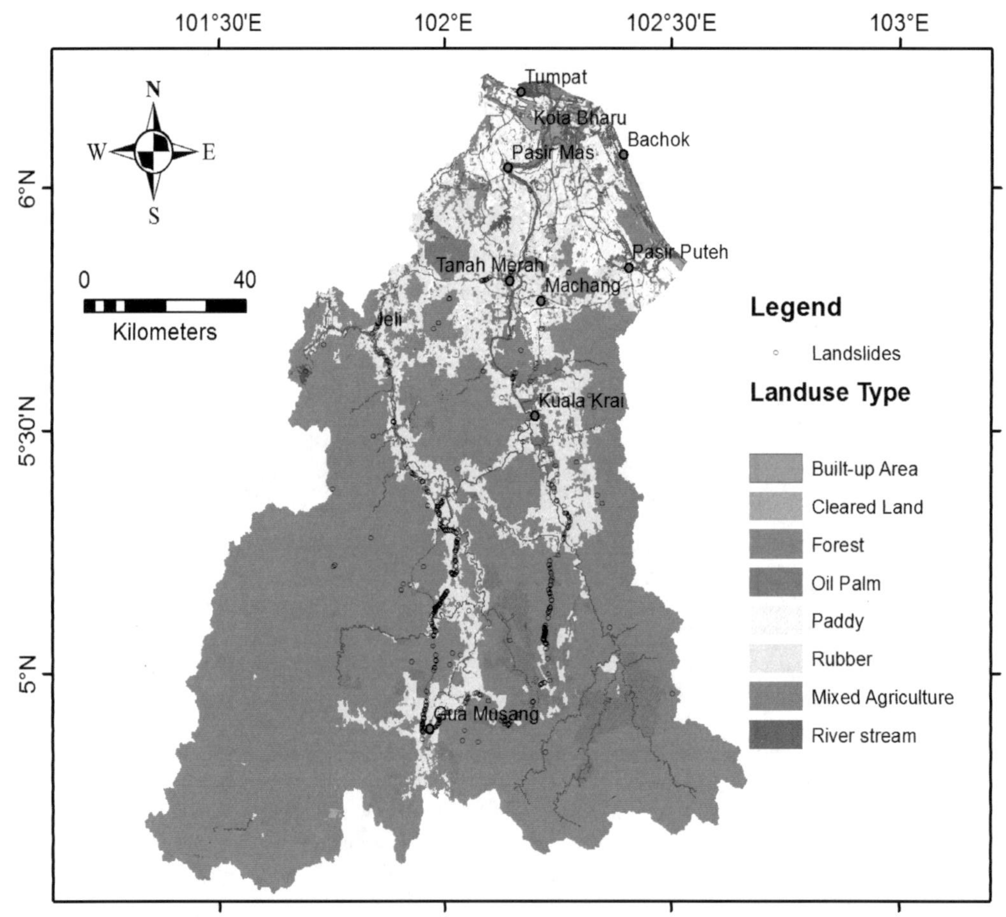

Figure 6. Landslides occurrences at different land use type in Kelantan state

Table 1. Matrix of AHP category shifting before and after flood episode

Before/After	LR	MR	FR	HR	VHR	Total (before)
LR	**523**	441	300	821	207	2292
MR	634	**1314**	216	982	520	3666
FR	576	105	**1179**	37	1736	3633
HR	238	100	57	**2066**	795	3256
VHR	200	81	105	49	**1073**	1508
Total (after)	2171	2041	1857	3955	4331	**14355**

4. Conclusion

This investigation indicate that the integration of satellite remote sensing data and GIS techniques is a applicable approach for landslide mapping and assessment in highly vegetated regions with tropical climate. Successful findings were achieved by combination of different sources of satellite remote sensing data and AHP approach using the 10 factors for landslide susceptibility mapping. Precipitation is the prime factor causing slope failure in the Kelantan state, followed by slope and distance to river factor, referred to weighting factor given by AHP with 0.5745 consistency ratio (CR). Based on the rainfall anomaly assessment, the 300mm precipitation in three days would trigger >15 landslides to occur. From the results, further action can be taken by relevant agencies of the Kelantan state to minimize flood impact to local residents in the future flooding event.

Acknowledgments

This study was conducted as a part of Transdisciplinary Research Grant Scheme (TRGS), Detecting, mapping, and characterization of landslides and landslide assessment in Kelantan River Basin during December 2014 Flooding using Satellite SAR Data- Vote no: R.J130000.7809.4L831, and Vote no: R.J130000.7809.4L837, Ministry of Higher Education (MOHE), Malaysia. We are thankful to the Universiti Teknologi Malaysia for providing the facilities for this investigation.

References

[1] Pour A B and Hashim M 2016 Application of PALSAR-2 remote sensing data for landslide hazard mapping in Kelantan river basin, Peninsular Malaysia *Int. Arch. Photogramm. Remote Sens. Spatial Inf. Sci.* **XLI-B8** 413

[2] Pour A B and Hashim M 2016 Application of PALSAR-2 remote sensing data for structural geology and topographic mapping in Kelantan river basin, Malaysia *IOP Conf. Ser.: Earth Environ. Sci.* **37** 012067

[3] Pour A B and Hashim M 2015 Integrating PALSAR and ASTER data for mineral deposits exploration in tropical environments: a case study from Central Belt, Peninsular Malaysia *International Journal of Image and Data Fusion* **6**(2) 170.

[4] Sanders B F 2007 Evaluation of on-line DEMs for flood inundation modeling *Advances in Water Resources* **30** 1831

[5] Pradhan B 2009 Flood susceptible mapping and risk area delineation using logistic regression *GIS and Remote Sensing Journal of Hydrology* **9**(2)

[6] Yalcin A 2008 GIS-based landslide susceptibility mapping using analytical hierarchy process and bivariate statistics in Ardesen (Turkey) Comparisons of results and confirmations Catena **72** 1

[7] Mardiana S, Ahmad A and Osman K 2000 Capability of RADARSAT data in monsoon flood monitoring. *GIS Development* 1

Spectral mineral mapping for characterization of subtle geothermal prospects using ASTER data

A J Abubakar[1], M Hashim*[1] and A B Pour[1, 2]

[1]Geoscience and Digital Earth Centre (Geo-DEC), Research Institute for Sustainability and Environment (RISE), Universiti Teknologi Malaysia (UTM)
[2]Korea Polar Research Institute (KOPRI), Songdomirae-ro,Yeonsu-gu, Incheon 21990, Republic of Korea

Email: mazlanhashim@utm.my

Abstract. In this study, the performance of ASTER data is evaluated for mapping subtle geothermal prospects in an unexplored tropical region having a number of thermal springs. The study employed a simple Decorrelation stretch with specific absorptions to highlight possible alteration zones of interest related to Geothermal (GT) systems. Hydrothermal alteration minerals are subsequently mapped using Spectral Angle Mapper (SAM) and Linear Spectral Unmixing (LSU) algorithms to target representative minerals such as clays, carbonates and AL-OH minerals as indicators of GT activity. The results were validated through field GPS survey, rock sampling and laboratory analysis using latest smart lab X-Ray Diffractometer technology. The study indicates that ASTER broadband satellite data could be used to map subtle GT prospects with the aid of an in-situ verification. However, it also shows that ASTER could not discriminate within specie minerals especially for clays using SWIR bands. Subsequent studies are aimed at looking at both ASTER and Hyperion hyperspectral data in the same area as this could have significant implications for GT resource detection in unmapped aseismic and inaccessible tropical regions using available spaceborne data.

1. Introduction

Globally, there is a serious energy concern and the clamour for sustainable alternative energy sources to forestall global warming and other unforeseen extreme climatic conditions. Geothermal (GT) resources, which are heat sources from the crust that can be harnessed for various uses, offers a reliable and sustainable alternative renewable energy not only in terms of power but other uses requiring less heat from the ground [1]. GT systems are associated with magmatic intrusions, which could serve as the source of heat. However, some GT systems are amagmatic and may occur in regions characterized by meteoric water rise within permeable crusts. In such regions hot fluids ascend to the surface as hotsprings, geysers and fumaroles [1]. GT systems may sometimes not have obvious surface manifestations but rather, subtle expressions such as hydrothermally altered rocks and minerals including clays, carbonates, silicates and calcites. These minerals are by-products of alteration as a result of metasomatism[2]. Blind and fossilized GT systems are characterized by these subtle imprints which are not easily identifiable using conventional techniques [2]. Remote sensing and spectroscopy offers a unique tool for synoptic detection of such prospective GT sites and have been successfully used in the prefeasibility stages of exploration by the detection of associated minerals as proxy [3]. However, previous studies have been focused in volcano-tectonic regions; there has been limited studies in unconventional aseismic settings with characteristically subtle GT features. There is the need for more studies in such regions and their identification using state of the art techniques of remote sensing and spectroscopy. This is especially true with the promising advances

Content from this work may be used under the terms of the Creative Commons Attribution 3.0 licence. Any further distribution of this work must maintain attribution to the author(s) and the title of the work, journal citation and DOI.
Published under licence by IOP Publishing Ltd

expected in the enhanced geothermal systems (EGS) which could make many regions of the world exploitable for GT renewable energy [1].

In this study, we assessed the applicability of the ASTER SWIR bands in discriminating subtle hydrothermal mineral signatures associated with GT systems as proxy for indirect characterization of hidden or fossilized systems, which are difficult to discern using only conventional survey techniques. The Wikki and Mawulgo thermal springs in Yankari Park area are studied.

2. Materials and Methods

2.1 Geology of the study area

The Yankari Park is located within Latitude 9.7500^{0}N, and Longitude 10. 5000^{0}E in Bauchi State, northeastern Nigeria (Fig. 1). Territorially it extends to about 2,244 square kilometres (866 sq. mi) [4]. Thermal springs in the Park include; Dimmil, Gwana,Mawulgo and Wikki. Geologically, it is within the Kerri formation characterized by Neogene to Mesozoic older sedimentary rocks which is composed of sandstone, silt stones, kaolinites, grits and clays [5].It is also part of the Benue Trough, bordered to the west by the basement complex crystalline rocks of the Jos plateau, to the northeast by the Biu plateau and to the southeast by the Adamawa highlands. Figure 1 shows geological map of Nigeria with Yankari Park.

2.2 ASTER remote sensing data

ASTER level 1T (AST_L1T, path/raw, 187/53) scenes with 25th January 2006 acquisition date covering the Yankari Park study area were obtained from the LPDAAC data pool. The ASTER level 1T is a Precision Terrain Corrected Registered At-Sensor Radiance that has been geometrically rectified, and turned to a north-up Universal Transverse Mercator (UTM) 32N projection with WGS 84 Datum. The data was atmospherically corrected using the internal average relative reflection (IARR) algorithm[6]. The image is thus converted from radiance to apparent reflectance to facilitate subsequent spectral analysis.

2.3 Data processing

A spatial subset was made of the ASTER scene focusing on Field surveyed areas around the Wikki and Mawulgo thermal springs and their proximate areas (Fig. 2). A decorrelation stretch (DCS) is first applied to the ASTER data to identify possible alteration areas of interest, this was done carefully with the specific minerals of interest in mind [6]. This included argillic alteration minerals including clays, carbonates, sulfates which are known to have diagnostic absorption features in the SWIR around (2.0 to 2.4 um), while limonitic alterations as a result of iron oxides and hydroxides show characteristic features within the VNIR around (0.3-1.6um) [7]. The focus of the study is to identify hydrothermal alteration zones associated with these minerals proximate to GT systems. In subsequent analysis, Spectral Angle Mapper (SAM) and Linear Spectral Unmixing (LSU) techniques are employed to target these alteration indicators and their pixel abundance on the ASTER data. The results are discussed below.

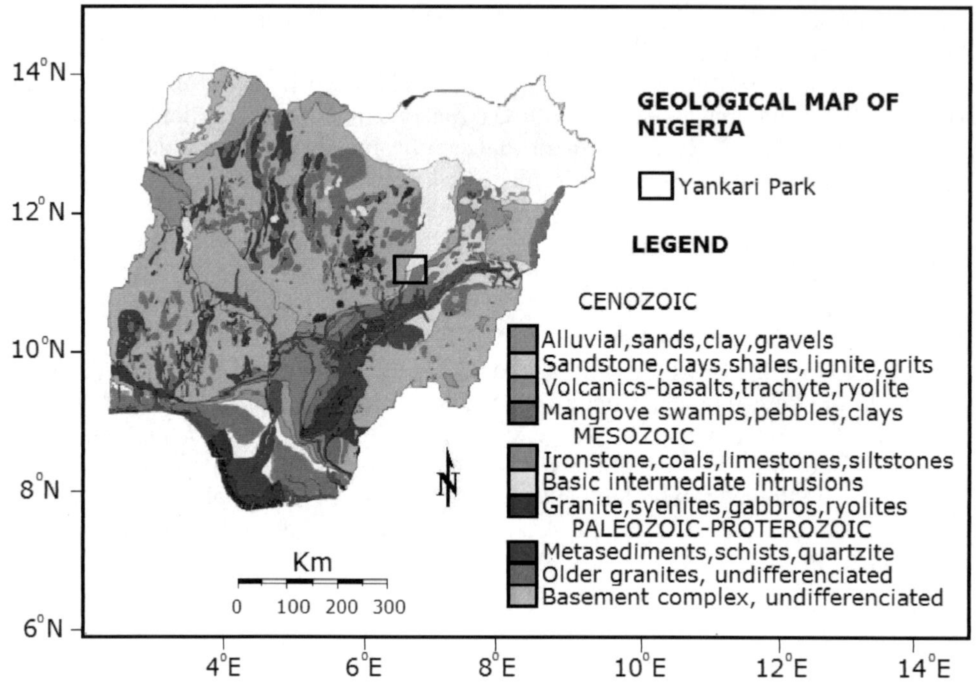

Figure 1. Geological Map of Nigeria showing Yankari Park location, modified from [5]

3. Results and discussion

3.1 Decorrelation Stretch (DCS)

The DCS is an image enhancement technique capable of highlighting image elements that are nearly invisible in true colour images by increasing the visualization of features [3] such as hydrothermal alteration zones of interest. It is employed in this study to visually heighten the spectral variations by removing the correlation among three input bands [8]. Figure 2 highlights red areas as possible argillic alteration zones, green as advanced argillic and blue as areas of carbonates[9]. The bands for DCS were carefully selected based on the absorption features of targeted alterations of interest.

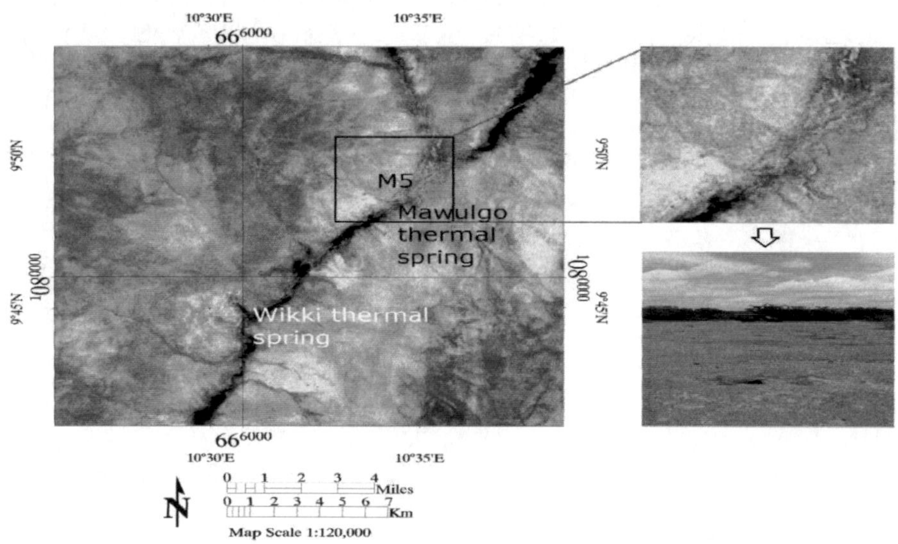

Figure 2. DCS bands 6, 5 and 8 in RGB and sampling site at M5 Mawulgo thermal spring

IOP Conf. Series: Journal of Physics: Conf. Series **852** (2017) 012024 doi:10.1088/1742-6596/852/1/012024

Many hydrothermal alteration minerals show diagnostic reflectance characteristics in the SWIR region. For instance, clay minerals show absorption within 2.20 um due to vibrational transition of hydroxyl ions and their bonding with Al-OH or Mg-OH cations while carbonates manifest within 2.35un [7]. Thus, the DCS was done using ASTER bands 6 (2.209um), band 5 (2.167) and band 8 (2.336). The DCS bands 6,5 and 8 were loaded as RGB. However, it is observed that the faint pink tones of the argillic alteration zones signifies weak alteration thus suggesting the need for more detailed analysis [6]. Subsequentlyanalysis is focused on M5 around Mawulgo where samples were obtained in the field and observed during field sampling to have clay alterations (Fig. 2).

3.2 Spectral Angle Mapper (SAM) Classification

SAM is a whole - pixel classification method used for comparing image spectra (an unknown) to a known reference spectra e.g. from spectral libraries, ASCII files etc. [3]. In this study, the reference spectra is chosen from USGS spectral library. The SAM techniques estimates the similarity between the image spectrum and the reference spectrum by calculating the angle between the two spectra while treating them as vectors in n-dimensional space [6]. To identify the argillic alteration zones that are indicated by the presence of clay minerals, five endmembers were selected from the spectral library (Fig. 3a). These included Kaolinite 1 and 2, Illite, Montmorillonite and Muscovite. A spectral subset of only SWIR bands was made and the selected endmembers are used as reference spectra with class colours as plotted in the Endmember collection spectra in Figure 3. Maximum radian angle of 0.100 was used. The result of the SAM is shown in Figure 3(b) that depicts Kaolinite as (red) pixels, Montmorillonite as (blue), and most of the pixels are classified as Illite and no pixel was classified as muscovite. The result indicates that the ASTER SWIR band could not discriminate within specie minerals since Kaolinite 1 and 2 were used in the analysis but only Kaolinite 1 is discriminated. The result however, identifies altered minerals of argillic alteration, which are of GT importance.

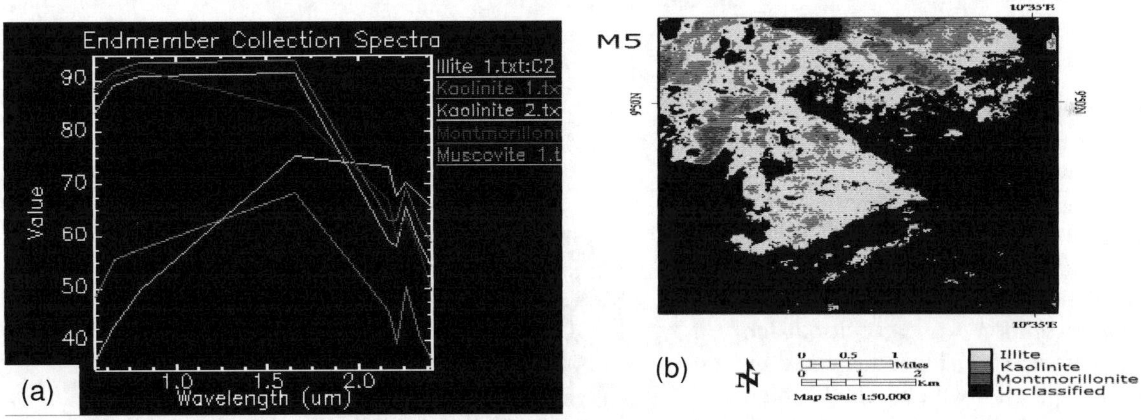

Figure 3. (a) Endmember collection spectra used as reference (USGS library spectra) for the analysis and (b) SAM rule image of M5 Mawulgo area showing SAM results for endmembers used as library spectra

3.3 Linear Spectral Unmixing (LSU)

The LSU algorithm is used to determine the actual proportion or abundance of pure endmember (class) materials within a pixel. The energy recorded by a remote sensing detector is a function of the amount of energy reflected or emitted by the materials within the Instantaneous Field Of View [7]. ENVI 5.1 was used to determine the distribution and relative abundance of the five indicator minerals as endmembers in the ASTER subset image. The LSU relies on the assumption that the reflectance at each pixel of the ASTER image is a linear combination of the reflectance of each endmember present within the pixel [9].

The result of LSU is depicted in Figure 4a, the image shows pixel abundance of Illite as cyan similar to the results in SAM, Kaolinite as red and Montmorillonite as blue. The sub-pixel LSU however, classified some areas initially of kaolinite in SAM, as Muscovite in (pink). It was observed that fewer pixels are classified as Kaolinite area at M5 which also corresponds to sampling sites where XRD results indicated presence of kaolinite and Quartz (see Figure 4b). It was however, observed that when Kaolinite 1 and 2 are loaded together with Montmorillonite as RGB, the LSU also classifies areas of Illite as Kaolinite indicating that the broad bands of ASTER is not effective in discriminating clay minerals with very close absorption features [10]. This signifies the need for using ASTER together witha sensor with higher spectral fidelity such as hyperspectral in detecting subtle GT related hydrothermal alteration minerals.

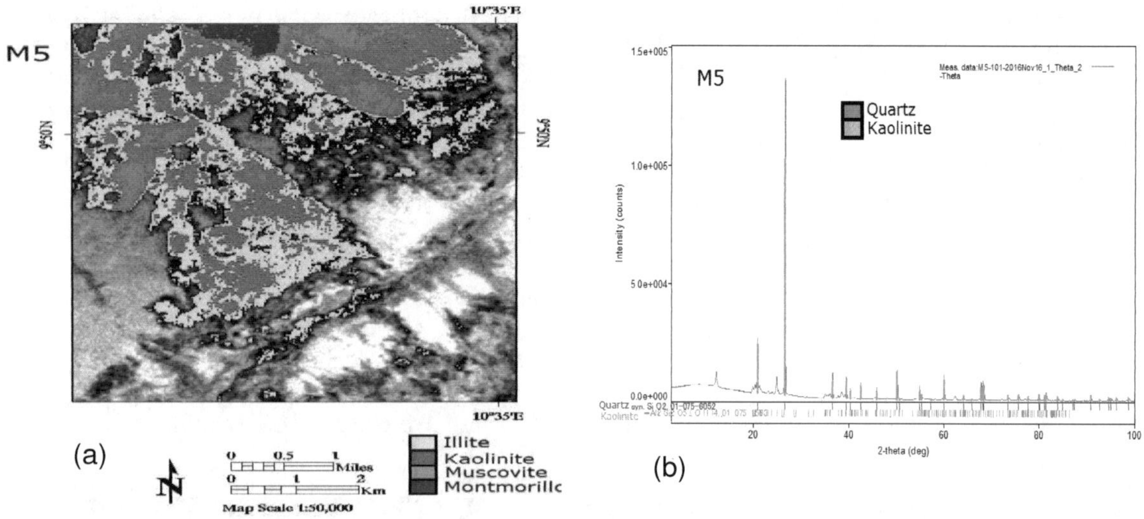

Figure 4. (a) LSU image map showing relative abundances of 4 endmembers Illite, Kaolinite, Muscovite and Montmorillonite on grayscale image at M5 and (b) X-Ray Diffraction results showing Quartz, Kaolinite mineral from altered rock samples at M5

The root mean square error (RMSE) image in Figure 6 shows the uncertainties in the mixing calculations and how well the least square solution was able to match the spectra [10]. This is indicated by bright pixels which show high values and high errors while the dark pixels indicate low errors and good fit [10, 11, 12]. The results of the RMSE (0.495) image indicates that areas around M5 where sampling was made was also classified sufficiently as shown by the dark pixels in the image area by the LSU algorithm which indicates the abundance of illite and Kaolinite with fewer pixels classified as montmorillonite and muscovite. The LSU thus better discriminates the alteration zones as compared to the SAM technique.

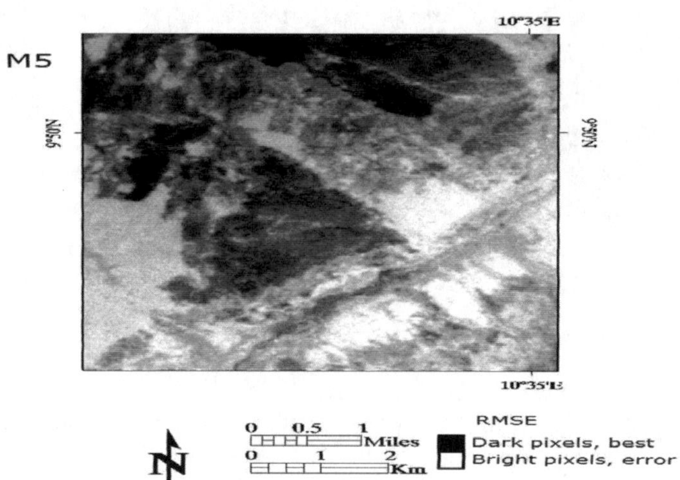

Figure 5. Root mean square error (RMSE) image showing bright as high error and dark as low error pixels in the LSU classification

4. Conclusion

This investigation shows the applicability of ASTER in discriminating subtle GT indicator minerals as proxy for identifying prospects and by narrowing targets for comprehensive surveys especially when backed with simple field validation. Alterations were identifiable around non-vegetated exposed areas at M5. However, despite ASTER SWIR ability to discriminate within specie alteration minerals, it is robust in identifying argillic and Hydroxyl-bearing alterations in general, which are very significant in identification of subtle GT surface mineralogical imprints. This could have implications for exploration of GT resources in unmapped regions characterized by subtle features with promising potentials.

Acknowledgments

We are thankful to the Universiti Teknologi Malaysia for providing the facilities for this investigation. Grant vote number R.J130000.7309.4B221 was supported registration fee for this paper. The authors wish to acknowledge Kaduna State University, Nigeria for providing the facilities for fieldwork study.

References

[1] Crosta A *et al.* 2003 Targeting key alteration minerals in epithermal deposits in Patagonia, Argentina, using ASTER imagery and principal component analysis *International journal of remote sensing* **24**(21) 4233

[2] Kratt C, Calvin W M and Coolbaugh M F 2010 Mineral mapping in the Pyramid Lake basin: Hydrothermal alteration, chemical precipitates and geothermal energy potential *Remote Sensing of Environment* **114**(10) 2297

[3] Vaughan R G et al. 2005 Surface mineral mapping at Steamboat Springs, Nevada, USA, with multi-wavelength thermal infrared images *Remote Sensing of Environment* **99**(1–2) 140

[4] Olokesusi F, Assessment of the Yankari Game Reserve, Nigeria: problems and prospects *Tourism Management* 1990 **11**(2) 153

[5] Kurowska E and S Krzysztof 2010 Geothermal exploration in Nigeria In *Proceedings World Geothermal Congress*

[6] Pour A B, Hashim M and Marghany M 2011 Using spectral mapping techniques on short wave infrared bands of ASTER remote sensing data for alteration mineral mapping in SE Iran *International Journal of Physical Sciences* **6**(4) 917

[7] Clark RN 1999 Spectroscopy of rocks and minerals, and principles of spectroscopy *Manual of remote sensing* **3** 3

[8] Rowan L C and J C 2003 Mars, Lithologic mapping in the Mountain Pass, California area using advanced spaceborne thermal emission and reflection radiometer (ASTER) data *Remote sensing of Environment* **84**(3) 350

[9] Pour A B and Hashim M 2011 Identification of hydrothermal alteration minerals for exploring of porphyry copper deposit using ASTER data, SE Iran *Journal of Asian Earth Sciences* **42**(6) 1309

[10] Van der Meer F 1996 Spectral mixture modelling and spectral stratigraphy in carbonate lithofacies mapping *ISPRS journal of photogrammetry and remote sensing* **51**(3) 150

[11] Pour A B and Hashim M 2011 ASTER, ALI and Hyperion sensors data for lithological mapping and ore minerals exploration *Springer Plus* **3**(1) 1

[12] Pour A Band M. Hashim 2015 Integrating PALSAR and ASTER data for mineral deposits exploration in tropical environments: a case study from Central Belt, Peninsular Malaysia *International Journal of Image and Data Fusion* **6**(2) 170

IOP Conf. Series: Journal of Physics: Conf. Series **852** (2017) 012025 doi:10.1088/1742-6596/852/1/012025

Application of ASTER SWIR bands in mapping anomaly pixels for Antarctic geological mapping

Amin Beiranvand Pour[1,2], Mazlan Hashim [1], Yongcheol Park[2]

[1]Geoscience and Digital Earth Centre (Geo-Dec) Research Institute for Sustainability and Environment (ROSE), Universiti Teknologi Malaysia (UTM)
[2]Korea Polar Research Institute (KOPRI) Songdomirae-ro,Yeonsu-gu, Incheon 21990, Republic of Korea

Email: beiranvand.amin80@gmail.com, mazlanhashim@utm.my

Abstract. Independent component analysis (ICA) was applied to shortwave infrared (SWIR) bands of ASTER satellite data for detailed mapping of alteration mineral zones in the context of polar environments, where little prior information is available. The Oscar II coast area north-eastern Graham Land, Antarctic Peninsula (AP) was selected to conduct a remote sensing satellite-based mapping approach to detect alteration mineral assemblages. Anomaly pixels in the ICA image maps related to spectral features of Al-O-H, Fe, Mg-O-H and CO3 groups were detected using SWIR datasets of ASTER. ICA method provided image maps of alteration mineral assemblages and discriminate lithological units with little available geological data for poorly mapped regions and/or without prior geological information for unmapped regions in northern and southern sectors of Oscar II coast area, Graham Land. The results of this investigation demonstrated the applicability of ASTER spectral data for lithological and alteration mineral mapping in poorly exposed lithologies and inaccessible regions, particularly using the image processing algorithm that are capable to detect anomaly pixels targets in the remotely sensed images, where no prior information is available.

1. Introduction

Remote sensing satellite data are commonly used for reconnaissance mapping of inaccessible regions, lithological mapping, structural analysis and mineral exploration in arid and semi-arid and tropical regions around the world [1, 2, 3].

In particular, remote sensing satellite imagery has high potential to provide a solution to overcome the difficulties and limitations associated with geological field mapping and mineral exploration in inaccessible region such Antarctic environments. The Advanced Spaceborne Thermal Emission and Reflection Radiometer (ASTER) is a high performance sensor for geological application and has VNIR subsystem with three bands between 0.52 and 0.86 μm (spatial resolution 15 m), SWIR subsystem with six bands from 1.6 to 2.43 μm, (spatial resolution of 30 m), and TIR subsystem with five band from 8.125 to 11.65 μm (spatial resolution of 90 m) [4]. Numerous research investigations used the ASTER data for geological application due to specific characteristics of ASTER bands for mineral exploration and lithological mapping [1, 2, 3].

Independent component analysis (ICA) is used on multispectral or hyperspectral datasets to transform a set of mixed, random signals into components that are mutually independent [4]. It is a statistical method for transforming an observed multidimensional random vector into components that are statistically as independent from each other as possible [5]. This transformation serves as a tool for blind source separation (BSS), where no prior information on the mixing is available [4, 5].

In this investigation, to map alteration mineral zone in inaccessible region such Antarctic environments, Independent Components Analysis (ICA) methods was applied to SWIR bands of ASTER data covering Oscar II coast area in eastern Graham, Antarctic Peninsula (AP).

2. Material and methods

2.1 Geology of the study area

Content from this work may be used under the terms of the Creative Commons Attribution 3.0 licence. Any further distribution of this work must maintain attribution to the author(s) and the title of the work, journal citation and DOI.
Published under licence by IOP Publishing Ltd

The Antarctic Peninsula (AP) contains a variety of poorly exposed lithologies and areas that have not been mapped directly providing an appropriate test of the use of image processing methods to multispectral remote sensing satellite data such as ASTER for alteration mapping in the polar context. In this research, the Oscar II coast area north-eastern Graham Land, Antarctic Peninsula (AP) was selected to conduct a remote sensing satellite-based mapping approach to detect poorly exposed lithological units and alteration mineral assemblages in the Antarctic environments.

Figure 1 shows geological map of the Oscar II coast area. The lithological units in this area consisting of (i) volcanic and volcaniclastic rocks (Mapple Formation); (ii) intrusive rocks; (iii) sedimentary rocks (Botany Bay and Trinity Peninsula Groups)) [6].

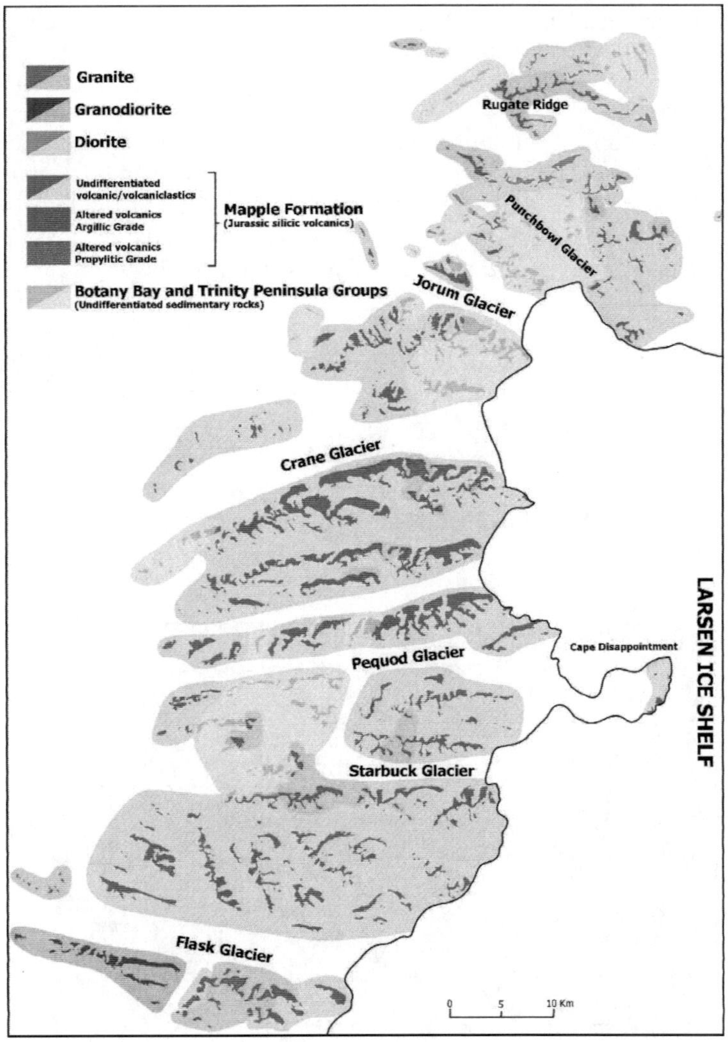

Figure 1. Geological map of the Oscar II Coast area. Extrapolated geological units are depicted as light coloured areas

2.2 Remote sensing data

The ASTER data used in this investigation were obtained from U.S. Geological EROS (http://glovis.usgs.gov/). An ASTER level 1 T (Precision Terrain Corrected Registered At-Sensor Radiance) scenes covering the Oscar II coast region north eastern part of Graham Land, Antarctic Peninsula (AP) were acquired on November 22, 2001. The average of cloud cover for the ASTER data used in this study is less than 10 %. The images were pre-georeferenced to UTM zone 20 South projection using the WGS-84 datum.

IOP Conf. Series: Journal of Physics: Conf. Series **852** (2017) 012025 doi:10.1088/1742-6596/852/1/012025

2.3 Data processing

Preprocessing of ASTER data was also required before analysis. Crosstalk correction was performed on the ASTER data sets used in this study. FLAASH atmospheric correction algorithm was applied on SWIR subsystems of the ASTER data sets. To minimize the reflectance effects of snow, ice and cloud areas from ASTER scene a masking procedure was used to restrict image processing algorithms to just regions of exposed rock.

ICA transformation can distinguish features of interest (defined target in this study for instance alteration mineral zones) even when they occupy only a small portion of the pixels in the image. ICA was applied to ASTER SWIR bands a datasets of the study region Forward IC rotation with sample X resize factor: 0.700000 and sample Y resize factor: 0.700000 was implemented to the datasets used in this study. IC parameter was adjusted as follows; change threshold: 0.00010000, maximum iterations: 100, maximum stabilization iterations: 100, contrast function: LogCosh and coefficient: 1.000000.

3. Results and discussion

In order to detail discrimination of alteration mineral assemblages and lithological units at regional scale, ICA was applied to ASTER SWIR bands, separately. IC bands contain anomaly pixels attributed to distinctive spectral features were assigned to RGB color combination for generating image map of alteration mineral assemblages in the study area. IC bands selected from ASTER SWIR dataset were used to generate FCC image maps for identification of alteration mineral assemblages in the study area.

Figure 2 shows FCC image map of ASTER full scene covering Oscar II coast area, IC5 (band 8 of ASTER: 2.295-2.365 μm), IC4 (band 7 of ASTER: 2.235-2.285 μm) and IC3 (band 6 of ASTER: 2.185-225 μm) were assigned to produce RGB image map. Anomaly pixels in this image map comprise spectral features related to Al-O-H, Fe, Mg-O-H and CO3 groups. Sericitically-altered rocks have a Al-OH absorption feature at 2.2 μm (the equivalent of band 6 of ASTER). Kaolinite shows Al-OH 2.165 μm and 2.2 μm absorption features (the equivalent of band 6 of ASTER). Propylitically-altered rocks contain Fe, Mg-O-H and CO3 2.31–2.33 μm absorption features (the equivalent of band 8 of ASTER) [3, 7, 8].

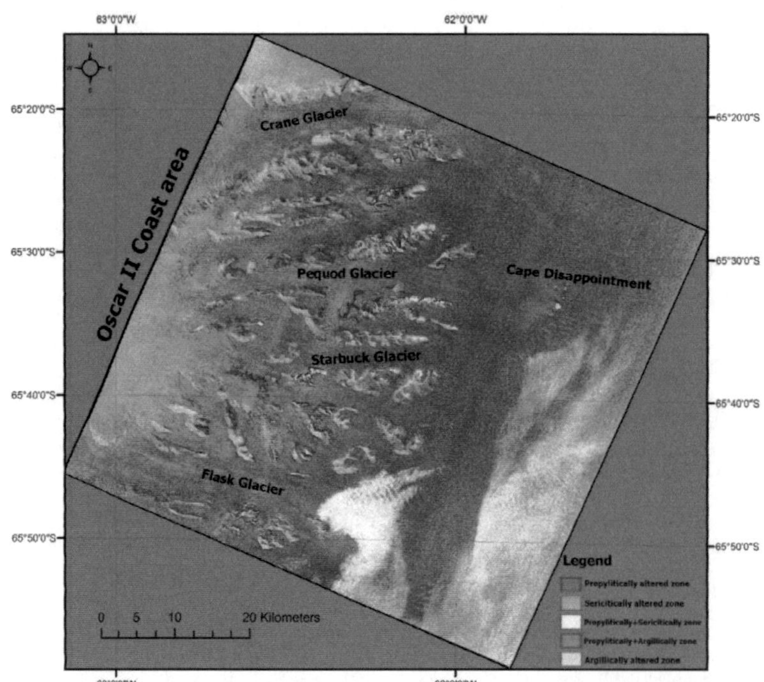

Figure 2. FCC image map (IC5, IC4 and IC3; SWIR bands) of ASTER full scene, Oscar II coast region

As shown in Fig. 2, FFC image map contains anomaly pixels exhibiting five colors. Each color class is indicative of surface abundance of different alteration mineral assemblages. Red color represents chlorite, epidote and calcite mineral group (propylitically-altered rocks). Green color is indicative of muscovite spectral features (sericitically-altered rocks). Aqua color is considered to show Kaolinite and alunite mineral assemblages (argillically-altered rocks), which is mixed with green color in many parts. Yellow color appearance indicates mixture of chlorite mineral group and muscovite (propylitically- and sericitically- altered rocks). Magenta color illustrates combination of chlorite, epidote and calcite mineral group (propylitically-altered rocks) and argillic alteration zone.

Comparison with geology map (see Fig. 1) of the Oscar II coast area and previous remote sensing study [9], indicate that the red pixels match chlorite and muscovite mineral assemblages. The green pixels match well with the muscovite high abundance zones. Aqua color pixels correspond with the Kaolinite and alunite high abundance regions. Yellow color pixels consist of muscovite-chlorite and Kaolinite-alunite mineral assemblages. Magenta color pixels coincide with chlorite-muscovite class. Furthermore, several anomaly pixels exhibiting distinctive alteration mineral assemblages were mapped in many parts of the Oscar II coast area (Fig. 2), which have been previously classified as low albedo or unassigned to any alteration mineral groups. Figure 3 displays FFC image map covering northern sector of the Oscar II coast.

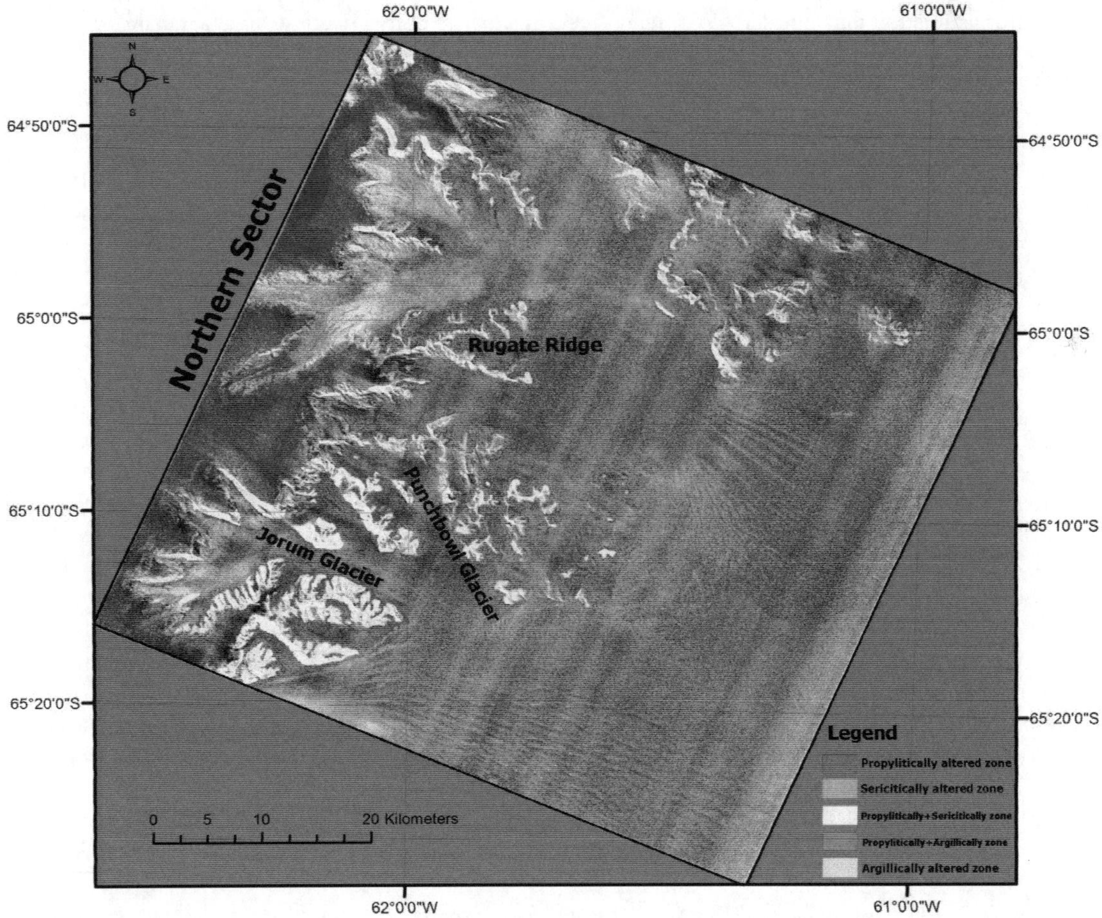

Figure 3. FCC image map (IC5, IC4 and IC3; SWIR bands) of ASTER full scene, northern sector of the Oscar II coast region

As shown in Fig. 3, anomaly pixels appear four dominant alteration mineral assemblages, including muscovite-chlorite and Kaolinite-alunite (yellow pixels), chlorite, epidote and calcite mineral group (red pixels), muscovite (green pixels) and chlorite- Kaolinite-alunite (magenta pixels). With reference to geology map of the northern part of the Oscar II Coast, undifferentiated sedimentary rocks (Botany Bay and Trinity Peninsula groups) and volcanic/volcaniclastic (Mapple Formation) located near Jorum and Punchbowl Glaciers are coincided with high abundance of yellow pixels and very low dissemination of red pixels.

However, intrusive rocks such as granite and diorite near Rugate Ridge are matched with high distribution of red, magenta and green pixels associated with low abundance of yellow pixels. Thus, poorly exposed lithologies detected in the northern part of Figure 3 might be dominated by intrusive rocks such as granite, granodiorite and diorite units and volcanic/volcaniclastic rocks. Propylitically- and sericitically- altered rocks and propylitically altered rocks are more distributed in the north western part of the image (Figure 3). However, north eastern part of the image dominates by propylitically- and sericitically- altered rocks and sericitically-altered rocks.

4. Conclusion

This investigation has indicated the application of SWIR bands of ASTER datasets for extrapolating satellite-based imagery from relatively mapped area such Oscar II coast area, in north-eastern Graham Land, Antarctic Peninsula (AP) into poorly mapped or unmapped domains. ICA mapped pixel related to Al-O-H, Fe, Mg-O-H and CO3 groups. ICA method provided image maps of alteration mineral assemblages and discriminate felsic to mafic lithological units with little available geological data for poorly mapped regions and/or without prior geological information for unmapped regions in northern sector of Oscar II coast area, Graham Land. The approach used in this study performed very well for alteration mineral mapping with little available geological data or without prior information of the study region.

Acknowledgments

This study was conducted as a part of Yayasan Penyelidikan Antartika Sultan Mizan (YPASM) research grant (Vote no: R.J130000.7309.4B221), Sultan Mizan Antarctic Research Foundation, Malaysia. We also would like to express our great appreciation to KOPRI (Korea Polar Research Institute) Asian Polar Science Fellowship Program 2016 for their great support during this research. This study was partially supported by KOPRI grant PE17050 and KOPRI's Asian Polar Science Fellowship Program. We are thankful to the Universiti Teknologi Malaysia for providing the facilities for this investigation.

References

[1] Pour A B and Hashim M 2011 The earth observing-1 (EO-1) satellite data for geological mapping, southeastern segment of the Central Iranian Volcanic Belt, Iran *International Journal of the Physical Sciences* **6**(33) 7638

[2] Pour A B and Hashim M 2014 ASTER, ALI and Hyperion sensors data for lithological mapping and ore mineral exploration *Springerplus* **3**(130) 1

[3] Pour A B and Hashim M 2015 Integrating PALSAR and ASTER data for mineral deposits exploration in tropical environments: a case study from Central Belt, Peninsular Malaysia *International Journal of Image and Data Fusion* **6**(2) 170

[4] Yamaguchi, Y I, Fujisada H, Kahle A B, Tsu H, Kato M, Watanabe H, Sato I and Kudoh M 2001 ASTER instrument performance, operation status, and application to Earth sciences *IEEE Transactions of Geosciences and Remote Sensing* 1215

[5] Haselwimmer C E, Riley T R and Liu J G 2013 Lithologic mapping in the Oscar II Coast area, Graham Land, Antarctic Peninsula using ASTER data *International Journal of Remote Sensing* **32**(7)

[6] Chang C I and Heinz D C 2000 Constrained subpixel target detection for remotely sensed imagery *IEEE Transactions on Geoscience and Remote sensing* **38**(3) 1144

[7] Harsanyi J C and Chang C I 1994 Hyperspectral image classification and dimensionality reduction: an orthogonal subspace projection approach *IEEE Transactions on Geoscience and Remote Sensing* **32**(4) 779

[8] Pournamdary M, Hashim M and Pour BA 2014a Application of ASTER and Landsat TM data for geological mapping of Esfandagheh ophiolite complex, southern Iran *Resource Geology* **64**(3) 233

[9] Pournamdary M, Hashim M and Pour B A 2014b Spectral transformation of ASTER and Landsat TM bands for lithological mapping of Soghan ophiolite complex, south Iran *Advances in Space Research* **54** (4) 694

IOP Conf. Series: Journal of Physics: Conf. Series **852** (2017) 012026 doi:10.1088/1742-6596/852/1/012026

Modelling airborne dispersion for disaster management

I A Musliman[1] and L Yohnny[1]

[1]Department of Geoinformation, Faculty of Geoinformation and Real Estate, Universiti Teknologi Malaysia, 81310 Skudai, Johor Bahru, Malaysia.

Email: ivinamri@utm.my, yohnniebabe@gmail.com

Abstract. Industrial disasters, like any other disasters, can happen anytime, anywhere and in any form. Airborne industrial disaster is a kind of catastrophic event involving the release of particles such as chemicals and industrial wastes into environment in gaseous form, for instance gas leakages. Unlike solid and liquid materials, gases are often colourless and odourless, the particles are too tiny to be visible to the naked eyes; hence it is difficult to identify the presence of the gases and to tell the dispersion and location of the substance. This study is to develop an application prototype to perform simulation modelling on the gas particles to determine the dispersion of the gas particles and to identify the coverage of the affected area. The prototype adopted Lagrangian Particle Dispersion (LPD) model to calculate the position of the gas particles under the influence of wind and turbulent velocity components, which are the induced wind due to the rotation of the Earth, and Convex Hull algorithm to identify the convex points of the gas cloud to form the polygon of the coverage area. The application performs intersection and overlay analysis over a set of landuse data at Pasir Gudang, Johor industrial and residential area. Results from the analysis would be useful to tell the percentage and extent of the affected area, and are useful for the disaster management to evacuate people from the affected area. The developed application can significantly increase efficiency of emergency handling during a crisis. For example, by using a simulation model, the emergency handling can predict what is going to happen next, so people can be well informed and preparations works can be done earlier and better. Subsequently, this application helps a lot in the decision making process.

1. Introduction

1.1. Airborne Dispersion

Air particles and gases exist as fluid in the environment; the substance has flexible form and is highly dynamic. Gases particles can take on any shape, the dispersion of the gas particles are dependent on internal factors as well as external factors. Dispersion of gases are evolved through three phases – rapid phase of internal slumping, intermediate entrainment phase and final phase of passive diffusion [3].

During the first phase – rapid phase of internal slumping, the movement of the gases are driven by the excessive hydrostatic forces due to high gas density compared to the surroundings. The dispersion pattern of the gas particles in this phase is in random direction, causing the particles to be pushed outwards from within. During the second phase – intermediate entrainment phase, the activity of the gas particles is directly influenced by the entrainment effects of the gas particles, in which the gases with lower density will be entrained into those with higher density. During the third phase – passive diffusion phase, the activity of the gas particles are fully dependent on the external forces such as wind

Content from this work may be used under the terms of the Creative Commons Attribution 3.0 licence. Any further distribution of this work must maintain attribution to the author(s) and the title of the work, journal citation and DOI.
Published under licence by IOP Publishing Ltd

and pressure [3]. The gas particles do not move on its own anymore in this stage, they remain stationary at the same spot until carried away. Various chemical substances are used in the industry for manufacturing and consumption purposes, some common names are such as naphthalene (C10H8), benzene (C6H6) and ammonia (NH3). These chemicals were commonly found in the manufacturing and electronic industry, naphthalene and ammonia are widely used and stored at Pasir Gudang industrial area, especially around the port area. These gases were stored in pressure tanks under high compressions, thus leakage can occur when the facilities were worn and old. Besides that, there are many other flammable and explosive gases stored in the industrial area such as propane and liquefied petroleum gas, these substance can be easily ignited and can cause explosions when heated, especially under hot and humid weather in tropical countries such as Malaysia. These chemicals are stored in storage tanks in abundance; the disaster can last up to few days and have a severe effect to the environment if it is not contained properly.

1.2. Modelling Air Particles Dispersion

Many gases are colorless and odorless, this makes the presence of the gas hard to be detected, and thus it is hard to predict the dispersion of the gas particles and the direction of the spread of the disaster. This makes the disaster management hard to be carried out as the emergency handling unit has to search for the whole area to identify the gases, the responses would be delayed as well in this case, causing the disaster handling to be slow and inefficient. In this case, modeling the dispersion of the air particles can be very useful for the emergency handling job. The simulation can be carried out using mathematical algorithm to predict the whereabouts of the gas particles and the most likely direction the spreading is heading to. Conventional methods such as Gaussian mathematical methods were widely used to model the dispersion of the liquid in the past; however this might not effectively model the dispersion of the gas particles as Gaussian models and Eulerian models assume the external variables in space and time to be global and constant all the time [7].

Lagrangian particle dispersion (LPD) is more suitable to be used to model the dispersion of air particles, this algorithm is able to compute dynamic input data such as wind speed and direction, which changes instantly in the real world [11][12]. Lagrangian particle models compute trajectories of a large number of so-called particles to describe the transport and diffusion of tracers in the atmosphere [1]. Many researches about simulation of gas leakage were done in the past, noticeably the most popular one was the Bhopal gas tragedy in India which occurred in 1984 [5][6][7][8]. These researches use numerical methods such as K - Closure meteorological model [7], Lagrangian model [9][10] and simulation models such as the Haddon matrix analysis [5] to simulate the aftermath of the disaster. By using the numerical models, the research can estimate the number of affected peoples and the properties and also the area covered by the leakage; these models are also useful to predict the factors and consequences of the event as well as the dispersion of the gases in graphical forms.

However, the researches in the past simulate the model in a mathematical coordinate system, there were no integration of GIS into past models so that the analysis were done in a projected coordinate system. So, there is a need to create a GIS integrated model to simulate situations related to location and environment.

1.3. Disaster Management

Disaster management is a set of precautionary procedures and protocols to be applied to avoid and to deal with disasters, it can be used as a guideline for pre-event preventions or post-event emergency handling. Disaster can be found almost everywhere, and it is often triggered by various factors, such as natural disasters, human mistakes and bad maintained facilities. One good example is gas leakages at industrial area due to corrosion of storage tanks by acidic substances. Industrial gas leakage can be fatal and can inflict permanent damage to the peoples and objects nearby, especially when the gases are highly toxic, flammable and highly reactive with other materials causing a chain effect to occur. Disaster management is important especially in medium and heavy industry area, these areas are always in larger scale and there are more chemical substances being stored, so the risk of disaster is higher and needs a better disaster management team to handle during an outbreak.

In Malaysia, the disaster risk indicator is classified into three categories; tier 1, 2 and 3 according to Department of Social Welfare Malaysia (JKM), these classes identifies the critical level of the disaster and to guide the decision makers to figure out suitable solution for the problems. The description of the classes is as shown in Table 1.

Table 1. Three categories of disaster risk indicator according to Department of Social Welfare, Malaysia (JKM)

Tier	Description
Tier 1	Local incident that is manageable and has no potential to spread. It is not very complex and may result in loss of life and property is small. This disaster did not affect the daily activities of the local population at large. Authorities at District Level have the ability to control and overcome this incident through agencies in the District Level with or without outside help are limited.
Tier 2	The more serious incident covers a large area or more than two (2) Districts and has the potential to spread. Likely to lead to loss of life and a lot property damage. This incident also destroy infrastructure and affect daily activities of life. Is more complex and complicate than Level 1 Disaster in term of search and rescue efforts. Need and be able to be operated by the Authority at State level without or with limited outside help
Tier 3	Disaster incident resulting from Level 2 is more complex or cover a larger area or more than two (2) states. Need and be able to be operated by the Authority at the Federal level without or with assistance from abroad.

1.4. History of Gas Leakage in Malaysia

Gas leakage is a common disaster that happens in Malaysia, there are various causes of gas leakage such as operational blunders, poorly maintained facilities and equipment, and natural disasters such as earthquake and thunder strike. Some of the larger scale airborne disasters are the Petroleum Sdn. Bhd. gasoline tank explosion and the ammonia gas leakage in Labis, Johor.

1.4.1. Petronas Gasoline Tank Explosion. Petronas is an oil and gas company which is located at Johor port in Pasir Gudang industrial area. This catastrophe happens on 24th April 2006 where eight of the gasoline storage tanks own by Petronas caught fire from thunder strike and black fumes appears at the area. The fire continue to burn for three days before being put out, it is one of the largest scale disaster in Pasir Gudang history. This disaster is a tier 3 disaster where neighbouring countries such as Singapore were helping to contain the fire. Lots of black fumes (carbon monoxide) due to the burning of gasoline had been released to the environment during the fire. The smoke can affect the health of residents that reside at that area, as well as pollute the rivers and crops nearby, causing losses to the nearby industries and affecting the operation of the nearby businesses. The smoke has spread to nearby village; Kampung Pasir Putih which is located north eastern of Pasir Gudang [4].

1.4.2. Labis Ammonia Leakage. Another large scale airborne disaster happens at Labis, Segamat, Johor where 300 tonnes of Ammonia (NH3) gas leaks to the surrounding areas from an illegal dumpsite of a company from Malacca. 300 families were evacuated from ground zero; Kampung Sungai Gatom and more than 100 factory employees were barred from entering the area to work as the company is located in that area. This incident has brought destructive aftermaths to the agricultural industry nearby the dumpsite where most of the plantations turned yellowish due to Ammonia poisoning [10]. It is difficult to contain the incident as the Ammonia gases are colorless, they were invisible to the naked eyes. One way to detect the presence of the Ammonia gas is by smelling because Ammonia gas is pungent in nature, so the response to the incident will be delayed because they need to wait for reporting from the peoples.

1.5. Modelling Air Particles in the Past and Today
Today, with the aid of advanced hardware capabilities and processing power, the modeling of the air particles has undergone a drastic improvement by integrating GIS technologies and mapping capabilities. Simulations can be carried out using live feed data from various sources and the processing power of hardware can cope with the huge amount of data exchange through the use of high bandwidth internet connections.

Therefore, the following question arises: How does today technology helps in improving emergency handling? Since the web platform and technologies has improved a lot after so many years, how can people utilize the technology to perform real time analysis and to disseminate information to the public in a quicker and more efficient way? By integrating GIS and web technology into emergency handling, the efficiency and the visualization of the emergency handling can be greatly improved. This allows the user to have a better understanding and accurate interpretation of the information as well as to disseminate the critical information to the public in the shortest delay. Therefore, integration of GIS and web technology undeniably could help improve emergency handling and to hasten the decision making process.

This paper is structured in the following order: first, a brief explanation of airborne dispersion and disaster management. Section 2 discuss on the modeling the airborne particles. Section 3 explains the algorithms used to perform the calculation of the dispersion of the air particles. The operation and output are presented in Section 4 and concluding remarks in Section 5.

2. Modeling Airborne Particle
To simulate the dispersion of air particles, an application prototype has been developed to compute the characteristics of air particles and the way air particles move around under the influence of wind.

Due to the dynamic characteristic of the wind, the dispersion of air particles in the air can be highly fluctuating. In order to model the wind, a simulation model which can accept dynamic parameters is needed, Lagrangian particle dispersion model is suitable to be used to simulate the movement of the air particles as the Lagrangian particle dispersion model calculates the next displacement of the air particles based on the previous position, so this model is suitable to be used in situation where highly dynamic data input is required.

The concept of the application is to make use of the web services from third party service providers such as Google and Yahoo! as the data provider so that the data input into the model can be updated by itself on real time and the changes to the data are handled automatically by the application.

2.1. System Design
The application prototype has a three tier infrastructure as shown in Figure 1, which is database, server and client side infrastructure. The application is distributed into three parts, so that the performance can fully utilize the hardware resources and web bandwidth.

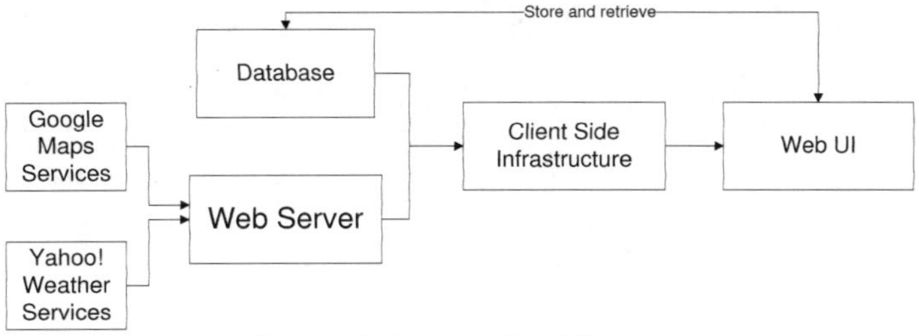

Figure 1. System Architecture

2.2. Web Server and Network Database

The application uses XAMPP hosting package to host the web system, XAMPP is an open source web hosting package that provides a variety of function and services such as Apache web server and MySQL database. The application uses the landuse data of Pasir Gudang, Johor and the Places-of-Interests (POI) data such as location of police station, hospital and express terminal to perform the analysis. The data geometry were stored in the MySQL Spatial database and loaded into the map view by using the Google Maps API map feature function. The landuse cover was created in the map view as polygon and the POI data was displayed in the map view as points. The geometric shape of the landuse data is as shown in Figure 2.

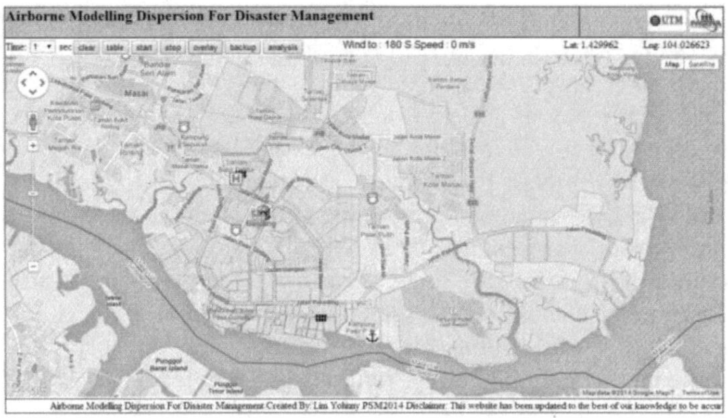

Figure 2. Landuse and POI data in Pasir Gudang, Johor

The attributes and the shapes of the landuse data and POI data were stored in the MySQL database as geometry data, the geometries of the polygons are stored as multipolygon spatial format which is supported natively in MySQL.

2.3. Application Flow Chart

The flow of the application is as shown in Figure 3.

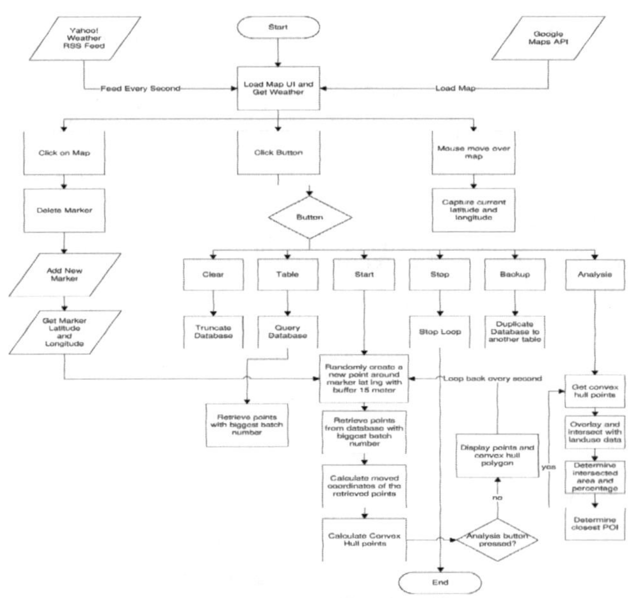

Figure 3. System application flowchart

2.4. Third Party Services and Data Feeds
The application prototype was developed on the web platform; the application makes use of existing Web Map Services (WMS) and weather RSS feeds as the data input for the parameters in real time. Google Maps API version 3 was used as the WMS map mashup of the application, this is the latest version of WMS from the browser machine company which has a stripped down architecture and is quicker in response to the mobile browsers. The weather RSS feeds were reading the data from Yahoo! Weather feeds by using its API, by using this method, the wind speed and direction can be updated every second. The Yahoo! Weather RSS feed is formatted to return the wind speed, bearing and wind direction.

There are a few analysis being carried out at the server side, this including the determination of the movement of the air particles under the influence of the wind and the conversion of the displacement of the particles from wind speed (meter per second) to the map view which is in decimal degree. The determination of the movement of the particles under the influence of wind speed is calculated by using trigonometry where the X and Y components forms part of the triangles and the speed can be calculated by using Pythagoras theorem. The condition of the algorithm is as shown in Figure 4.

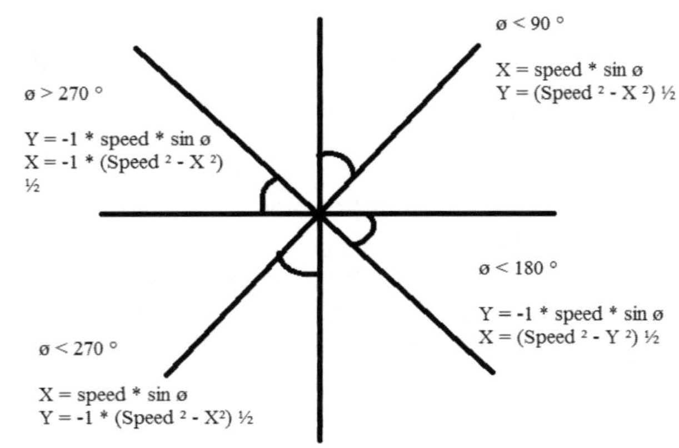

Figure 4. Trigonometry to determine wind component

The unit of wind components is in meter per second; this means that the wind components are in metric units, metric units cannot be directly integrated into the Google Maps view which is using decimal degrees (Coordinate System: World Geodetic System 1984 (WGS84)). In order to integrate the wind components into the map view, a conversion has to be carried out to convert the wind vector components (speed in north-south and east-west direction) from metric units into decimal degree; this requires conversion from metric coordinate system Malaysian Rectified Skew Orthomorphic (MRSO) to Malaysian Revised Triangulation (MRT) to WGS84.

2.5. Modelling Gas Leakage
The application applied mathematical algorithm to model the dispersion of the air particles and the direction the particles are heading to. The mathematical algorithm used was the Lagrangian particle dispersion model.

In order to model the scenario of the gas leakage, the application would randomly populate gas clouds (point feature) into the map view in a 15 meters radius buffer from the incident point (disaster source). This is to simulate the movement of the gas particles from the first to phase of dispersion, the rapid phase of internal slumping and the intermediate phase of entrainment of the particles which will

IOP Conf. Series: Journal of Physics: Conf. Series **852** (2017) 012026 doi:10.1088/1742-6596/852/1/012026

push the particles outwards into a certain distance in all direction, so the particles are distributed randomly at this stage. The particles will then move to other directions based on the wind parameters feed in by Yahoo! Weather. To model this movement, the application utilizes the Lagrangian particle dispersion model to calculate the individual displacement of the particles in the map view.

3. Applying the Algorithms

3.1. Lagrangian Dispersion Model (LPD)

The Lagrangian particle dispersion model was designed to compute the individual displacement of the gas particles along range and mesoscale environment [2]. The algorithm simulates the dispersion and the trajectories of the air particles by using the carrier vector components. The dispersions are calculated according to:

$$X(t + \Delta t) = X(t) + U(t) * \Delta t \quad (1)$$

$$Y(t + \Delta t) = Y(t) + V(t) * \Delta t \quad (2)$$

where,
X = longitudinal position of the air particle
Y = latitude position of the air particle
t = the current time
Δt = change of time
U = the east-west vector component (speed) of the wind
V = the north-south component (speed) of the wind

The Lagrangian particle dispersion model makes use of the wind component to simulate the trajectories of the air particles. The position of the particles at time t + Δt is calculated based on the previous position of the particles at time t. Thus this algorithm is suitable to be used to model the dispersion of the air particles as the calculation uses different parameter from time to time, this is suitable to the characteristics of the wind which is dynamic and changes all the time. The movement of the particles is as shown in Figure 5, the particles move to bearing 30 degrees as follow to the wind direction.

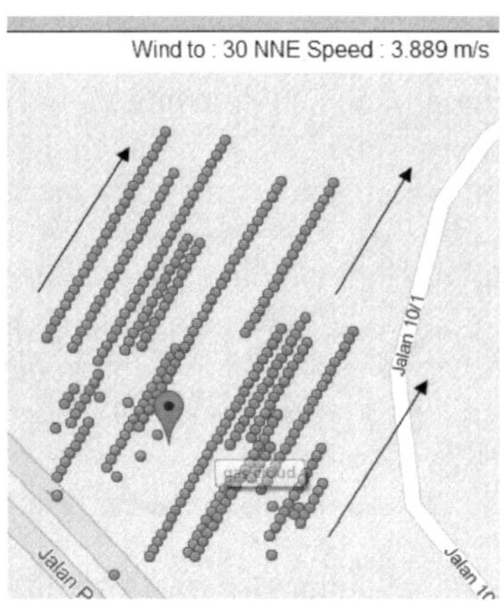

Figure 5. Movement of air particles in LPD

The movement of the particle in close up would be show the evolution of the air particles from one position to the next position. Each particle is retained and the next position of the air particle under the influence of the wind is calculated using the Lagrangian particle dispersion model.

IOP Conf. Series: Journal of Physics: Conf. Series **852** (2017) 012026 doi:10.1088/1742-6596/852/1/012026

3.2. Convex Hull

Convex hull is a mathematical algorithm used to calculate the minimum bounding polygon that contains points X given a set of point X. This means that the convex hull algorithm is used to identify the outermost points of a given set of points, the convex points and connect the points to form a polygon which is able to contain all the other points in the polygon.

The convex points were calculated according to a logical loop where the algorithm loops through all the points in the given point set. The algorithm would then determine whether the point is the point located at the left most position as it loops through the set of points as shown in Figure 6(a).

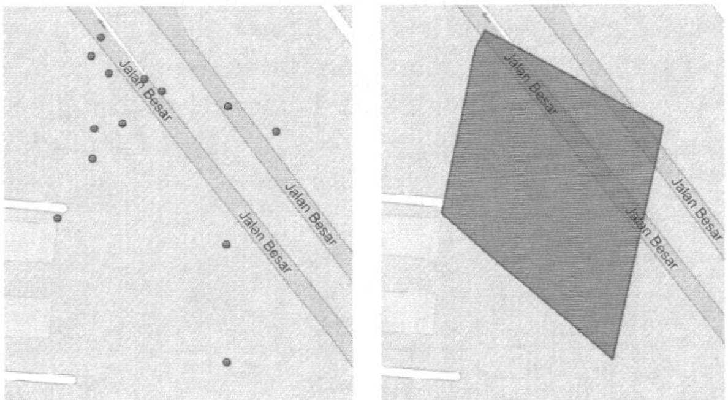

Figure 6. (a) Left: Given a set of point. (b) Right: The convex polygon

The points will be looped through one by one to determine the left most points. By looping through the points, the algorithm determine the minimum and maximum latitude and longitude position of the points, if it has either a maximum or a minimum position, then it is a convex point. The result of the convex hull is as shown in Figure 6(b).

3.3. Performing GIS Analysis (Intersection) using MySQL Spatial

The position of each of the nodes of the convex polygon was sent to the database server to perform an intersection between the convex polygon and the landuse polygon layer. The coordinates of the points send were structured into a multipolygon format as shown in Figure 7 to integrate with the MySQL Spatial syntax.

```
MULTIPOLYGON(((103.91327776263
1.4351883190923,103.91327905747085
1.4350334250699057,103.91330370048
1.4346968706534,103.9142149557
1.4350539911868,103.91460819316
1.4352793731575,103.91508216362
1.4358569928478,103.91550265241
1.4371085971566,103.91532149251
1.4372297262572,103.9150346561
1.4373135287365,103.91368583431
1.4358619104029,103.91327776263 1.4351883190923)))
```

Figure 7. Multipolygon data representation in MySQL Spatial format

The intersection is carried out by using the following SQL Spatial syntax:

```
SELECT intersects (GeomFromText("the convex polygon points"),
GeomFromText("ogc_geom column")) FROM zone;
```

4. Result and Validation

4.1. Application Output

The analysis of the application is to determine the coverage area of the air particle over the Pasir Gudang area to determine the effects of the disaster to the surrounding environment. The analysis involves intersecting the convex hull polygon to the landuse cover data of Pasir Gudang and the POI data. In order to determine the effect of the disaster to the surrounding environment, the convex hull polygon was overlaid on top of the landuse cover, then an intersection process was carried out to retrieve the intersected pieces of land polygon, this results give us the affected area of the Pasir Gudang industrial area. Following the intersection, the application will also calculate the extent of the affected area (in km^2) and the percentage of the affected area compared to the total landuse parcel.

The application also produce a suggestion to the user the nearest evacuation point from the incident area, the application calculates the distance of each POI from the incident area and return the nearest one to the user. The dispersion of the air particles follows the wind direction, the wind component is 3.333m/s to bearing 40 NNE as shown in Figure 8(a), Figure 8(b) and Figure 8(c), we can see that the dispersion of the air particles flow from the source point towards the north east direction, following the trajectories of the wind.

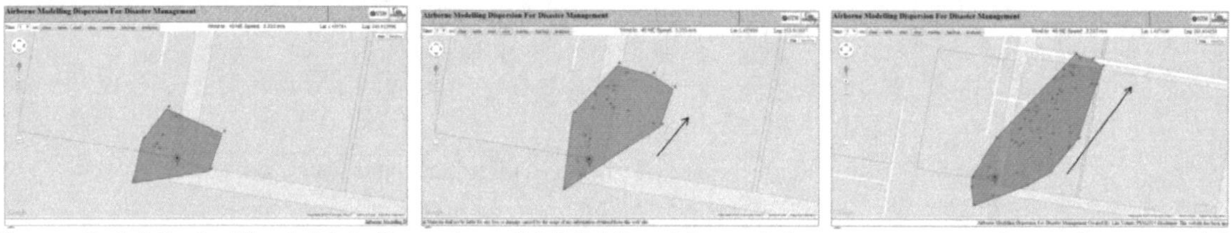

Figure 8. (a) Left: The initial dispersion, (b) Middle: Air particles moves towards north east and (c) Right: Air particles moving further up north east

4.2. Airborne Dispersion Validation

In order to validate the results of the simulation model, comparison of results from the developed prototype has been made with the results obtained from an online simulation application called HYSPLIT Air Dispersion Model developed by National Oceanic and Atmospheric Administration (NOAA). The HYSPLIT dispersion model uses Lagrangian Particle Dispersion model to compute the dispersion of the air particle as well [13]. The result from HYSPLIT program is an hour long of processing data, which is simulated using the archived weather data and then exported to ESRI Shapefile along with a GIF image map. The GIF image of the result is shown in Figure 9a and the shapefile is opened in QGIS software as shown in Figure 9b.

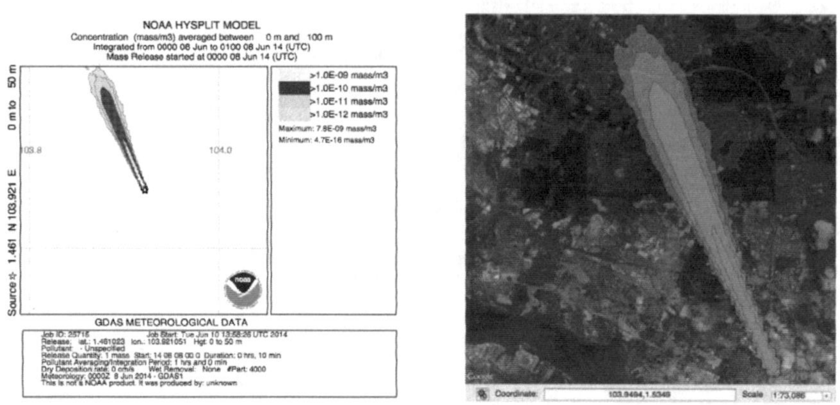

Figure 9. (a) Left: GIF map results from HYSPLIT and (b) Right: HYSPLIT result in vector format (Shapefile) overlay with Google Maps imagery in QGIS

IOP Conf. Series: Journal of Physics: Conf. Series **852** (2017) 012026 doi:10.1088/1742-6596/852/1/012026

For comparison, set of air particle points (from LPD model) have been generated using the developed prototype simulation model for the period of one hour, the results were overlaid on top of the HYSPLIT result. The distance covered by the HYSPLIT result is 14.9km while the result generated by the prototype simulation model is 14.2km, thus the results were identical to each other up to 95%. Figure 10 shows the overlaid comparison of both results for 1 hour duration.

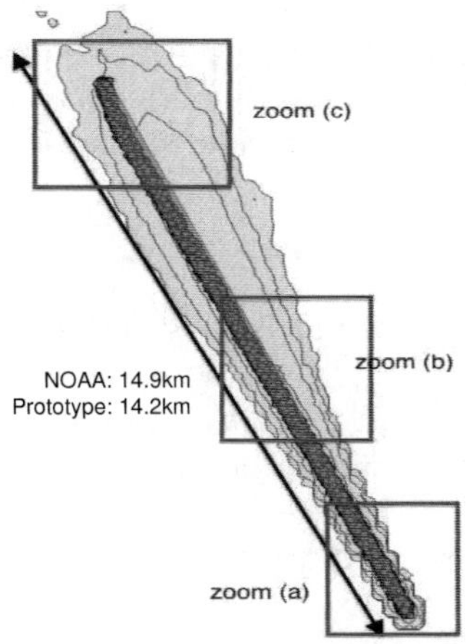

Figure 10. Comparison of results from HYSPLIT (NOAA) and the developed prototype

For visual inspection, the validation result is divided into three zones, namely the zoom (a) zone as shown in Figure 11(a) with high concentration gases, zoom (b) zone as shown in Figure 11(b) where the concentration starts to spread outward and zoom (c) as shown in Figure 11(c) zone where the gases are in various concentrations.

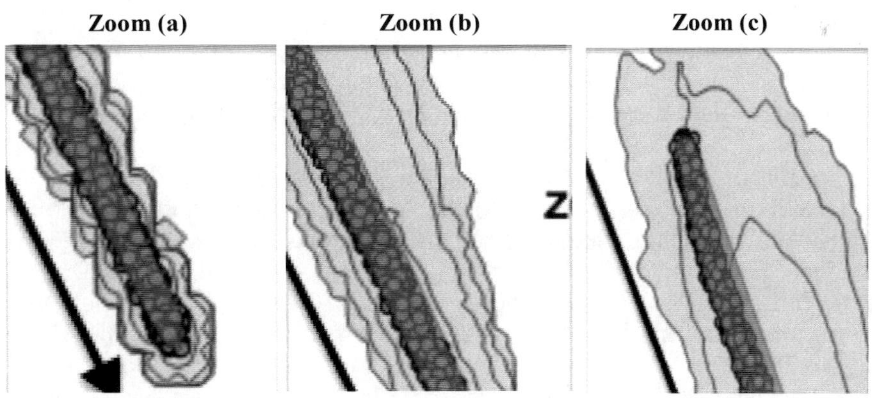

Figure 11. (a) Left: Zoom (a), the concentration of the gases at ground-zero, (b) Middle: Zoom (b), concentration of the gases start to decline and the dispersion pattern appears to be diverged outwards and (c) Right: Zoom (c), dispersion pattern diverged outwards and gases are in various concentration.

In Figure 11(a), the result of the model is same as the result produced by NOAA, this is because the concentration of the gases are the same. While in Figure 11(b) the concentration of the gases start to decline and the dispersion pattern appears to be diverged outwards. In Figure 11(c) the dispersion pattern of the results are not the same because the NOAA model takes into consideration the decline in gas concentration; while the developed prototype model only simulate the air particles in the same concentration throughout the period, so the results appear to be in constant width without diverging outwards. However, the results of both models appear to be quite similar to each other.

IOP Conf. Series: Journal of Physics: Conf. Series **852** (2017) 012026 doi:10.1088/1742-6596/852/1/012026

5. Conclusion

In conclusion, the dispersion of the air particles in the environment is usually hard to be traced, so it is important to have modeling system to predict the trajectories of the air particles in the air. Through the use of the modeling system, the authorities can be well informed of the direction of dispersion of the air particles and affected area of Pasir Gudang. This can help a lot in decision making to save lives and to evacuate peoples to safety during a disaster outbreak. Hopefully by using this application, the emergency handling in Pasir Gudang industrial area can be improved and more life can be saved. Besides that, the model can also be applied to other field besides airborne disaster such as to determine the dispersion of air pollution and dispersion of radioactive materials from factories, as long as the subject is airborne and can be easily carried away by wind.

References

[1] Stohl A, Forster C, Frank A, Seibert P and Wotawa G 2005 Technical Note: The Lagrangian Particle Dispersion Model FLEXPART Version 6.2 *Atmospheric Chemical Phys.* **5** 2461

[2] Stohl A, Hittenberger M and Wotawa G 1998 Validation of The lagrangian particle dispersion model FLEXPART against large scale tracer experiment data. *Atmospheric Environment* 32 (24) 4245

[3] Eidsvik KJ 1980 A model for heavy gas dispersion in the atmosphere *Atmospheric Environment* **14** 769

[4] Harian Metro Online (29 April 2006) Petir Musnah 8 Tangki Minyak

[5] Eckerman I 2005 The Bhopal gas leak: analyses of causes and consequences by three different models *J. Loss Prevention in the Process Industries* **18** 213

[6] Havens J, Walker H and Spicer T 2012 Bhopal atmospheric dispersion revisited *Journal Hazardous Materials* **233 234** 33

[7] Sharan M, McNider R T, Gopalakrishnan S G and Singh M P 1994 Bhopal gas leak: a numerical simulation of episodic dispersion *Atmospheric Environment* **29**(16) 2061

[8] Gopalakrishnan S G and Sharan M 1997 A lagrangian particle dispersion model for marginally heavy gas dispersion *Atmospheric Environment* 31 (20) 3369

[9] The Star Online (12 January 2006) 300 Tonnes of Toxic Waste Emitting Ammonia Fumes

[10] Leelőssy Á, Molnár F, Izsák F, Havasi Á, Lagzi I and Mészáros R 2014 Dispersion modeling of air pollutants in the atmosphere: a review *Open Geosciences* **6**(3) p 257 doi:10.2478/s13533-012-0188-6

[11] Berbekar E, Harms F, Leitl B, Boris J, Moses A, Obenschain K, Patnaik G, Fischer S, and Storm K 2016 Hazmat dispersion modeling in support of urban emergency response for the fire brigade in Hamburg, Germany *2016 IEEE Symposium on Technologies for Homeland Security (HST)* Waltham MA 1

[12] Bartzis J G, Efthimiou G C and Andronopoulos S 2015 Modelling short term individual exposure from airborne hazardous releases in urban environments *Journal of Hazardous Materials* 300182

[13] Stein A F, Draxler R R, Rolph G D, Stunder B B, Cohen M D and Ngan F F 2015 NOAA's HYSPLIT atmospheric transport and dispersion modeling system *Bulliten American Meteorological Society* **96** 2059

IOP Conf. Series: Journal of Physics: Conf. Series **852** (2017) 012027 doi:10.1088/1742-6596/852/1/012027

Remote sensing assessment of absorbing aerosol over Peninsular Malaysia from OMI onboard Aura satellite

K C Tan*, H S Lim and M Z Mat Jafri

School of Physics, Universiti Sains Malaysia, 11800 Penang, Malaysia

*E-mail: kctan@usm.my

Abstract. The observation of aerosol index derived from the Ozone Monitoring Instrument (OMI) on board the Dutch-Finnish Aura satellite with spatial resolution 1° x 1° have been analyzed over Peninsular Malaysia for 2013-2015, from June to September, respectively. The results show significant spatial and temporal variabilities in aerosol index with higher values during June 2013 and September 2015. On the other hand, the aerosol index does not show significant differences between the Peninsular Malaysia for the remaining study duration. The high positive aerosol index values over Southern Peninsular Malaysia clearly reveal the ultraviolet absorbing nature of smoke particles affecting the area during Indonesia forest fire, associated with the Southwest monsoon season. The spatial distribution of aerosol index has been analyzed using monthly OMI/Aura data obtained from the NASA-operated Giovanni. The result shows that the satellite measurements can measure and observe the increase of the aerosol index over different regions.

1. Introduction

The report by the Intergovernmental Panel on Climate Change had recognized that atmospheric aerosols affect the climate, energy balance and hydrological cycle [1]. Besides, the atmosphere consists of many aerosols which affect the atmospheric process such as chemical processes, visibility, air quality, precipitation, and cloud formation. Aerosol particles interact with solar radiation by absorbing and scattering it, resulting from direct effect of aerosols on solar radiation and subsequent climate change [2]. The indirect effect is that they can influence on formation and properties of clouds by acting as cloud condensation nuclei (CCN) and affect the microphysical properties and formation of clouds [3]. Hence, the studies of aerosol processes and properties can give an obvious view on the factors that involved in local and global climate change.

Recently, there is growing of researches have reporting the use of satellite measurements to investigate the indirect effects of temporal and spatial patterns of aerosols, at both regional and global scales [4]. Meanwhile, satellite remote sensing is an important tool for examine the radiative effects and global aerosol budget that aerosols have on climate [5]. In theory, satellite remote sensing allow users to assess the spatial distribution and properties, and it also provide major benefit by allowing synoptic and complete mapping of a large area in a single snapshot [6]. Satellite remote sensing has been successfully applied in deducing the origin, spatial and temporal distribution of absorbing aerosols [7], and in constraining biomass-burning emissions [8].

In particular, the ultraviolet aerosol index provides a useful tool in satellite remote sensing that indicates the presence of elevated absorbing aerosols in the troposphere [7]. The absorbing aerosol index has an advantage that capable to detect aerosols over land due to the low surface albedo of land within ultraviolet wavelength. The spectral comparison between two ultraviolet wavelengths is used to

Content from this work may be used under the terms of the Creative Commons Attribution 3.0 licence. Any further distribution of this work must maintain attribution to the author(s) and the title of the work, journal citation and DOI.
Published under licence by IOP Publishing Ltd

calculate aerosol index. The existing of spectral contrast is caused by the difference in the characteristic of absorbing aerosols and other phenomena such as Rayleigh scattering, surface reflectance, gaseous absorption and cloud and aerosol scattering [9].

The magnitude of aerosol index depends on the height of aerosol layer, viewing geometry, aerosol type (single scattering albedo), and aerosol optical depth [10]. Negative values of aerosol index are assigned by pure scatterers while positive values of aerosol index are assigned by desert dust and smoke [7]. Nevertheless, the other features such as spectral slopes in the surface reflectance among the reference wavelengths and high altitude clouds also bring the negative value of aerosol index [11]. In addition, the presence of clouds will give the near zero values of aerosol index [12]. Whilst aerosol index values greater than 1.0 represent the absorbing aerosols such as dust or soot particles [13].

2. Ozone Monitoring Instrument (OMI)

The O_3 Monitoring Instrument (OMI), a contribution of the Netherlands and Finland to the National Aeronautics and Space Administration' (NASA) Aura mission, is flown on the Aura spacecraft. Aura is part of NASA's long term Earth Observation System (EOS) mission and was launched on 15 July 2004, from Vandenberg Air Force base in California. The OMI is the first of a new generation of space-borne spectrometers that combine a high spatial resolution (13×24 km^2 at nadir) with daily global coverage [14]. Measuring the trend in stratospheric O_3 as well as detecting tropospheric O_3 on regional scales is top priorities among the science objectives for Aura satellite [15].

OMI is used to measure the reflected solar radiation in the ultraviolet and visible part in the spectral range between 270 nm to 500 nm, by using two channels with a spectral resolution of about 0.5 nm. The light entering the telescope is pseudodepolarized using the scrambler and then split into two channels, which are ultraviolet channel and the visible channel.

3. Study Area

Peninsular Malaysia is located in Southeast Asia within the latitudes of 1° to 7° N and the longitudes of 99° to 105° E (south of Thailand, north of Singapore, and east of Sumatra). Peninsular Malaysia has an area of approximately 131,587 km^2 (Figure 1) and an estimated population of 23 million.

Peninsular Malaysia enjoys a humid tropical climate throughout the year; the weather is warm and humid, and the temperature ranges from 20 °C to 32 °C [16]. Mountainous topography and complex land-sea interactions significantly influence the tropical climate. Intra-seasonal and inter-decadal fluctuations, such as the El Niño Southern Oscillation, the Indian Ocean Dipole, and the Madden Julian Oscillation, significantly influence Malaysia's inter-annual climate variability. April to May and July to August are the months with the highest average temperatures, and the lowest average temperatures occur from November to January.

In Peninsular Malaysia, the average monthly humidity falls between 70% and 90%, varying by location and month. Peninsular Malaysia experiences two rainy seasons throughout the year due to the effects of the northeast monsoon (NEM) season from November to February and the southwest monsoon (SWM) season from May to August [17]. In most areas, maximum rainfall typically occurs near the end of the year during the early NEM. The lowest monthly rainfall occurs in February, while the highest monthly rainfall occurs in December [18]. These monsoons are associated with the contribution of many regional pollutant sources that affect atmospheric parameters and the amount of pollutants transported to Malaysia [19].

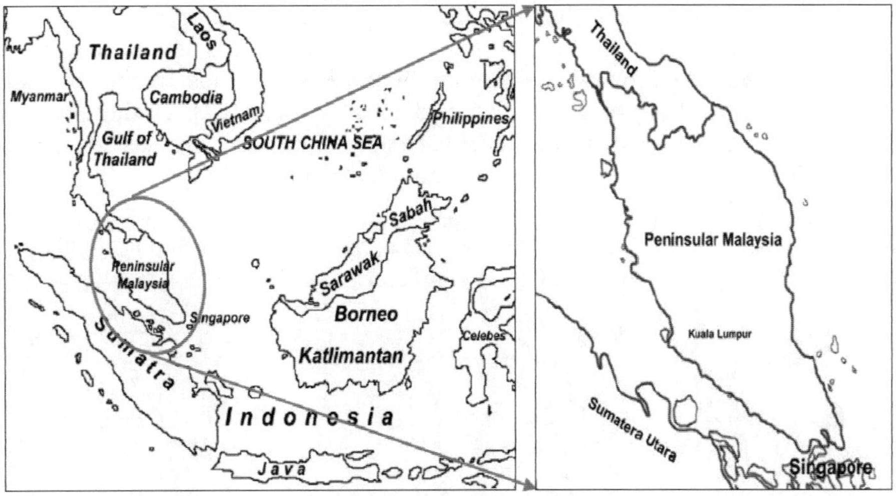

Figure 1. Geographical features of the study area.

4. Acquisition and Specification

Observations of the earth through remote sensing techniques require a variety of diverse platforms and instruments, ranging from hand-held, close-range spectrometers to images and instruments onboard satellites that orbit many thousands of kilometers above the earth [20]. Various types of sensor data have different advantages and disadvantages. Different types of sensors are suitable for certain studies and/or research purposes. Hence, understanding and updating information on all of these sensors on various types of available satellites is indispensable for remote-sensing applications.

From OMI, ultraviolet aerosol index used to detect the presence of ultraviolet absorbing aerosol such as soot and dust. Ultraviolet aerosol index is calculated based on the contrast method in the ultraviolet region, due to the small absorption of ozone in this region. Thus, it is the difference between model calculations and observations of non-absorbing and absorbing spectral radiance ratios. For OMI, aerosol index (AI) defined as [21]:

$$AI = 100[\log_{10}(I_{360}/I_{331})_{\text{measured}} - \log_{10}(I_{360}/I_{331})_{\text{calculated}}] \quad (1)$$

5. Methodology

This research has been carried out for twelve months' data from June to September 2013 till 2015. To better access and analyze the impacts of the forest fire over study area, the maps of aerosol index obtained from NASA-operated Giovanni. The daily time averaged maps of ultraviolet aerosol index (OMTO3dv003) were utilized for this purpose. In addition, the mean wind vector also obtained as a supportive data, in order to give an idea on how the transportation of wind can affects the value of aerosol index.

6. Data Analysis and Results

Fig. 2-5 show the time averaged of monthly coverage of ultraviolet aerosol index over study area for June-September, 2013-2015. The OMI/Aura resolution of reanalyses for the region covering Peninsular Malaysia used in this study is (1° log x 1° lat). The ultraviolet aerosol index spatial patterns were almost same for each month, except on June 2013 and September 2015. Average value of ultraviolet aerosol index for every single month is about 0.333-1 (except on June 2013 and September 2015). Whilst for June 2013 and September 2015, the average values for ultraviolet aerosol index range from 0.333-1.667. Aerosol index shows positive values for ultraviolet absorbing particles such as volcanic ash, desert dust, and smoke whereas negative values for non-absorbing particles such as sulphate aerosols and sea-salt particles [21].

Previously, other researcher had shown increasing aerosol index greater than 1.7 in biomass burning plumes during June, 2013 for the Pelalawan District [21]. Biomass burning plumes have carbonaceous aerosols, which exhibit different radiative properties, both absorbing and non-absorbing. Black carbons give the absorbing nature of carbonaceous aerosols [22]. During June 18th-July 1st 2013, an intense haze event occurred in Indonesia that brings significant impacts to other neighbouring countries, such as Singapore, Brunei, Malaysia, and southern Thailand [21].

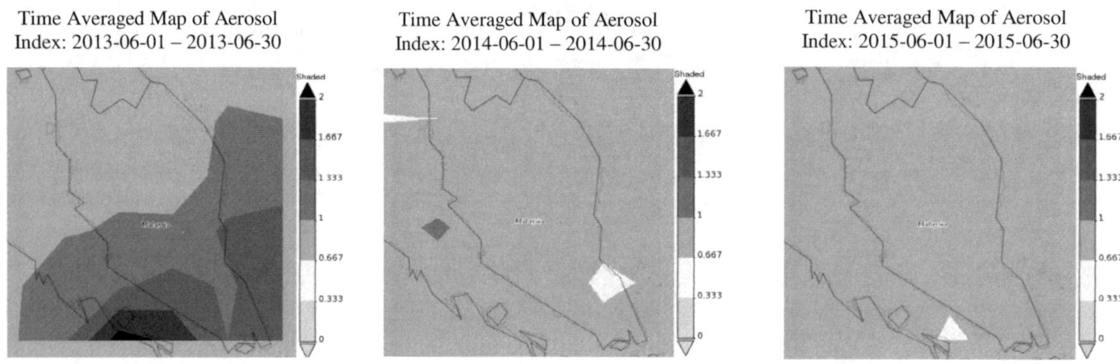

Figure 2. Spatial and temporal variations of monthly averaged aerosol index over Peninsular Malaysia for (June) 2013 to 2015.

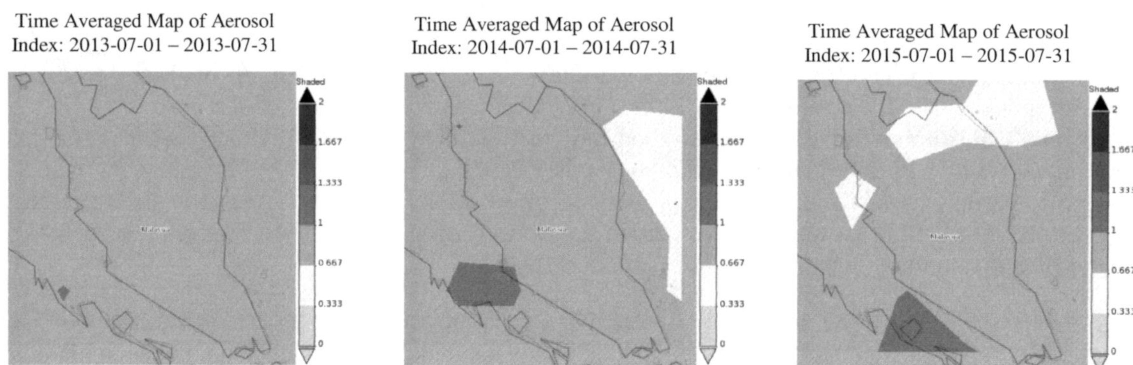

Figure 3. Spatial and temporal variations of monthly averaged aerosol index over Peninsular Malaysia for (July) 2013 to 2015

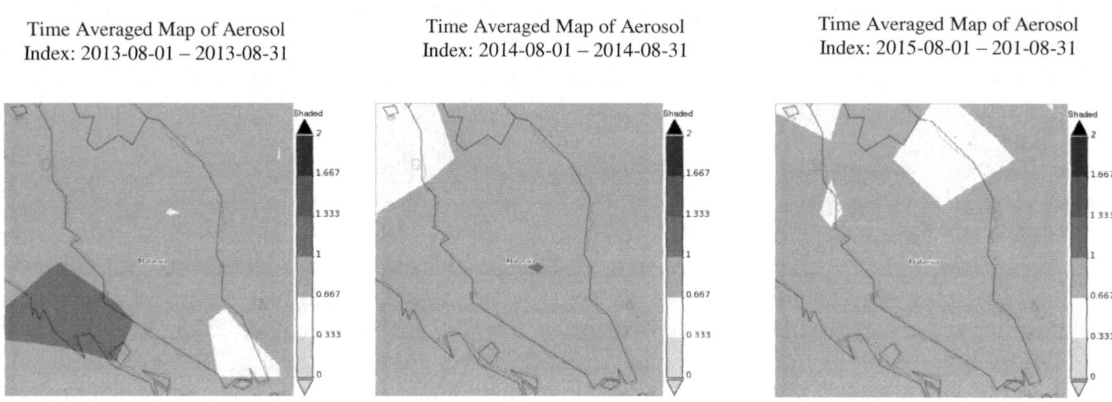

Figure 4. Spatial and temporal variations of monthly averaged aerosol index over Peninsular Malaysia for (August) 2013 to 2015

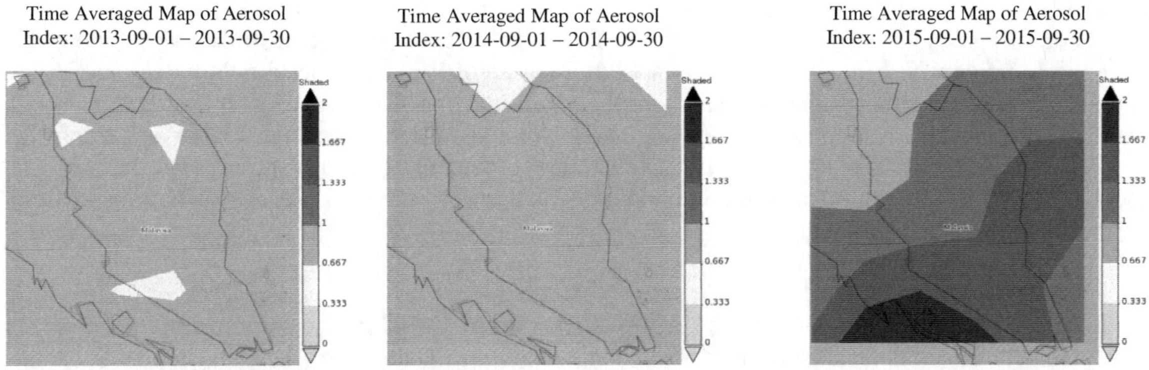

Figure 5. Spatial and temporal variations of monthly averaged aerosol index over Peninsular Malaysia for (September) 2013 to 2015

The mean synoptic chart of the wind vector (Fig. 6) was also developed for the region. The results showed that the monthly wind patterns for the June-September (Southwest monsoon) season followed the dominant wind direction as a result of the monsoon wind. When the Intertropical Convergence Zone (ITCZ) moves northward across continental Southeast Asia into the Northern Hemisphere, the Southwest monsoon brings marine air masses from the Indian Ocean to Peninsular Malaysia and experiences the Southwest monsoon season. Figure 6 shows vector components of mean wind systems at 1000 mb for the SWM or summer monsoon (June to September) from 2013 to 2015. The wind vector plots from the analysis data clearly define the ITCZ area.

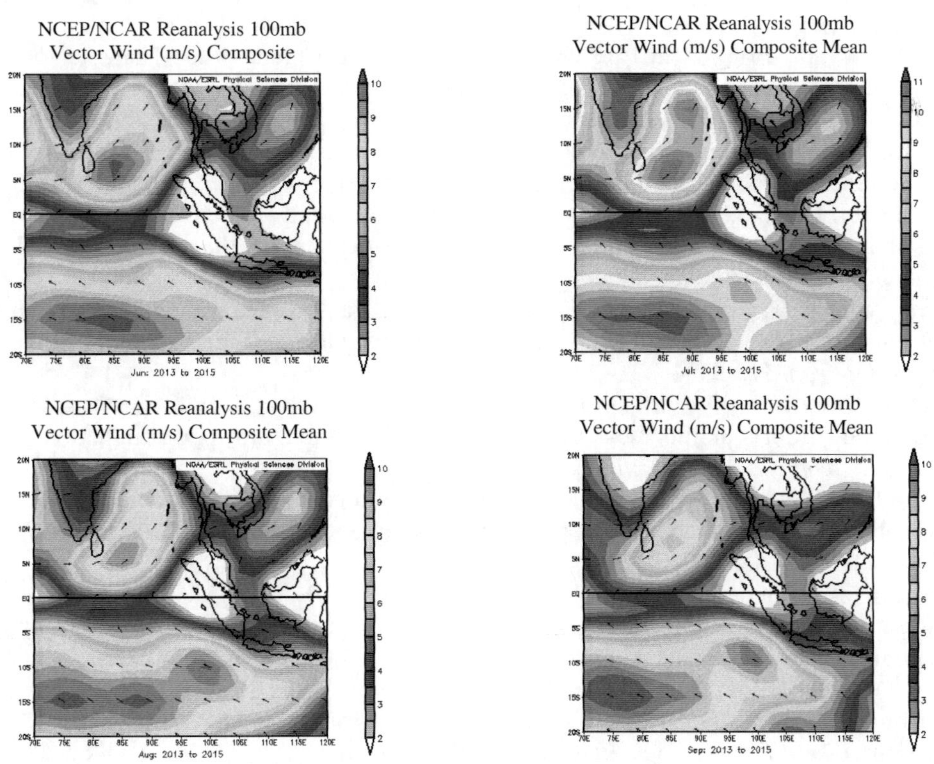

Figure 6. The mean wind vector at 1000 mb over Peninsular Malaysia region for (June – September) from 2013 to 2015

7. Conclusion

In this study, we present OMI's monthly observation data of aerosol index from June-September 2013-2015 in Peninsular Malaysia. The detail analyzes over study area successfully identify both the spatial and temporal variations, especially in examining aerosol index. The ultraviolet aerosol index spatial patterns were almost same for each month, except on June 2013 and September 2015. During June 2013 haze event in Indonesia, it released huge amount of aerosols in the region and bring the significant impacts on aerosol index over the study region. The result shows that the satellite measurements are efficiently measure the increase of aerosol index of the absorbing aerosol and transportation over different regions.

Acknowledgements

The authors gratefully acknowledge the National Aeronautics and Space Administration (NASA) Goddard Earth Sciences Data Information and Services Center (DISC) for the provision of the OMI data and images used in this paper, were acquired using GES-DISC interactive online visualization and analysis infrastructure (Giovanni). This project was carried out using the financial support from the short-term grant, Satellite Remote Sensing of Total Column Ozone over Peninsular Malaysia, account number: 304/PFIZIK/6313260.

References

[1] IPCC (Intergovernmental Panel in Climate Change), Climate Change 2013 The Physical Science Basis: Contribution of Working Group 1 to the Fifth Assessment Report of the Intergovernmental Panel on Climate Change

[2] Lin C Y, Zhao C, Liu X, Lin N H, and Chen W N 2014 Modelling of long-range transport of Southeast Asia biomass-burning aerosols to Taiwan and their radiative forcings over East Asia *Tellus* **66** 1

[3] Chuang M, Lee C, Chou C C, Engling G, Chang S, Chang S, Sheu G, Lin N, Sopajaree K, Chang Y and Hong G 2016 Aerosol transport from Chiang Mai, Thailand to Mt. Lulin, Taiwan – Implication of aerosol aging during long-range transport *Atmos. Environ.* **137** 101

[4] Matheson M A, Coakley J A and Tahnk W R 2005 Aerosol and cloud property relationships for summertime stratiform clouds in the northern Atlantic from advanced very high resolution radiometer observations *J. Geophys. Res.* **110** D24204. doi:10.1029/2005JD006165

[5] Tripathi S N, Dey S, Chandel A, Srivastava S, Singh R P, and Holben B N 2005 Comparison of MODIS and AERONET derived aerosol optical depth over the Ganga Basin, India *Ann. Geophys.* **23** 1093

[6] Kosmopoulos P G, Kaskaoutis D G, Nastos P T, and Kambezidis H D 2008 Seasonal variation of columnar aerosol optical properties over Athens, Greece, based on MODIS data *Remote Sens. Environ.* **112** 2354

[7] Torres O, Bhartia P K, Herman J R, Ahmad Z and Gleason J 1998 Derivation of aerosol properties from satellite measurements of backscattered ultraviolet radiation: theoretical basis *J. Geophys. Res.* **103** 17099

[8] Duncan B N, Martin R V, Staudt A C, Yevich R and Logan, J A 2003 Interannual and seasonal variability of biomass burning emissions constrained by satellite observations *J. Geophys. Res.* **108**(D2) 4100

[9] De Graaf M, Stammes P, Torres O and Koelemeijer R B A 2005 Absorbing aerosol index: Sensitivity analysis, application to GOME and comparison with TOMS *J. Geophys. Res.-Atmos.* **110** art. No. D01201

[10] Torres O, Bhartia P K, Herman J R, Sinyuk A, Ginoux P and Holben B N 2002 A long-term record of aerosol optical depth from TOMS observations and comparison to AERONET measurements *J. Atmos. Sci.* **59** 398

[11] Baddock M C, Bullard J E and Bryant R G 2009 Dust source identification using MODIS: A comparison of techniques applied to the Lake Eyre Basin, Australia *Remote Sens. Environ.* **113** 1511

[12] Hsu N C, Herman J R, Torres O, Holben B N, Tanre D, Eck T F, Smirnov A, Chatenet B, and Lavenu F 1999 Comparisons of the TOMS aerosol index with sun-photometer aerosol optical thickness: Results and applications *J. Geophys. Res.-Atmos.* **104** 6269

[13] Washington R, Todd M, Middleton N J and Goudie A S 2003 Dust-storm source areas determined by the total ozone monitoring spectrometer and surface observations *Ann. Assoc. Am. Geogr.* **93** 297

[14] Levelt P F, van der Oord G H J, Dobber M R, Mälkki A, Visser H, de Vries J, Stammes P, Lundell J O V and Saari H 2006a The Ozone Monitoring Instrument *IEEE T. Geosci. Remote* **44** 1093

[15] Levelt P F, Hilsenrath E, Leppelmeier G W, van den Oord G H J, Bhartia P K, Tamminen J, de Haan J F and Veefkind J P 2006b Science objectives of the ozone monitoring instrument *IEEE T. Geosci. Remote* **44**(5) 1199

[16] Omar D 2009 Urban form and sustainability of a hot humid city of Kuala Lumpur *J. Soc. Sci.* **8** 353

[17] Wong C L, Venneker R, Uhlenbrook S, Jamil A B M and Zhou Y 2009 Variability of rainfall in Peninsular *Hydro. Earth Sys. Sci.* **6** 5471

[18] Varikoden H, Samah A A and Babu C A Spatial and temporal characteristics of rain intensity in the Peninsular Malaysia using TRMM rain rate *J. Hydro.* **387** 312

[19] Pochanart P, Kreasuwun J, Sukasem P, Geeratithadaniyom W, Tabucanon M S, Hirokawa J, Kajiii Y and Akimoto H 2001 Tropical tropospheric ozone observed in Thailand *Atmos. Environ.* **35** 2657

[20] Tan K C, Lim H S and Mat Jafri M Z 2016 Prediction of column ozone concentrations using multiple regression analysis and principal component analysis techniques: A case study in peninsular Malaysia *Atmos. Pollut. Res.* **7** 533

[21] Vadrevu K P, Lasko K, Giglio L and Justice C 2014 Analysis of Southeast Asian pollution episode during June 2013 using satellite remote sensing datasets *Environ. Pollut.* **195** 245

[22] Andreae M O and Gelencsér A 2006 Black carbon or brown carbon? The nature of light-absorbing carbonaceous aerosol *Atmos. Chem. Phys.* **6** 3131

Measurement of Radon Concentration in Selected Houses in Ibadan, Nigeria

M R Usikalu[1], V Olatinwo[2], M Akpochafor[2], M A Aweda[2], G Giannini[3] and V Massimo[3]

[1]Department of Physics, Covenant University, Ota, Ogun State, Nigeria
[2]Department of Radiobiology and Radiotherapy, College of Medicine, Lagos, Nigeria
[3]Department of Physics, Istituto Nazionale di Fisica Nucleare (INFN), Trieste, Italy

Email: moji.usikalu@covenantuniversity.edu.ng

Abstract. Radon is a natural radioactive gas without colour or odour and tasteless. The World Health Organization (WHO) grouped radon as a human lung carcinogen. For this reason, there has been a lot of interest on the effects of radon exposure to people all over the world and Nigeria is no exception. The aim of this study is to investigate the radon concentration in selected houses in three local government areas of Ibadan. The indoor radon was measured in both mud and brick houses. Fifty houses were considered from the three Local government areas. A calibrated portable continuous radon monitor type (RAD7) manufactured by Durridge company was used for the measurement. A distance of 100 to 200 m was maintained between houses in all the locations. The living room was kept closed during the measurements. The mean radon concentration measured in Egbeda is 10.54 ±1.30 Bqm^{-3}; Lagelu is 16.90 ± 6.31 Bqm^{-3} and Ona-Ara is 17.95 ± 1.72 Bqm^{-3}. The mean value of the annual absorbed dose and annual effective dose for the locations in the three local government areas was 0.19 $mSvy^{-1}$and 0.48 $mSvy^{-1}$respectively. The radon concentration for location 10 in Ono-Ara local government exceeded the recommended limit. However, the overall average indoor radon concentration of the three local governments was found to be lower than the world average value of 40 Bqm^{-3}. Hence, there is need for proper awareness about the danger of radon accumulation in dwelling places.

1. Introduction

Radon is a natural radioactive gas that is derived from the decay of ^{238}U. It exists ubiquitously, small in degree to be noticed but can only be measured using a detector.^{238}Uranium is available in soil and rocks in limited quantity. It slowly disintegrates to other compounds like radium, which later turns to radon. Part of the radium migrates to the soil surface and finds their ways to the air, while other part remains at the sub-surface and penetrates the ground water [1, 2]. Admit the radioisotopes that add to general background radiation, radon posed major risk to human health. About 55% of the yearly radiation dose incurred by the general public can be traced to it [3]. Alpha particles are heavy particles and can only move a short distance. It cannot permeate through our skin but it can be inhaled and settled in the lung tissue. The energy deposition in lungs is highly localized and this will likely elevate the risk of lung cancer [4]. The International Agency for Research on Cancer (IARC) and the World Health Organization (WHO) categorized it as a human lung carcinogen [5, 6]. The ICRP suggests that all the countries of the world should carry out radon survey to find radon prone areas [7]. Due to dearth of information about the concentration and harmful effects [8, 9] of radon in Nigeria, lot of people are working and living in places unaware of the dangers posed to their lives. Concentration of radon gas in dwellings has been found to depend on geological condition, meteorological condition, construction materials and ventilation [10]. The buildings in the study areas were built from cement, concrete, mud, brick, granite etc, thus it is nearly impossible not to detect activity of radon in the areas. Also, there is a large variability in indoor concentration levels between different countries, even between areas in the same country, due to variations in the climatic parameters, geology of the subsoil, and building characteristics [11]. ^{222}Rn exposure levels in large buildings such as commercial buildings, schools and multiunit residential structures may be different from exposure patterns in

Content from this work may be used under the terms of the Creative Commons Attribution 3.0 licence. Any further distribution of this work must maintain attribution to the author(s) and the title of the work, journal citation and DOI.
Published under licence by IOP Publishing Ltd

detached houses due to difference in operation going on in the building [3]. There is no specific regulation or safe limit value in Nigeria concerning indoor radon levels in either homes or workplaces. Therefore, it is necessary to take measurement with the RAD 7 radon detector and other passive detector [12] to measure radon levels inside homes. Hence, the aim of this study is to investigate the concentration of radon in homes (both mud and brick houses) in Egbeda, Lagelu and Ona-Ara local government areas of Ibadan, Oyo State, Nigeria and the results will be used to estimate the annual effective dose for the homes dwellers.

2. Materials and Method

Ibadan is located on latitude 7.39° N and longitude 3.9° E, southwestern Nigeria. It has mud houses in rural areas and brick houses in both rural and urban area. The elevation of the town ranges from 150 m in the valley to 275 m above sea level. It has eleven local governments out of which three were selected for this study. They are: Egbeda, Lagelu and Ona-ara local Government Area. Egbeda headquarter has an area of 191 km^2 and a population of 281,573 according to the 2006 census. Iyana Offa is the headquarter town of Lagelu. It has an area of 338 km^2 with an estimated population of 147,957. Akanran is the headquarter town of Ona-Ara local government. It has an area of 290 km^2 and population of 265,059. The choice of selecting the three local governments is because of mud houses and brick houses situated at the locations. There are more mud houses in Egbeda and Ona-Ara local Government than Lagelu local Government. In this study, indoor radon was measured in both mud and brick houses. A calibrated portable continuous radon monitor type RAD7 manufactured by Durridge company was used. The gas to be measured enters in diffusion mode through a large surface of glass filter into an ionization chamber. The radon progeny products were prevented while only the gaseous ^{222}Rn can enter the ionization chamber. A personal computer (PC) Software (CAPTURE) allows a graphic presentation and calculation of the average concentration in the measured period. The measurements were carried out in fifty houses from the three locations: twenty were taken from Lagelu local Government (all brick houses), twenty five were taken from Ona- Ara local Government (fifteen mud houses and ten brick houses) and five mud houses were taken from Egbeda local Government. 100 to 200 m was maintained from one house to the other in all the locations. For all runs, the RAD7 was positioned on the table and on the floor in the center of the living room (Figure 1). The living room was kept closed during the measurements. The average radon concentration was calculated by the system in RAD7 for every cycle (20 min), displayed on LCD and a printer printed out the reading.

2.1 Dose Estimation

The indoor radon concentrations measured in each location of the study were used to estimate the annual effective doses and annual absorbed dose. The annual absorbed dose was calculated using equation 1.

$$D\,(mSvy^{-1}) = C_{Rn}\, D_c\, F\, H\, T \qquad (1)$$

where D is the annual absorbed dose ($mSvy^{-1}$); C_{Rn} is the radon concentration (Bqm^{-3}); D_c is the dose conversion factor ($9.0 \times 10^{-6} mSvh^{-1}$ per Bqm^{-3}); F is the equilibrium factor (0.4); T is hours in a year (8760); H is the occupancy factor (0.4). The annual effective dose was calculated by using equation 2.

$$E\,(mSvy^{-1}) = DW_R\, W_T \qquad (2)$$

where E is the annual effective dose; D is the annual absorbed dose ($mSvy^{-1}$); W_R is the radiation weighting factor for alpha particles (20) according to [1]; W_T is the tissue weighting factor for the lung (0.12) according to [1].

3. Results and Discussion

The radon concentration measured, estimated annual effective dose and annual absorbed in the study areas are presented in Table 1-3. The concentration of radon was found to vary from one location to the other. Figure 1 is the pictorial presentation of the radon concentration in the three local government areas. Table 1 showed the measured concentrations in Ona-Ara local government, the

IOP Conf. Series: Journal of Physics: Conf. Series **852** (2017) 012028 doi:10.1088/1742-6596/852/1/012028

concentration of the radon was found to have the highest value in location 10 (which is mud house) with the mean radon concentration of 42.54 ± 1.32 this was found to exceed the recommended limit of 40 Bqm^{-3}. This may be because the measurement was done in the night when all the windows were closed and this increased the radon concentration level as they were trapped due to decrease in the cross ventilation.

Table 1. Indoor Radon Concentration, annual absorbed dose and annual effective dose of Ona-Ara Local Government mud and brick houses

Location	Age of the building (yr)	Mean radon concentration (Bqm^{-3})	Annual absorbed dose ($mSvy^{-1}$)	Annual effective dose ($mSvy^{-1}$)
1	30+	38.09 ± 20.43	0.48	1.15
2	30+	8.76 ± 0.01	0.11	0.26
3	30+	8.74 ± 0.02	0.11	0.26
4	30+	11.03 ± 0.29	0.14	0.34
5	30+	18.28 ± 1.18	0.23	0.55
6	20+	14.58 ± 0.94	0.18	0.43
7	25+	17.00 ± 0.48	0.21	0.50
8	30+	8.76 ± 0.10	0.11	0.26
9	30+	37.79 ± 7.04	0.48	1.15
10	30+	42.54 ± 1.32	0.54	1.30
11	30+	33.62 ± 0.90	0.42	1.01
12	30+	8.67 ± 0.07	0.11	0.26
13	30+	10.77 ± 0.12	0.14	0.34
14	30+	9.81 ± 0.04	0.12	0.29
15	25+	8.78 ± 0.01	0.11	0.26
Mean (Mud)		**18.48 ± 0.58**	**0.23**	**0.56**
16	10+	18.23 ± 5.10	0.23	0.55
17	20+	8.78 ± 0.00	0.11	0.26
18	15+	15.71 ± 0.01	0.20	0.48
19	10+	16.81 ± 1.72	0.21	0.50
20	10+	20.66 ± 0.16	0.26	0.62
21	10+	17.64 ± 0.14	0.22	0.53
22	8+	13.27 ± 0.66	0.17	0.41
23	10+	17.12 ± 1.39	0.22	0.53
16	10+	18.23 ± 5.10	0.23	0.55
17	20+	8.78 ± 0.00	0.11	0.26
18	15+	15.71 ± 0.01	0.20	0.48
19	10+	16.81 ± 1.72	0.21	0.50
20	10+	20.66 ± 0.16	0.26	0.62
21	10+	17.64 ± 0.14	0.22	0.53
20	10+	20.66 ± 0.16	0.26	0.62
21	10+	17.64 ± 0.14	0.22	0.53
22	8+	13.27 ± 0.66	0.17	0.41
23	10+	17.12 ± 1.39	0.22	0.53
24	15+	19.78 ± 0.67	0.25	0.6
25	15+	23.66 ± 0.11	0.3	0.72
Mean (Brick)		**17.17± 1.72**	**0.22**	**0.52**

Note: Locations 1 - 15 are mud houses while 16 to 25 are brick houses.

Table 2. Indoor Radon Concentration, annual absorbed dose and annual effective dose of Egbeda Local Government Area mud houses

Location	Age of the building (yr)	Mean radon concentration (Bqm^{-3})	Annual absorbed dose ($mSvy^{-1}$)	Annual effective dose ($mSvy^{-1}$)
1	30+	13.46 ± 4.43	0.17	0.71
2	30+	9.27 ± 0.50	0.12	0.29
3	30+	8.76 ± 0.01	0.11	0.26
4	30+	11.39 ± 1.56	0.14	0.34
5	30+	9.80 ± 0.01	0.12	0.29
Average		**10.45 ± 1.30**	**0.13**	**0.38**

Table 3. Indoor Radon Concentration, annual absorbed dose and annual effective dose of Lagelu Local Government brick houses

Location	Mean radon concentration (Bqm^{-3})	Annual absorbed dose ($mSvy^{-1}$)	Annual effective dose ($mSvy^{-1}$)
1	24.73 ± 7.80	0.31	0.74
2	18.54 ± 9.30	0.23	0.55
3	3.67 ± 4.45	0.05	0.12
4	19.99 ± 8.96	0.25	0.60
5	17.35 ± 4.81	0.22	0.53
6	18.50 ± 8.86	0.23	0.55
7	12.42 ± 5.70	0.17	0.41
8	28.76 ± 1.44	0.36	0.86
9	4.79 ± 4.41	0.06	0.14
10	14.28 ±12.88	0.18	0.43
11	15.64 ± 7.98	0.20	0.48
12	19.29 ± 1.29	0.24	0.58
13	15.72 ± 0.01	0.20	0.48
14	13.27 ± 0.56	0.17	0.41
15	18.83 ± 9.02	0.24	0.58
16	18.87 ± 9.02	0.24	0.58
17	14.30 ±12.85	0.18	0.43
18	14.26 ± 5.95	0.18	0.43
19	17.95 ± 4.07	0.23	0.55
20	26.81 ± 6.92	0.34	0.82
Average	**16.90 ± 6.31**	**0.21**	**0.51**

The overall mean value of the indoor radon concentration in Ona-Ara was 17.95 Bqm^{-3}. In Ona-Ara it was found that radon concentration measured in some mud houses and the mean was higher when compared to those obtained in brick houses. Table 2 displayed measured concentration in Egbeda local Government. The highest radon concentration in Egbeda was obtained in location 1 with a mean value of 13.46 ± 4.43 Bqm^{-3} while the lowest value was obtained in location 3 with the average value of 8.76 ± 0.01 Bqm^{-3}. The highest value may be attributed to type of ventilation in location 1 the window is small and there is very low ventilation as a result of this, the radon concentration levels accumulate, while in location 3 there is a good ventilation in which the radon concentration mixed with the air to reduce the radon concentration level. The overall mean value of the indoor radon concentration in Egbeda was 10.45 Bqm^{-3}.

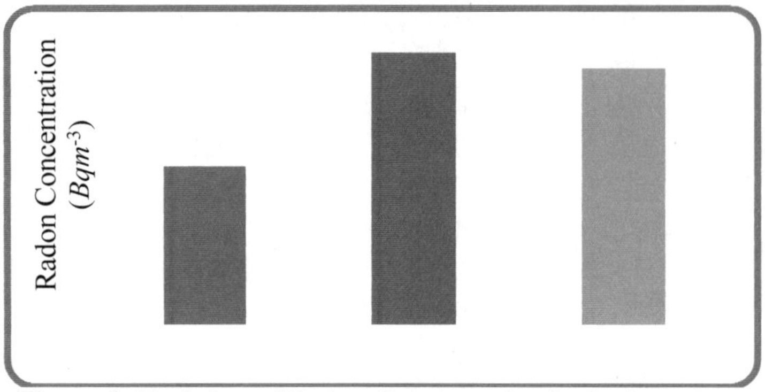

Figure 1. Variation of Radon Concentration in the Locations

Table 3 displayed the measured concentration in Lagelu local Government, the highest radon concentration was found in location 8 with the mean concentration of 28.76 ± 1.44 Bqm^{-3} while the lowest value was in location 3 with the average value of 3.67 ± 4.45 Bqm^{-3}. The high value of radon concentration in location 8 may be due to the age of the building, due to crack on the ground which is likely to be one of the causes of increase in the concentration. Also in location 3, the reading were taken at the first floor of the building, the elevation of the building can also contribute to the decrease in radon concentration level. The overall mean value of indoor radon concentration in Lagelu was 16.90 Bqm^{-3}.

The radon concentration obtained in Ono-Ara local government was quite high when compared to the other two local governments which may be as result of the materials used in the construction of the houses and the mode of the construction of the building as the sizes of the window in this location was small compared to other locations. Meanwhile, the overall average radon concentration from the three (3) local Government Areas in Ibadan is 15.10 ± 3.11 Bqm^{-3}, which is lower than the world average value of 40 Bqm^{-3} [4, 13]. The radon concentration obtained in these locations compared well with studies conducted by [14] where the concentration measured was 52 Bqm^{-3}, 28 Bqm^{-3} and 17Bqm^{-3} for Japan, Pakistan and Kuwait respectively. The correlation coefficient of Lagelu brick and Egbeda mud houses; Egbeda mud and Ona-Ara brick houses; Lagelu brick and Ona-Ara mud houses and Lagelu brick and Ona-Ara brick houses values are 0.45, 0.09, -2.26 and -0.28 respectively. The negative means the correlation coefficient between them are extremely low. The correlation coefficient value between the three locations is -5.53 and this means there is no correlation between them. The mean annual absorbed dose calculated was 0.13 $mSvy^{-1}$, 0.23 $mSvy^{-1}$ and 0.21 $mSvy^{-1}$for Egbeda, Ona-Ara and Lagelu respectively. Since the annual effective dose depends on mean indoor radon concentrations, the location with highest mean indoor radon concentration also has the highest value of the effective dose. The estimated mean annual effective was of 0.38 $mSvy^{-1}$, 0.54 $mSvy^{-1}$ and 0.51 $mSvy^{-1}$for Egbeda, Ona-Ara and Lagelu respectively. Locations 1, 9, 10 and 11 exceeded the recommended limit of 1 $mSvy^{-1}$. However, the average annual effective dose obtained for the three study areas is below 1 $mSvy^{-1}$ the world average value for normal background radiation.

The results obtained from RAD7detector showed that, the average radon concentration for the 3 local government areas surveyed in Ibadan was 15.10 ± 3.11 Bqm^{-3} which is lower when compared to the measurement reported by [9] conducted in some Secondary Schools in Oke-Ogun area of Oyo State with average radon concentration of 45 ± 27 Bqm^{-3}. This study revealed that proper ventilation in a building reduces the level of radon that will be recorded in a particular building, the more the ventilation the more the air enters into the room, the lower the radon level. The lifestyle of the individual living in this building is also important in some of these places people smoked cigarratte which increased the radon level. The floor of the room equally plays some role as observed here. The rooms in ground floor have higher value than those in first floor. The age of a building is equally important in the concentration obtained.

4. Conclusion

The first radon measurement in Lagelu, Egbeda and Ona-Ara Local Government Areas, Ibadan Southwestern Nigeria was conducted using RAD7. The mean value of indoor radon concentration of the three local governments was found to be 15.10 ± 3.1 Bqm^{-3} which is lower than the recommended limit of 40 Bqm^{-3}. The average annual effective dose estimated for this study was found to be 0.48 $mSvy^{-1}$which is lower than the recommended limit of 1 $mSvy^{-1}$. Therefore, the study provides preliminary data about indoor radon levels in the three local governments in Ibadan as a baseline study for a wider future national survey in the whole Ibadan city.

Acknowledgements

The authors are grateful to Covenant University and Professor O.O Oni for making RAD7 available for the measurement. We thank the houses that participated in the study.

References

[1] Bodansky D., Maurice A., Robkin and David R. S. 1989 Indoor Radon and its Hazards. *Univ. of Washington Press Seattle and London, second edition*

[2] Patrick RichonYann Klinger, Paul Tapponnier, Chen-Xia Li, Jerome Van Der Woerd, Fre´de´ric Perrier 2010 Measuring radon flux across active faults: Relevance of excavating and possibility of satellite discharges. **45** 211–218.

[3] George, A.C 2009 World history of radon research and measurement from the early 1900's to today *26th Avenue Bayside NY 11360 US,* pp.201-207.

[4] UNSCEAR (United Nations Scientific Committee on the Effects of Atomic Radiation) 2000 Sources and Effects of Ionizing Radiation.*Scientific Annexes*

[5] Darby, S., Hill, D., Auvinen, A., Barros-Dios, J.M., Baysson, H., Bochicchio, F. 2005 A radon in homes and risk of lung cancer: collaborative analysis of individual data from 13 European case-control studies. *British MedicalJournal*, **330**: 223-227.

[6] International Commission on Radiological Protection (ICRP) 2010. Lung Cancer Risk from Radon and Progeny and Statement on Radon. *Ann.* **40**(1).

[7] Ji Young Yoon, Jung-Dong Lee, So Won Joo and DaeRyong Kang 2016 Indoor radon exposure and lung cancer: a review of ecological studies. Annals of Occupational and Environmental Medicine **28**:15 DOI 10.1186/s40557-016-0098-z.

[8] Nasir T. and Shah M., 2012 Measurement of Annual Effective Doses of Radon from Drinking Water and Dwellings by CR-39 Track Detectors in Kulachi City of Pakistan. *Journal of Basic & AppliedSciences*.**8**: 528-536.

[9] Obed R.I., Ademola A.K., Vascotto M., Giannini G., 2011 Radon measurements by nuclear track detectors in secondary schools in Oke-Ogun region, Nigeria. *Journal of Environmental Radioactivity*, **102**: 1012-1017.

[10] Jung K.W, Won Y.J, Kong H.J, Oh C.M, Cho H, Lee D.H. 2015 Cancer statistics in Korea: incidence, mortality, survival and prevalence in 2012. Cancer Res Treat, **47**(2):127–41.

[11] Yu K.N., Young, E.C.M., Stokes, M.J. and Tang K.K. 1998 Radon properties in offices. *Health Phys*, ***75***: 159-164.

[12] Papastefanou C., 2008 Radioactive Aerosols. Aristotle University of Thessaloniki.Atomic & Nuclear Physics Laboratory. Greece.

[13] Ali K. Alsaedi, B.A. Almayahi, A.H. Alasadi 2013 Cement ^{222}Rn and ^{226}Ra concentration measurements in selected samples from different companies. Asian journal of natural & applied sciences, **2**: 2186 –8468.

[14] Shafik S. Shafik, Aamir A. Mohammed. 2013 Measurement of Radon and Uranium Concentrations and Background Gamma Rays at the University of Baghdad -Jadiriyah Site. International Journal of Application or Innovation in Engineering & Management **5**: 2319 – 4847.

Validation of Sea levels from coastal altimetry waveform retracking expert system: a case study around the Prince William Sound in Alaska

N H Idris[a,b], X Deng[c] and N H Idris[a]

[a]Department of Geoinformation, Faculty of Geoinformation and Real Estate, Universiti Teknologi Malaysia, Skudai 81310, Johor Darul Takzim, Malaysia.
[b]Geoscience and Digital Earth Centre, Research Institute for Sustainability and Environment, Universiti Teknologi Malaysia, 81310, Johor Bahru, Malaysia.
[c]School of Engineering, The University of Newcastle, University Drive, Callaghan 2308, New South Wales, Australia.

[*]E-mail: nurulhazrina@utm.my

Abstract This paper presents the validation of Coastal Altimetry Waveform Retracking Expert System (CAWRES), a novel method to optimize the Jason satellite altimetric sea levels from multiple retracking solutions. The validation is conducted over the region of Prince William Sound in Alaska, USA, where altimetric waveforms are perturbed by emerged land and sea states. Validation is performed in twofold. First, comparison with existing retrackers (i.e. MLE4 and Ice) from the Sensor Geophysical Data Records (SGDR), and second, comparison with in-situ tide gauge data. From the first validation assessment, in general, CAWRES outperforms the MLE4 and Ice retrackers. In 4 out of 6 cases, the value of improvement percentage (standard deviation of difference) is higher (lower) than those of the SGDR retrackers. CAWRES also presents the best performance in producing valid observations, and has the lowest noise when compared to the SGDR retrackers. From the second assessment with tide gauge, CAWRES retracked sea level anomalies (SLAs) are consistent with those of the tide gauge. The accuracy of CAWRES retracked SLAs is slightly better than those of the MLE4. However, the performance of Ice retracker is better than those of CAWRES and MLE4, suggesting the empirical-based retracker is more effective. The results demonstrate that the CAWRES would have potential to be applied to coastal regions elsewhere.

1. Introduction

An accurate coastal sea level measurement has been a great demand by the scientific community for various applications. For examples, the accuracy of 1 mm/year is desired for measuring sea level rises and of 10 cm is required for detecting eddies in the East Australian Current system. Nowadays, local coastal forecast systems such as BLUElink Ocean Model Analysis and Prediction System [OceanMAPSv1.0b; 1] are already exploiting the coastal altimetry data for operational applications. The parameter of sea level anomaly (SLA) is derived quantitatively from satellite altimetry observations. However, within 200 m isobaths, the altimetry SLAs are not assimilated in the system due to land contamination in coastal altimetry signals, and lower accuracy of geophysical and atmospheric corrections [1].

Issues regarding the altimetry data over coastal have becomes an overwhelm discussion by the scientists worldwide. They realized the needs of specific treatments for recovering the altimetry data over coastal, so that the accuracy of the data is as good as the open ocean. Specific treatment is needed to correct the corrupted altimetry signals due to the land contamination [cf. 2, 3]. This can be performed by 'retracking' waveform that applies the coastal retrackers [e.g. 4, 5-10] to correct the estimation of geophysical parameters (i.e. SLA, significant wave height and wind speed). Waveform retracking has been conducted over global oceans to improve the accuracy of the altimetry measurements. When retrieving SLAs near coastal, attention is also needed when applying the corrections of sea states (e.g. inverse barometer and sea state bias) and of atmospheres (e.g. wet and dry tropospheric, and ionospheric) because they are less accurate due to high variability of the ocean

Content from this work may be used under the terms of the Creative Commons Attribution 3.0 licence. Any further distribution of this work must maintain attribution to the author(s) and the title of the work, journal citation and DOI.
Published under licence by IOP Publishing Ltd

signals and land contamination [11]. The state-of-the-art about retracking and geophysical corrections can be found in the book *Coastal Altimetry* by Vignudelli [12].

Innovative ideas is continuously developing to provide alternative approach to better accuracy of altimetry data from corrupted waveforms near coastal. These include modification of the standard Brown model [e.g. 13] to fit the noisy coastal waveforms, retracking of multiple waveforms [e.g. 7, 9, 14] to exploit the inter-waveform properties, retracking of reduced gate of waveforms [e.g. 8, 15, 16, 17] to exclude the non-ocean signals, and retracking using multiple retrackers based on the expert system [e.g. 18, 19-21] to select the optimal retracker.

The Coastal Altimetry Waveform Retracking Expert System (CAWRES) developed by Idris and Deng [6] is designed to optimize the estimation of SLAs by selecting the optimal retracker via a fuzzy expert system, and to provide a seamless transition from the open ocean to the coasts (or vice versa), when switching retrackers, via a neural network approach. Through the system, the fuzzy expert system is exploited to improve the selection method of the retrackers by integrating information about the physical features of waveforms and the statistical features of retracking results. Validation of the CAWRES retracked SLAs over the region of the Great Barrier Reef in Australia has been reported in Idris [2]. It shows that the SLAs from CAWRES generally outperform those from conventional methods. The retracked SLAs have satisfactory agreement with in-situ tide gauges.

In this paper, the validation of CAWRES is performed over the region of the Prince William Sound in Alaska, USA. The region is surrounded by steep and glaciated mountains, rough coastal sea states due to the notoriously stormy seas, and a complex hydrological system of freshwater from rivers and glaciers. The validation protocol is twofold: 1) comparison with existing retrackers from SGDR product and geoidal heights, and 2) comparison with tide gauge data. The comparison with tide gauge evidences to finding the accuracy and precision of the SLAs, while with the geoidal height, only the precision can be estimated.

2. Study Area and Data

The CAWRES has been applied to waveforms in the region of Prince William Sound (Figure 1). Waveforms in this area are highly perturbed by the mountainous terrain, varying ocean depth, complicated coastal geometry, and rough coastal sea states due to the notoriously stormy seas, and a complex hydrological system of freshwater from rivers and glaciers.

The Prince William Sound is a small (~100 km^2) semi-closed sea located in the northeast corner of the Gulf of Alaska. It is connected to the Northern Gulf of Alaska via two major passages: Hinchinbrook Entrance and Montague Strait, and has complex bottom topography and coastal orography. The maximum water depth, which is not shown, is about 800 m. To some extent, Prince William Sound also has the character of an estuary due to the strong runoff from snowmelt along the shoreline, especially in late summer/early autumn. The coastline is convoluted with many islands and fjords, several of which contain tidewater glaciers. It is surrounded by the Chugach Mountains, which reach a height of 4,300 m, and contain the most extensive system of valley glaciers in North America. With a shoreline length of about 6,900 km and a tidal range of 6-8 m, Prince William Sound has an enormously varied shoreline habitat of reefs, rocks, mud flats, eelgrass beds, wetlands, and cobble beaches [22]. Thus, altimetry data in this region can capture diverse patterns because of the mountainous terrain, the notoriously stormy seas, and a complex hydrological system of freshwater from rivers and glaciers.

The data used are Jason-1 and Jason-2/OSTM during the tandem mission. The Ku-band 20-Hz 104-sample waveform data are from January 2009 to December 2011, which corresponds to cycles 262 to 370 of Jason-1, and cycles 19 to 143 of Jason-2. Waveforms along one ascending pass (123) and two descending passes (28, and 104) of Jason-1, and two ascending passes (47, and 123) and one descending pass (28) of Jason-2 are investigated in the area of Prince William Sound (Figure 1).

In producing SLAs, environmental and geophysical corrections from SGDR products, and the DTU10 mean sea surface are applied to the altimeter range. The wet and dry tropospheric corrections are from the European Centre for Medium-Range Weather Forecasts (ECMWF) numerical prediction models, and ionospheric correction is from the General Ionospheric Model Map. The more accurate instrumental radiometer wet correction and dual frequency ionospheric correction are not used because of coastal contamination effects. The ocean tidal signals are removed using a pointwise tide modelling (Idris et al. 2014) rather than the global ocean tide model such as FES2004 and GOT4.8

because it better resolves the tidal signals in the study region. The sea state bias correction is not applied because it is not appropriate for waveforms near coasts. It is applied neither to coastal data, to avoid additional error, nor to open ocean data, to keep consistency of datasets in the area. The corrections have been interpolated from 1 Hz to 20 Hz.

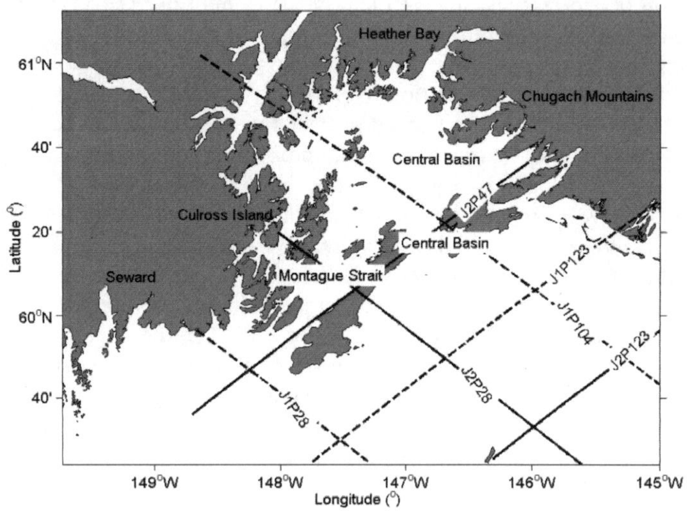

Figure 1. Jason-1 and Jason-2/OSTM satellite passes in Prince William Sound, Alaska. The red star shows the tide gauge station.

The quality and consistency of sea levels derived from the CAWRES is compared with in-situ measurements of geoid height and tide gauges. The geoidal height is based on the Earth Gravitational Model 2008 (EGM2008) with 2.5 minute resolution (http://earth-info.nga.mil/GandG/wgs84/gravitymod/egm2008/egm08wgs84.html). The tide gauge data are from the University of Hawaii Sea Level Centre (http://ilikai.soest.hawaii.edu/uhslc). It is the hourly sea level data from Seward (60.072°N, 149.3°W) stations (see Figure 1). The assessment with tide gauge merits to finding the accuracy and precision of the SLA estimates, while with the geoidal heights, only the precision can be computed.

3. Validation of Retracked Sea Level from CAWRES

The quality of retracked sea levels from CAWRES is assessed by comparing the results with other existing retrackers available from the SGDR product (Section 3.1). The accuracy of the results are also compared with tide gauge data (Section 3.2).

3.1 Comparison with Existing Retrackers in the SGDR Product

Comparison between the CAWRES retracked sea surface heights (SSHs) with existing retrackers from the SGDR product is performed. The retrackers from SGDR product are the MLE4 and Ice retrackers. The parameter of SSHs is used to enable comparison with geoid height. It is realized that both datasets are relative to difference ellipsoid. However, conversion of reference ellipsoid is not performed because the impact on the analysis is assumed to be insignificant, as the STD and IMP are computed over a short SSH profiles (~10 km from the coastline) or small areas.

Three assessments are carried out: 1) the standard deviation of difference (STD) between retracked sea level with respect to geoid height and the improvement of percentage (IMP) of the sea levels; 2) the STD between 20 Hz retracked SLAs and its average 1 Hz SSHs, hereafter called 'the noise STD'; and 3) the percentage of reasonable sea levels after removing outliers with predefined editorial criterion.

The IMP can be computed using Equation 1 [23],

$$IMP = \frac{\sigma_{raw} - \sigma_{retracked}}{\sigma_{raw}} \times 100 \quad (1)$$

where σ_{raw} and $\sigma_{retracked}$ are the standard deviation of the difference between raw SSHs and geoid heights, and retracked SSHs (e.g. CAWRES or Ice) and geoid heights, respectively. The raw SSHs are the SSHs retracked by the MLE-4 retracker because it is the standard retracking solution for the ocean.

Table 1 and Table 2 show that CAWRES improves the precision of the MLE4 SSHs over the region in 4 out of 6 passes. The IMPs are improved up to 33% for Jason-1 and 54% for Jason-2. However, deterioration in precision is found in pass 104 of Jason-1 and pass 47 of Jason-2, where the CAWRES retracked SSHs have negative IMPs (-7% and -22.2%, respectively). This suggests that the MLE4 retracked SSHs have a better precision than those of the CAWRES in both passes. Although deteriorations are recorded in both passes, results in Figure 2 show that the CAWRES recovers more (up to 70%) data than the MLE4 retracker. The percentage near the coast is extremely large (up to 70%), suggesting that the CAWRES retrieves more data than those of the MLE4 retracker. This also indicates that the CAWRES has extended the SLAs much closer to the coastline than those of the MLE4 retracker.

Table 1. STDs and IMPs between the Jason-1 retracked SSHs and the geoid heights.

Pass	MLE4 retracker	CAWRES		No. of points
	STD (cm)	STD (cm)	IMP (%)	
28	34	31	8.8	25,020
104	43	46	-7	53,362
123	46	31	32.6	62,262

Table 2. STDs and IMPs* between the Jason-2 retracked SSHs and the geoid heights.

Pass	MLE4 retracker	CAWRES		Ice retracker		No. of points
	STD (cm)	STD (cm)	IMP (%)	STD (cm)	IMP (%)	
28	65	30	**53.8**	54	16.9	54,487
47	45	55	-22.2	69	-53.3	36,120
123	12	11	**8.3**	15	-25	38,019

*The highest IMPs are indicated by bold numbers.

When compared to SGDR Ice retracker (Table 2), the IMPs (STDs) of CAWRES are always higher (lower), suggesting that CAWRES retracked SSHs are less noisy than those of Ice retracker. Along passes 47 and 123, Ice retracked SSHs show worse performance when compared to MLE4 retracked SSHs with -53.3% and -25% of IMPs.

Results in Table 3 show the number of reasonable SSHs and noise STDs of retrackers over the Prince William Sound region. When compare with SGDR retrackers, CAWRES presents the best performance in producing valid observations up to 84% and 85% for Jason-1 and Jason-2, respectively. The reasonable SSHs retrieved by the MLE4 are much less as ~54% and 71% of total dataset, respectively, while by the Ice retracker are ~72% for Jason-2. The same results can also be observed from Figure 2. This suggests that CAWRES is an effective method to enhance the spatial resolution of coastal altimetry data.

Beside the increment in the validity of the measurements, CAWRES also has the lowest noise STD when compared to the SGDR retrackers (Table 3). For Jason-1, CAWRES produces slightly smaller noise STD (46 cm) than the MLE4 retracker (48 cm), while for Jason-2, their values are similar (62 cm). The noise STD of Ice retracker is slightly higher (63 cm) than those of the CAWRES and MLE4 retrackers. However, it is realized that the difference in the noise STD of all retrackers is insignificant. Table 3 tells that, in general, CAWRES can spatially recover more SSHs than MLE4 (see also Figure 2) and Ice retrackers, and meanwhile achieves similar data quality to MLE4 and Ice other retrackers. This, therefore, confirms that CAWRES outperforms the SGDR retrackers over the tested region.

Table 3. Number of reasonable SSHs and noise STD of retrackers over the Prince William Sound region

Jason-1 missions			
Retracker	**Total number of SSHs**	**Numbers of reasonable SSHs (%)**	**Noise STD (cm)**
CAWRES	139,380	117,165 (84.06)	46
MLE4	139,380	75,145 (53.91)	48
Jason-2 missions			
CAWRES	135,680	115,221 (84.92)	62
MLE4	135,680	96,562 (71.17)	62
Ice1	135,680	97,668 (71.98)	63

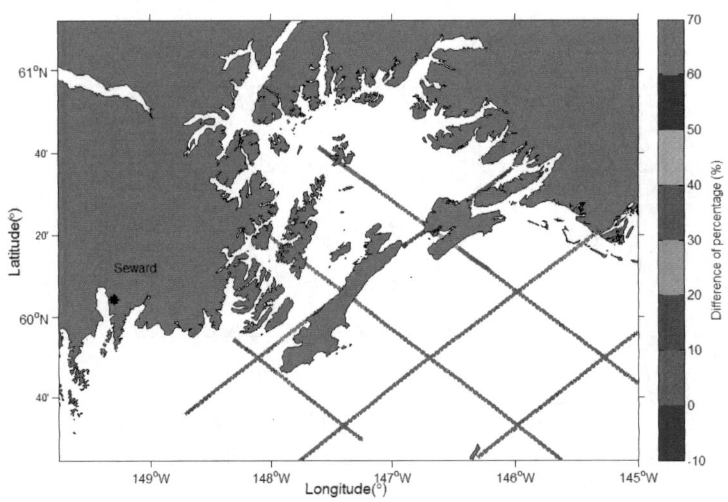

Figure 2. Spatial plot of the difference of SLA data recovered (in percentage) by the CAWRES and the MLE4 retracker. A positive value means the CAWRES retrieves more data than the MLE4 retracker

3.2 Comparison with Tide Gauge Data

In comparison with tide-gauge data, the performance of CAWRES and SGDR retrackers is assessed by computing the mean values of temporal correlation and root mean square (RMS) error (Table 4) in the area near (<100 km) the Seward station. They are computed from Jason-1 (J1), Jason-2 (J2) and the combination of Jason-1 and Jason-2 (J1+J2) data. These values are not computed for the Jason-1 Ice retracker because it is unavailable from the SGDR products. Figure 3 shows the spatial plots of temporal correlation and RMS error of the retrackers over the region.

Results from Table 4 and Figure 3 show that there is a satisfactory agreement between the retracked SLAs from the CAWRES and the tide gauge. The mean value of the temporal correlations (RMS errors) of the CAWRES is slightly higher (lower) than that of the MLE4 retracker, suggesting that the SLAs from CAWRES are more accurate than those of the MLE4 retracker.

When comparing with the Ice retracker, it is seen that the temporal correlation (0.94) of Ice retracker is better than those of the CAWRES (0.86) and MLE4 retracker (0.85). It is noted that no significant difference in their RMS errors (~7 cm). This is an astonishing result as the Ice retracker is intended for hydrology and cryosphere applications, and not for Open Ocean or coastal applications. It, however, outperforms the MLE4 of an ocean retracker. A similar result has been found in the Arctic ocean by Jain et al. (2015), where the performance of the Ice retracker is found superior than the physical-based (i.e. MLE4) retracker in comparison with the tide gauge data. In the cryospheric areas, like in the Arctic and the Prince William Sound, the physical-based retrackers may be ineffective because the ocean usually have a significant presence of sea ice, thus affecting the accuracy of estimated SLAs.

From Table 4, it is realized that the accuracy of Jason-2 retracked data is better than that of Jason-1 data. This is partly due to a stable mispointing angle of Jason-2, while mispointing angle of Jason-1 is severe. That is, the Jason-1 slope of waveform trailing edge departed from the value expected for a nadir-pointing instrument [24], thus affecting the accuracy of the estimations.

The plot of the SLA time series from different retracking methods and tide gauge at nine locations around Prince William Sound is shown in Figure 4 (refer to Figure 3a for the distribution of the locations), along with their correlations and RMS errors. The retracked SLAs from the Ice retracker are not shown in Figure 4a to Figure 4e because they are unavailable from the Jason-1 SGDR products.

Based on results in Figure 4, in general, the performance of both CAWRES and MLE4 retrackers are almost equal with slight differences in the value of correlations and RMS errors. It is seen that in some cases (Figure 4a, b, and e), the correlation and RMS error of the MLE4 retracker are slightly better than those of the CAWRES, and vice versa (Figure 4c, g and h). However, the accuracy of Ice retracker is always better than those of the other two retrackers (Figure 4g-i), proving that it is the optimal solutions for retrieving accurate SLAs over the tested region.

The results shows that the CAWRES provides retracked SLAs that agree with tide gauge data at almost all locations with correlations $\geq$0.57 and RMS errors $\leq$23 cm, except for locations d and f (Figure 4d and Figure 4f), where the CAWRES shows disagreements. In such complicated locations, the CAWRES as well as the SGDR retrackers fail to produce accurate SLAs, mainly due to the extreme corruption of waveforms and data corrections (e.g., tides, wet and dry tropospheric delays). The SLAs retracked by the MLE4 are totally missing in Figure 4d, indicating that the MLE4 cannot fit waveforms at that coastal location. The SLAs near the coast retracked by the MLE4 (e.g., Figure 4c) and Ice retrackers (e.g., Figure 4f) are sometimes discarded as the waveforms are highly perturbed by emerged lands and coastal sea states.

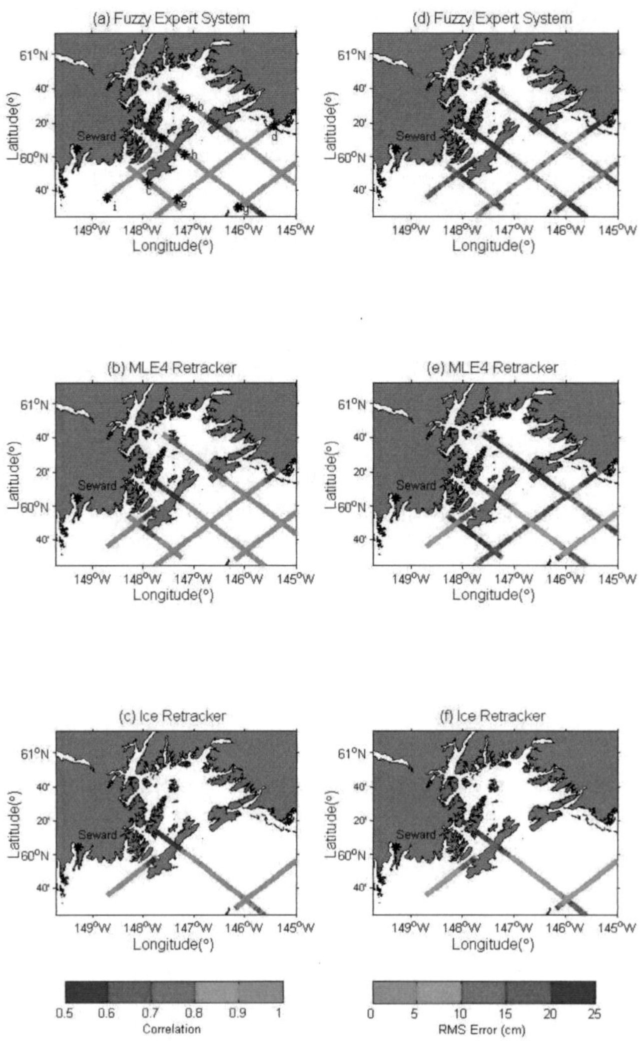

Figure 3. Spatial plots of the temporal correlation (left panels) and RMS error (right panels) of the sea level anomalies from the CAWRES and the Seward tide gauge station. The SLA time series at locations a-i (black dots in a) is shown in Figure 4.

Table 4. Mean of temporal correlation and RMS error in the area near (<100 km) the Seward tide gauge station

Mean temporal correlation						
CAWRES			MLE4			Ice
J1	J2	J1+J2	J1	J2	J1+J2	J2
0.84	0.86	0.87	0.60	0.85	0.67	0.94
Mean RMS error (cm)						
7	7	12	30	7	24	7

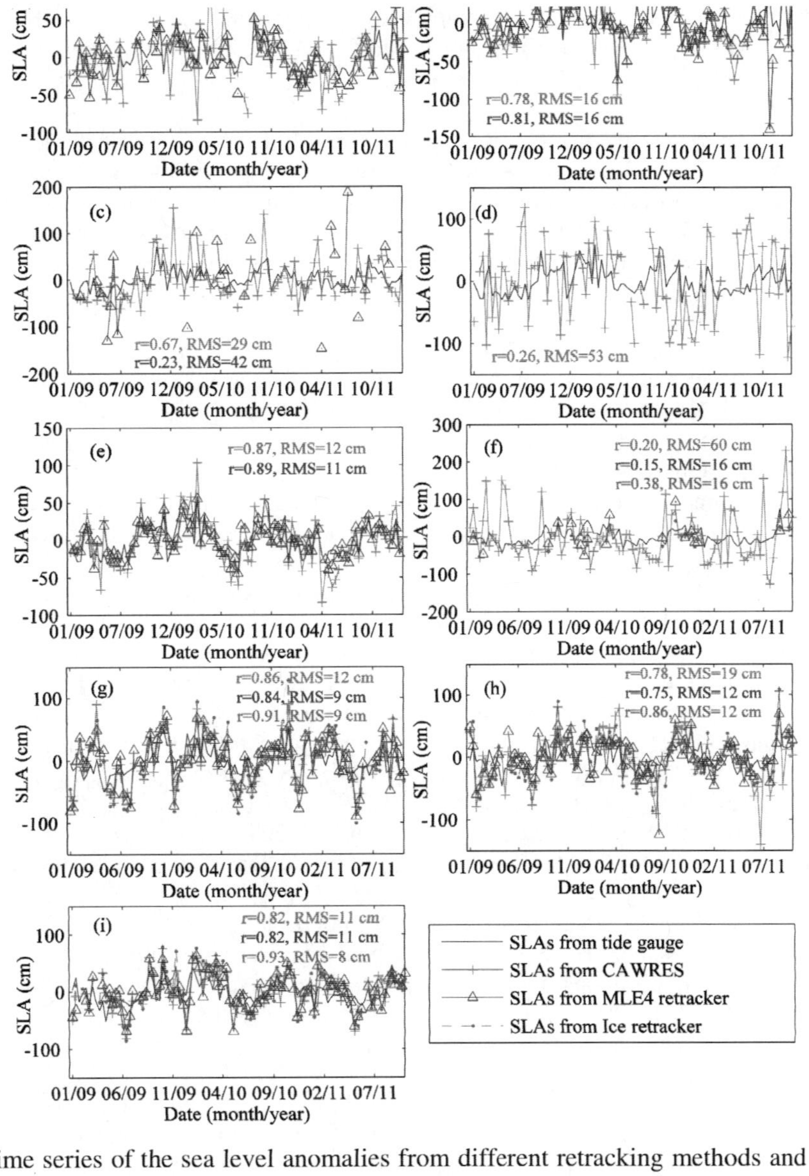

Figure 4. Time series of the sea level anomalies from different retracking methods and tide gauge at nine locations around the Prince William Sound (see Figure 3a). The retracked SLAs shown in a-e are from the Jason-1 mission, while the retracked SLAs shown in f-i are from the Jason-2 mission. The retracked SLAs from the Ice retracker are not shown in a-e because they are unavailable from the Jason-1 SGDR product. The retracked SLAs from the MLE4 retracker are unavailable in d. The temporal correlation (r) and RMS error of for each retrackers are also shown in each subplots.

4 Conclusions

The results over the tested regions emphasize that the retracked sea levels from the CAWRES are consistent with those of the geoid height and tide gauges. It reduces the STD of the MLE4 retracked sea levels by up to 15 cm for Jason-1 and 35 cm for Jason-2. It recovers 12-30% more data than the MLE4 and Ice retrackers over the regions. Although CAWRES improves the spatial resolution of the coastal altimetry data over those regions, sometimes, the quality of the retrieved SLAs is poor. Therefore, the CAWRES products should be used with caution. Analysis with tide gauge indicates that the retracked SLAs from Ice retracker outperform the CAWRES and MLE4 retracker, suggesting that the empirical-based retracker is more effective when over the ocean with significant presence of sea ice. With respect to the standard MLE4 retracker, CAWRES produces more accurate and precise SLAs. Current research is undertaken to compare the CAWRES SLAs with the Regional Ocean Model System (ROMS) over the Prince William Sound to identify their consistency.

Acknowledgements

The research is supported Fundamental Research Grant Ministry of Higher Education Malaysia (Grant No. 4F776). We would like to acknowledge the Archiving, Validating, and Interpretation of Satellite Oceanography (AVISO) data team for kindly providing Jason-2 data, and the University of Hawaii for providing tide gauge data. Thanks to Dr. Wang Xiaochun (Jet Propulsion Laboratory, California Institute of Technology, USA) for his helps and suggestions.

References

[1] Taylor A, Brassington G B and Nader J 2010 *Assessment of BLUElink OceanMAPSv1.0b againsts coastal tide gauges*, in *CAWCR Technical Report No. 030*

[2] Idris N H, Deng X and Andersen O B 2014 The importance of coastal altimetry retracking and detiding: a case study around the Great Barrier Reef, Australia *Int. Journal of Remote Sensing* **35**(5): p 1729-1740

[3] Scozzari A, Gomez-Enri J, Vignudelli S and Soldovieri F 2012 Understanding target-like signals in coastal altimetry: experimentation of a tomographic imaging technique *Geophysical Reseach Letters* **39** L02602

[4] Guo J Y, Gao Y G, Hwang C W and Sun JL 2010. A multi-subwaveform parametric retracker of the radar satellite altimetric waveform and recovery of gravity anomalies over coastal oceans *Sci China Earth Sci* **53** 610

[5] Halimi A, Mailhes C, Tourneret J Y and Thibaut P 2011 A new model for peaky altimetric waveforms *in Geoscience and Remote Sensing Symp. IGARSS IEEE International* p 2825

[6] Idris NH and Deng X 2012 Coastal waveform retracking for sea surface height estimates: a fuzzy expert system approach in *20 Years of Progress in Radar Altimetry Esa Publication*: Venice Italy

[7] Kuo C Y, Kao H C, Lee H, Cheng K C and Lin L C 2012 Assessment of Radar Waveform Retracked Jason-2 Altimetry Sea Surface Heights Near Taiwan Coastal Ocean *Marine Geodesy* **35**(2) 188

[8] Lee H, Shum C K, Yi Y, Braun A and Kuo C Y 2008 Laurentia crustal motion observed using TOPEX/POSEIDON radar altimetry over land *Journal of Geodynamics* **46**(3-5) 182-193

[9] Quartly G 2011 Hyperbolic pretracker: a means to filter waveform data in *OSTST meeting* San Diego, USA

[10] Thibaut P, Severini J, Mailhes C, Tourneret J Y, Bronner E and Picot N 2010 A multi-peak model for peaky altimetric waveforms in *4th Coastal Altimetry Workshop* Porto, Portugal

[11] Obligis E, Desportes C, Eymard L, Fernandes J, Lazaro C and Nunes AL 2011 Tropospheric corrections for coastal altimetry in *Coastal altimetry* ed S Vignudelli et al. (Springer: Berlin) chapter 6 pp 147

[12] Vignudelli S, Kostianoy AG, Cipollini P and Benveniste J 2011 Coastal altimetry (Berlin: Springer)

[13] Thibaut P, Poission J C, Halimi A, Mailhes C, Tourneret J Y, Boy F and Picot N 2011 A review of CLS retracking solutions for coastal altimeter workshop, in *5th Coastal Altimetry Workshop* San Diego USA

[14] Thibaut P, Poission J C, Ollivier A, Bronner E and Picot N 2009 Singular value decomposition applied on altimeter waveforms in *OSTST meeting* Seattle USA

[15] AVISO 2010 Coastal and Hydrology Altimetry produc*t (PISTACH) handbook* p58 (Toulouse France)

[16] Bao L, Lu Y and Wang Y 2009 Improved retracking algorithm for oceanic altimeter waveforms Progress *Natural Science* **19**(2) 195

[17] Passaro M, Cipollini P, Vignudelli S, Quartly G D and Snaith H M 2014 ALES: A multi-mission adaptive subwaveform retracker for coastal and open ocean altimetry *Remote Sensing of Environment* **145** 173

[18] Berry PAM 2000 Topography from land radar altimeter data: Possibilities and restrictions *Physics and Chemistry of the Earth Part A: Solid Earth and Geodesy* **25**(1) 81

[19] Deng X and Featherstone WE 2006 A coastal retracking system for satellite radar altimeter waveforms: Application to ERS-2 around Australia *Journal of Geophysical Research* **111** C06012

[20] Freeman J and Berry PAM 2006 A new approach to retracking ocean and coastal zone multi-mission altimetry in *15 Years of Progress in Radar Altimetry* Venice Italy

[21] Idris NH and Deng X 2013 An iterative coastal altimetry retracking strategy based on fuzzy expert system for improving sea surface height estimates in *2013 IEEE International Geoscience and Remote Sensing Symposium IGARSS* IEEE International p 2954

[22] Schoch GC, Chao Y, Colas F, Farrara J, McCammon M, Olsson P, and Singhal G 2011 An ocean observing and prediction experiment in Prince William Sound, Alaska *Bulletin American Meteorological Society* **92**(8) p 997

[23] Hwang C, Guo J, Deng X, Hsu HY and Liu Y 2006 Gravity anomalies in coastal waters from retracked Geosat/GM altimetry: comparison with shipborne and airborne gravity data *Journal of Geodesy* **80**(4) 204

[24] Amarouche L, Thibaut P, Zanife O Z, Dumont J P, Vincent P and Steunou N Improving the Jason-1 ground retracking to better account for attitude effects *Marine Geodesy* **27** 171

IOP Conf. Series: Journal of Physics: Conf. Series **852** (2017) 012030 doi:10.1088/1742-6596/852/1/012030

Investigation to determine the vulnerability of reclaimed land to building collapse using near surface geophysical method

O O Adewoyin*[1], E O Joshua[2], M L Akinyemi[1], M Omeje[1] and E S Joel[1]

[1]Department of Physics, College of Science and Technology, Covenant University, Ota, Ogun State, Nigeria
[2]Department of Physics, Faculty of Science, University of Ibadan, Ibadan, Oyo State, Nigeria

*E-mail: segadot@yahoo.com

Abstract. Adequate knowledge of the geology and the structures of the subsurface would assist engineers in the best way to carry out constructions to avoid building collapse. In this study, near surface seismic refraction method was used to determine the geotechnical parameters of the subsurface, the results obtained were correlated with the result of borehole data drilled in the study area. The results of seismic refraction method delineated mostly two distinct layers with the first layer having the lower geotechnical parameters. It was observed that in the first layer, the Young's modulus ranged from 0.168 to 0.458 GPa , shear modulus ranged between 0.068 and 0.185 GPa, the bulk modulus ranged between 0.106 and 0.287 GPa while the bearing capacity ranged from 0.083 to 0.139 MPa. On the other hand, in the second layer, the Young's modulus ranged between 3.717 and 7.018 GPa, shear modulus ranged from 1.500 to 2.830 GPa while the bulk modulus ranged from 2.383 to 4.449 GPa. Significantly, the formation of the second layer appeared to be more competent than the first layer, therefore engineering construction in this geological setting is recommended to be founded on the second layer at depth ranging between 7 and 16 m.

1. Introduction

The alarming rate of building failure has necessitated the knowledge about the structure of the subsurface especially in area where lands are recovered from water bodies for the purpose of construction. This type of land is known to contain geologic formations [1], which are mechanically unstable for founding an engineering structure [2]. Therefore, adequate approach of investigation must be employed in order to study this type of terrain giving consideration to the high number of casualties that would be recorded and the worth of resources that would waste if the result obtained does not properly represent the condition of the subsurface [3,4]. Geophysical techniques are in high demands lately because of their relevance in hydrological and geotechnical investigations [1,5]. They have the advantage that they are capable of providing information about the spatial and/or temporal distribution of the features of the subsurface [6,7]. Prominent among the geophysical methods used for geotechnical studies is the seismic refraction method [8,9,10]. This technique is highly useful in many applications such as engineering, environmental, groundwater, hydrocarbon, and industrial-mineral exploration [11].It is used to characterize the subsurface structure and geological condition. This method divides the subsurface structure into different layers and it gives information on the engineering parameters of each layer and their thicknesses [12]. Seismic refraction method operates on the principle that the speed of propagation of seismic energy varies with the medium of propagation. It is on this basis that the first arrival is of utmost importance in the analysis of seismic data because it is

Content from this work may be used under the terms of the Creative Commons Attribution 3.0 licence. Any further distribution of this work must maintain attribution to the author(s) and the title of the work, journal citation and DOI.
Published under licence by IOP Publishing Ltd

IOP Conf. Series: Journal of Physics: Conf. Series **852** (2017) 012030 doi:10.1088/1742-6596/852/1/012030

used to determine the distribution of underground seismic velocities. It provides 2-dimensional profiles including depth and distance that simplified the characterization of relatively large volumes of the subsurface [3].

Various geomechanical properties can be measured by the propagation of seismic energy through soil deposits. Some of the soil properties that could be obtained by this method include Young's modulus, shear modulus, bulk modulus and Poisson's ratio. All these geotechnical parameters require the knowledge of both p and s-waves to determine their values. This method has been used to obtain in-depth information on different geological formation, previous studies revealed that this method can be successfully applied for site characterization [3,13]. Martinez and Mendoza [7] were able to show that this method would give quality image of the subsurface for site characterization. In a recent development, Grelle and Guadagno [14] established the applicability of seismic refraction method for groundwater level determination. This method has also been used to determine the ground role motion in soil. Seismic refraction method has been found to be very useful in the characterization of the subsurface for geotechnical engineering applications [14]. The advantage of seismic refraction method in seepage study was noted by Osazuwa and Chinedu [15]. Percussion drilling test is often conducted to obtain physical information on the geology of the study area in order to corroborate the results obtained by geophysical methods [16,17].

In order to prevent loss of lives and cost suffered in the cases of building collapse, it is important to understand the subsurface characteristics and the geologic condition of reclaimed land before they are recommended for construction purposes. Therefore, it is on this note that seismic refraction method was engaged in the present study to determine the condition of the subsurface and to correlate the result with borehole lithologic log of the area in order to determine the depth to the most competent layer in the area of study.

2. Description and Geological Setting

The study area is located in Lagos Island area of Lagos State, southwestern Nigeria Coastal zone, a zone of coastal creeks and lagoons developed by barrier beaches associated with sand deposition [18]. The topography is gentle with elevation averaging about 2 m above mean sea level. This area is situated within the Nigerian sector of the Benin-basin, and near the eastern margin of the basin. The geological formations of the study area are composed of sediments laid down under fluviate, lascustrine and marine environments. These sediments grade into one another and vary widely in lateral extent and thickness [18,19]. In this region, six lithostratigraphic formations are present, which are Abeokuta, Ewekoro, Akinbo, Oshosun, Ilaro and Benin Formations (arranged sequentially from the oldest to the youngest. The local geology is predominantly Coastal Plain Sands which consist of poorly sorted clayey sands, reddish mud/mudstone, clay and sand lenses, and sandy clay with lignite of Miocene to Recent (Figure 1).

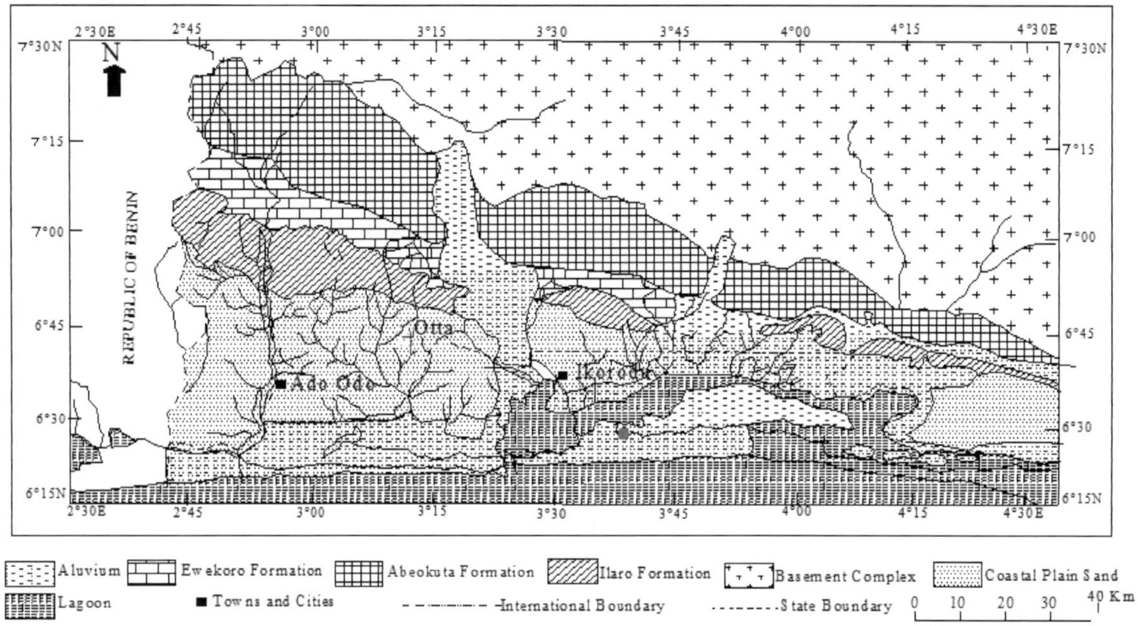

Figure 1. Geological map of the study area, with the area of study marked in red (Modified after [19])

3. Methodology

Seven seismic refraction profiles were surveyed in order to obtain adequate data for this study. Each profile length extended between 50 and 150 m as a result of space. The inter-geophone spacing of 2 m was used and the shot-to-first geophone spacing was also 2 m. A total number of 24 geophones were used on every 50 m profile length. A 24 Channel ABEM MK 6 Terraloc seismogram [20] was used for the data acquisition and energy source engaged was a 15 kg sledge hammer and a metallic base plate. Multiple shots were taken at each location and the data stacked about 8 times in order to increase the data quality. On each 50 m profile length, shots were taken at the offset location, between the sixth and the seventh geophones, between the twelfth and thirteenth geophones and mid-point between the eighteenth and nineteenth geophones and finally at 2m after the last geophone. This was done in order to adequately cover the refractor layer in the subsurface and to improve the quality of the image. High consideration was given to possible sources of noise such as machinery, daily human activities, traffic and other similar factors so as to improve the signal-to-noise ratio. From the seismic refraction method, p-wave velocity of each layer was used to determine the some engineering parameters for the two layers using some theories that exist in literature [21].

A borehole was drilled to a depth of 30 m in the study area using Shell and Auger percussion boring techniques with a Pilcon Wayfarer rig to bore through the overburden in accordance with the code of practice for soil investigations [2,13,22]. This is used to produce a borehole lithologic log for the surveyed area which formed a basis for the correlation of the results of seismic refraction method.

4. Results and discussion

The first arrival travel times of all recorded traces were picked, and then the picked travel times were inverted using a software code called seisImager. The seismic tomography section revealed two geologic layers in the study area (Figure 2). The p-wave velocity of the first layer ranged between 340 and 554 m/s while in the second layer, the velocity was between 588 and 2012 m/s. Other geotechnical parameters were determined, the Young's modulus varied between 0.168 and 0.458 GPa and it ranged between 0.518 and 7.071 GPa. Relatively, the bulk modulus ranged between 0.106 and 0.287 GPa in the first layer while the values lied between 0.326 and 4.455 GPa in the second layer. In another measurement, the shear modulus ranged between 0.068 and 0.185 GPa in the first layer while it was between 0.210 and 2.862 GPa in the second layer. Further, the allowable bearing capacity was discovered to vary between 0.083 and 0.139 MPa while it was between 0.149 and 0.593 MPa in the second layer. The depth of the first layer was about 8 m from the topsoil while in the second layer the thickness measured was about 11 m. Geotechnically, the geologic formation with higher values of geotechnical parameters are considered for engineering construction purposes as was observed in the second layer.

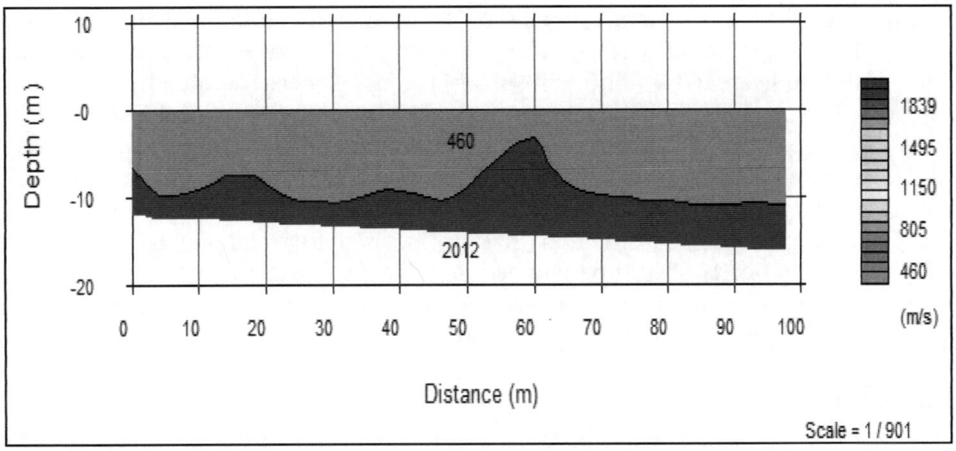

Figure 2. 2D seismic refraction section of the study area, indicating the number of layers and the values of the p-wave velocities

The result revealed that the first layer is predominantly composed of unconsolidated geologic formation which was responsible for the lower values of geotechnical parameters observed. This result was established by the borehole lithologic log of the study area which revealed the geologic formation from the topsoil to a depth of about 1.5 m as a geologic formation that is composed of peaty clay (soft geomaterials). This layer was underlain directly by wet sand deposit that is about 2.0 m in thickness. The two earlier mentioned geologic deposits were later underlain by clay and sandy clay of soft to medium-dense consistencies to a depth of 7 m in the subsurface. This type of depositional pattern is typical of a zone of coastal creeks and barrier beaches that characterize the study area. The second layer was found to have higher geotechnical parameters than the first layer. The borehole lithologic log revealed that the geologic formation in this layer is composed of sandy clay of medium dense to dense consistency. The geologic formation in this layer was found to be more competent than the first layer as this layer was found to possess high shear strength and low compressibility potential.

Geologically, the study area is located in the alluvium deposit, which is a prominent formation of a coast line area. The composition of the first layer is more recent in age of deposition than the second layer, which explained why the first layer was characterised by unconsolidated geologic formation. The weak geologic formation observed in the first layer might also be because there was no direct pressure from other geologic formation on top of this layer that could assist to hasten the consolidation of the soil deposit in this layer. Also, the second layer was found to be composed of more consolidated geological formation than the first layer, which might be because of the age of deposition, which allowed it to be compressed by the presence of other recent deposit lying on it. The pressure exerted on this layer by other geological formation lying on it helped to improve its geotechnical capability. It is worthy of mentioning that the results obtained in the present study correlated with the results of Pegah and Liu [23], in Manjil city of northern Iran, which is a similar terrain to the study area.

Therefore, the depth to the most competent geologic formation was found to be between 7 and 16 m corresponding to consolidated layer in the subsurface. This is the portion of the subsurface recommended for siting the foundation for the purpose of engineering construction. This study revealed that in any reclaimed land having similar geology, as in the study area engineering construction that is designed to provide safety and stand the test of time should be founded on the depth from the second layer.

5. Conclusion

In this study near surface seismic refraction method was used to explore the geological condition and engineering parameters of the subsurface. The observed results were correlated with the borehole lithologic log of the study area. Two different subsurface layers were revealed by the seismic method in the study area. The engineering parameters were observed to be higher in the second layer than in the first layer. The second layer corresponded to a consolidated layer on the borehole lithologic log obtained from the borehole drilled in the area, which is found to be sandy clay of medium to dense consistency. This layer was found at a depth between 7 and 16 m. It is therefore recommended that the foundation of any engineering construction that is intended to stand the test of time and offer protection to lives and property should be founded on the consolidated layer. This study revealed that in any part of the world with similar geological formation, the layers beneath the first layer should be considered for the foundation of engineering structures.

Acknowledgements

The authors wish to appreciate the management of Covenant University for providing financial support for the success of this research endeavour.

References

[1] Pueyo-Anchuela O, Casas-Sainz A M, Soriano M A and Pocovi-Juan A 2011 Geophysical techniques applied to urban planning in complex near surface environments. Examples of Zaragoza, NE Spain *Physics and Chemistry of the Earth* **36** 1211

[2] Sopaci E and Akgun H 2015 Geotechnical assessment and engineering classification of the Antalya tufa rock, southern Turkey *Engineering Geology* **197** 211

[3] Azwin I N, Saad R and Nordiana M 2013 Applying the seismic refraction tomography for site characterization *APCBEE Procedia* **5** 227

[4] Mohamed A M E, Abdel Hafiez H E and Taha M A (2013a) Estimating the near-surface site response to mitigate earthquake disasters at the October 6th city, Egypt, using HVSR and seismic techniques *National Research Institute of Astronomy and Geophysics* **2** 146

[5] Cardarelli E, Cercato M and De Donno G 2014 Characterization of an earth-filled dam through the combined use of electrical resistivity tomography, P- and SH- wave seismic tomography and surface wave data *Journal of Applied Geophysics* **106** 87

[6] Prasad B R, Venkateswarlu N, Prasad A, Murthy A S N and Sateesh T 2010 Basement configuration of on-land Kutch basin from seismic refraction studies and modeling of first arrival travel time skips *Journal of Asian Earth Sciences* **39** 460

[7] Martinez K and Mendoza J A 2011 Urban seismic site investigations for a new metro in central Copenhagen: Near surface imaging using reflection, refraction and VSP methods *Physics and Chemistry of the Earth* **36** 1228

[8] Mohamed A M E, Abu El Ata A S A, Azim F A and Taha M A (2013b): Site-specific shear wave velocity investigation for geotechnical engineering applications using seismic refraction and 2D multi-channel analysis of surface waves *National Research Institute of Astronomy and Geophysics* **2** 88

[9] Fabien-Ouellet G and Fortier R 2014 Using all seismic arrivals in shallow seismic investigations *Journal of Applied Geophysics* **103** 31

[10] Rumpf M and Tronicke J 2014 Predicting 2D geotechnical parameter fields in near-surface

sedimentary environments *Journal of Applied Geophysics* **101** 95
[11] Khalil M H and Hanafy S M 2008 Engineering applications of seismic refraction method; a field example at Wadi, Warden, Northeast Gulf Of Suez, Sinai, Egypt *Journal of Applied Geophysics* **65** 132
[12] Engelsfeld T, Sumanovac F and Krstic V 2011 Classification of near-surface anomalies in the seismic refraction method according to the shape of the time-distance graph: A theoretical approach *Journal of Applied Geophysics* **71** 149
[13] Das B M 2007 Principles of foundation engineering 6th Edition Thomson p 90
[14] Grelle G and Guadagno F M 2009 Seismic refraction methodology of groundwater level determination: "water seismic index" *Journal of Applied Geophysics* **68** 301
[15] Osazuwa I B and Chinedu A D 2008 Seismic refraction tomography imaging of high-permeability zones beneath an earthen dam, in Zaria area Nigeria *Journal of Applied Geophysics* **66** 44
[16] Alkroosh I S, Bahadori M, Nikraz H and Bahadori A 2015 Regressive approach for predicting bearing capacity of bored piles from cone penetration test data *Journal of Rock Mechanics and Geotechnical Engineering* 1-9 (article in press)
[17] Feng S, Gao K, Chen Y, Li Y, Zhang L M, Chen H X 2016 Geotechnical properties of municipal solid waste at Laogang landfill, China *Waste Management* (article in press)
[18] Adepelumi A A and Olorunfemi M O 2000 Engineering geological and geophysical investigation of the reclaimed Lekki, Peninsula, Lagos, South west Nigeria *Bulletin of Engineering, Geology and The Environment* **58** 125
[19] Durugbo E U and Aroyewun R F 2012 Palynology and paleoenvironments of the upper Araromi Formation, Dahomey Basin, Nigeria *Asian Journal of Earth Sciences* **5** 50
[20] ABEM Instrument AB 1996 Equipment Manual for TERRALOC MK6 Software Version 2.21 Bromma Sweden
[21] Tezcan S S, Ozdemir Z and Keceli A 2009 Seismic technique to determine the allowable bearing pressure for shallow foundations in soils and rocks *Acta Geophysica* **57**(2) 1
[22] Hunt R E 2005 Geotechnical engineering investigation handbook 2nd edition Taylor and Francis 1
[23] Pegah E and Liu H 2016 Application of near-surface seismic refraction tomography and multichannel analysis of surface waves for geotechnical site characterization: a case study *Engineering Geology* **208** 100

IOP Conf. Series: Journal of Physics: Conf. Series **852** (2017) 012031 doi:10.1088/1742-6596/852/1/012031

Extraction of shoreline changes in Selangor coastal area using GIS and remote sensing techniques

S N Selamat[1], K N Abdul Maulud[1,2*], O Jaafar[2] and H Ahmad[2]

[1]Earth Observation Centre, Institute of Climate Change (IPI), Universiti Kebangsaan Malaysia, 43600 UKM, Bangi, Selangor Malaysia
[2]Department of Civil & Structural Engineering, Faculty of Engineering & Built Environment, Universiti Kebangsaan Malaysia, 43600 UKM, Bangi, Selangor Malaysia

*Email: knam@ukm.edu.my

Abstract. Nowadays, coastal zones are facing shoreline changes that stemming from natural and anthropogenic effect. The process of erosion and accretion will affect the physical environment of the shoreline. Therefore, the study of shoreline changes is important to identify the patterns of changes over time. The rapid growth of technology nowadays has facilitated the study of shoreline changes. Geographical Information System (GIS) alongside Remote Sensing (RS) technology is a useful tool to study these changes due to its ability to generate information, monitoring, analysis and prediction of the shoreline changes. Hence, the future projection of the trend for a specific coastal area can be done effectively. This study investigates the impact of shoreline changes to the community in Selangor area which mainly focus on the physical aspects. This study presents preliminary result using satellite image from SPOT 5 to identify the shoreline changes from the year 1984 to 2013 at Selangor coastal area. Extraction of shoreline from satellite image is vital to analyze the erosion and accretion along the shoreline area. This study shows that a shoreline change for the whole area is a categorized as a medium case. The total eroded and accretion of Selangor area from 1984 to 2013 is 2558 hectares and 2583 hectares respectively. As a result, Kapar, Jugra, Telok Panglima Garang and Kelanang are categorized as high risk erosion area. Shoreline changes analysis provides essential information to determine on the shoreline changes trends. Therefore, the results of this study can be used as essential information for conservation and preservation of coastal zone management.

1. Introduction

Phenomena of sea level rise (SLR) constitute as an effect of climate change. Based on the Intergovernmental Panel on Climate Change [1] envisaged that global mean SLR may increase between 0.26 m and 0.82m by the 2100 year. Predicted of climate change will greatly give high risk to coastal population [2]. One of the main factors that contribute towards the shoreline changes is sea level rise phenomena. According to National Hydraulic Research Institute of Malaysia (NAHRIM), sea level of Peninsular Malaysia will increase with the range of 0.253m to 0.517m by the year 2100 (from the year 2010 as a baseline) or 2.7mm to 7.0mm/ year. An increase of sea level rise has been recognized as one of the major threats to societies around the world [3].

Coastal zone is one of the most complex ecosystems that has a large number of non-living and living resources [4,5]. Many series of physical processes such as tidal flooding, sea level rise, land subsidence, erosion, and sedimentation, thus that area was categorized as a high impact of environment natural processes. According to [6], over 70% of the worlds coastal are experiencing

Content from this work may be used under the terms of the Creative Commons Attribution 3.0 licence. Any further distribution of this work must maintain attribution to the author(s) and the title of the work, journal citation and DOI.
Published under licence by IOP Publishing Ltd

coastal erosion and this presents a serious hazard to many coastal regions. The coastal zone of the world exposed to danger due to several factors, among which is the development of industry, trade, tourism, migration and human population growth in coastal areas. [7]. Shoreline zone is considered as one of the zones that face the most dynamic processes in the coastal zone by [8,9]. Monitoring of shoreline is a significant task in environmental protection and national development [5], in which, extraction of shoreline is the fundamental study of necessity by [10]. Shoreline expert faced one of the obstacles to managing coastal environment changes due to the uncertainty of climate prediction [11,12]. Shoreline changes that stemming from natural and anthropogenic effect pose a threat to countries with high population and socio economic activities in coastal are such as Malaysia.

Nowadays, application of Geographical Information System (GIS) and satellite remote sensing data are being used in storing, extraction, analyze and mapped the shoreline changes. Application of GIS integrates the location of spatial information with its attribute information. GIS and remote sensing technology are able to calculate the historical rate of changes. Historic maps, aerial photographs, beach profiles, topographic and bathymetric surveys can be used to qualitatively and quantitatively evaluate the spatial and temporal changes in coastal areas [13]. Using updated technology of GIS and remote sensing, several techniques have been introduced to extract shoreline and detect shoreline changes from satellite imagery [14,15,16].

This study will be carried out analysis of satellite imagery SPOT 5 and topographic map to investigate the shoreline changes at several Selangor coastal area that clarify as erosion and accretion that has occurred. The main agenda of the study was to determine the shoreline changes in Selangor coastal area and the analysis of this study for these areas can be used for further investigation.

2. Material and method

Malaysia is a coastal nation rich with natural resources and biodiversity such as marine life and wildlife. Malaysia covers an area of 329, 750 square kilometers (km^2) with a coastal zone transverse along 4809 kilometers and the country is divided into two (2) landmasses which are being separated by the South China Sea. Peninsular Malaysia is located in the west covers an area of 131, 590 km^2. Whereas, the island of Borneo Sabah and Sarawak each covering an area of 73.711 km^2 and 124.449 km^2 [17]. Figure 1 shows the shoreline of the Malaysia coastal area. Malaysia coastal zone consists of broad sandy beaches such as those on the East Coast of Peninsular and muddy coastal areas in the west.

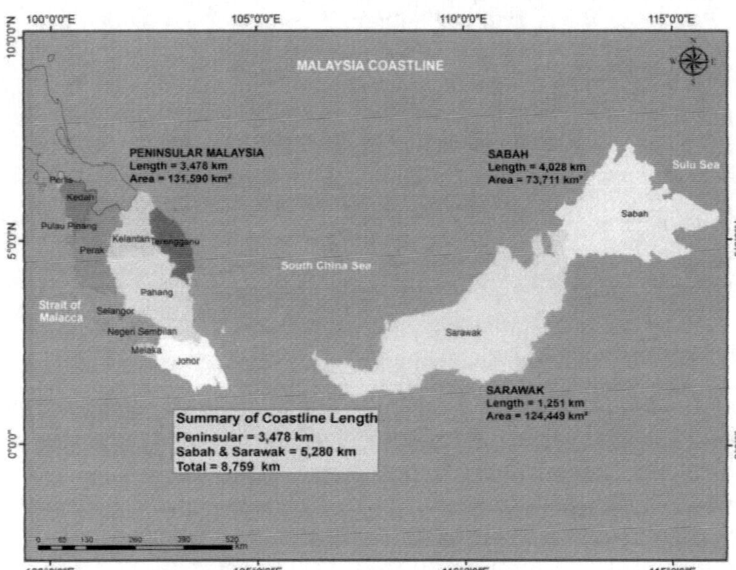

Figure 1. The shoreline of Malaysia area

As shown in the figure, most of Malaysia shoreline is occupied by important industries, agricultural, construction, fisheries, tourism and major city/ports are coast which contributes significantly to Malaysia's economic growth. The study area is located along the Selangor coastal area. This area is located on the west coastal of Peninsular Malaysia. Selangor's coastal is one area that located at the west coast of Peninsular Malaysia. Selangor coastal area covers the shoreline along Kuala Langat toward Sabak Bernam. The coastal areas along 276 km which are from Kuala Langat and Sabak Bernam exposed to potentially catastrophic hazards such as coastal erosion, flooding and technological hazards. According to *Rancangan Fizikal Zon Persisiran National* (RFZPPN), the boundary of shoreline in Malaysia is divided into management units, which peninsular coastal zone consist of 17 management units. Every management unit is being determined by the administration, use of land, and ecological characteristic. The study area and three management units are shown in Figure 2.

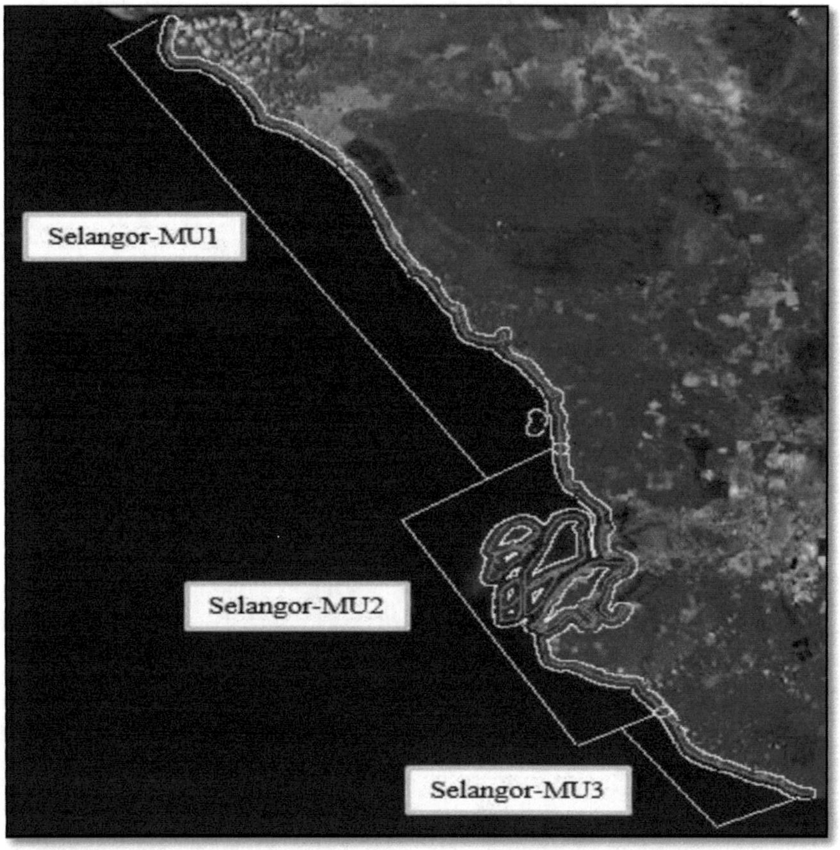

Figure 2. Study area and management units from Kuala Langat to Sabak Bernam [18]

Every management units consist of their own shoreline area where the shoreline is divided by district. According to RFZPPN, part of the management unit covers the southern of Perak and the northern of Negeri Sembilan. Management Unit 1 (MU1) is starting from Sungai Perak at Perak to northern border sub-district of Kapar, Management Unit 2 (MU2) starting from northern border sub-district of Kapar to southern border sub-district of Kelanang, and Management Unit 3 (MU3) along the coastal zone from southern border sub-district of Kelanang to southern border sub-district of Jimah at Negeri Sembilan. Table 1 shows the details of sub-district which include the management units.

Table 1. Management unit area according to RFZPPN 2012

Management Unit (MU)	Shoreline area	District	Sub-district
MU1	Sungai Bernam-Northern border sub-district of Kapar	Sabak Bernam Kuala Selangor	Bagan Nakhoda Omar, Panchang Bedena, Pasir Panjang, Tanjong Karang, Hujong Permatang, Kuala Selangor, Pasangan, Api-Api and Jeram.
MU2	Northern border sub-district of Kapar-Southern border sub-district of Kelanang	Kuala Selangor Kelang	Kapar, kelang, jugra, telok panglima garang and kelanang.
MU3	Southern border sub-district of Kelanang - Southern border district of Kuala Langat	Kuala Langat	Jugra, Telok Paglima Garang, Bandar, Morib, and Batu.

In this study, SPOT 5 satellite imagery and Topographic map acquired from 1984 to 2013 are used to extract the shoreline changes along the Selangor coastal area. Table 2 indicates the data sources used for the shoreline extraction. Projection system that used in this study is Rectified Skew Orthomorphic (RSO) projection system.

Table 2. Data acquisition

Type of data	Year	Scale/Resolution
Topographic map	1984	1: 50 000
SPOT 5	2013	2.5 meter
SPOT 5	2013	2.5 meter
SPOT 5	2013	2.5 meter

Figure 3 illustrate spatial model process generated by using ArcGIS software to extract the eroding and accreting shoreline changes. In this study, shoreline topographic map on year 1984 is used as a baseline whilst the extraction of shoreline on year 2013 from SPOT 5 satellite imagery as an investigation analysis. Thus, based on the overly of baseline and satellite imagery data it is able to identify the erosion and accretion changes.

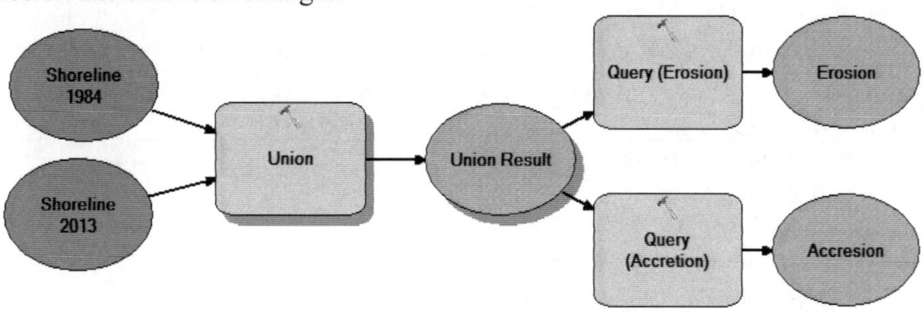

Figure 3. Spatial Model to classify the coastal change

3. Result and Discussion

The shoreline changes phenomena will occur as a two events either erosion or accretion. Erosion phenomena is a process that describe the condition of coastal area that destruction effect from activities of sea water, while accretion is an additional material caused by the removal of material from river. The focus of this research is extraction of shoreline changes at Selangor coastal area with the total length is 276 km from Kuala Langat to Sabak Bernam. Figure 4 shows the map of erosion and accretion area at management unit 1 (MU1) which is from Sungai Bernam to northern border sub-district of Kapar.

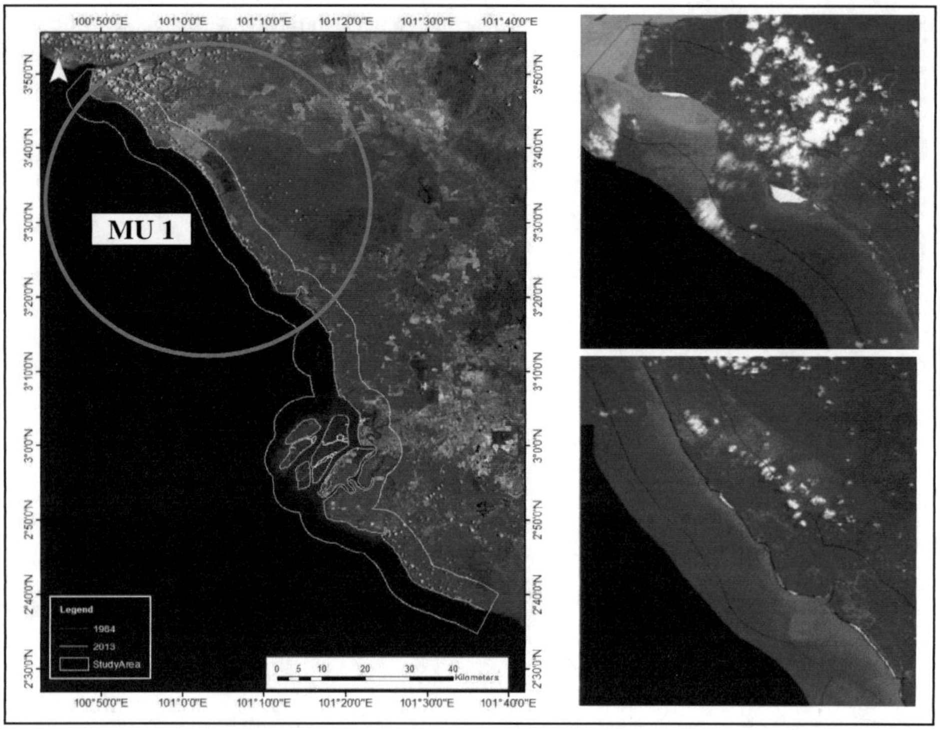

Figure 4. Map of erosion and accretion along Management Unit 1 (MU1)

The purposed of this study is to investigate the shoreline changes between different years. The shoreline change in different parts of the coastal area is determined by the process of swash and backwash of sediments due to seasonal current movements [19]. Therefore, the changes rate is differences based on their specific area and the shoreline change process is due to dynamic process (see Table 3).

Table 3. Eroding and accretion shoreline in Selangor coastal area on 1984 to 2013

Management Unit (MU)	Location	Shoreline Length (km)	Eroding (Ha)	Accreting (Ha)
MU1	Sungai Bernam - Northern border sub-district of Kapar.	114.727	813	960
MU2	Northern border sub-district of Kapar– Sothern border sub-district of Kelanang	293.675	1,549	1,523
MU3	Southern border sub-district of Kelanang - Southern border district of Kuala Langat	29.858	196	100
	Total	438,260	2,558	2,583

During year 1984 to 2013 about 2558 hectra was eroded from the area, and the value of accrating is 2583 hectra. Table 2 illustrate the rate of eroding at Selangor coastal area from year 1984 to 2013. Based on the table below, the highest erosion area occurred at MU2 compare to the MU1 and MU3, meanwhile erosion in MU1 is more than MU3. Figure 5 shows the map for management unit 2 (MU2) which from which is from northern border sub-district of Kapar to southern border border sub-district of Kelanang, while Figure 6 represent the changes for management unit 3 (MU3) which from Southern border sub-district of Kelanang to Southern border district of Kuala Langat respectively.

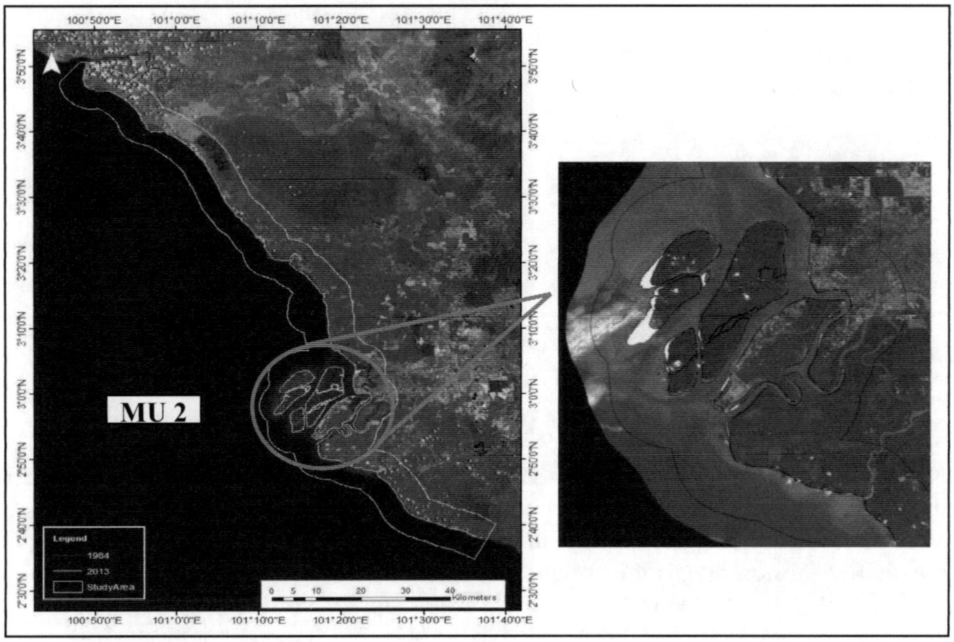

Figure 5. Map of erosion and accretion along Management Unit 2 (MU2)

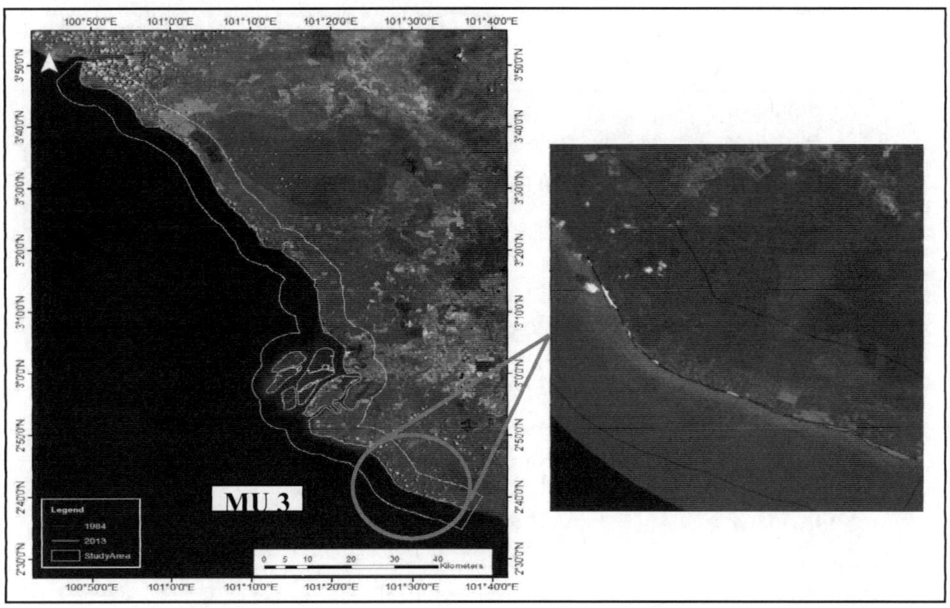

Figure 6. Map of erosion and accretion along Management Unit 3 (MU3)

Based on the Table 3, shoreline in MU2 is the highest erosion compare with MU1 and MU3, meanwhile erosion in MU1 is more than MU3. The persentage of eroding and accrating is shown in Figure 7. Based on the result, the higher eroding area occurred at MU2 and the lower eroding area occurred at MU3 which is 60% and the 8 % respectively. Where as, the higher accration area occurred at MU2 and the lower accration occurred at MU3 in 59% and 4% respectively. As a result, MU2 indicates facing the higher erosion rate and clasify as the high risk erosion area.

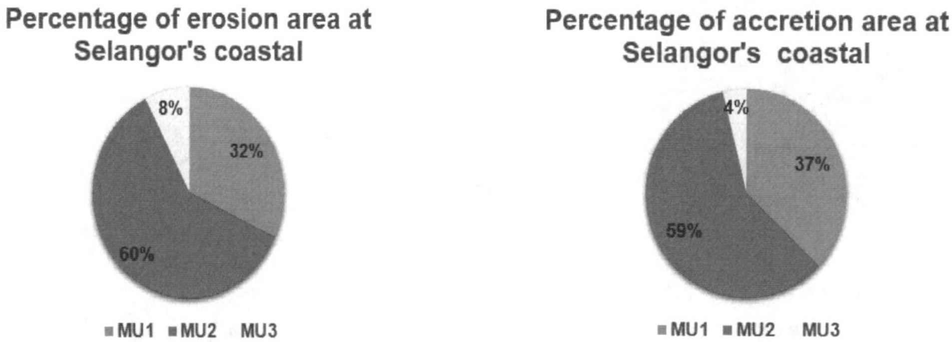

Figure 7. Percentage of erosion and accretion area at Selangor coastal

Table 4 shows the rate of erosion and accretion for MU1, MU2 and MU3. The erosion and accretion rate have calculated in meter unit and per year. The erosion rate for this study varies between 2.0 to 18.00 per year, while the accretion rate fluctuated between 3.0 to 100 meters per year. Shoreline in MU2 is the highest erosion rates compare with MU1 and MU3, meanwhile rate of erosion in MU1 is more than MU3. MU2 get the highest erosion rate where applicable 17.79 meters per year compare with the highest erosion at MU2 is 14.41 meter and MU3 only 10.28 meter per year.

Table 4. The rate of erosion and accretion per year

Management Unit (MU)	Location	Erosion (m)	Accretion (m)	Rate per year	
				Erosion	Accretion
MU1	Sg.Selangor	289		9.97	
	Sg.Beting Kepah	159		5.48	
	Bagan Sg Burong	418		14.41	
	Sg. Panchang Bedena		1548		53.38
	Bagan Sg Tengkorak		313		10.79
MU2	Sg.Ular	516		17.79	
	Sg. Langat	493		17.00	
	Sg.Tiram Besar	287		9.90	
	Sg.Labu Lemak		240		8.28
	Pulau Tengah		2896		99.86
MU3	Sg.Gaping	149		5.14	
	Sg. Kudang	109		3.76	
	Sg.Pulai	298		10.28	
	Sg. Kanchong		206		7.10
	Sg.Tengkah		108		3.72

The highest rate of accretion also in MU2 compared with MU1 and MU3. MU2 get the highest accretion rate where applicable 99.8 meters per year compare with the highest accretion at MU2 is 53.38 meter and MU3 only 7.10 meter per year. Pulau Tengah located at MU2 is constituted the highest rate of accretion area where the island is a mangrove swamp that helps increase the rate of accretion per year.

4. Conclusion

This study focused on the extraction of shoreline changes using GIS and remote sensing technique to determine the impact of shoreline changes due to the physical environment. The analysis of shoreline change assessment using SPOT 5 satellite imagery exposes that the study area has experienced the high rate of erosion and accretion along the different coastal zone during 1984 to 2013. Monitoring and management of the coastal zone are very complex because it needs to focus on all aspects that influence to the disaster of the coastal zone. Therefore, Coastal zone management has been divided into sub-sections that call as a management unit and it was divided into small units according to natural coastal processes and land use activity. Every development that planned at coastal areas must comply with the policies and negotiations as *Konvensyen Antarabangsa berkaitan Tanah Lembap Berkepentingan Antarabangsa atau Konvensyen Ramsar, Rancangan Fizikal Negara 2, Rancangan Fizikal Zon Persisiran pantai Negara (RFZPPN),* and International Union for Conservation of Nature (IUCN). Each policies and negotiation have their own aspects because different soil type and different physical activities contributed the different impact of the coastal zone. The coastal development plan must comply with the conditions that have been stated and each application should be accompanied by an Environment Impact Assessment (EIA) report that represents preliminary report on the activities to be carried out on water quality. Consequently, development planning at coastal zone is important to ensure that area is not threatened. This study denotes as a preliminary study to investigate the shoreline changes in Selangor coastal area. As a recommendation, further research should be conducted by using the statistical technique for other critical location throughout Malaysia due to its effectiveness for prediction of shoreline changes.

Acknowledgements

The authors gratefully acknowledge the Earth Observation Center, Institute of Climate Change, Universiti Kebangsaan Malaysia for sharing the satellite data Selangor year 2013. This work was supported by The Research Fund by Research University Grants (TRGS/1/2015/UKM/5/1).

References

[1] Physical T and Basis S 2013 Climate Change 2013
[2] Gornitz V, Couch S and Hartig E K 2002 Impacts of sea level rise in the New York City metropolitan area *Glob. Planet. Change* **32** 61
[3] Nicholls R, Wong P, Burkett V, Codignotto J, Hay J and McLean R 2007 Coastal systems and low-lying areas. In: Climate Change 2007: impacts adaptations and vulnerability (eds. Parry M, C999anziani O, Palutikof J, Linden, J P and Hanson C) (Cambridge: Cambridge University Press) pp 315
[4] Costanza R, Arge R, Groot R De, Farber S, Grasso M, Hannon B and Sutton P 1997 The value of the world's ecosystem services and natural capital *Nature* **387**(May) 253
[5] Aedla ., Dwarakis, G S and Reddy D V 2015 Automatic shoreline detection and change detection analysis of netravati-gurpurrivermouth using histogram equalization and adaptive thresholding techniques *Aquatic Procedia* **4**(Icwrcoe) 563
[6] Addo K A and Kufogbe K S 2011 Quantitative analysis of shoreline change using medium resolution satellite imagery in Keta, Ghana *Marine Scioence* **1**(1) 1
[7] Nayak S 2000 Critical issues in coastal zone management and role of remote sensing In: Subtle Issues in Coastal Management *Indian Institute of Remote Sensing* Dehradun pp 77
[8] Bagli S 2016 Pan-European Coastline from Landsat Etm + Images, (January 2003)

[9] Mills J P, Buckley S J, Mitchell H L, Clarke P J and Edwards S J 2005 A geomatics data integration technique for coastal change monitoring *Earth Surface Processes and Landforms* **664** 651

[10] Rasuly A, Naghdifar R and Rasoli M 2010 International society for environmental information sciences *2010 annual conference monitoring of caspian sea coastline changes using object-oriented techniques* **2**(5) 416

[11] Norman B 2009 Planning for coastal climate changes "an insight into international and national approaches" Victorian government *Department of Planning and Community Development and Department of sustainability and Environment* Melbourne

[12] Tam L 2009 Sea level rise and the future of the bay area *J. Urbanist* **487** 9

[13] Islam A, Hossain S, Hasan T and Murshed S 2014 Shoreline changes along the Kutubdia Island, South East *Bangladesh J. Sci. Res.* **27**(1) 99

[14] Chen L C and Rau J Y 1998 Detection of shoreline changes for tideland areas using multi-temporal satellite images *Inter. J. Remote Sensing* **19**(17) 3383

[15] Lipakis M and Chrysoulakis N 2005 Shoreline extraction using satellite imagery *BEACHMED-e/OpTIMAL* 81

[16] Salghuna N N and Bharathvaj S A 2015 Shoreline change analysis for northern part of the coromandel coast *Aquatic Procedia* **4** (Icwrcoe) 317

[17] Department of Irrigation and Drainage Malaysia 2000

[18] Federal Department of Town and Country Planning Peninsular Malaysia 2012 Ministry of Urban Wellbeing, Housing and Local Government *National Physical Plan-2*

[19] Sheeja P S and J A G A 2016 Application of digital shoreline analysis system in coastal erosion assessment *International Journal of Engineering Science and Computing* **6(**6) 7876

IOP Conf. Series: Journal of Physics: Conf. Series **852** (2017) 012032 doi:10.1088/1742-6596/852/1/012032

Estimation of lidar ratios during haze events in Penang, Malaysia

W Y Khor*, M Z Mat Jafri and H S Lim

School of Physics, Universiti Sains Malaysia, Penang, Malaysia

*Email: kwy13_phy071@student.usm.my

Abstract. Three haze events occurred in Malaysia during the years of 2014 and 2015. Humidity and visibility parameters are obtained and used to estimate the Lidar ratio (LR) values during these three haze events using the algorithm generated. The possible aerosol types during the haze were also evaluated. The haze event that happened from August to October 2015 was the worst among the three cases with a maximum LR value of 177 sr. The LR values also range from 50 sr to 145 sr. The possible aerosol type during the haze events are highly light absorbing particles such as biomass burning aerosols. The LR values are much larger compared to other literature which may be due to overestimation. The assessed LR values are consistent with the changes of the number of hotspots in neighbouring countries during the haze events.

1. Introduction

Haze is a seasonal issue every year in Malaysia. This is caused by the monsoon seasons and seasonal burning in neighbouring countries of Malaysia, such as Vietnam, Indonesia and many more. In year 2014 and 2015, Malaysia was struck with three separate events of haze during different time periods. A moderate haze episode occurred between February and March 2014 whereas transboundary haze pollution occurred during the Southwest monsoon from June to October 2014. Year 2015 had only one occurrence of haze from August to September 2015 which was more devastating than the previous two haze events [1].

The API levels for these three haze events were very high which preached the unhealthy level of more than 200. The visible atmospheric condition was also very hazy with very low visibility. To further learn about the atmospheric conditions during these three time periods, the lidar ratios (LR) values can be assessed to further invert lidar signals and extract information. The possible aerosol types can be known based on the values of the LR [2]. Therefore, the LR values during these three time periods are estimated and presented in this paper.

2. Study Area

The island of Penang is located in the northwestern region of Peninsular Malaysia and lies within latitudes 5.2°N to 5.5°N and longitudes 100.15° E to 100.43°E (Figure 1). The lidar used is the Raymetrics LB100-ESS-D200 Eye-Safe Scanning Lidar system. It is placed on the rooftop of the School of Physics, USM, Penang at a latitude of 5.40N and a longitude of 100.30E.

Content from this work may be used under the terms of the Creative Commons Attribution 3.0 licence. Any further distribution of this work must maintain attribution to the author(s) and the title of the work, journal citation and DOI.
Published under licence by IOP Publishing Ltd

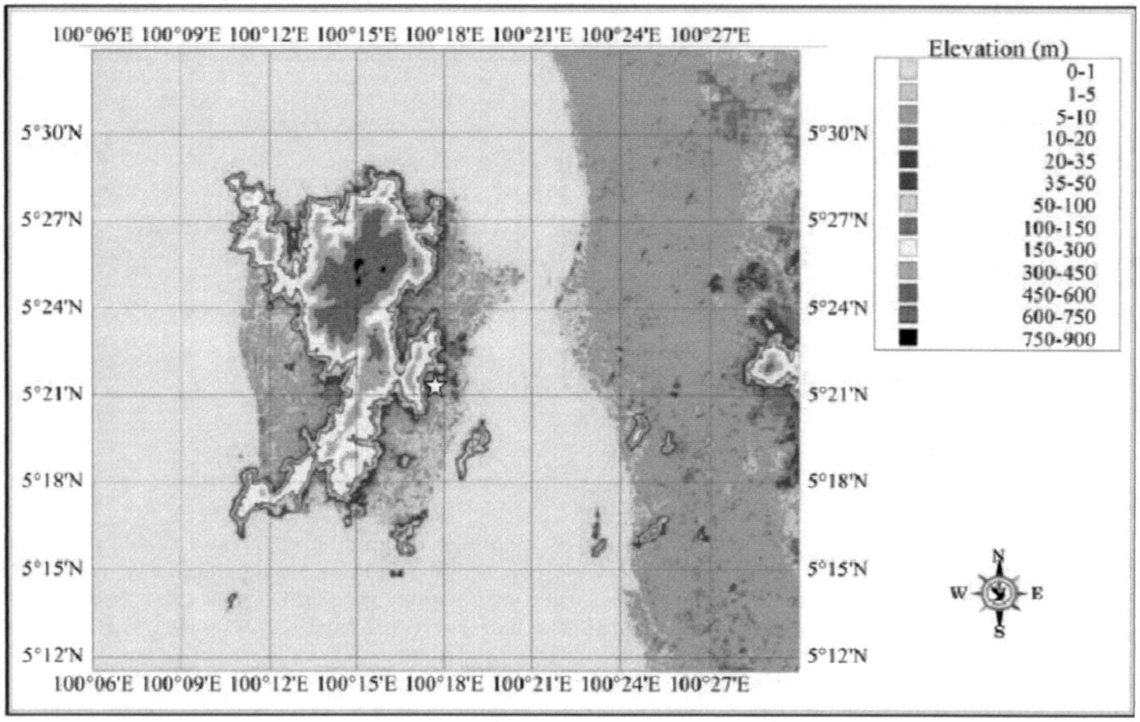

Figure 1. Topographic map of Penang region with elevation in meters. The five edge star represents the Lidar location (USM)

3. Data Acquisition and Processing

An algorithm for estimating the LR values is generated based on multiple regressions. The final algorithm is as shown in Eq (1):

$$LR = a0 + a1\ (\ log10\ H) + a2\ (\ H\ {*}V) \qquad (1)$$

The parameters required for implementing in the equation for estimating LR are humidity (H) and visibility (V). These two parameters are average values obtained daily. Once the H and V values for the three time periods of haze is obtained, the LR values during the haze events in year 2014 and year 2015 was generated based on the algorithm. Graphs are plotted to display the values of the LR. The LR values are plotted according to the 3 haze event that has occurred separately in the years 2014 and 2015.

4. Results and Discussion

Table 1 shows the maximum, minimum, mean and median values of LR during the three haze events respectively. The maximum LR value is highest for the Aug – Oct haze event in 2015 with 177 sr while the lowest was in Feb – Mar 2014 with 123 sr. The average LR was highest for the haze in 2015 with 97 sr while the other two haze events had 79 sr and 73 sr respectively. It can be implemented that the haze event was the worst from August to October 2015 compared to the previous two cases in year 2014.

It should be noted that although the haze event in year 2015 was reported to be from August to September, the haze event in Penang seemed to occur for a longer period which is from August to October 2015. This may be due to the fact that the pollution particles have lingered in the atmosphere during the inter-monsoon season of October. Lack of wind and rain this month has prolonged the lifetime of aerosols in the atmosphere from deposition and removal from the atmosphere through rain and other natural atmospheric processes.

Table 1. Maximum, minimum, mean and median values of lidar ratios during three haze events (Feb – Mar 2014, June – Oct 2014, Aug – Oct 2015)

Lidar ratios, sr	Feb-Mar14	June-Oct14	Aug-Oct15
Max	123	141	177
Min	58	46	48
Mean	79	73	97
Median	77	69	88

Figure 2 shows the LR values for these three haze events during the respective period of occurrence. February to March 2014 haze displays LR values in the general range of 60 to 95 sr. Only two extreme values of LR values are found. On the other hand, the second haze event from June to October 2014 contains LR values within the range from 50 sr to 100 sr. 9% of the values are greater than 105 sr. The third haze event from August to October 2015 showed LR values which are more scattered over a very large range of LR values from 50 sr to 145 sr. 11% of the LR values during this period were larger than 160 sr.

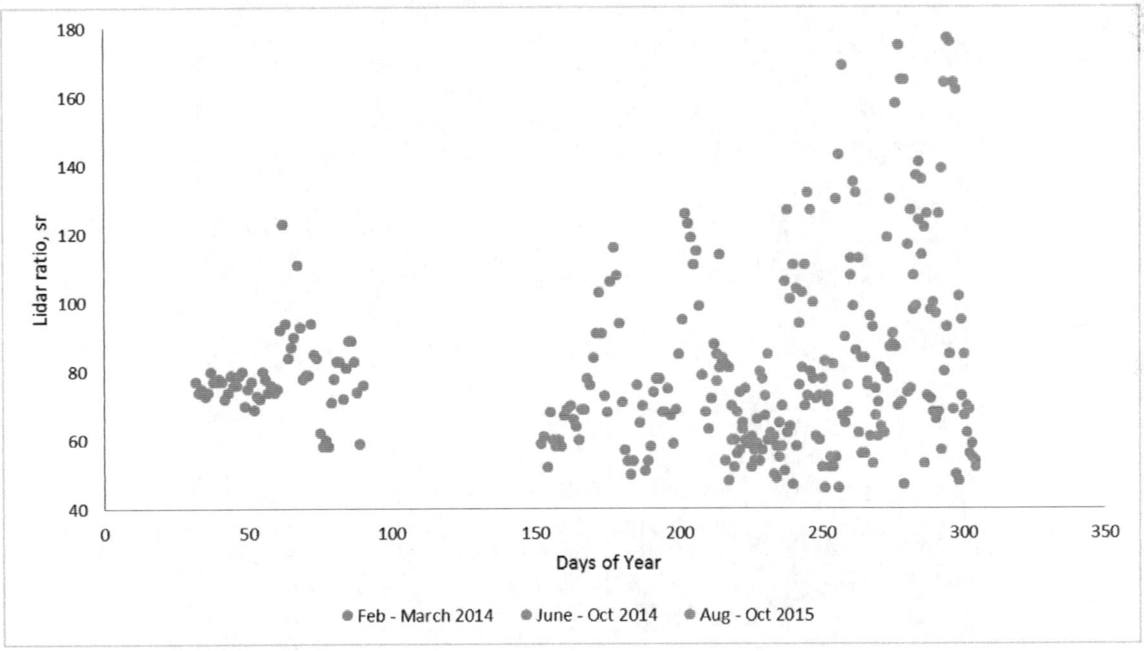

Figure 2. Lidar ratios of the three haze events (February to March 2014, June to October 2014 and August to October 2015)

Among these three haze events, the most serious case is no other than the one that happened in year 2015. It contains the highest values of LR and the LR value range is very large. The other two haze events were less serious compared to this haze event.

Based on the LR values, the possible aerosol types during the haze events are highly light absorbing fine particles which are possibly biomass burning aerosols and fine particulate matter particles. LR values within the range from 50 sr to 80 sr are highly light absorbing particles as reported by these authors [3], [4]. It is also reported that Malaysia is one of many Southeast Asian countries that has particulate matter pollution from vehicle and industrial emissions, including Indonesian peatland fire during the southwest monsoon (dry) season from June to October [5]. The LR values assessed in this study are very high which is not highly likely to occur. An observation which took place in India during haze also showed a maximum LR value of 110 sr only for 355-nm wavelength [6]. The LR values are only estimations and may be over-estimated.

The high values of LR in year 2015 compared to year 2014 can also be proved by the high number of hotspots during the period of August to October 2015 compared to the same period in year 2014. This is shown in figure 3. As shown, during the haze event in February to March 2014, the number of hotspots is much less compared to the other two periods. The second haze event that happened from June to October 2014 is less serious compared to the one in year 2015. This is proven by the number of hotspots that is less saturated compared to that in year 2015. A backward trajectory on the 15th of September 2015 shows that biomass burning aerosols from Indonesia would be advected to Penang.

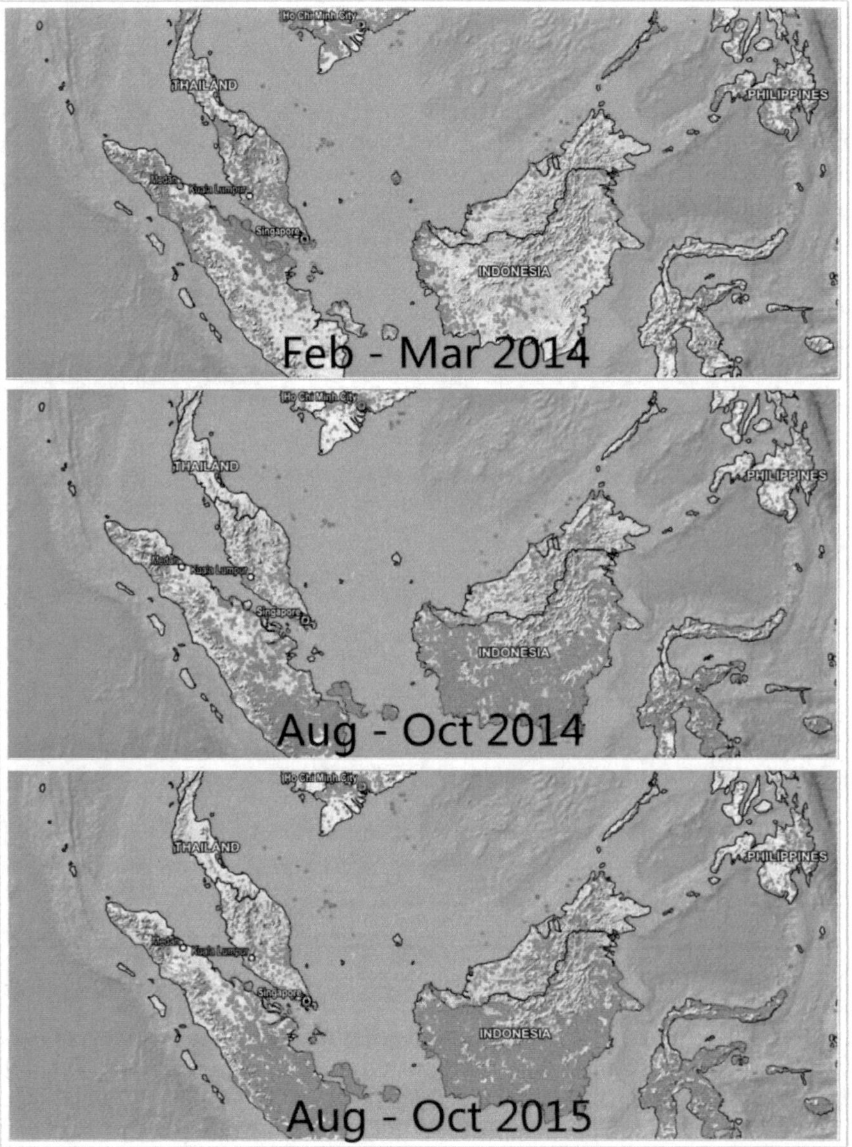

Figure 3. Fire hotspots during February to March 2014 (Top), August to October 2014 (Middle) and August to October 2015 (Bottom)

The number of hotspots for the same period of time is shown here which is from August to October to show the difference of the number of hotspots in the same months but different year. The LR values of in the months of August to October is much higher in year 2015 than in year 2014 which is the same for the number of hotspots during these three months. With the high number of hotspots, it would also mean that the density of smoke and biomass burning is high which attributes to a high number of biomass burning aerosols. These fine particles of biomass burning would induce the increment of the LR values. This means that the assessed LR values would be quite accurate for each haze event. The assessed LR values would be proportional to the number of hotspots surrounding Penang.

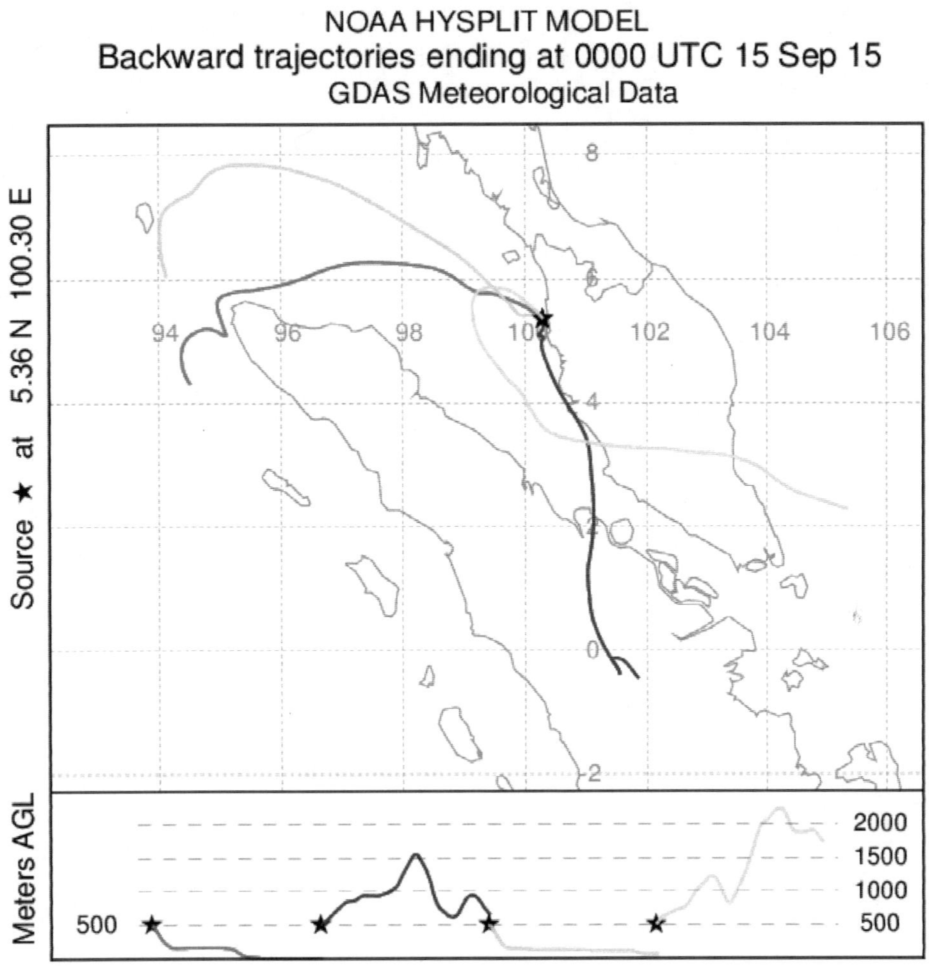

Figure 4. Backward trajectory on 15 September 2015 with 4 trajectories generated every 72 hours from the same location (Penang)

5. Conclusion

The most serious haze event is found to be from August to October 2015 with a large LR value range from 50 sr to 145 sr and a maximum LR value of 177 sr. The other two cases were less serious compared to this haze event. The maximum values of LR for February to March 2014 and for June to October 2014 are 123 sr and 141 sr respectively. Possible aerosol types during the haze events are biomass burning aerosols and fine particulate matter particles. The LR values estimated are very much larger than LR values found in literature which may be overestimated. The LR values for the haze events changes with relation to the number of hotspots surrounding Penang.

Acknowledgments
The authors gratefully acknowledge the financial support under the ERGS grant 203/PFIZIK/6730051, Fundamental Research Grant Scheme (FRGS) 203/PFIZIK/6711352 - Investigation Of Aerosol Characteristics And Distribution In Malaysia By Using Aeronet And Lidar Data, ScienceFund 305/PFIZIK/613615 - Environmental Effects And Its Influence of Increased Green House Gasses In Peninsular Malaysia and RUI grant (Individual) 1001/PFIZIK/811228 - Investigation Of The Impacts Of Summertime Monsoon Circulation To The Aerosols Transportation And Distribution In Southeast Asia Which Can Lead To Global Climate Change.

References

[1] Department of Environment Malaysia (DOE) 2015 Chronology of haze episodes in Malaysia pg 3: https://www.doe.gov.my/portalv1/info-umum/info-kualiti-udara/kronologi-episod-jerebu-di-malaysia/319123

[2] Müller D, Ansmann A, Mattis I, Matthias T, Wandinger U, Althausen D and Pisani G 2007 Aerosol-type-dependent lidar ratios observed with Raman lidar *J. Geophys. Res.: Atmospheres* **112** D16

[3] Novakov T, Bates T S and Quinn P K 2000 Shipboard measurements of concentrations and properties of carbonaceous aerosols during ACE 2 Tellus *Ser.* B **52** 228

[4] Franke K, Ansmann A, Müller D, Althausen C, Wagner F and Scheele R 2001 One-year observations of particle lidar ratio over the tropical Indian Ocean with Raman lidar *Geophys. Res. Lett.* **28** 4559

[5] Fujii Y, Mahmud M, Tohno S, Okuda T and Mizohata A 2016 A Case Study of PM 2.5 Characterization in Bangi, Selangor, Malaysia during the Southwest Monsoon Season *Aerosol Air Qual. Res.* **16** 2685

[6] Franke K, Ansmann A, Müller F, Althausen D, Venkataraman C, Reddy M S, Wagner F and Scheele R 2003 Optical properties of the Indo-Asian haze layer over the tropical Indian Ocean *J. Geophys. Res.* **108**(D2) 4059, doi:10.1029/2002JD002473.

Application of 2D electrical resistivity imaging and cone penetration test (CPT) to assess the harzadous effect of near surface water on foundations in Lagos Nigeria

O O Adewoyin[1,*], E O Joshua[2], M L Akinyemi[1], M Omeje[1] and E S Joel[1]

[1]Department of Physics, College of Science and Technology, Covenant University, Ota, Ogun State, Nigeria
[2]Department of Physics, Faculty of Science, University of Ibadan, Ibadan, Oyo State, Nigeria

*E-mail: segadot@yahoo.com

Abstract. Adequate information on the condition of the subsurface is very important for site evaluation for engineering purposes. In this study two dimensional (2D) geoelectrical resistivity survey and cone penetration tests were conducted to study the hazardous effect of excess near surface water on the foundation of building in a reclaimed land located at Victoria Island area of Lagos State. The results of the inverted 2D geoelectrical resistivity data revealed three distinct geoelectrical layers characterized by low to moderate electrical resistivity of 2.23 and 129 Ωm and 9.46 to 636 Ωm respectively. The topsoil is characterized by wet sandy soil, which is underlain by sandy clay and banded at the below by a geologic formation of low resistivity which is suspected to be clay. The clay material may be responsible for the excess water retention observed in the area. The CPT method on the other hand revealed a geological formation of low resistance to penetration between 2-3 kg/cm^2 from the topsoil to a depth of 7 m, which may be the effect of excess water in the near surface. This study revealed that the foundation of building may not be founded directly on the soil in any reclaimed land as this may result in collapse as a result of upward migration of water to the near surface.

1. Introduction

Lack of appropriate studies of the subsurface is one of the factors contributing to the frequent building collapse in our society of late. Therefore, adequate studies of the subsurface [1] will assist building developers on the right choice of building materials, type of foundation and the design that matches a particular geological setting [2][3]. Several approaches have been used for the success of foundation investigations among which are the conventional and modern geophysical and geotechnical methods, particularly electrical resistivity technique and cone penetration test [4][5]. This combination has become presently very reliable and popular techniques to investigate the subsurface underground features for different applications, such as environmental, groundwater potential study, engineering and geotechnical, investigation [6][7][8].

The electrical resistivity method is non invasive, cost effective and fast to conduct, it also measures the lateral and vertical variation of resistivity with depth, the acquired resistivity data is used to produce an inverted 2D image of the surveyed area[9]. The cone penetration test on the other hand measures the resistance of the subsurface to cone penetration, which is used to determine the strength or the bearing capacity of the subsurface [10][8][11]. This method is capable of a single point investigation but it is essential to ground-truth the result obtained from other geophysical methods. Although, the method is invasive but it is not destructive, it is a very fast geotechnical method and

Content from this work may be used under the terms of the Creative Commons Attribution 3.0 licence. Any further distribution of this work must maintain attribution to the author(s) and the title of the work, journal citation and DOI.
Published under licence by IOP Publishing Ltd

IOP Conf. Series: Journal of Physics: Conf. Series **852** (2017) 012033 doi:10.1088/1742-6596/852/1/012033

very economical [12][13][14][15]. In the present study, 2D resistivity method using Wenner array electrode configuration and cone penetration test were conducted in order to determine the effect of excess near surface water to engineering construction.

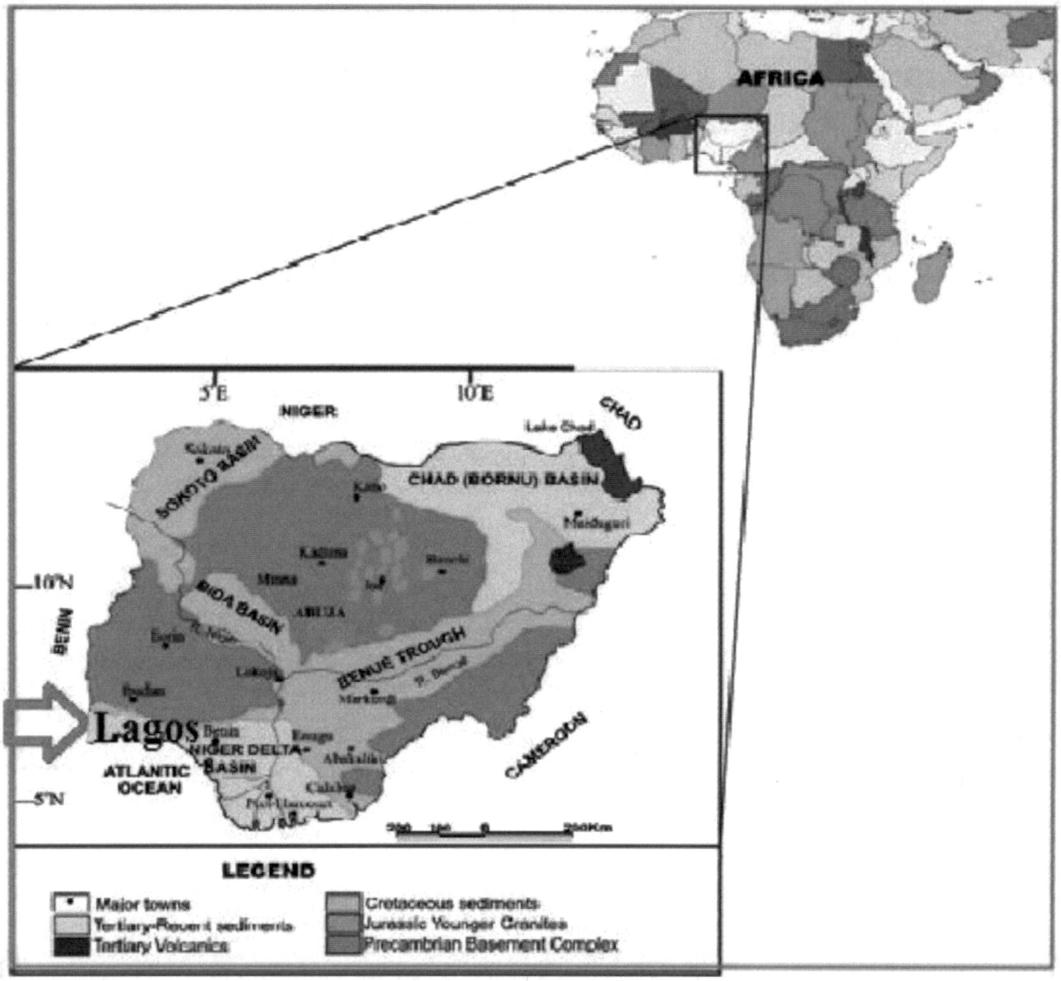

Figure 1. Location map of the studied area

2. Field description

2.1. Geological setting

The study area is located at Eti-Osa local government in the southeastern part of Lagos State (Figure 1). It lies between latitudes $6^0\ 30^I\ 37^{II}$ and $6^0\ 30^I\ 18^{II}$ N and longitude $3^0\ 36^I\ 3^{II}$ and $3^0\ 35^I\ 34^{II}$ E in South West Nigeria. This area is in the zone of coastal creeks and lagoons developed by barrier beaches associated with sand deposition. It is situated in the Nigeria sector of the Benin-basin and near the eastern margin of the basin. The geological formation of the study area is composed of sediments that are typical of the marine environments, which is an intercalation of sand and clay (Figure 2). These sediments also grade into one another and vary widely in both lateral extent and thickness [16][17]. The choice of this study area is because the area is reclaimed from water bodies and it is always logged with water.

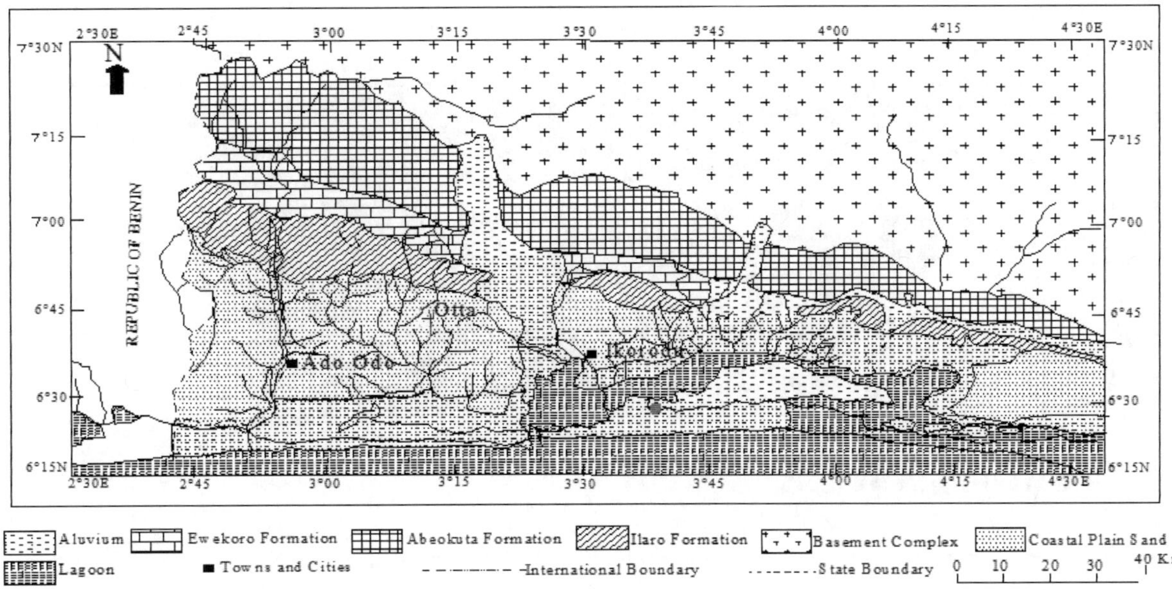

Figure 2. Simplified geological map of the study area, with the study area marked in red

3. Data acquisition and processing

2D electrical resistivity survey was conducted in the study area using an ABEM Terrameter (SAS 1000/4000) [18]. Four 2D profiles were surveyed with profile length ranging between 85 and 250 m as a result of accessibility, 5m electrode spacing was used. The 2D data was processed using RES2DINV software. The RES2DINV inversion code uses nonlinear optimization technique that automatically determines the inverse model of the 2D resistivity distribution of the subsurface [12]. The RES2DINV code subdivides the subsurface into a number of rectangular blocks on the basis of the distribution and volume of the data measured, which also depends on the electrode spacing, positions of electrodes and the data level [13]. The inversion was subjected to least-squares inversion technique with standard least-squares constraint, which reduces the square of the difference between the observed and the computed apparent resistivity. The cone penetration test was conducted with the use of a 2.5 ton Shell and Auger equipment. Two cone tests were conducted at two different locations in the study area. In this method a cylindrical penetrometer with a conical tip was driven into the ground at a constant rate. During penetration, the forces acting on the cone tip and on the shaft behind the tip are measured. For this study a standard Dutch cone (60° apex angle and base area of 10 cm^2) was used. The investigation of each location was concluded when the anchors that held the CPT rig firmly to the ground began to pull out of the ground. The cone resistance data obtained by this technique was plotted against the depth.

4. Results and discussion

Electrical resistivity profiling is used to acquire information on the lateral and vertical variation of resistivity with depth. The results of the inverted 2D electrical resistivity data showed three distinct geoelectrical layers which were delineated in the study area (Figure 3).

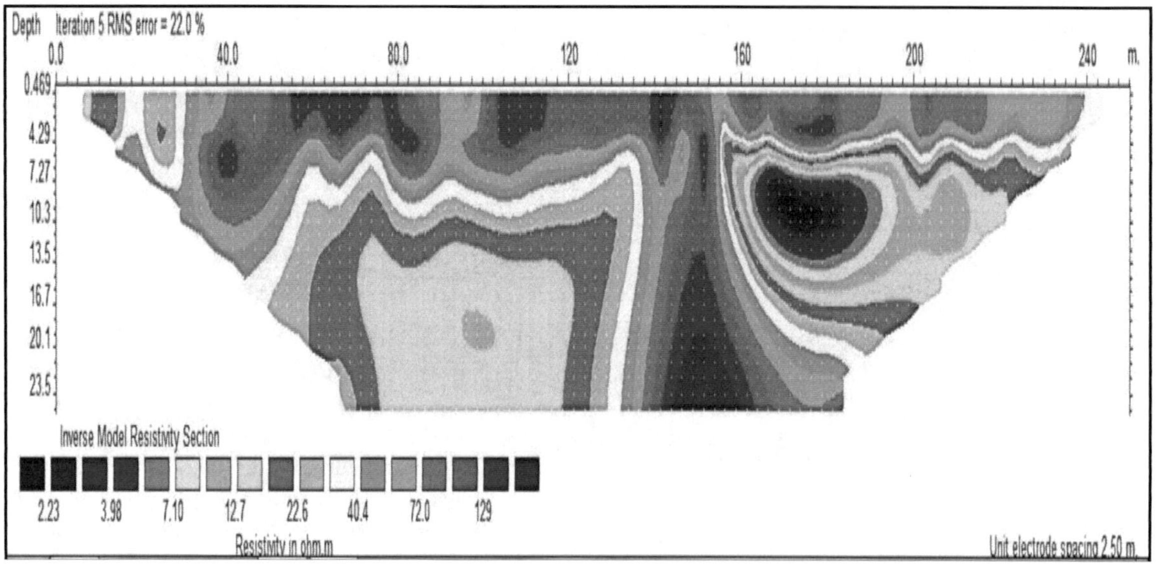

Figure 3. 2D inverse resistivity model conducted in the study area

The lithology of the delineated layers was dependent on all other available information from boreholes, hand-dug wells, known geology and previous studies. The first geoelectric layer represents the top soil with thickness varies from 0.5 to 7 m and has high resistivity value of about 150 Ωm corresponding to wet sand. The second geoelectric layer has a moderate resistivity values that ranged between 30 and 70 Ωm with an average thickness of 6 m representing sand clay. The third geoelectric layer depicted very low resistivity values that ranged between 2 and 10 Ωm with an average thickness of about 8 m corresponding to clay layer in the study site. Within the depth of investigation for foundation at the study area, the 2D electrical resistivity image revealed variations in electrical resistivity with clay layer of appreciate thickness in the subsurface. Also, there is a channel for upward migration of water from the subsurface noticed in the 2D image. Clay materials have poor porosity and this explains why the study area is always logged with water, which may be very harzadous to the foundation of the building. The result of the cone penetration test revealed the material from the topsoil to a depth of about 7 m to be of low resistance of about 2-3 kg/cm^2 (Figure 4). The geologic formation with this type of resistance value may be regarded as a weak geomaterial and of low bearing capacity. The low bearing capacity observed in the first region of the subsurface may be as a result of the influx of water into the geomaterial which is retained by the presence of clay material in the subsurface.

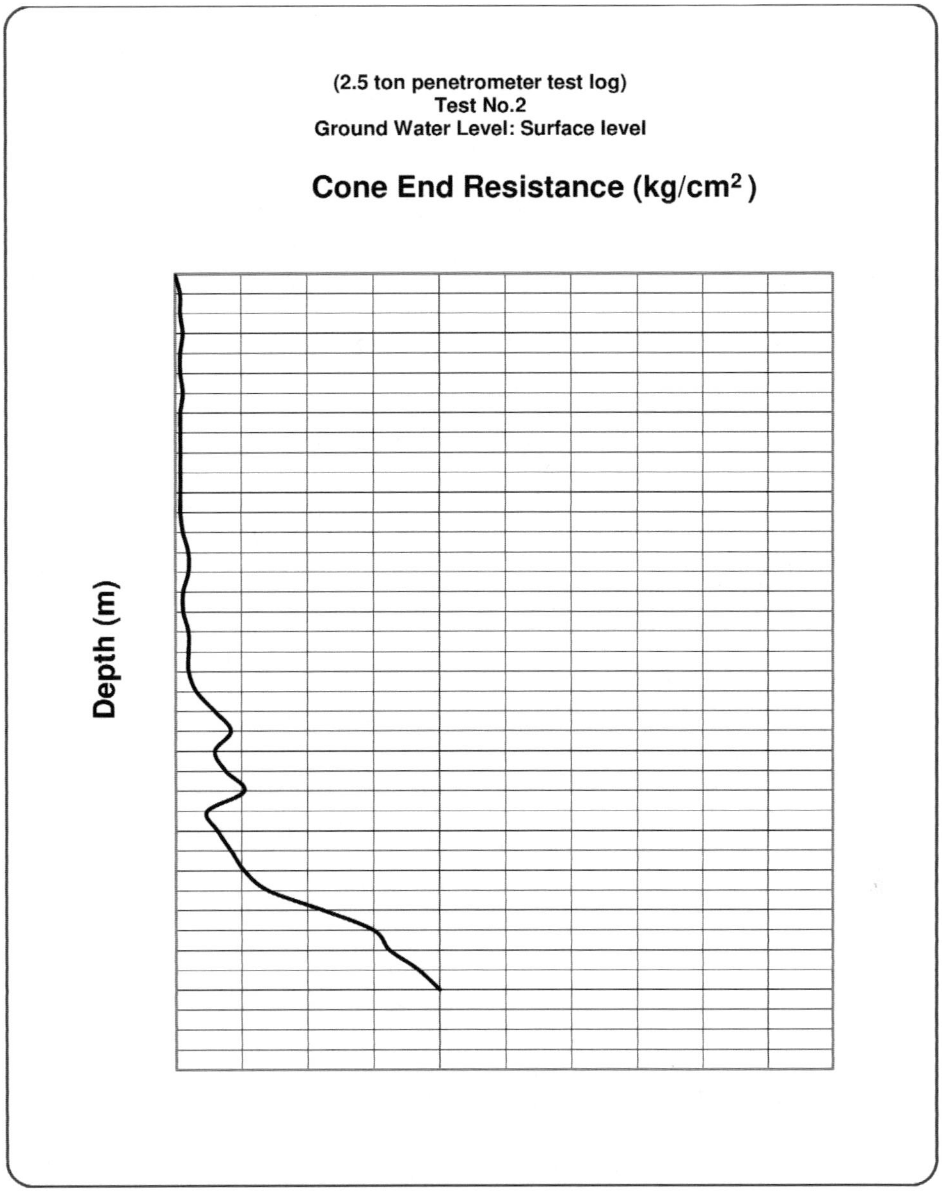

Figure 4. Result of cone penetration test conducted in the study area.

5. Conclusions

From the geologic information, interpretation of resistivity data and cone penetration test results conducted in the study area, it can be concluded that there are three distinct geoelectric layers, which are (from top to bottom); wet sand (150 ohm.m), sandy clay (30 to 70 ohm.m) and clay layer (2-10 ohm.m). Within the depth of investigation for foundation, there are no faults and cavities but a channel that allows for upward migration of water to the near surface. The geomaterial from the topsoil to a depth of about 7 m is composed of soft geologic formation, which may be the influence of water retained in the body of the soil. Some sort of soil improvement or pile foundation is recommended in order to mitigate the direct impact of the foundation on the soil.

References

[1] Rainone M L, Rusi S and Torrese P 2015 Mud volcanoes in central Italy: subsoil characterization through a multidisciplinary approach *Geomorphology* **234** 228
[2] Ayolabi E A, Enoh I J E and Folorunsho A F 2012 Engineering site characterisation using 2-D and 3-D tomography *Earth Science Research* **2**(1) 133
[3] Cassidy R, Comte J, Nitsche J, Wilson C, Flynn R and Ofterdinger V 2014 Combining multi scale geophysical techniques for robust hydro-structural characterization in catchments underlain by hardrock in post-glacial regions *Journal of Hydrology* **517** 715
[4] Muhammad Y K 2013 Engineering geophysical study of unconsolidated topsoil using shallow seismic refraction and electrical resistivity techniques *Journal of Environment and Earth Science* **3**(8) 120
[5] Poenaru A 2016 Correlations between cone penetration test and seismic dilatometer marchetti test with common laboratory investigations *Energy Procedia* **85** 399
[6] Busato L, Boaga J, Perruzo L, Himi M, Cola S, Bersan S and Cassiani G 2016 Combined geophysical surveys for the characterization of a reconstructed river embankment *Engineering Geology* **211** 74
[7] Heon-Joon P, Derk-Moon K, Ki-Seng K, Hee-Yoon A, Dong-Soo K 2012 Noninvasive geotechnical site investigation for stability of cheomseongdae *Journal of Cultural Heritage* **13** 98
[8] Perri M T, Boaga J, Bersan S, Cassiani G, Cola S, Deiana R, Simonini P and Patti S 2014 River embankment characterization: The joint use of geophysical and geotechnical techniques *Journal of Applied Geophysics* **110** 5
[9] Perrone A, Lapenna V and Piscitelli S 2014 Electrical resistivity tomography technique for landslide investigation: A review *Earth-Science Reviews* **135** 65
[10] Loperte A, Soldovieri F, Palombo A, Santini F and Lapenna V 2016 An integrated geophysical approach for water infiltration detection and characterization of Morte Cotugno rock-fill dam (southern Italy) *Engineering Geology* **211** 162
[11] Sarma V S 2014 Electrical resistivity (ER), self potential (SP), induced polarization (IP), spectral induced polarization (SIP), and electrical resistivity tomography (ERT) prospection in NGRI for the past 50 years-A brief review *J. Ind. Geophys. Union* **18**(2) 245
[12] Aizebeokhai A P, Olayinka A I and Singh V S 2010 Application of 2D and 3D geoelectrical resistivity imaging for engineering site investigation in a crystalline basement terrain, Southwestern Nigeria *Environmental Earth Science* **61**(7) 1481
[13] Loke M H 2001 Electrical imaging surveys for environmental and engineering studies: a practical guide to 2D and 3D surveys Available at http:// www.geoelectrical.com [Accessed: 14/01/2013]
[14] Oyedele K F, Oladele S and Adedoyin O 2011 Application of geophysical and geotechnical methods to site characterization for construction purposes at Ikoyi, Lagos, Nigeria *Journal of Earth Sciences and Geotechnical Engineering* **1**(1) 87
[15] Sopaci E and Akgun H 2015 Geotechnical assessment and engineering classification of the Antalya tufa rock, southern Turkey *Engineering Geology* **197** 211
[16] Adepelumi A A and Olorunfemi M O 2000 Engineering geological and geophysical investigation of the reclaimed Lekki, Peninsula, Lagos, South west Nigeria *Bulletin of Engineering, Geology and The Environment* **58** 125
[17] Adepelumi A A, Ako B D, Ajayi T R, Afolabi O and Omotoso E J 2009 Delineation of salt water intrusion into the fresh water aquifer of Lekki Peninsula, Lagos, Nigeria *Environmental Geology* **56** 927
[18] ABEM Instrument AB 2006 Equipment Manual for ABEM Terammeter SAS 1000/4000

Design of a Dual C slot Reflectarray with Enhanced Phase Range Performance

H I Malik[1], M Y Ismail[1], M Amin[2] and F Mohamed[3]

[1] Universiti Tun Hussein Onn Malaysia (UTHM)
[2] Institute of Space Technology (IST), Islamabad, Pakistan
[3] Universiti Kebangsaan Malaysia (UKM)

Email: yusofi@uthm.edu.my

Abstract. This paper deals with the design, fabrication and performance evaluation of a dual C slot compact reflectarray element. It has been demonstrated that the progressive phase distribution of array elements can be achieved by minor variation of C slot widths, without changing the entire patch dimensions. A comparison between measured and simulated results confirms wider phase range coverage of 340° and a 10% bandwidth of 70 MHz have been achieved. A good comparison between simulated and measured results has been exhibited by comparing three different slot widths of the proposed design.

1. Introduction

Reflectarray antennas are a moderate choice between the bulky parabolic dish reflectors and expensive phased array antennas. They have proved their significance as effective antennas in radar systems and long range communications. Reflectarrays were introduced in 1963 as a bulky waveguide structure, fed by a distant spatial source [1]. The potential of reflectarray was not gauged well at that time but the with the advent of printed circuit board (PCB) technology, new design schemes have been presented combining the prime features of parabolic dish reflectors and phased array antennas. Today's reflectarray antenna consists of a planar surface with printed array of radiating elements, illuminated by a feed source. The phases of elements are controlled in such a way to form a planar wavefront at far field distance from array [2].

Different design schemes for the reflectarray unit cells have been presented in the literature to control the phase distribution of individual elements on the periodic reflectarray antenna. Recent techniques for phase distribution include stub loaded microstrip elements, elements of variable sizes and embedded slot configuration in rectangular patches [3-5]. It has also been demonstrated that different ring shaped elements such as the circular spit ring, concentric split ring square and split ring with variable rotation can also be used to control the phase behavior of the reflectarray elements [6-8]. However the phase range of the split ring element is observed to be compromised. Moreover the introduction of a gap in the structure greatly decreases the gradient of the phase curve and reducing the linear phase region of the phase curve. Thus, it is leading to phase errors in a periodic array.

This work presents a novel X-band dual C slot reflectarray antenna. The proposed slot configuration offers a wider phase range coverage and bandwidth performance over the X-band frequency range. The phases of individual elements can be easily controlled by changing the width of embedded slot configuration, without any changes in the entire element size. The design was fabricated over Rogers Duroid 5880 substrate of 0.508 mm thickness. Waveguide simulator technique

Content from this work may be used under the terms of the Creative Commons Attribution 3.0 licence. Any further distribution of this work must maintain attribution to the author(s) and the title of the work, journal citation and DOI.
Published under licence by IOP Publishing Ltd

was used to measure the scattering parameters of the fabricated design. A good agreement was found between the measured and simulated results.

2. Design Considerations

The simulation designs were carried out using CST Microwave Studio. Two C slots were inserted in a conventional rectangular microstrip patch element as a mirror image of each other. The proposed element design has been presented in figure 1(a). Where h_s is the height of the slot while l_{arm} is the length of the arms of C slot. Suitable boundary conditions were applied in CST with electric and magnetic boundaries in order to realize an infinite reflectarray.

The orientation of the E-field excitation is shown in figure. 1(b) by an E-field vector. Figure. 2 shows two patch elements that were fabricated for waveguide simulator measurements. The inter element spacing 'd' was kept half wavelength at 10 GHz to avoid the problem of grating lobes. The width of the slots was varied from 0.6 to 1 mm for three different values, to monitor the effect over scattering parameters. Table 1 summarizes the design specification of double C slot element at X-band frequencies.

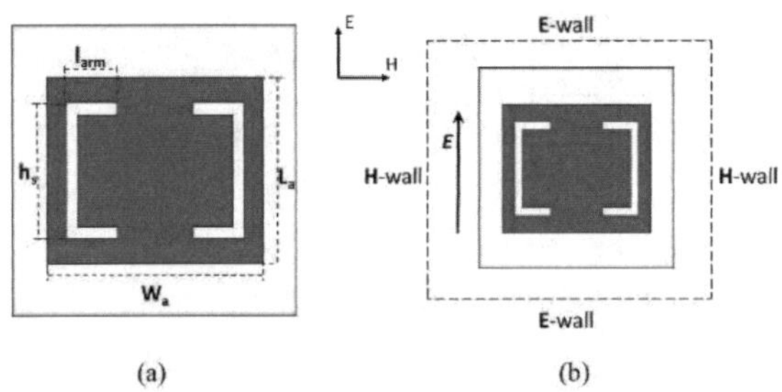

Figure 1. Proposed element design for reflectarray element (a) A dual C slot element (b) Applied boundary conditions in CST MWS

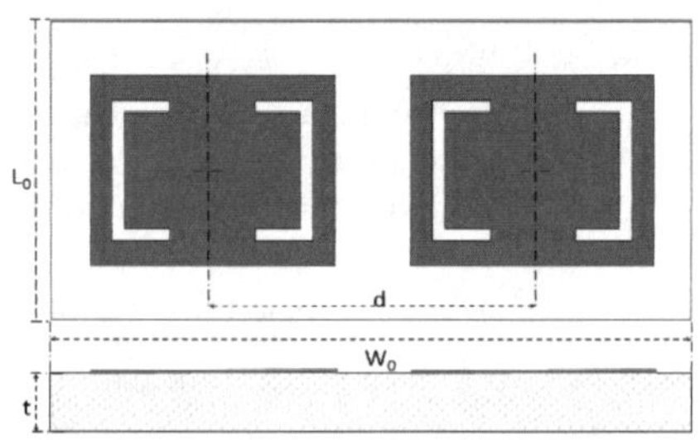

Figure 2. Design of a two element sample for waveguide simulator measurement

Table 1. Design specification for double C slot element for X-band operation

Parameter	Dimension (mm)
L_a	9.3
W_a	11.4
l_{arm}	2.3
h_s	7
W_o	40
d	15
t	0.508
L_o	23

The surface current distribution of the patch element is shown in figure. 3. The surface current distribution shows a major flow of charge carriers between the upper and lower radiating edges of the patch. While travelling between the edges a lot of charge carriers follow the spacing between the slots and the non-radiating edges (mentioned by green shade in figure. 3). There is an even distribution of the charge carriers between the arms of the C slots and charge carriers follow the edges of the C slot, thus increasing the length of the arms will increase the distance travelled by the charge carriers. This will result in decrease in resonant frequency. Moreover the width of the C slot in the patch also affects the flow of the charge carriers, increasing the width decreases the resonant frequency.

In order to examine the dominant mode inside the cavity between the patch and ground plane, E-field monitors were simulated at the resonant frequency. The results have been presented in figure. 4.

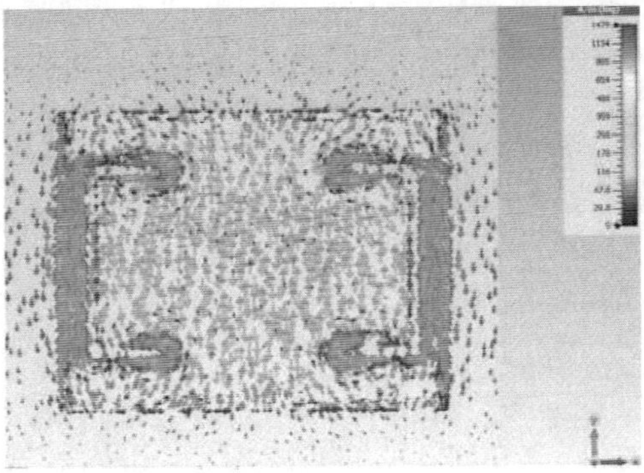

Figure 3. Surface current distribution of a dual C slot element configuration

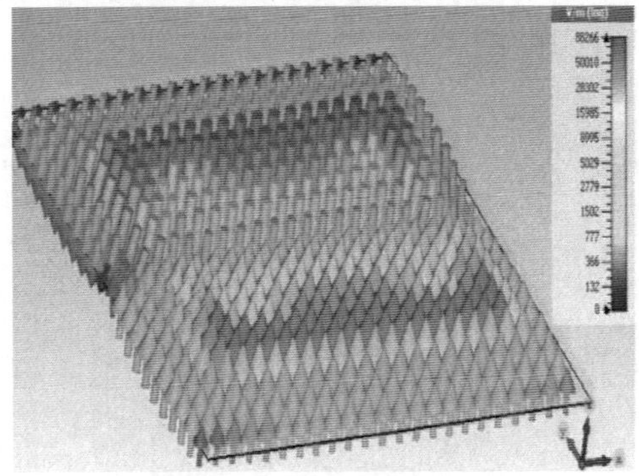

Figure 4. Field Configuration (mode) for the proposed design inside the cavity

The orientation of the E-field lines slows a similar trend as presented for TM01 mode in [9]. At TM01 mode shows a maximum electric field intensity of 88.2 mV/m inside the cavity. This shows that the insertion of C slots does not affect the dominant mode of operation of patch element.

3. Measurement and Considerations

Samples with two unit elements were fabricated over a Rogers Duroid 5880 substrate, with a thickness of 0.508 mm using photolithic etching process. Elements with three different width of C slots were fabricated to validate the proposed design. An X-band waveguide simulator was used in order to realize the effect of an infinite periodic reflectarray. The waveguide simulator technique applies perfect electric and magnetic boundaries for unit reflectarray element to determine its reflection phase and loss before embedding the element into the array [10]. Fabrication tolerances were incorporated in the simulated results after a thorough examination of fabricated samples under a microscope. Figure 5(a) shows the fabricated samples with three different slot sizes. Three samples for each slot width were fabricated and tested in order to ensure repeatability of results. Figure 5(b) shows the measurement setup for the arrangement of the waveguide simulator attached to a vector network analyser via coaxial cable.

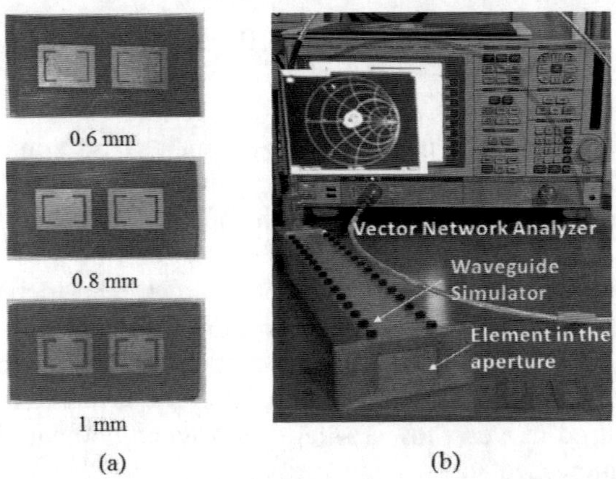

Figure 5. (a) Fabricated samples of dual C slot element with variable slot widths (b) Measurement set up for scattering parameter measurements, waveguide simulator attached to a vector network analyzer

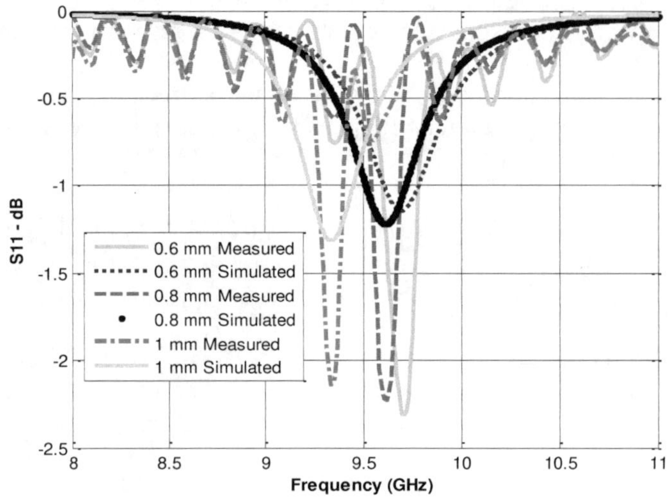

Figure 6. Comparison of simulated and measured reflection loss measurements for 0.6, 0.8 and 1 mm slot width samples

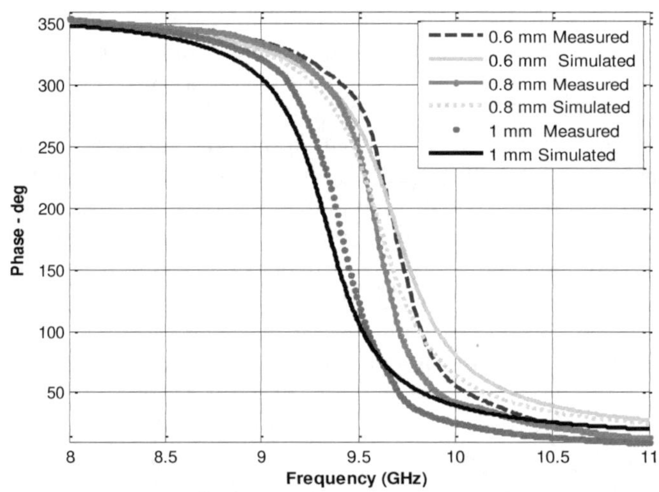

Figure 7. Comparison of simulated and measured reflection phase measurements for 0.6, 0.8 and 1 mm slot width samples

A comparison between scattering parameter measurements and simulated results is presented in figure. 6. The ripples present in the measured results might be due to limitation of the measurement set up. The results demonstrate that with the increase in the slot width from 0.6 to 1 mm the resonant frequency decreases from 9.7 to 9.34 GHz. It can be seen that the loss for the simulated designs is below -1.4 dB while for measured results the loss is -2.3 dB. The increase in loss might be due to losses of the measurement set up such as the insertion loss of connectors and the coaxial cable. The measured results show a 10% bandwidth of 50, 70 and 73 MHz for 0.6, 0.8 and 1 mm slot width elements respectively. Figure. 7 shows the reflection phase measurements of fabricated elements. It can be seen that the measured curves show a steep trend when compared to simulated phase results. This may be due to the increased loss of the measured results than simulated results. In order to analyse the phase curves of measured samples a Figure of Merit (FOM) has been defined in equation 1 [11].

$$FOM = \frac{\Delta\varphi}{\Delta f} \tag{1}$$

where $\Delta\varphi$ is the reflection phase in degrees, and Δf is the frequency range covered. The FOM is calculated in °/MHz. In order to verify the wide phase range of the proposed design, a comparison was also done with the reflection phase of a conventional rectangular patch element with similar substrate properties. Figure 8 shows the reflection phase curves for dual C slot element and rectangular patch element. FOM was calculated for both the designs to compare the static phase region.

Table 2 summarizes the performance analysis of the proposed design configuration. It has been shown that the bandwidth of dual C slot element decreases with increase in slot width. The 10% bandwidth reduces from 73 MHz to 50 MHz as the slot width is increased from 0.6 to 1 mm respectively.

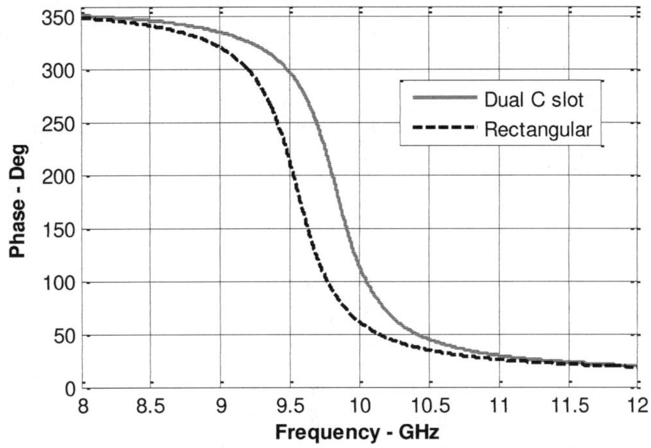

Figure 8. Comparison between reflection phase curves between conventional rectangular patch element and dual C slot embedded elements

Table 2. Performance comparison between rectangular patch and dual C slot element

Slot Width	Bandwidth (MHz)		FOM
	10%	20 %	deg/MHz
Rectangular patch	78	120	0.52
0.6 mm	73	105	0.82
0.8 mm	70	96	0.70
1 mm	50	87	0.55

Moreover, it can be seen from Table I that FOM decreases from 0.82 to 0.55 deg/MHz with slot width increase from 0.6 to 1 mm respectively. The FOM for a rectangular patch is 0.52 with a 20 % bandwidth of 120 MHz, however dual C slots of 0.6 mm provides an FOM of 0.82 with a 20 % bandwidth of 105 MHz. The-decrease in the bandwidth is due to slight increase in the loss of the element. Thus a dual C slot configuration offers a better phase gradient as compared to conventional rectangular patch.

4. Conclusion

A novel rectangular microstrip reflectarray element embedded with dual C slots has been presented for X-band frequency applications. It has been demonstrated that a small variation in the width of C slots results in a significant change in the resonant frequency of the element. The results show that by changing the slot width from 0.6 mm to 1 mm the resonant frequency varies from 9.7 to 9.34 GHz. Moreover the proposed design with a slot width of 0.6 mm is shown to offer wider phase range coverage of 340° along with efficient bandwidth performance. The proposed design scheme offers a 20% bandwidth improvement from 87 to 105 MHz with the decrease in slot width from 0.6 to 1 mm. A Figure of Merit (FOM) has been defined in order to analyse the gradient of phase curves. Comparison of phase gradient with a conventional microstrip patch shows wider phase coverage for dual C slot embedded configuration.

Acknowledgement
The work was funded by GPPS Grant (VOT466), RACE Grant (VOT1119) and Research and Development (R&D) Fund awarded by Ministry of Higher Education Malaysia.

References

[1] Berry, G D G, Malech, R G and Kennedy W A 1963 The Reflectarray Antenna. *IEEE Trans. Antennas and Propagation* AP-11
[2] Huang, J and Encinar, J A 2008 Reflectarray Antennas *IEEE Press*, John Wiley
[3] Metzler, T A 1995 Stub Loaded Microstrip Reflectarrays *International Symposium on Antenna and Propagation Society* **1** 574
[4] Pozar, D M and Metzler T M 1993 Analysis of a Reflectarray Antenna using Microstrip Patches of Variable Size *Electronic Letters* **29(8)** 657
[5] Ismail M Y and Inam M 2010 Performance Improvement of Reflectarrays Based on Embedded Slot Configurations *Progress in Electromagnetic Research C* **14** 67-78.
[6] Misran N 2004 Design Optimisation of Ring Elements for Broadband Reflectarray Antennas *PhD Thesis, Faculty of Engineering*, The Queen's University Belfast, Ireland.
[7] Yusop S H, Misran N, Islam M T and Ismail M Y 2009 Analysis of Concentric Split Ring Square Reflectarray Element for Bandwidth Enhancement *Proc. International Conference on Space Science and Communication pp* 62
[8] Han C M and Chang K 2003 Ka-band reflectarray using ring elements *Electronic Letters* **39**(6)
[9] Balanis C A 2005 Antenna Theory *3rd Ed Wiley*
[10] Ismail MY and Inam, M 2010 Analysis of design optimisation of bandwidth and loss performnace of reflectarray antennas based on material properties *Modern Applied Sciences CCSE* **4(1)**
[11] Hannan P W and Balfour M A 1965 Simulation of phased-array antenna in waveguide *IEEE Transactions on Antennas and Propagation* **13(3)** 342

IOP Conf. Series: Journal of Physics: Conf. Series **852** (2017) 012035 doi:10.1088/1742-6596/852/1/012035

Design and analysis of dual U slot reflectarray antenna for X-band applications

M Y Ismail[1], H I Malik[1] and M Amin[2]

[1] Universiti Tun Hussein Onn Malaysia (UTHM)
[2] Institute of Space Technology (IST), Islamabad, Pakistan

Email: yusofi@uthm.edu.my

Abstract. A novel design of a dual frequency single layer reflectarray antenna is presented for X-band frequency applications. Dual U slots embedded on conventional rectangular microstrip reflectarray have been designed to attain a dual frequency operation. A detailed analysis of the effect on surface current distributions with the introduction of dual U slots is presented. Moreover a parametric study on the variation of significant dimensions of the design have been carried out and analysed thoroughly using a commercially available CST computer model. Proposed design configurations were fabricated above a 0.508 mm thick substrate of Rogers Duroid 5880. The dual U slot configuration offers a significant dual frequency behavior at 8.54 and 11.56 GHz with 10% bandwidth improvement of 47 and 56 MHz with a reflection loss of -4.54 and 4.11 dB respectively.

1. Introduction

Parabolic dish reflectors have been widely used particularly for high gain and efficient bandwidth applications. They have demonstrated efficient performance for their long range communication effectiveness and radar applications, but the bulkiness of size and the rotary mechanism needed for the paraboloid, increases the complexity and cost. A reflectarray antenna on the other hand has a light weight flat surface with strategic arrangement of distributed elements. The flat surface reflects the incident electromagnetic energy of a feed antenna in such a way that a planar wavefront is formed at a far field region from the antenna [1].

Reflectarray antennas were introduced to replace the widely used parabolic dish reflectors [2]. The planar surface of the reflectarray is spatially fed by a feed antenna usually a horn antenna. The spherical beam front of horn antenna is converted into a planar beam front by strategic phase distribution of the elements over the array [3]. The first presented design of reflectarray was a bulky arrangement of open ended wave guides [2]. The idea though was not advanced further at that time, due to sheer size of the waveguide array. In the past few years, reflectarrays have significantly evolved with the advent of printed circuit board technology (PCB) in which printed reflectarrays emerged as compact and easily fabricated antennas compared to parabolic dish reflectors. However due to intrinsic low bandwidth, microstrip reflectarray antennas face a critical challenge of a narrow bandwidth and high Q factors [1].

Microstrip reflectarrays pose a narrow bandwidth generally less than 5%. In order to improve the bandwidth performance of microstrip reflectarray, different techniques have been presented in the literature. In the past, different techniques have been reported to realize dual frequency reflectarrays such as the use of split ring slots, variable dual band gaps embedded in conventional rectangular patches, split ring square elements and by adding phase delay lines to microstrip reflectarray elements [4-7]. Multilayer structures have also been reported for bandwidth improvement of reflectarrays. It

Content from this work may be used under the terms of the Creative Commons Attribution 3.0 licence. Any further distribution of this work must maintain attribution to the author(s) and the title of the work, journal citation and DOI.
Published under licence by IOP Publishing Ltd

has been demonstrated that two and three layer structures can also be used for bandwidth improvement and for realizing a dual frequency of operation [8-10]. However the addition of phase delay lines to microstrip patches leads to undesired spurious radiation losses which adds up to increased cross polarization levels in arrays. Additionally, the use of multilayer structures for dual band of operation also increases the complexity of the design. The two or three layer structure needs precise alignment

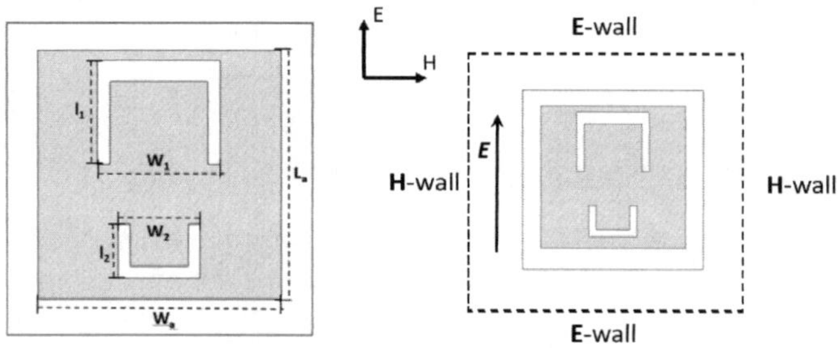

(a) (b)

Figure. 1. Simulated designs of dual U slot element configuration (a) Simulated design and (b) Boundary conditions of a dual U slot unit cell

of layers to achieve fruitful results. Moreover the increase in the substrate height also may result in generation of unwanted surface waves that increase the loss of array. The individual reflectarray element reflection phase is required to cover entire range of 360°, however due to fabrication tolerances this requirement is not fulfilled and leads to phase errors in an array. To avoid this problem dual resonance elements have also been reported in order to achieve a phase shift greater than 360°, thus avoiding phase errors [11].

This paper presents a novel dual U slot configuration for dual resonance operation in X-band. Dual U slots embedded in a conventional rectangular patch element were designed. Significant parameters of the design were also swept over suitable frequency ranges to analyse the effect over the two resonances. Fabricated samples were tested for scattering parameters using a waveguide simulator in order to cater the effects of an infinite array [12-13]. The results demonstrate a good agreement between the simulated and fabricated element designs.

2. Design Considerations

Simulation designs of the proposed dual frequency reflectarray element were done using CST MWS. The proposed dual U slot element design is shown in figure 1(a). By using controlled boundary conditions for a dual U slot unit reflectarray element, perfect electric and magnetic walls have been realized to obtain an infinite periodic reflectarray.

Table 1. Dual U slot element dimensions on Rogers Duroid 5880

Parameter	Dimension (mm)
W_a	12
L_a	8.8
l_1	3
W_1	4
W_2	6
l_2	1

The boundary conditions for the unit element are shown in figure 1(b). The orientation of the incident electric field is shown by the E-field vector while the dotted line shows the electric and magnetic walls. The dimension of the U slot elements were selected in such a way that dual frequencies can be realized in the X-band frequency range The U slot elements were printed above a Rogers Duroid substrate of 0.508 mm thickness, a dielectric permittivity of 2.2 and a loss tangent of 0.0009. The simulated dimensions for the patch element are tabulated in Table 1.

3. Parametric Analysis

In order to fully understand the behavior of the proposed design, different parameters of the elements were studied. The variation of the selected parameters were studied for X-band frequency range to monitor the effect on the reflection phase and the reflection loss of element. The selection of the dimensions was done such that the resonances remain in the X-band of operation. The variation of selected dimensions is tabulated in Table II, Where L_a is the length of the antenna and l_1 is the length of the arms of bigger U slot.

Table 2. Dimensions and ranges for dual U slot configuration parametric analysis

Parameter	Range (mm)
L_a	8.8 – 10
l_1	2.5 – 3

Reflection loss of the element over this range was simulated and compared with the surface current distribution of the element to analyse the results. The length of the antenna L_a was varied from 8.8 to 10 mm. Sweeping the element length (L_a), results in decrease of frequency for both the upper and lower resonances of the element. Figure 3 shows the effect of patch length on the reflection loss of both resonant frequencies. As shown in figure 2, by increasing the patch length L_a from 8.8 to 10 mm, the lower and upper resonant frequencies are observed to be shifted by 470 and 600 MHz respectively. Moreover the loss is shown to offer an alternate effect for both resonances. The loss for lower resonance decreases from -3.58 to -2.54 dB while for higher resonance the loss increases from -2.67 to -3.35 dB. Since the surface currents for both the resonances lie in between the top and bottom edges of patch, so increasing length of the patch will increase the current paths for both the resonances, thus reducing the resonant frequencies.

As shown in figure 3, the variation of the U slot arms significantly affects both upper and lower resonant frequencies. The increase of arm length results in decrease in both resonant frequencies. As the length is increased from 2.5 to 3.2 mm, the lower frequency is observed to decrease from 8.4 to

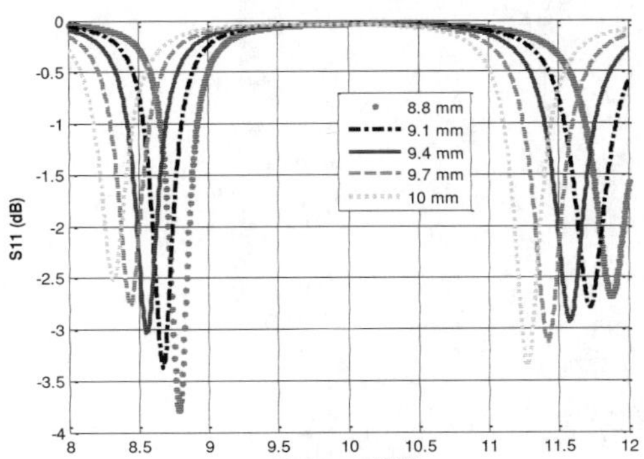

Figure 2. Effect of vertical length L_a of the patch on the resonance behavior for the range $8.8 \leq L_a \leq 10$ mm

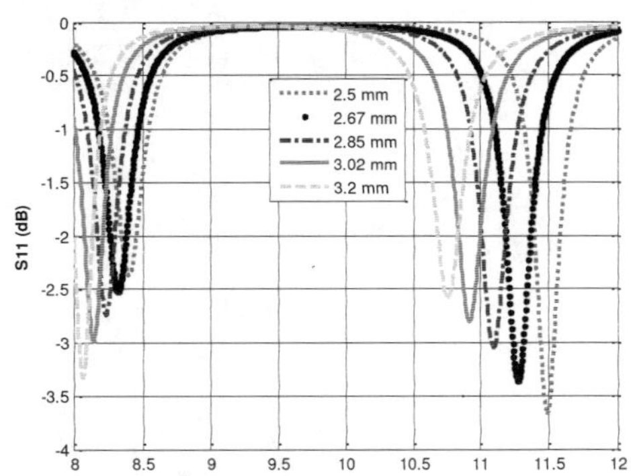

Figure 3. Effect of arm length (l_1) variation for the range $2.5 \leq l_1 \leq 3.2$ mm

IOP Conf. Series: Journal of Physics: Conf. Series **852** (2017) 012035 doi:10.1088/1742-6596/852/1/012035

8.06 GHz, whereas upper resonant frequency tends to decrease from 11.49 to 10.74 GHz respectively. This effect is due to the decrease in the path length of the charge carriers that exists between the lower edges of the path element around the arm lengths (l1) and the upper edge of element. Additionally it is observed that, the loss for this parametric sweep follows alternate trends for both the resonances. For the lower resonance the loss increases from -2.63 to -3.2 dB, whereas the loss of the upper resonance decreases from -3.67 to -2.57 dB respectively. This behavior shows that both the length of the patch element and the length of the arms of U slot should be monitored in order to control the resonance behavior of the patch element.

4. Measurements

In order to validate the reliability of simulated results, dual U slot configurations were fabricated above Rogers Duroid 5880 substrate of 0.508 mm thickness. The proposed designs were fabricated using photolithic etching technique. The ground plane of the elements were extended to fit in the aperture of the waveguide. Before development of full array individual elements of the array needs to be characterized for their reflection phase and loss. Waveguide simulator technique provides realization of an infinite reflectarray with perfect magnetic and electric boundaries [12]. Fabricated elements were placed inside the aperture of the waveguide simulator to record the scattering parameters. Figure 4 shows the fabricated sample, waveguide simulator and the measurement set up for the measurement of reflection loss.

Figure 4(a) shows the fabricated dual U slot elements over Rogers Duroid 5880. As shown in figure 4(b) an X-band adapter connected to a tapered waveguide to form a waveguide simulator. Multiple samples of single design were fabricated and measured to ensure the repeatability of the results. Dimensional errors were introduced after the fabrication of the samples due to limitation of the fabrication process. In order to cater those dimensional errors, dimensions of the samples and the inserted U slots were measured again using a microscope. The simulated results were then updated with the fabricated dimensions for correction.

Figure 4(c) shows the measurement set up for scattering parameter measurements. The measurement set up consists of a waveguide simulator over the X-band frequency operation. A vector network analyzer (VNA) was used to measure the scattering parameters of the elements under test. Waveguide and the vector network analyzer are connected together by a coaxial cable. The vector network analyzer was calibrated for the X-band and was tested for short, open and matched loads to remove any discrepancies of the set up. The fabricated elements were then placed one by one in the aperture of the waveguide to record the reflection loss and the reflection phase of the elements. Scattering parameters for the proposed design are presented in Figure 5.

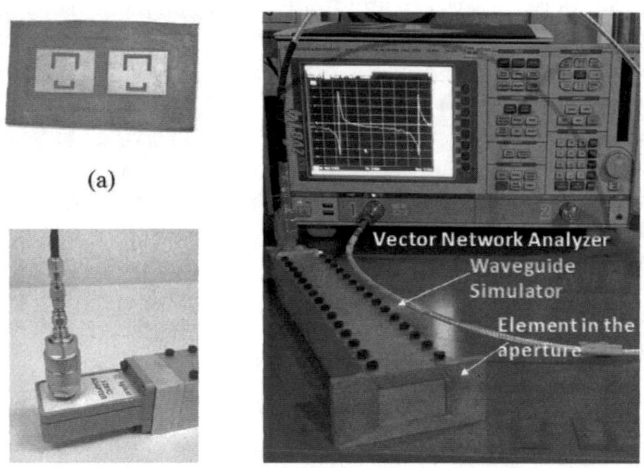

(b) (c)

Figure 4. Measurement setup for scattering parameters (a). Fabricated sample for dual U slot configuration. (b) An X-band adapter connected to a tapered waveguide (c) Measurement set up with a waveguide simulator attached to VNA for scattering parameter measurements

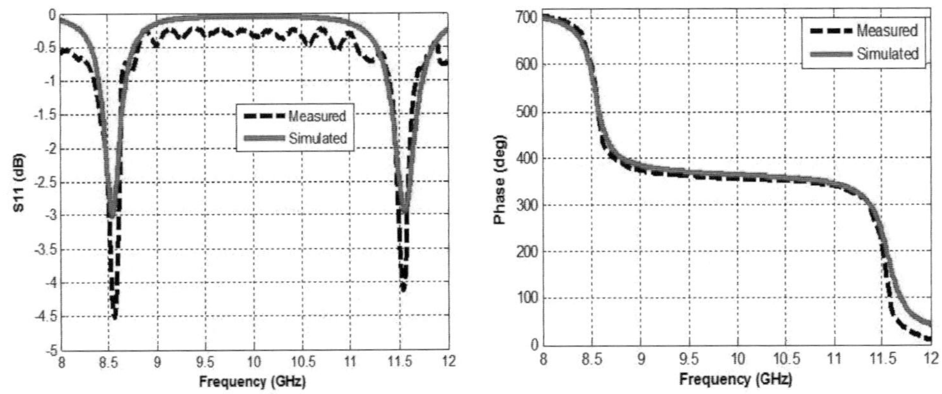

(a) (b)

Figure 5. Simulated and measured results for 0.508 mm thick substrate (a) Reflection loss (b) Reflection Phase

The results shown in figure 5 demonstrate that a good agreement of a dual frequency operation is realized at 8.52 and 11.51 GHz respectively as shown in figure 5(a). Moreover a reflection loss of -4.5 dB and -4.11 dB have been achieved. The simulated results are shown by red line while the measured results for dual U slot element is shown by dotted black line. The ripple present in the measured results may be due to deviation of the waveguide simulator from its ideal behavior. The results show that the bandwidth performance of 47 and 56 MHz have been achieved for upper and lower resonances, with a reflection loss of -4.5 and -4.11 dB respectively.

Figure 5(b) shows the reflection phase of the fabricated element. Due to the dual resonance behavior of the proposed element, reflection phase should cover a phase range of 720 deg ideally. The reflection phase of simulated design covers a phase range of 670 degrees. However fabricated sample covers a phase range of 690 degrees. The increase in the phase coverage comes at cost of increased loss of the element. In order to compare the linear static region of the phase curves, a Figure of Merit (FOM) is defined as a ratio of reflection phase covered to the change in frequency and can be mathematically expressed as.

$$FOM = \frac{\Delta\varphi}{\Delta f} \tag{1}$$

where $\Delta\varphi$ the phase range is covered by the reflection phase curve and Δf is the change in resonant frequencies in MHz. The FOM is measured as deg/MHz. FOM is calculated for both the resonances. A summary of the scattering parameter results is tabulated in Table 3.

Table 3. Comparison between measured and simulated results for Dual U slot reflectarray element

	Resonant Frequency (GHz)	Reflection Loss (dB)	Phase Range (Deg)	Bandwidth (MHz)	Figure of Merit (Deg/MHz)
Simulated	8.54 | 11.56	-3.01 | -2.94	670	65 | 80	1.03 | 0.91
Measured	8.58 | 11.52	-4.55 | -4.56	690	47 | 56	1.33 | 1.45

Table 3 shows the summarized results of the simulated and measured designs for dual resonance behavior. The results depicted in Table 3 show that, there is an increase in the measured loss when compared to simulated design, which is due to the loss of measurement set up and fabrication limitations. The phase range covered by the fabricated samples is slightly more than the simulated. This may be due to the increased loss of the fabricated samples. A good bandwidth performance is shown by both the samples, however the reduction of bandwidth compared to simulated is due to increased loss. Figure of merit shows that increase in loss increases the slope of the phase curve thus FOM for fabricated samples is more than that of simulated designs.

5. Conclusion
A novel dual U slot microstrip reflectarray configuration has been presented for dual frequency operation. By embedding two U slots of variable sizes in a conventional rectangular patch, a phase range greater than 360° can be achieved which provides an improved bandwidth performance. A detailed parametric study for significant dimensions of the design shows that both the resonances can be adjusted conveniently with in X-band operation. A 13 % bandwidth improvement has been achieved by dual U slot configuration compared to conventional rectangular patch element. Moreover a Figure of Merit (FOM) defined shows that a phase range greater than 360 deg can be attained at a gradient of 1.33 and 1.45 for the higher and lower resonance of 8.58 and 11.52 GHz respectively.

Acknowledgement
The work was funded by GPPS Grant (VOT466), RACE Grant (VOT1119) awarded by Ministry of Higher Education Malaysia.

References
[1] Huang J and Encinar J A 1963 *Reflectarray Antennas*. IEEE Press, John Wiley, 2008.
[2] GDG. Berry, RG Malech, and WA Kennedy, The Reflectarray Antenna. *IEEE Trans. Antennas and Propagation* AP-11
[3] Ismail M Y, Inam M, Zain A F Mand Misran N 2013 Mathematical Model for Progressive Phase Distribution of Ku-band Reflectarray Antennas. *International Journal of Communication Science and Engineering* 7(12) 105
[4] Fuentes J A O, Montero J A, Lopez J I M, Cuevas J R and Martynyuk A E 2016 Dual-Frequency Reflectarray Based on Split Ring Slots. *IEEE Antennas and Wireless Propagation Letters.* **99** 1
[5] Ismail M Y and Sulaiman N H 2011 Enhanced bandwidth reflectarray antenna using variable dual gap *Proc. International Conference on Instrumentation, Communication, Information Technology and Biomedical Engineering (ICIC-BME)* pp 92
[6] Yusop S H, Misran N, Islam M T and Ismail M Y 2013 Analysis of concentric split ring square reflectarray element for bandwidth enhancement *Proc. International Conference on Space Science and Communication* pp 62
[7] Chunhui H, Yunhua Z and Qingshan Y 2016 Single-layer reflectarray antennas with improved bandwidth by attaching phase-delay lines *10th European Conference on Antennas and Propagation (EuCAP)* Davos, Switzerland p 1
[8] Encinar J A 1996 Design of a Dual Frequency Reflectarray using Microstrip Stacked Patches of Variable Sizes. *Electronics Lett.* **32** (12) 1049
[9] Encinar J A and Zornoza J A 2003 Broadband Design of Three Layer Printed Reflectarrays. *IEEE Transactions on Antenna and Propagation* **51**(7) 1662
[10] Encinar J A and Zornoza J A 2004 Three Layer Printed Reflectarrays for Contoured Beam Space Applications. *IEEE Transactions on Antennas and Propagation* **52**(5) 1138
[11] Tong Cai, Guang- Ming, Xiao-Fei Zhang 2016 Compact dual-resonance element with low phase sensitivity for offset reflectarray antennas *IEEE Antenna and Propagation Letters* pp 99
[12] PW Hannan and MA Balfour 1965 Simulation of phased-array antenna in waveguide *IEEE Transactions on Antennas and Propagation* **13**(3) 342
[13] Stockmann J and Hodges R 2005 The Use of Waveguide Simulator to Measure the Resonant Frequency of Ku-band Microstrip Arrays *IEEE International Symposium on Antenna and Propagation Society* **1**A 417

IOP Conf. Series: Journal of Physics: Conf. Series **852** (2017) 012036 doi:10.1088/1742-6596/852/1/012036

Transparent solar antenna of 28 GHz using transparent conductive oxides (TCO) thin film

N I Mohd Ali[1,2], N Misran[1], M F Mansor[1] and M F Jamlos[3]

[1]Department of Electrical, Electronic and System Engineering, Faculty of Engineering and Built Environment, Universiti Kebangsaan Malaysia, 43600, Bangi, Selangor, Malaysia
[2]Faculty of Engineering Technology, Universiti Malaysia Perlis, Kampus UniCITI Alam, Sungai Chuchuh, 02100, Padang Besar, Perlis, Malaysia
[3]School of Computer and Communication Engineering, Universiti Malaysia Perlis, Kampus Tetap Pauh Putra, 02600, Arau, Perlis, Malaysia

E-mail: nurizzati@siswa.ukm.edu.my, bahiah@ukm.edu.my, m.mansor@ukm.edu.my faizaljamlos@unimap.edu.my

Abstract. This paper presents the analysis of 28GHz solar patch antenna using the variations of transparent conductive oxides (TCO) thin film as the radiating patch. Solar antenna is basically combining the function of antenna and solar cell into one device and helps to maximize the usage of surface area. The main problem of the existing solar antenna is the radiating patch which made of nontransparent material, such as copper, shadowing the solar cell and degrades the total solar efficiency. Hence, by using the transparent conductive oxides (TCO) thin film as the radiating patch, this problem can be tackled. The TCO thin film used is varied to ITO, FTO, AgHT-4, and AgHT-8 along with glass as substrate. The simulation of the antenna executed by using Computer Simulation Technology (CST) Microwave Studio software demonstrated at 28 GHz operating frequency for 5G band applications. The performance of the transparent antennas is compared with each other and also with the nontransparent patch antenna that using Rogers RT5880 as substrate, operating at the same resonance frequency and then, the material that gives the best performance is identified.

1. Introduction

Throughout the years, the communication systems are rapidly developed. Typical antenna and microwave circuit are made of thick metal layers as conducting element and placed on a dielectric substrate, and then a thick metal layer as ground plane placed on the bottom of it. Recently, the communication system demanding antenna is with low profile characteristic. It is to minimize if not, eliminate the visual impact of antenna in the system [1, 2, 3]. Hence, transparent antenna has caught researcher attentions as one of the solutions. By implementing this feature, an antenna design can be more forgiven in terms of size, installation, and structure. There are other applications that also beneficial from this features, for instance, solar antenna.

Transparent Conducting Oxides (TCO) film is a thin film that is conductive electrically and transparent optically [4, 5]. This remarkable combination of transparency and conductivity leads to great development of electronic components and telecommunication devices, such as touch screens, solar cells, and electrodes for displays, as it can reduce the visual impact and flexibility in terms of installations.

Content from this work may be used under the terms of the Creative Commons Attribution 3.0 licence. Any further distribution of this work must maintain attribution to the author(s) and the title of the work, journal citation and DOI.
Published under licence by IOP Publishing Ltd

Solar antenna is basically combining two applications in one device and it is mostly used on satellite as it helps to maximize the usage of satellite's surface area, which is the main concern. The main problem of the solar antenna that have been designed using non-transparent material [6, 7, 8, 9, 10] and semi-transparent material [11, 12, 13, 14, 15, 16] are including bulky, have complicated design, and also experiencing shadowing effect that causing the degradation of total solar efficiency.

In this paper, transparent solar patch antennas for 5G band application with operating frequency of 28 GHz are designed. In order to investigate the performance, instead of using Copper, several types of TCOs have been used as the conducting patch of the antenna which is ITO, AgHT-4, AgHT-8, and FTO. The performance of the antennas are compared to each other and to the reference antenna which is nontransparent solar patch antenna at the same resonance frequency that made of copper and Rogers RT5880 as the patch and substrate respectively.

2. Design

The antenna design used in this paper is the basic rectangular microstrip patch antenna as shown in Figure 1 and based on the design specification in Table **1**. It comprises of two parts which are the antenna part and the solar part. Figure 1(b) shows the design of the patch as the radiating element and (c) shows the structure of the solar cell which consists of three main layers, anode lattice, silicon, and cathode. The material used in this design for the anode and cathode layers are silver and aluminum respectively.

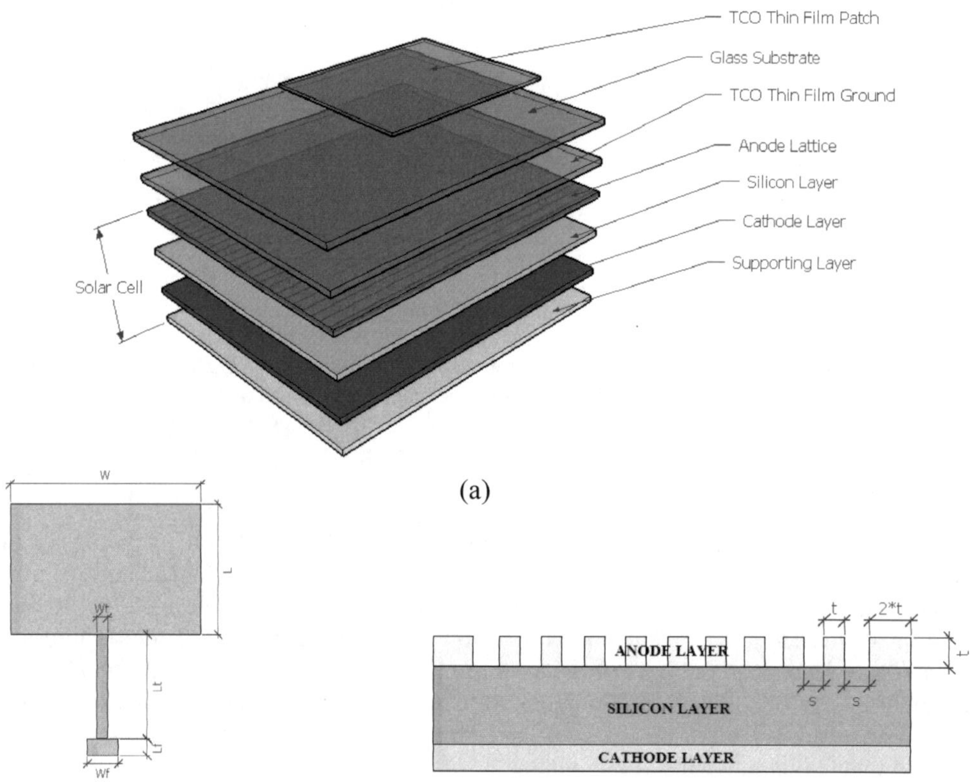

(a)

(a) (b)

Figure 1. (a) Structure of designed antenna (b) top view of the patch design and (c) cross section of solar cells

Table 1. Design specification

Operating freq.	Gain	Bandwidth	Rad. efficiency
28 GHz	>4 dB	>1.5 GHz	>55%

For the nontransparent solar antenna which for reference purposes, Copper material used as the ground and the radiating patch material, whereas Rogers RT5880 used as the substrate with dielectric constant, ε_r and thickness of 2.2 and 0.254 mm respectively. Table 2 shows the optimized dimension of the design.

Table 2. Optimized dimensions of nontransparent antenna

Parameter	Length (mm)
L	3.357
W	4.235
W_t	0.2
L_t	1.8046
W_f	0.7826
L_f	1.958
s	2
t	0.1

There are four transparent solar antennas have been designed and will be discussed in this paper. Glass materials have been used as the transparent substrate with dielectric constant, ε_r of 4.82 and thickness of 1.1mm. As for the radiating patch and ground layer, TCO thin film are used and varied to Indium Tin Oxide (ITO) films, fluorine-doped tin oxide (FTO), conductive silver-coated thin film, AgHT-4, and AgHT-8 with sheet resistance, R_S of 10Ω/sq, 24Ω/sq, 4.5Ω/sq, and 8Ω/sq respectively. Equation (1) below show the relationship between the sheet conductivity, σ and sheet resistance, R_S [1, 17, 18, 19].

$$R_S = \frac{\rho}{t} \text{ and } \sigma = \frac{1}{\rho} \tag{1}$$

Table 3 shows the optimized width and length of each transparent antenna. Even though the properties of the conducting materials are varied, the optimized dimensions of all four transparent antennas are similar as the same substrate used. This is in line with the equation from [20] that shows the dimensions of a rectangular patch antenna are guided mainly by the properties of the dielectric substrate.

Table 3. Optimized dimension of transparent antennas

PARAMETERS	ITO	FTO	AgHT-4	AgHT-8
Length of patch (mm)	1.594	1.57	1.57	1.57
Width of patch (mm)	3.90	3.90	3.90	3.90

3. Result and Analysis

3.1. Return Loss and Bandwidth

Figure 2 shows the frequency response of transparent and non-transparent antenna with resonance frequency of 28GHz, and Table 4 shows the results summary of return loss and bandwidth of the design.

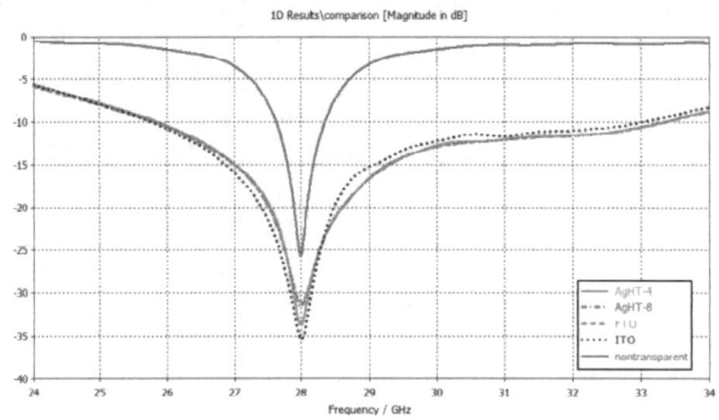

Figure 2. Frequency Response

Table 4. Return Loss and Bandwidth Results.

Conducting material	PEC	ITO	FTO	AgHT-4	AgHT-8
Return loss	-25.729	-35.437	-31.123	-33.833	-31.573
Bandwidth (GHz)	0.7395	7.2808	7.5964	7.4912	7.584
Bandwidth (%)	2.64%	26.00%	27.13%	26.75%	27.09%

As predicted, the bandwidth of the transparent and non-transparent antenna shows great difference due to its substrate thickness which is more than 4 times thicker than the non-transparent. Transparent antenna that gives the highest bandwidth is the FTO whereas ITO gives the best performance in terms of return loss.

3.2. Surface Current Density

Results as depicted in Figure 3 shows the surface current of each element of the conducting material. As foreseen, copper gives the highest maximum surface current with 230.4 A/m as copper have the highest conductivity. Same goes to the transparent antenna, maximum current density of ITO is the highest with 125.1 A/m as it have the highest conductivity compare to the other thin films.

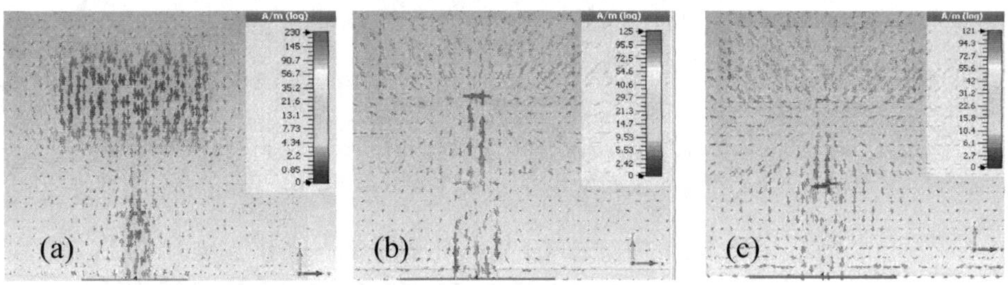

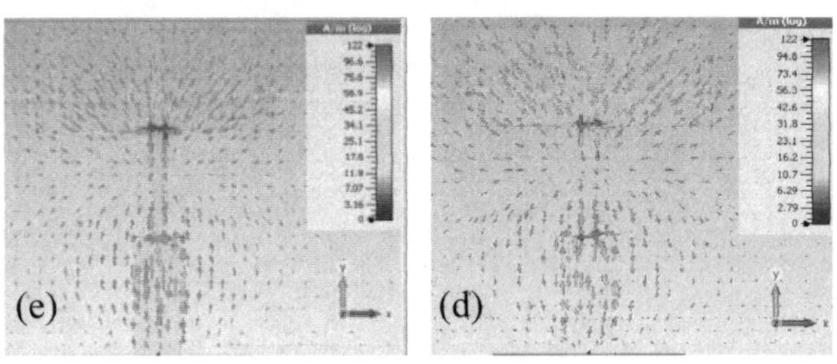

Figure 3. Surface current density of (a) non-transparent (Copper) (b) ITO, (c) FTO, (d) AgHT-4 and (e) AgHT-8

IOP Conf. Series: Journal of Physics: Conf. Series **852** (2017) 012036 doi:10.1088/1742-6596/852/1/012036

3.3. Gain, Radiation Pattern, VSWR, and Radiation Efficiency

Figure 4 shows the radiation pattern of the non-transparent antenna and all transparent antenna on both $\phi=90°$ and $\theta=90°$ planes. From the figures, it shows that all antennas have directional radiation pattern on both planes.

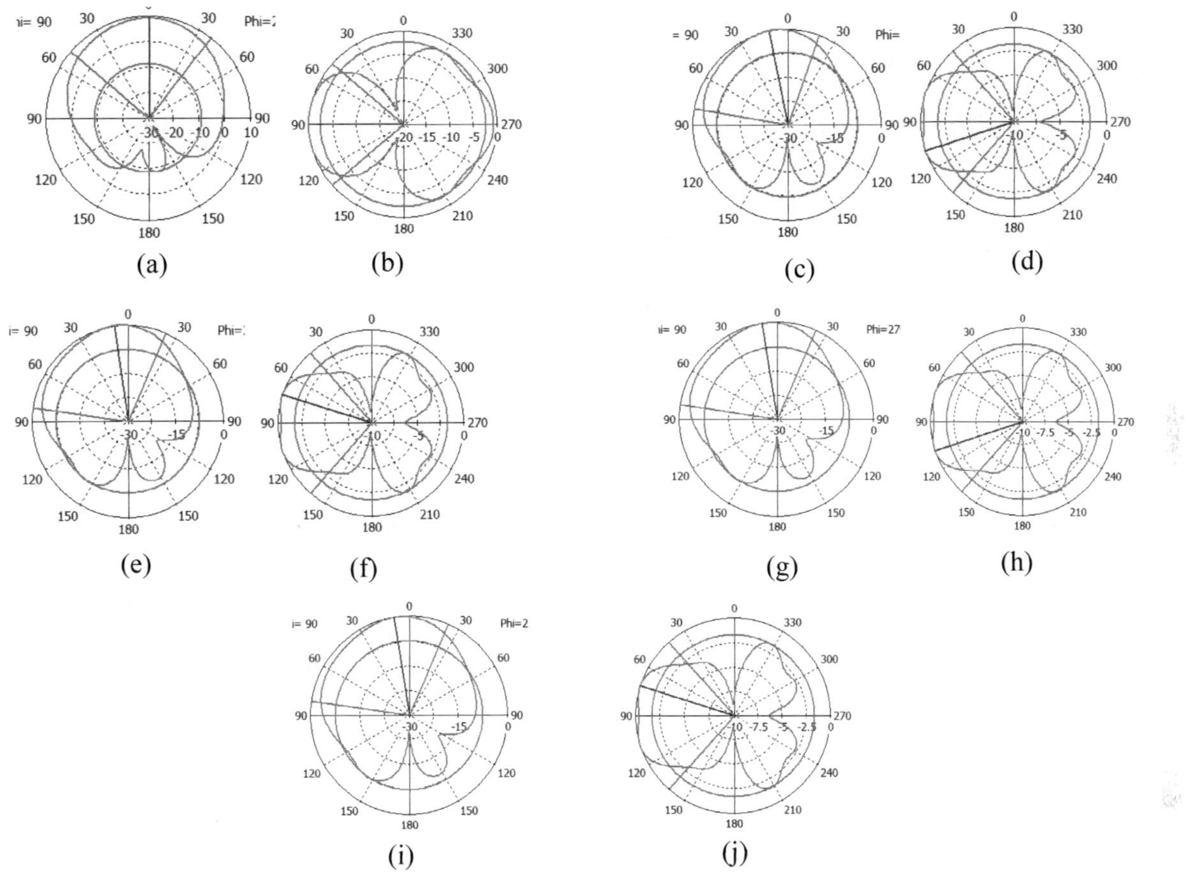

Figure 4. Radiation Pattern of (a) non-transparent antenna on ϕ = 90° planes and (b) ϕ = 90° planes, (c) ITO on ϕ = 90° planes (d) ITO on ϕ = 90° planes, (e) FTO on ϕ = 90° planes (f) FTO on ϕ = 90° planes, (g) AgHT-4 on ϕ = 90° planes (h) AgHT-4 in ϕ = 90° planes, (i) AgHT-8 in ϕ = 90° planes and (j) AgHT-8 in ϕ = 90° planes.

Table 5 shows the summarization of gain, VSWR, and the radiation efficiencies. As shown in the table, all transparent antennas give similar results of gain, but ITO gives slightly better value than others, and then followed by AgHT-4. The VSWR and radiation efficiency of all antennas are satisfactory which are nearly to 1 and higher than 50 percent respectively. But again, ITO gives the best efficiency compare to the other transparent antennas.

Table 5. Gain, VSWR, and radiation efficiency results

Conducting material	**PEC**	**ITO**	**FTO**	**AgHT-4**	**AgHT-8**
Gain (dB)	7.066	4.883	4.263	4.412	4.299
VSWR	1.109	1.102	1.057	1.042	1.054
Rad. Efficiency (%)	86.49%	81.05%	73.08%	75.54%	73.72%

4. Conclusion

Transparent solar antenna of 28 GHz operating frequency has been designed with radiating patch and ground material are made of Transparent Conductive Oxides thin film which varied to ITO, FTO, AgHT-4, and AgHT-8. These antennas are then compared to the nontransparent solar antenna and to each other. From the result obtained, it shows that transparent antenna can be designed using TCO thin film as all designs give satisfactory results and fulfilling the design specification. As for comparison, transparent antenna that made of ITO thin film gives the best performance with highest current density, gain, and efficiency but with minor differences compared to each other. Transparent antenna also shows great performance in terms of bandwidth when comparing to the nontransparent antenna with more than 20 percent of differences. It is due to higher thickness of substrate used for the transparent antenna.

References

[1] Hautcoeur J,Colombel F and Himdi M 2013 Large and Optically Transparent Multilayer for Broadband H-Shaped Slot Antenna *IEEE Antennas and Wireless Propagation Letters* **12** 933

[2] Hautcoeur J, Talbi L and Hettak K 2014 60 GHz optically transparent microstrip antenna made of meshed AuGL material *IET Microwaves, Antennas & Propagation* **8**(13) 1091

[3] Sun G, Muneer B and Zhu Q 2014 A study of microstrip antenna made of transparent ito films," *AP-S International Symposium (Digest) (IEEE Antennas and Propagation Society)* 1867

[4] Augarten Y, Wringley A, Rau U and Pieters B E 2016 Calculation of the TCO Sheet Resistance in thin film modules using electroluminescence imaging *IEEE 43rd Photovoltaic Specialists Conference (PVSC)* Oregon USA

[5] George J, Haley R, Pate B, Rozeveld S, Krafft M, Bernius M and Yeung S 2009 Optimization of electrical and optical properties in multilayer TCO thin film structures *34th IEEE Photovoltaic Specialists Conference (PVSC)* Philadelphia USA

[6] Conchubhair O O , McEvoy P and Amm M J 2012 Integration of Inverted-F Antenna with Solar Cell Substitute," in *Antennas & Propagation Conference* Loughborough UK

[7] Lim E H, Leung K W, Su C C and Wong H Y 2010 Green antenna for solar energy collection *IEEE Antennas and Wireless Propagation Letters* **9** 689

[8] Roo-Ons M J, Shynu S V, Seredynski M, Ammann M J, McCormack S J and Norton B 2010 Influence of solar heating on the performance of integrated solar cell microstrip patch antennas *Solar Energy* **84** (9) 1619

[9] Yurduseven O, Smith D and Pearsal N 2012 Design of A Highly Efficient Wideband Suspended Solar Array Antenna *Proceedings of the 2012 IEEE International Symposium on Antennas and Propagation* Chicago IL

[10] Shynu S V, Roos-Ons M J, Ruvio G and Ammann M J 2008 A microstrip printed dipole solar antenna using polycrystalline silicon solar cells *2008 IEEE Antennas and Propagation Society International Symposium* San Diego CA

[11] Yurduseven O, Smith D, Pearsall N and Forbes I 2012 A Transparent Solar Patch Antenna for 2.4/2.5 GHz WLAN-WiMAX Applications *2nd International Symposium on Environment-Friendly Energies and Applications (EFEA)* Newcastle upon Tyne UK

[12] Roo-Ons M J, Shynu S V, Ammann M J, McCormack S J and Norton B 2011 Transparent Patch Antenna on a-Si tTin-film Glass Solar Module," *Electronics Letters* **47**(2) 85

[13] Yurduseven O, Smith D, Pearsall N, Forbes I and Johnston D 2012 A meshed multiband solar patch array antenna *Antennas & Propagation Conference* Loughborough, UK

[14] Yurduseven O, Smith D and Elsdon M 2014 A transparent meshed solar monopole antenna for UWB applications *The 8th European Conference on Antennas and Propagation (EuCAP 2014)*

The Hague

[15] Sheikh S and Shokooh-Saremi M 2015 Rectangular Meshed Patch Antenna Integrated on ilicon Solar Cell *23rd Iranian Conference on Electrical Engineering* Tehran, Iran

[16] Nashad F, Foti S, Smith D, Elsdon M and Yurduseven O 2016 Development of transparent patch antenna element integrated with solar cells for Ku-band satellite applications in *Antenna & Propagation Conference (LAPC)* Loughborough

[17] F. Colombel X, Castel M., Himdi G *et al.* 2009 Ultrathin metal layer, ITO film and ITO/Cu/ITO multilayer towards transparent antenna *IET Science, Measurement and Technology* **3**(3) 229

[18] T. Peter T I, Yuk R, Nilavalan and Cheung S W 2011 A novel technique to improve gain in transparent uwb antennas *Antennas & Propogation Conference* Loughborough UK

[19] Lee M -H J and Collier R J 2005 The sheet resistance of thin metallic films and stripes at both dc and 130 GHz *IEEE Transactions on Instrumentation and Measurement* **54** 2412

[20] Balanis C A 2005 Microstrip Antenna *Antenna Theory: Analysis and Design, 3rd Edition*, SA John Wiley & Sons, Inc p 811

[21] Yurduseven O, Smith D, Pearsall N *et al.* 2012 A meshed multiband solar patch array antenna *Antennas and Propagation Conference (LAPC)*, Loughborough UK

[22] Rappaport T S, Sun S, Mayzus R, Zhao H and Azar Y 2013 Millimeter wave mobile for 5G cellular: It will work *IEEE Access* **1** 335

IOP Conf. Series: Journal of Physics: Conf. Series **852** (2017) 012037 doi:10.1088/1742-6596/852/1/012037

The analysis of the algorithms of the complex optimal estimates interpolation in tasks of satellite navigation

A I Perov[1], O V Chernoyarov[1,2,*], E N Boldenkov[1] and A N Faulgaber[3]

[1]Department of Radio Engineering Systems, National Research University "MPEI", 14 Krasnokazarmennaya Street, Moscow, Russia
[2]International Laboratory of Statistics of Stochastic Processes and Quantitative Finance, National Research Tomsk State University, 36 Lenin Avenue, Tomsk, Russia
[3]Department of Radio Engineering Systems, Moscow Technical University of Communications and Informatics, 8a Aviamotornaya Street, Moscow, Russia

*Email: chernoyarovov@mpei.ru

Abstract. For the tasks of satellite navigation, we conduct the synthesis of the interpolation algorithms within the fixed interval and in the fixed point, when the complex processing of measurements of the range and Doppler frequency is implemented. The simulation results are provided.

1. Introduction

The requirements for the present radio navigation satellite systems (RNSS), such as GPS and GLONASS, are constantly increasing. There are various ways to improve the system characteristics, such as accuracy, reliability, integrity. One approach to the system improvement is to incorporate on-board equipment for inter-satellite measurements (OEIM) into the system [1]. Such equipment carries out off-line testing of the orbits parameters of navigation space vehicles (NSVs), with the terrestrial command-instrumentation facility unemployed.

There are made great demands of the inter-satellite instrumentation concerning the NSVs reciprocal movement measurements accuracy. Specifically, for pseudo-delay they are up to tens of centimeters and for pseudo-velocity – less than ten millimeters per second. In order to achieve such accuracy, it is necessary to apply the optimal tracking algorithms for signals parameters measuring [2]. The information interchange between NSVs revealing the measured parameters of the reciprocal movement occurs in the period of communication sessions. Here the transmitted data are related to the some fixed point of time $t_{ms} \in [t_k, t_k + T]$, which is situated within the last time measuring span. Thus, there is a problem of generation of the smoothed (interpolated) estimates of pseudo-range and pseudo-velocity at the specified relevant point in time t_{ms}, based on the measurements obtained into the set interval $[t_k, t_k + T]$. With very high precisions of the generated estimates required, for the solution of this problem it is necessary to apply the optimal interpolation algorithms. Common approaches to the synthesis of the interpolation algorithms in case of the linear problem are described in [3-6]. In [7] there are specified optimal filtering and interpolation algorithms for nonlinear dynamic model and linear observations. Some mathematical aspects of the interpolation algorithms stability are discussed in [8]. The present paper is focused on the study of the interpolation algorithm in the context of the inter-satellite measurements for the GLONASS system.

Content from this work may be used under the terms of the Creative Commons Attribution 3.0 licence. Any further distribution of this work must maintain attribution to the author(s) and the title of the work, journal citation and DOI.
Published under licence by IOP Publishing Ltd

2. The problem statement

The equipment for inter-satellite measurements is intended for determining the relative ranges and velocities between 24 NSVs of the GLONASS system, distributed in three orbital planes in normal mode. As measuring signals the phase-shift keyed signal [1, 9] are used in inter-satellite measuring system. Phase-shift keying is implemented by pseudorandom sequences, known as the ranging codes. As within GLONASS system, the measuring signals are used simultaneously for the navigating information transfer realized by additional digital phase modulation. Thus, OEIM signal can be mathematically presented as

$$S(t) = AG_{RC}(t-\tau)G_{DM}(t-\tau)\cos[\omega_0(t-\tau)+\phi_0], \quad (1)$$

where A is the signal amplitude, $G_{RC}(t)$ is the ranging code, $G_{DM}(t)$ is the digital message, ω_0 is the signal carrier frequency, τ is the signal pseudo-delay, ϕ_0 is the initial phase. We presuppose that the signal (1) is distorted by Gaussian white noise $n(t)$ with one-sided spectral density N_0 and designate the power-to-noise ratio P_c/N_0 (in Hz) as q_{c/n_0}.

The inter-satellite measuring system cycle is 20 s and is divided into for intervals on 5 s allocated for the operation of the corresponding satellite constellations. The transmission session of one satellite is about 4.5 s, subject to the time required for switching the reception/transmission modes. During the same time interval, the reception goes on by others NSVs of the signal informing about the pseudo-range and pseudo-velocity estimates currently generated. It is necessary to output the information on the measured pseudo-range and pseudo-velocity at the point of time $t_{ms} \in [0,4.5]$ s (for example, $t_{ms} = 4$ s). We now assume a task of the synthesis of optimal estimates of pseudo-range and pseudo-velocity matching the point of time $t_{ms} \in [0,4.5]$, according to the results of the observations obtained over the time interval $t_{ms} \in [0,4.5]$ s.

3. The synthesis of the current optimal estimates

3.1. Generalities

In the optimal filtering theory the required estimates belong to the class of the interpolated estimates [3, 10]. The problem of interpolation of some process $x(t)$ according to the observations $y(t)$ over the time interval $[0,t]$ is to form the estimate $\hat{x}_{\nu|t}$ for the time moment $\nu \in [0,t]$. In certain interpolation algorithms such estimate is formed through additional processing of the current estimates $\hat{x}_{t|t}$, i.e. the estimates which have been obtained during the observations accessible at a present point of time. We will start with briefly considering the algorithms for the generation of optimal estimates referring to the problem of inter-satellite measurements (see [1, 2] in more details).

In [1, 2] it is shown that the range change between NSVs moving in their orbits within the time intervals with the duration of 5 s is sufficiently approximated by the second order polynomial

$$R(t) = R_0 + V_0 t + a_0 t^2/2$$

with the following key parameters: maximum range $R_{\max} = 51000$ km, minimum range $R_{\max} = 10000$ km, maximum relative velocity $V_{\max} = 6200$ m/s, maximum relative acceleration $a_{\max} = 3.9$ m/s^2.

We designate the received signal envelope delay as τ and introduce a state vector $\mathbf{x}_\tau$ [1] for its description, so

$$\tau_k = \mathbf{C}\mathbf{x}_{\tau,k}, \quad (2)$$

where $\mathbf{C} = \|1 \quad 0 \quad 0\|$, $\mathbf{x}_{\tau,k}$ is the vector described by the equation

$$\mathbf{x}_{\tau,k} = \mathbf{F}\mathbf{x}_{\tau,k-1} + \mathbf{G}\xi_{\tau,k-1}, \tag{3}$$

In Eq. (3) the designations are: $\mathbf{F} = \begin{Vmatrix} 1 & T & 0 \\ 0 & 1 & T \\ 0 & 0 & 1 \end{Vmatrix}$, $\mathbf{G} = \begin{Vmatrix} 0 \\ 0 \\ 1 \end{Vmatrix}$ and $\xi_{\tau,k-1}$ is the generating discrete Gaussian white noise with zero mathematical expectation and a certain dispersion D_{ξ_τ}.

Taking into account the high requirements to the system accuracy and the high volatility of the estimated processes, in order to generate the optimal current estimates of pseudo-range and pseudo-velocity, there should be applied the principles of a complex filtration, all necessary information acquired from both envelope and phase of the received signal. Therefore, as in [2], we consider OEIM tracing systems having the structure presented in Fig. 1 and including the components described below.

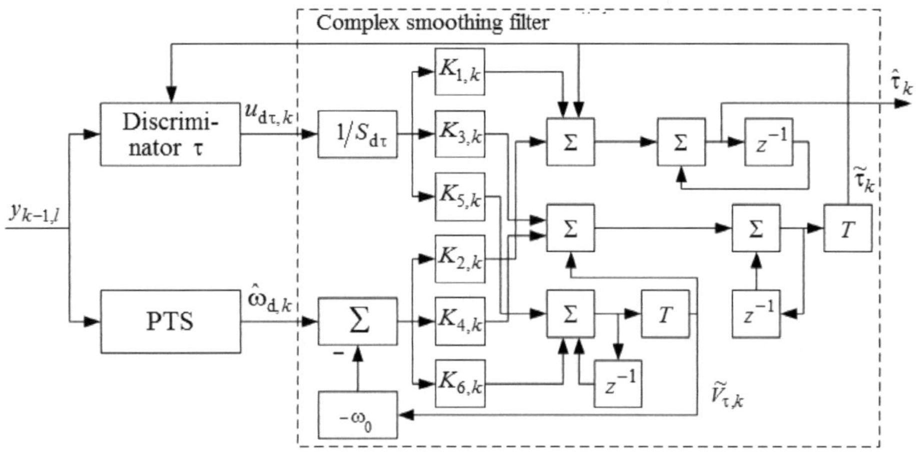

Figure 1. Block diagram of the complex signal delay tracing system

3.2. *Autonomous signal phase tracking system (PTS)*

The PTS discriminator is described by the expression

$$u_{\mathrm{d}\phi,k}(\phi_k) = -Q_k \,\mathrm{th}(I_k),$$

where $\mathrm{th}(\cdot)$ is the hyperbolic tangent, I_k, Q_k are the signals at the correlator output of the OEIM receiver:

$$I_k = \frac{A}{\sigma_n^2}\sum_{l=1}^{M} y(t_{k-1,l})G_{RC}(t_{k-1,l} - \tilde{\tau}_k)\cos(\omega_0 t_{k-1,l} + \tilde{\phi}_k),$$

$$Q_k = \frac{A}{\sigma_n^2}\sum_{l=1}^{M} y(t_{k-1,l})G_{RC}(t_{k-1,l} - \tilde{\tau}_k)\sin(\omega_0 t_{k-1,l} + \tilde{\phi}_k),$$

and

$$y(t_{k-1,l}) = S(t_{k-1,l}) + n_{k-1,l}.$$

Here $n_{k-1,l}$ is the discrete Gaussian white noise with the dispersion $\sigma_n^2 = N_0/2T_d$, T_d is the sampling interval, $MT_d = T$ is the accumulation time at the correlator, $\tilde{\tau}_k$ is the reference correlator signal delay, $\tilde{\phi}_k$ is the reference correlator signal phase.

As the received signal phase is linearly related to signal delay, i.e. $\phi_k = -\omega_0\tau$, we describe it by the equations that are similar to Eqs. (2), (3)

$$\phi_k = \mathbf{C}\mathbf{x}_{\phi,k}, \qquad \mathbf{x}_{\phi,k} = \mathbf{F}\mathbf{x}_{\phi,k-1} + \mathbf{G}\xi_{\phi,k-1}$$

with the state vector $\mathbf{x}_{\phi,k} = \left\| x_{1\phi,k} \quad x_{2\phi,k} \quad x_{3\phi,k} \right\|^{\mathrm{T}}$. Here $\xi_{\phi,k-1}$ is the discrete Gaussian white noise with the certain dispersion $D_{\xi,\phi}$, and other designations coincide with the ones introduced in Eqs. (2), (3). It should be noted that the component $x_{2\phi,k}$ of the vector $\mathbf{x}_{\phi,k}$ is the Doppler translation, i.e. $x_{2\phi,k} \equiv \omega_{\mathrm{d},k}$.

The filter equations in PTS have the form of

$$\hat{\mathbf{x}}_{\phi,k} = \tilde{\mathbf{x}}_{\phi,k-1} + \mathbf{D}_{\phi,k}\mathbf{C}^{\mathrm{T}} u_{\mathrm{d}\phi,k-1}\left(\tilde{\phi}_k\right), \qquad \tilde{\mathbf{x}}_k = \mathbf{F}\hat{\mathbf{x}}_{\phi,k-1},$$
$$\mathbf{D}_{\phi,k}^{-1} = \left(\mathbf{F}\mathbf{D}_{\phi,k-1}\mathbf{F}^{\mathrm{T}} + \mathbf{G}D_{\xi,\phi}\mathbf{G}^{\mathrm{T}}\right)^{-1} + \mathbf{C}^{\mathrm{T}}\mathbf{C}/\tilde{\sigma}_n^2 \qquad (4)$$

with the initial conditions $\hat{\mathbf{x}}_{\phi,0} = \hat{\mathbf{x}}(0)$, $\mathbf{D}_{\phi,0} = \mathbf{D}_\phi(0)$. Here $\tilde{\sigma}_n^2 = \left(1 + 1/2q_{c/n_0}T\right)/2q_{c/n_0}T$ is the dispersion of the equivalent observation noise of the phase discriminator.

3.3. Complex signal delay tracking system (DTS)

Unlike PTS, the DTS is complex one. The delay discriminator is described by the equation

$$u_{\mathrm{d}\tau,k}(\phi_k) = \left(I_{E,k} - I_{L,k}\right)\mathrm{th}(I_k),$$

where

$$I_{E,k} = \frac{A}{\sigma_n^2}\sum_{l=1}^{M} y(t_{k-1,l})G_{RC}\left(t_{k-1,l} - \tilde{\tau}_k + \Delta\tau/2\right)\cos\left(\omega_0 t_{k-1,l} + \tilde{\phi}_k\right),$$

$$I_{L,k} = \frac{A}{\sigma_n^2}\sum_{l=1}^{M} y(t_{k-1,l})G_{RC}\left(t_{k-1,l} - \tilde{\tau}_k - \Delta\tau/2\right)\cos\left(\omega_0 t_{k-1,l} + \tilde{\phi}_k\right),$$

and $\Delta\tau$ is the time shift between reference correlator signals.

When designing the complex DTS filter, the equivalent linear observations of the delay discriminator are used [1, 10]:

$$\tilde{y}_{\tau,k} = \tau_k + \tilde{\eta}_{\tau,k}.$$

Here $\tilde{\eta}_{\tau,k}$ is the noise of equivalent observations approximated by discrete Gaussian white noise with the dispersion $D_{\tilde{\eta}} = \Delta\tau^2/4q_{c/n_0}T$.

We now write down the observations of the Doppler frequency estimate $\hat{\omega}_{\mathrm{d},k} \equiv \hat{x}_{2\phi,k}$ formed into PTS as follows

$$\tilde{y}_{V_\tau,k} = \hat{\omega}_{\mathrm{d},k} = \omega_{\mathrm{d},k} + \tilde{\eta}_{V_\tau,k} = -\omega_0 V_{\tau,k} + \tilde{\eta}_{V_\tau,k},$$

where $V_\tau = d\tau/dt$, $\tilde{\eta}_{V_\tau,k}$ can be taken by the discrete Gaussian white noise with the dispersion $D_{\tilde{\eta}_{V\tau}} = 4\sigma_{\tilde{n}}^2 (6\Delta f_{\text{PTS}}/5)^2$, as shown in [1], and Δf_{PTS} is the PTS bandpass.

We introduce the following observation

$$\tilde{\mathbf{y}}_k = \begin{Vmatrix} \tilde{y}_{\tau,k} \\ \tilde{y}_{V_\tau,k} \end{Vmatrix} = \mathbf{H}\mathbf{x}_{\tau,k} + \tilde{\boldsymbol{\eta}}_{\tau,k},$$

where $\mathbf{H} = \begin{Vmatrix} 1 & 0 & 0 \\ 0 & -\omega_0 & 0 \end{Vmatrix}$, $\tilde{\boldsymbol{\eta}}_{\tau,k}$ is the vector process with the dispersion $\mathbf{D}_{\tilde{\boldsymbol{\eta}}_\tau} = \begin{Vmatrix} D_{\tilde{\eta}_\tau} & 0 \\ 0 & D_{\tilde{\eta}_{V\tau}} \end{Vmatrix}$, $\mathbf{x}_{\tau,k}$ and is the state vector determined by Eq. (2).

The current estimate-fitting procedure for obtaining $\hat{\tau}_{k|k}$ and optimal filter equations of the vector $\mathbf{x}_{\tau,k}$ are written down in the form of

$$\hat{\tau}_{k|k} = \mathbf{C}\hat{\mathbf{x}}_{\tau,k|k}, \qquad \hat{\mathbf{x}}_{\tau,k|k} = \tilde{\mathbf{x}}_{\tau,k|k} + \mathbf{K}_{\tau,k|k}\left(\tilde{\mathbf{y}}_k - \mathbf{H}\tilde{\mathbf{x}}_{\tau,k|k}\right), \tag{5}$$

$$\mathbf{K}_{\tau,k|k} = \mathbf{D}_{\tau,k|k}\mathbf{H}^{\mathrm{T}}\mathbf{D}_{\tilde{\boldsymbol{\eta}}_\tau}^{-1}, \qquad \tilde{\mathbf{x}}_{\tau,k|k} = \mathbf{F}\hat{\mathbf{x}}_{\tau,k-1|k-1}, \tag{6}$$

$$\tilde{\mathbf{D}}_{\tau,k|k-1} = \mathbf{F}\mathbf{D}_{\tau,k-1|k-1}\mathbf{F}^{\mathrm{T}} + \mathbf{G}D_{\xi_\tau}\mathbf{G}^{\mathrm{T}}, \tag{7}$$

$$\mathbf{D}_{\tau,k|k}^{-1} = \tilde{\mathbf{D}}_{\tau,k|k-1}^{-1} + \mathbf{H}^{\mathrm{T}}\mathbf{D}_{\tilde{\boldsymbol{\eta}}_\tau}^{-1}\mathbf{H}, \tag{8}$$

where $\mathbf{K}_{\tau,k|k} = \begin{Vmatrix} K_{1,k} & K_{2,k} \\ K_{3,k} & K_{4,k} \\ K_{5,k} & K_{6,k} \end{Vmatrix}$ is the weight matrix. In Eqs. (5)-(8) index $k|k$ designates that the corresponding estimate (parameter) is formed at k-th moment of time, according to the observations received up to the k-th moment inclusive. Further, we would name such estimate as the filtration estimate, i.e. obtained as a result of a current filtration of information processes.

Block diagram of the signal DTS described by Eqs. 5-8 is shown in Fig. 1. Here it is taken into account that

$$\tilde{y}_{\tau,k} - \tilde{\tau}_k = u_{\mathrm{d}\tau,k}/S_{\mathrm{d}\tau},$$

where $S_{\mathrm{d}\tau}$ is the slope of characteristic of the delay discriminator which is described by the formula (6).

PTS is described by Eqs. (4), it represents a classic signal phase tracking system with third-order astaticism [1].

4. The interpolation algorithms

4.1. Generalities

In optimal filtering theory there are a number of certain interpolation algorithms differing by the problem statement [3, 4, 10]: interpolation within the fixed interval, interpolation in the fixed point and interpolation with the fixed delay.

The algorithm for the optimal interpolation within the fixed interval presupposes that the observation interval is fixed, and the moment of time, for which it is necessary to obtain the estimate, varies.

The algorithm for the optimal interpolation in the fixed point implies that the moment of time is specified for which it is necessary to obtain the estimate, but observation continues after this time, and the estimate of parameters at this point keeps on getting more and more precise. The first two algorithms can be used for the solution of the considered problem. In the third algorithm we see that there is a time delay between the estimate and the observations being received, and this case is not studied here.

4.2. Optimal interpolation within the fixed interval
Let us assume that it is necessary to obtain the estimate of parameters for a time t_ν. We designate the current estimate $\hat{\mathbf{x}}_{\tau,k|k}$ that has been interpolated at the moment of time t_ν as $\hat{\mathbf{x}}_{\tau,\nu|k}$. Then the interpolation equations can be written down in the form of

$$\hat{\mathbf{x}}_{\tau,\nu|k} = \hat{\mathbf{x}}_{\tau,\nu|\nu} + \hat{\mathbf{K}}_\nu \left(\hat{\mathbf{x}}_{\tau,\nu+1|k} - \mathbf{F}\hat{\mathbf{x}}_{\tau,\nu|\nu}\right), \tag{9}$$

$$\hat{\mathbf{K}}_\nu = \mathbf{D}_{\tau,\nu|\nu}\mathbf{F}^{\mathrm{T}}\mathbf{D}^{-1}_{\tau,\nu+1|\nu+1}, \tag{10}$$

$$\hat{\mathbf{D}}_{\tau,\nu|k} = \mathbf{D}_{\tau,\nu|\nu} + \hat{\mathbf{K}}_\nu \left(\hat{\mathbf{D}}_{\tau,\nu+1|k} - \mathbf{D}_{\tau,\nu+1|\nu+1}\right)\hat{\mathbf{K}}^{\mathrm{T}}_\nu, \tag{11}$$

where $\hat{\mathbf{x}}_{\tau,\nu|\nu}$ is the optimal filtration estimate of the parameters formed according to Eq. (4), $\mathbf{D}_{\tau,\nu|\nu}$ is the dispersion of the filtering error of the vector $\hat{\mathbf{x}}_{\tau,\nu|\nu}$ determined by Eqs. (7), (8), and $\hat{\mathbf{D}}_{\tau,\nu|k}$ is the dispersion of the error of the interpolated estimate.

In order to obtain the interpolated estimate at the step ν, it is necessary, according to Eqs. (5)-(8), to carry out an optimal filtration to the end of measurements (until the step with the number *k*), then, referring to the initial conditions $\hat{\mathbf{x}}_{\tau,k|k}$, $\mathbf{D}_{\tau,\nu|\nu}$, to solve Eqs. (9)-(11) in reverse time up to the moment t_ν. Results of numerical simulation of the algorithm (9)-(11) are presented below.

4.3. Optimal interpolation in the fixed point
For the considered task, the optimal interpolation algorithm in the fixed point can be written down using the general equations [10] and gets the form of

$$\hat{\mathbf{x}}_{\tau,\nu|k} = \hat{\mathbf{x}}_{\tau,\nu|k-1} + \hat{\mathbf{K}}_{\nu|k}\left(\hat{\mathbf{x}}_{\tau,k|k} - \tilde{\mathbf{x}}_{\tau,k|k-1}\right), \tag{12}$$

$$\hat{\mathbf{K}}_{\nu|k} = \hat{\mathbf{K}}_{\nu|k-1}\mathbf{D}_{\tau,k-1|k-1}\mathbf{F}^{\mathrm{T}}\tilde{\mathbf{D}}^{-1}_{\tau,k|k-1}, \tag{13}$$

$$\hat{\mathbf{D}}_{\tau,\nu|k} = \hat{\mathbf{D}}_{\tau,\nu|k-1} - \hat{\mathbf{K}}_{\nu|k}\mathbf{K}_{\tau,k|k}\mathbf{H}\tilde{\mathbf{D}}_{\tau,k|k-1}\hat{\mathbf{K}}^{\mathrm{T}}_{\nu|k}, \tag{14}$$

where $\mathbf{K}_{\tau,k|k}$ are the coefficients of the optimal filter determined by Eq. (6).

In order to solve the specified equations, it is necessary to solve the filter equations (5)-(8) up to the moment t_ν inclusive. Then the initial conditions $\hat{\mathbf{x}}_{\tau,\nu|\nu} = \hat{\mathbf{x}}_{\tau,\nu|\nu}$, $\hat{\mathbf{K}}_{\nu|\nu} = \mathbf{I}$ should be referred to them, where $\mathbf{I}$ is the unit matrix, $\hat{\mathbf{D}}_{\tau,\nu|\nu} = \mathbf{D}_{\tau,\nu|\nu}$ and after that we pass to the joint solution of the filter equations (9)-(11) and the interpolation equations (12)-(14), until the termination time of observations t_k comes. As a result, the required interpolated estimate $\hat{\mathbf{x}}_{\tau,\nu|k}$ will be obtained.

5. The simulation results of the interpolation algorithms

The interpolation algorithms (9)-(11), (12)-(14) were simulated when operating within the OEIM structure. Besides, the filtering estimate $\hat{\mathbf{x}}_{\tau,\nu|\nu}$ was additionally controlled, so that the measurements following the time point t_ν were not used with it.

In Figs. 2 and 3 there are shown the dependences of mean square errors (MSEs) of the estimates of range and velocity for the interpolation algorithm within the fixed interval. These dependences had been obtained by statistical averaging over 200 realizations. In Figs. 4 and 5 the similar dependences are traced for the interpolation algorithm in the fixed point, they had been obtained by statistical averaging over 50 realizations, the number of the realizations reduced due to huge computing costs. It should also be noted that in Figs. 4, 5 the simulation results are given for the case when the "position of the fixed point" changes from 1 s to 4.5 s.

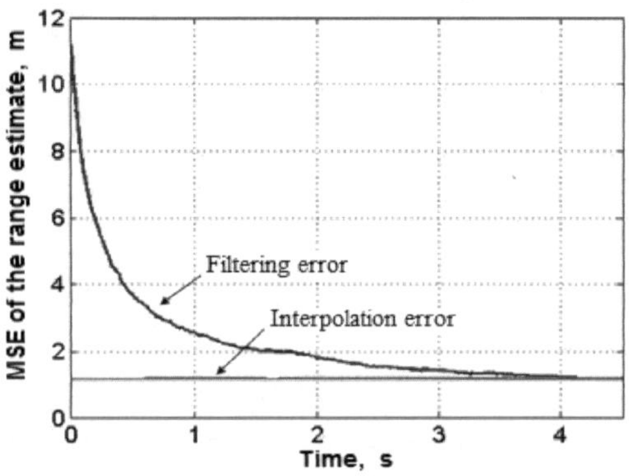

Figure 2. The dependences of mean square errors of the range filtering and range interpolation within the fixed interval

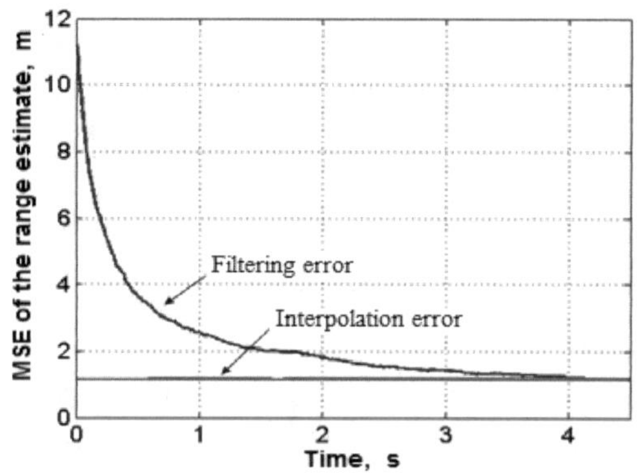

Figure 3. The dependences of mean square errors of the velocity filtering and velocity interpolation within the fixed interval

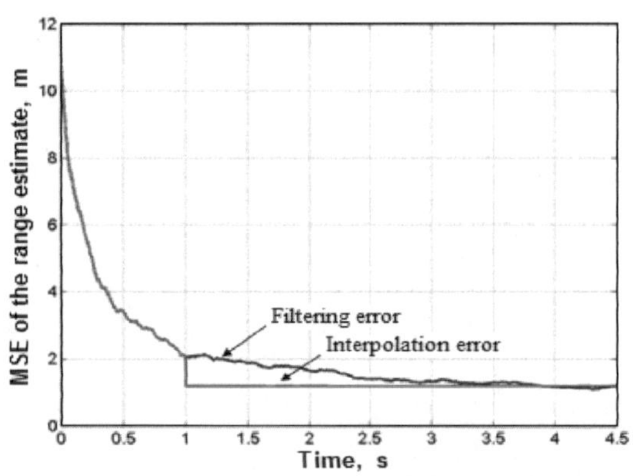

Figure 4. The dependences of mean square errors of the range filtering and range interpolation in the fixed point

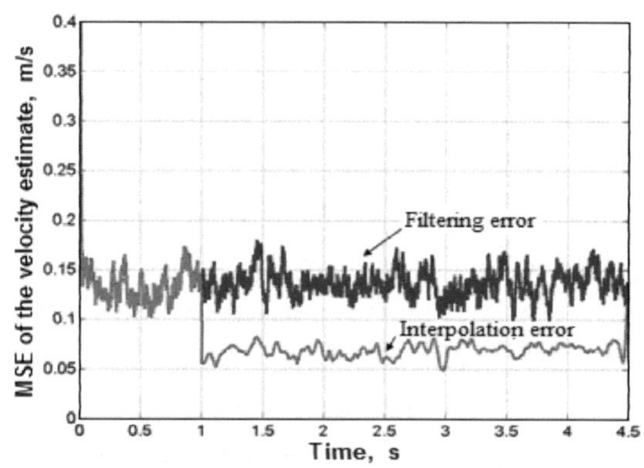

Figure 5. The dependences of mean square errors of the velocity filtering and velocity interpolation in the fixed point

From the presented dependences it follows that, if compared with the filtering estimates, the interpolated estimates of the range demonstrate a gain in accuracy, the more time has passed since the end of the observation interval for which the corresponding estimate is formed, and this gain may make up to 2 and even more times.

The functional efficiency of the algorithms for the interpolation within the fixed interval and in the fixed point is almost identical. From the point of view of practical implementation, the algorithm for the interpolation in the fixed point is preferred, since in this case interpolation and filtration are carried out simultaneously. At the same time, the algorithm for the interpolation within the fixed interval requires that filtration should be conducted at first, so that the intermediate variables could be stored, and only then one could proceed to solving the interpolation equations in reverse time.

6. Conclusion

Thus, by applying the optimal information processes filtering theory, for a problem of satellite navigation, the optimal interpolation equations are obtained within the fixed interval and in the fixed point. The specified equations realize the complex processing of the measurements of navigation radio signal delay and Doppler translation. By means of simulation modeling, it is shown that the synthesized interpolation algorithms provide a gain in accuracy of the interpolated estimates of range over the results provided by the filtering estimates, and that such gain may be 2 times and more. A gain in accuracy of the interpolated estimates increases with the interpolation interval length. It is also demonstrated that the functional efficiency of interpolation algorithms within the fixed interval and in the fixed point is practically identical. Therefore, from the point of view of practical implementation, the algorithm for the interpolation in the fixed point can be recommended, as it is simpler in realization technically.

Acknowledgments

The reported study was supported by Russian Science Foundation (research project No. 14-49-00079).

References

[1] Perov A I and Kharisov V N 2005 *GLONASS. Principles of Construction and Functioning* [in Russian] (Moscow: Radiotekhnika)

[2] Perov A I and Boldenkov E N 2005 The synthesis of signal parameters tracking algorithms into the on-board equipment for inter-satellite measurements [in Russian] *Radiotekhnika* **7** 15

[3] Sage A P and Melse J L 1972 *Estimation Theory with Application to Communication and Control* (New York: McGraw-Hill)

[4] Brown R G and Hwang P Y C 2012 *Introduction to Random Signals and Applied Kalman Filtering*, (New York: Wiley)

[5] Najim M 2010 *Modeling, Estimation and Optimal Filtration in Signal Processing* (New Jersey: Wiley-ISTE)

[6] Grewal M S and Andrews A P 2008 *Kalman Filtering: Theory and Practice Using MATLAB* (New York: Wiley)

[7] Briers M, Doucet A and Maskell S 2010 Smoothing algorithms for state-space models *Annals of the Institute of Statistical Mathematics* **62** 61

[8] Bolzern P, Colaneri P and De Nicolao G 2004 On discrete-time H^{∞} fixed-lag smoothing *IEEE Trans. on Signal Processing* **52** 132

[9] Litvinenko V P, Litvinenko Yu V, Matveev B V, Chernoyarov O V and Makarov A A 2016 The new detection technique of phase-shift keyed signals *Proc. Int. Conf. on Applied Mathematics, Simulation and Modelling (AMSM2016)* (Paris: Atlantis Press) pp 355

[10] Perov A I 2003 *Statistical Radio System Theory* [in Russian] (Moscow: Radiotekhnika)

IOP Conf. Series: Journal of Physics: Conf. Series **852** (2017) 012038 doi:10.1088/1742-6596/852/1/012038

Efficient Pricing Technique for Resource Allocation Problem in Downlink OFDM Cognitive Radio Networks

O B Abdulghafoor[1,*], M M R Shaat[2], M Ismail[3], R Nordin[3], T Yuwono[4] and O N A Alwahedy[1]

[1]Electrical and Electronic Engineering, Al-Madinah International University-Malaysia
[2]Centre Tecnològic de Telecomunicacions de Catalunya (CTTC), Barcelona-Spain
[3]Electrical, Electronic and System Engineering Dept., University Kebangsaan Malaysia, Bangi, Malaysia
[4]Electrical Engineering Dept. University Islam Indonesia, Yogyakarta, Indonesia

*E-mail: omar.bahjeet@mediu.edu.my

Abstract. In this paper, the problem of resource allocation in OFDM-based downlink cognitive radio (CR) networks has been proposed. The purpose of this research is to decrease the computational complexity of the resource allocation algorithm for downlink CR network while concerning the interference constraint of primary network. The objective has been secured by adopting pricing scheme to develop power allocation algorithm with the following concerns: (i) reducing the complexity of the proposed algorithm and (ii) providing firm power control to the interference introduced to primary users (PUs). The performance of the proposed algorithm is tested for OFDM- CRNs. The simulation results show that the performance of the proposed algorithm approached the performance of the optimal algorithm at a lower computational complexity, i.e., $O(N\log N)$, which makes the proposed algorithm suitable for more practical applications.

1. Introduction

Studies carried out by the Federal Communications Commission (FCC) have shown that conventional spectrum allocation technique is becoming insufficient to address the rapid development of wireless technologies, and there is a need for the development of an open spectrum allocation methodology to compensate spectrum underutilisation [1]. To meet this important demand, dynamic spectrum access (DSA) and cognitive radio (CR) have been proposed as smart techniques to resolve the problems related to the fixed spectrum approach. CR can be defined as a system, in which wireless devices have the capability to sense the spectrum which is belong to primary user (PU) and opportunistically access any unused bands based findings. However, the generated interference from CR-to-PU band must be at or below a defined threshold. OFDM technique is seen as a promising candidate for cognitive radio networks (CRNs) due to its reliability and flexibility in allocating the available resources among CRs [2-4]. In OFDM-based CRNs, the CR and PU band exist side by side, which results in mutual interference. This mutual interference is considered as a preventive factor that affects the performance of both networks [5]. The resource allocation problem in downlink non-cognitive radio networks has been widely examined [e.g., 6, 7]. However, the use of a conventional water-filling algorithm is not sufficient in case of CRNs, and an additional constraint must be proposed in the optimisation problem to control any negative effects of generated interference from CR to PU.

The problem of resource allocation in the context of multicarrier CRNs (MC-CRNs) has garnered much interest in the literature; see for example [9-10]. In contrast, game theory and supermarket theory have been used to provide efficient spectrum sharing among CRs and PUs. This is because such

Content from this work may be used under the terms of the Creative Commons Attribution 3.0 licence. Any further distribution of this work must maintain attribution to the author(s) and the title of the work, journal citation and DOI.
Published under licence by IOP Publishing Ltd

techniques can properly define interaction and competition among players [11]. Moreover, there are a number of studies adopted game and supermarket theory in MC-CRNs to address the problem of resource allocation, as in [12-15]. In [12], a spectrum monopoly-market scenario based on non-cooperative game theory has been proposed for OFDM-based CRNs. The authors used a non-cooperative auction game to determine the optimal power for CRs in a leasing market scenario. However, convergence of the proposed game to the Nash equilibrium (NE), is a bit slow. A more complex scenario in MC-CRNs based on non-cooperative games has been considered in [13]. Instead of single-cell scenario, the authors proposed a resource allocation algorithm in multiuser, multi-cell MC-CRNs that proves to be a notable contribution compared to other studies in the literature. Resource allocation in MC-CRNs based on market games has been considered in [14]. Unlike [13], this study includes simple optimization problem and adopts Blotto game to allocate resources among players while produces fair allocation in both uplink and downlink with fast convergence to the NE point.

In this paper, we consider more practical downlink set-up in OFDM-CRNs by adopting pricing scheme in the surplus function. The main objective is to maximize the downlink capacity of the CRNs at a lower computational complexity compared to that in [5]. Furthermore, the main contribution of this paper is that the pricing scheme adopted to provide flexibility in managing harmful interference to PUs rather than increasing the revenue of PUs. Hence, CRs have more freedom to utilize both active and non-active PU bands, considering that the total interference is at or below the defined threshold.

The remainder of this paper is organized as follows. In Section II, the system model, including the OFDM-based CR, system setting and interference model is presented. In Section III, we develop problem formulation. The downlink market model based on pricing scheme is illustrated in Section IV and includes subcarrier allocation algorithm, market power allocation, Nash equilibrium concept and the proposed power allocation algorithm. Simulation results of the proposed algorithm and the comparison study are given in Section V. Finally, we conclude this paper in Section VI.

2. System Model

2.1. OFDM-based Cognitive Radio: System Setting

In this paper, a hybrid network in an underlay spectrum-sharing scenario has been considered. A network composed of one CRN-based OFDM coexists with primary user network (PUN) in the same geographical area, and sharing same spectrum, as shown in Figure 1 (a). Single-cell, OFDM-based CR scenario has been assumed in this research. CRN consists of M CRs, denoted by $m = \{1,2,..., M\}$, sharing licensed band with L PUs. Moreover, the spectrum band is divided into N subcarrier, denoted by $N = \{1,2,..., N\}$, with a bandwidth represented by Δf. The distribution of the frequency among users is supposed to be a side-by-side distribution, as shown in Figure 1(b).

The frequency bands, denoted as $B_1^{PU}, B_2^{PU}, ..., B_L^{PU}$, are utilised by PUs, i.e., active band, while the remainder of the bands represent the non-active bands, where PUs are not transmitting. The channel gains of the links shown in Figure 1 can be defined as follows $g_{i,l}^{c}$ denotes channel gain of interference link from CRBS to the lth PU on the *i*th subcarrier. The superscript (^{c}) refers to CR. The channel gain of interference link from the lth PU's transmitter to the *m*th CR over the *i*th subcarrier is denoted by $y_{l,m,i}^{p}$ where the superscript (^{p}) refers to PU. Finally, $h_{i,m}$ denote the *i*th subcarrier fading gain from the CRBS to the *m*th user

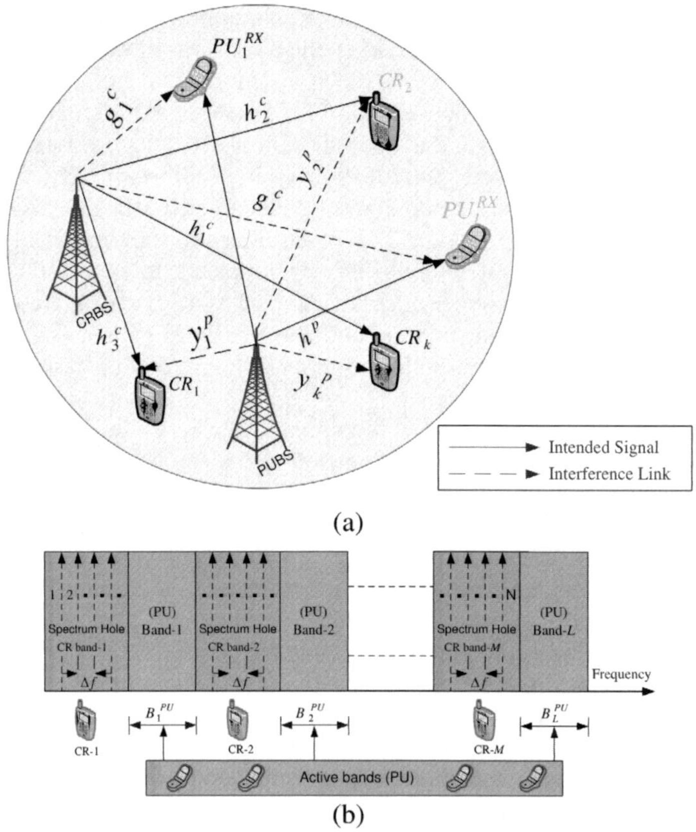

Figure 1. System model: (a) Example of coexistence among PUs and CRs and (b) Frequency distribution of the active and non-active primary bands

2.2. Interference Model

2.2.1 Interference generated by CR's signal

A common problem in CR hybrid networks is the generated interference from CRs to PUs. Hence, interference must be well controlled to avoid an unacceptable performance degradation in primary user network (PUN) to achieve feasible coexistence between CRs and PUs. The total available bandwidth is divided into N OFDM subcarriers, as shown in Figure 1 (b). The power spectrum density of the ith OFDM subcarrier can be defined as [9, 10]

$$\phi_i(f) = p_i T\left(\frac{\sin \pi f T}{\pi f T}\right), \tag{1}$$

where T is the OFDM symbol duration and p_i is the total power produced by the ith subcarrier. The interference generated by the mth CR on the ith subcarrier to the lth PU, denoted by $I_{m,i}^{PU_l}(d_i, p_i)$, can be defined as the integral of power spectrum density (PSD) of the ith subcarrier across the lth PU band. Mathematically speaking, the interference introduced by the CR's signal can be modelled according to [9]

$$I_{m,i}^{PU_l}(d_i, p_i) = \int_{d_i - B_l/2}^{d_i + B_l/2} \left|g_{i,l}^{c}\right|^2 \phi_i(f)df = P_i \Omega_i^l, \tag{2}$$

where d_i is the distance between the ith subcarrier and the lth PU's spectrum band, i.e., B_l, and Ω_i^l is the interference component of the ith subcarrier.

2.2.2 *Interference generated by PU signal*

The interference generated by the lth PU to the *m*th CR on the *i*th OFDM subcarrier, denoted by $I_{l,i}^{CR_m}$, can be defined as the integral of the PSD of the lth PU over the *i*th subcarrier which can be modelled according to [9]

$$I_{l,i}^{CR_m} = \int_{d_i-\Delta f/2}^{d_i+\Delta f/2} \left| y_{l,i}^{p} \right|^2 \phi(f)_l \, df, \tag{3}$$

where $\phi_l(f)$ is the power spectrum density of the lth PU signal

3. Problem Formulation

3.1. *Protecting PUs*

CRs are allowed to share the spectrum with Pus in underlay scenario, and this could result in poverty in the performance of PUN The expected degradation can be avoided by introducing stringent interference and power constraints. Moreover, adopting pricing techniques can also be used to maintain good QoS in primary network. Let I_{th} represents the maximum tolerable interference by PU and that $\tilde{P}$ is the upper limit power budget constraint for CR across all subcarriers. Let $s_{i,m}$ be the subcarrier indicator allocation, where $s_{i,m} = 1$ if and only if the given subcarrier belongs to the *m*th user and $s_{i,m} = 0$ otherwise. Moreover, the aggregate interference at the lth PU can be mathematically defined according to (4).

$$\sum_{m=1}^{M}\sum_{i=1}^{N} s_{i,m} p_i^m \Omega_i^l \le I_{th}, \forall l \in \{1,2,...,L\}, \tag{4}$$

Furthermore, the transmission power for all CRs in the networks should be less than the maximum power budget of CRBS, i.e., $\tilde{P}$. Therefore, the total power constraint can be represented mathematically as (5)

$$\sum_{m=1}^{M}\sum_{i=1}^{N} s_{i,m} p_i^m \le \tilde{P}, \tag{5}$$

If both interference and power constraints are satisfied, then CRs can use the spectrum of PUN and thus increase the spectral efficiency

3.2. *The Optimization Problem*

Assuming that p_i^m is the transmit power of the *m*th CR on the *i*th OFDM subcarrier. Therefore, the signal-to-interference-plus-noise ratio (SINR) for the *i*th subcarrier can be formulated as

$$\gamma_i^m = \frac{p_i^m \left| h_{i,m} \right|^2}{\delta_{AWGN} + \sum_{l=1}^{L} I_{l,i}^{CR_m}} = \frac{p_i^m \left| h_{i,m} \right|^2}{I^{Tot}}, \tag{6}$$

The total interference can be calculated according to $I^{Tot} = \delta_{AWGN} + \sum_{l=1}^{L} I_{l,i}^{CR_m}$, where δ_{AWGN} is the mean variance of the additive white Gaussian noise (AWGN) and $I_{l,i}^{CR_m}$ is the interference introduced by the lth PU to the *m*th CR on the *i*th subcarrier. Furthermore, the interference factor, i.e., $I_{l,i}^{CR_m}$, which is presented in (3), is assumed to be the superposition of a large number of independent random variable components, i.e., $\sum_{l=1}^{L} I_{l,i}^{CR_m}$. Thus, using central limit theorem, the interference factor ($I_{l,i}^{CR_m}$) can be model as AWGN. Note that this assumption is used different works in the literature [4, 9].

Therefore, the transmission rate of the ith subcarrier utilised by the mth CR can be modelled as follows:

$$R_i^m = \Delta f \log_2 (1+\gamma_i^m). \quad (7)$$

The main objective is to maximise the total capacity of the CRN by considering interference constraint in (4) and total power constraint in (5). Hence, the problem of resource allocation in OFDM-based downlink CRNs can be formulated as follows:

$$\text{P1}: \max_{p_i^m} (R_i^m),$$

$$\text{s.t.} \begin{cases} C1: s_{i,m} \in \{0,1\}, & \forall i, m, \\ C2: \sum_{m=1}^{M} s_{i,m} \le 1, & \forall i, \\ C3: \sum_{m=1}^{M} \sum_{i=1}^{N} s_{i,m} p_i^m \Omega_i^l \le I_{th}, \forall l \in \{1,2,..., L\}, \\ C4: \sum_{m=1}^{M} \sum_{i=1}^{N} s_{i,m} p_i^m \le \tilde{P}, \\ C5: p_i^m \ge \overline{P} & \forall n \in \{1,2,..., N\}. \end{cases} \quad \text{(P1)}$$

C2, in (P1) ensures that any given sub-channel should be allocated to one and only one user, and $\overline{P}$ in C5 refers to the minimum transmission power for CR. The optimisation problem in (P1) includes both binary variables, e.g., C1 in (P1), and continuous variables. Thus, it is a hard problem to solve. Moreover, the maximum data rate in downlink set-up is achieved if each subcarrier is allocated to the user with the best channel gain for that subcarrier, as shown in [7]. Moreover, in the downlink CRN, the CRBS has a common interference factor for all of the available CRs in the network, i.e., the value of the interference component (Ω_i^l) is independent of the CR. Consequently, the proof presented in [7] is valid for the CRN scenario as well

4. Downlink Market Model

4.1. Definition and Subcarrier Allocation

The joint solution to the problem of resource allocation (RA) in multicarrier-based CRNs is, in general, an NP-hard problem. Instead, the problem can be decomposed and solved in two steps, subcarrier and power steps, by many suboptimal algorithms presented in the literature, e.g., [16]. Once the subcarrier is allocated in the first step, the problem of RA can then be virtually addressed as a single-user optimisation problem. A common subcarrier allocation algorithm is adopted, similar to [5], to allocate the subcarrier to CRs, as shown in Algorithm 1 (Table 1).

Table 1. Algorithm 1 for subcarrier to CR allocation

	Algorithm 1. Subcarrier to CR Allocation
1-	Initialisation Step:
	Set $s_{i,m} = 0 \quad \forall i, m$
2-	SA Step:
	For i = 1 to N do
	$m = \arg\max_{m} \{h_{i,m}\};\ s_{i,m} = 1$
	End

Moreover, the channel gain can be determined from the subcarrier allocation steps according to

$$h_i = \sum_{m=1}^{M} s_{i,m} h_{i,m} \tag{8}$$

4.2. Market Power Allocation (MPA)

In power allocation sub-problem, the pricing technique is adopted from supermarket theory to further simplify (P1). This can be achieved by relaxing interference constraint in (P1) and letting the pricing scheme in the utility function manage the interference generated to PUs. The market power allocation (MPA) can be defined as follows MPA$\langle M, \{a_i^{\text{ction}}\}, \{S^c(\bullet)\}\rangle$, where $M = \{1,2,...,M\}$ is a finite set of decision makers (i.e., players or CR nodes); $\{a_i^{\text{ction}}\}$ is the action space, i.e., power to subcarrier allocation; and $S^c(\bullet)$ is the surplus function linked to each CR node in the network, where the superscript (^{c}) refers to CR. Assuming that the distributed power strategy over the available subcarriers, is a compact convex set bounded by minimum and maximum power budgets denoted by $\hat{P}_i = [\overline{P}, \tilde{P}]$. Mathematically speaking, the strategy space can be modelled as

$$\hat{P}_i = \left\{ \boldsymbol{P}_i : \sum_{i=1}^{N} p_i \le \tilde{P},\ \overline{P} \le p_i \le, \tilde{P}, \forall i \in N \right\}. \tag{9}$$

Without loss of generality, we assume that $\overline{P} = 0$.

4.3. Design of Surplus Function

A commonly chosen surplus function in market theory is the transmission rate for a given link in the network. The surplus function can be described as a function of action chosen by player m on the *i*th subcarrier, i.e., $(\mathbf{a}_i^{\text{ction}})$, and actions chosen by all of the players in the game, except those of player *m*, i.e., $(\mathbf{a}_{-i}^{\text{ction}})$. Mathematically speaking, the surplus function can be formulated as follows:

$$S_{i,m}^{c}(\boldsymbol{p}_{i,m}, \boldsymbol{p}_{-i,m}) = \sum_{i=1}^{N} \log_2\left(1 + \tfrac{\gamma_i^m}{I_i^{Tot}}\right) - p_i^m \sum_{l=1}^{L} \alpha_l \Omega_i^l, \tag{10}$$

where $S_m^c(\bullet)$ is the net surplus function for the *m*th player, which can be described as the difference between the rate of the user and the pricing function. Knowing that α_l is the price-control factor that manages the generated interference to the PUs. Accordingly, pricing technique has been used to decrease the complexity of (P1) by relaxing interference constraint (C3). By adopting Algorithm 1 and surplus function in (10), problem (P1) can be reformulated as follows:

$$\text{P2}: \max_{\hat{p}_i} (S_{i,m}^{c}),$$
$$\text{s.t.} \begin{cases} C1: \sum_{i=1}^{N} \hat{p}_i \le \tilde{P}, \\ C2: \hat{p}_i \ge \overline{P} \quad \forall i \in \{1,2,...,N\}. \end{cases} \tag{P2}$$

Any solution to the above market model is a solution that obtains the NE.

Theorem 1: At least one NE exists in the market model, i.e., (P2), if it satisfies the following conditions:

1) The action strategy profile (i.e., $\hat{P}_i$) is a closed, bounded, convex subset.
2) The surplus function $S_{i,m}^{c}(\bullet)$ is a continuous, quasi-concave function over the strategy set.

Proof: It is straightforward to prove that the above conditions are satisfied for the following reasons:

1) Because the strategy profile, as defined in (9), is defined by a minimum and maximum power, the first condition is readily satisfied.
2) To prove that the second condition is also satisfied, we have to show that the given price-based utility function is quasi-concave in the strategy profile.

Lemma 1: The surplus function given in (10) is concave in the defined power strategy profile.
Proof: To show that the second condition is true, we have to solve the following set of equations: $\frac{\partial^2 S^c_{i,m}}{\partial^2 p^m_i} < 0$. In the following, we have $\frac{\partial^2 S^c_{i,m}}{\partial^2 p^m_i} = -\frac{|h_{i,m}|^2}{\left(I^{Tot} + p_{i,m}|h_{i,m}|^2\right)^2}$.

which is <0.

Because both conditions have been proved, the NE exists in the proposed MPA.

4.4. Proposed Power Algorithm
In this section, optimal power allocation is derived according to the following proposition.
Proposition 1: If power is allocated to subcarriers by maximising (P2), then the optimal power strategy across the subcarriers is given by

$$\hat{p}_i = \left[\frac{1}{\varphi\left(\lambda + \sum_{l=1}^{L} \alpha_l \Omega_i^l\right)} - \frac{I^{Tot}}{|h_i|^2}\right]^+ \tag{11}$$

Proof: The solution of (P2) can be easily found by adopting the Lagrangian technique. Thus, the Lagrangian related to (P2) can be defined as follows:

$$\Gamma(\hat{p}_i) = \sum_{i=1}^{N} \log_2\left(1 + \frac{\hat{p}_i |h_i|^2}{I^{Tot}}\right) - \hat{p}_i \sum_{l=1}^{L} \alpha_l \Omega_i^l - \lambda\left(\sum_{i=1}^{N} \hat{p}_i - \tilde{P}\right) + \sum_{i=1}^{N} \hat{p}_i \beta_i \tag{12}$$

where β and λ are non-negative Lagrange multipliers. Because $\beta_i \hat{p}_i = 0$ and $\beta_i \geq 0$, we obtain $\hat{p}_i = 0$. Therefore, the optimal solution for the power allocation can be written as follows:

$$\hat{p}_i = \left[\frac{1}{\lambda + \sum_{l=1}^{L} \alpha_l \Omega_i^l} - \frac{I^{Tot}}{|h_i|^2}\right]^+ \tag{13}$$

where $[\bullet]^+ = \max(0,\bullet)$, λ is the Lagrangian multiplier. The power strategy algorithm in (13) utilises a single Lagrangian multiplier, in contrast to the use of the two Lagrangian multipliers in [5]. Thus, the complexity of the power strategy algorithm is reduced, which is one of the advantages behind adopting pricing mechanism. The Lagrangian multiplier (λ) can be calculated by substituting (13) into (P2-C1), i.e., $C1: \sum_{i=1}^{N} \hat{p}_i \leq \tilde{P}$, accordingly the Lagrangian multiplier (λ) can be determined according to (14).

$$\lambda = \frac{|N|}{\psi_l \left(\widetilde{P} + \sum_{i=1}^{N} \frac{I^{Tot}}{|h_i|^2} \right)} - \sum_{l=1}^{L} \alpha_l \Omega^{avg} \tag{14}$$

where ψ_l is the factor that compensates for replacing interference component by the average interference component, which can be modelled as follows, $\psi_l = \sum_{i=1}^{N} (\frac{\Omega_i^l I^{Tot}}{|h_i|^2}) + I_{th}$. Base on the above derivations and analysis, the MPA algorithm is proposed to determine the optimal power, which is denoted by $\hat{p}_i$. Furthermore, the procedure of MPA is shown in Algorithm 2 (Table 2).

Table 2. Algorithm 2 for Market Power Allocation (MPA) algorithm

Algorithm 2. MPA Algorithm
a) Initialisation step: $N, M, \widetilde{P}, \overline{P}, \Omega_i^l, I^{Tot}$
b) Set the value of α_l such that the interference constraint is satisfied.
c) Call Pricing-based Water-filling Algorithm
• Find λ according to (14)
• Set $\hat{p}_i$ according to (11)
d) Declare the power $\hat{p}_i$.
End

5. Simulation Results

5.1. Simulation Setting

The simulations are implemented according to the system model shown in Figure 1. The multicarrier networks consist of four (4) CRs, i.e., M = 4, and the total number of subcarriers is assumed to be $N = 16$. The values of Δf, $\widetilde{P}$ and I_{th} are assumed to be 0.3125 MHz, 1 W and 1 mW, respectively. The thermal noise is assumed to be 10^{-6}. The subcarrier fading gains h and g are assumed to be the outcome of independent, identically distributed (i.i.d) Rayleigh distributed random variables with unity mean. Furthermore, the channel gains are assumed to be perfectly known by the CRBS. In addition, all the results have been compiled over 1000 iterations.

5.2. Deployment Model and Simulation Results

The simulations are implemented according to the general system model shown in Figure 1. However, to enable a fair comparison, only one active PU band is assumed, i.e., L=1 in the deployment setting, as shown in Figure 2. An underlay coexistence scenario has been considered in this deployment, as defined in Figure1. Thus, the CRs are allowed to transmit over both active and non-active bands considering that the generated interference to PU must be less than the defined threshold I_{th}.

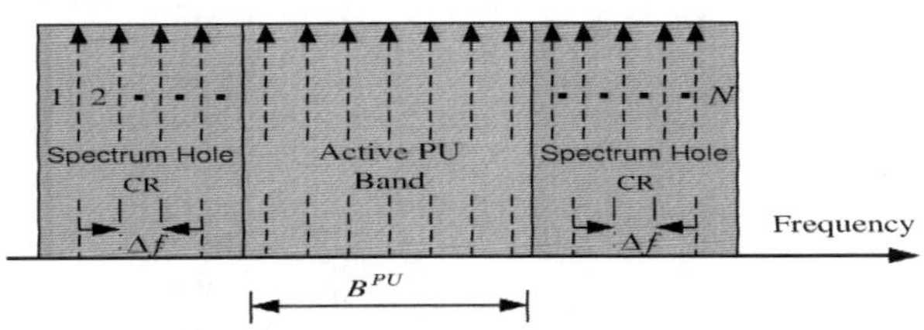

Figure 2. Deployment Scenario

The achievable capacity of CRs against the interference induced to PU band in the given deployment scenario using **MAP** and **PI algorithm** for different interference constraints is plotted in Figure 3. Note that Figure 3 shows for a given interference threshold, the proposed **MPA** algorithm almost approaches the **optimal** algorithm scheme. Moreover, the capacity of the **MPA** algorithm is slightly less than that of the **PI-algorithm** scheme. The net interference induced to PUs using the **MPA** and **PI algorithm** is plotted in Figure 4.

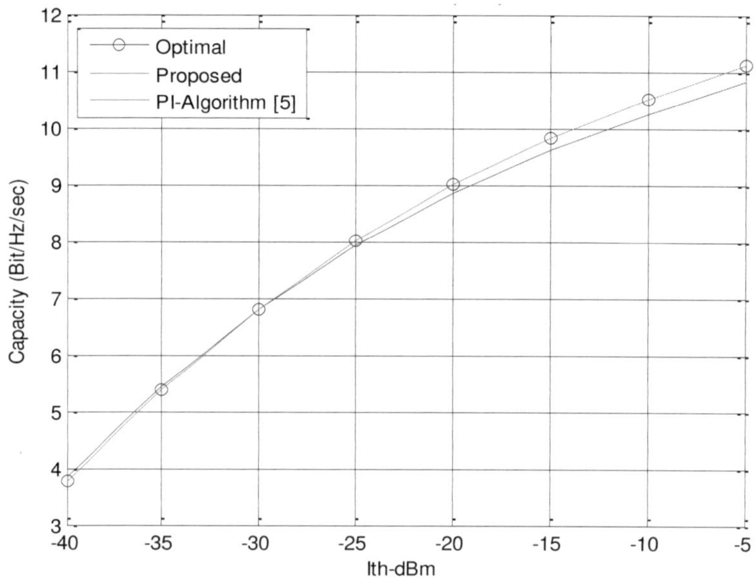

Figure 3. Achieved capacity vs. allowable interference threshold for OFDM-CRNs

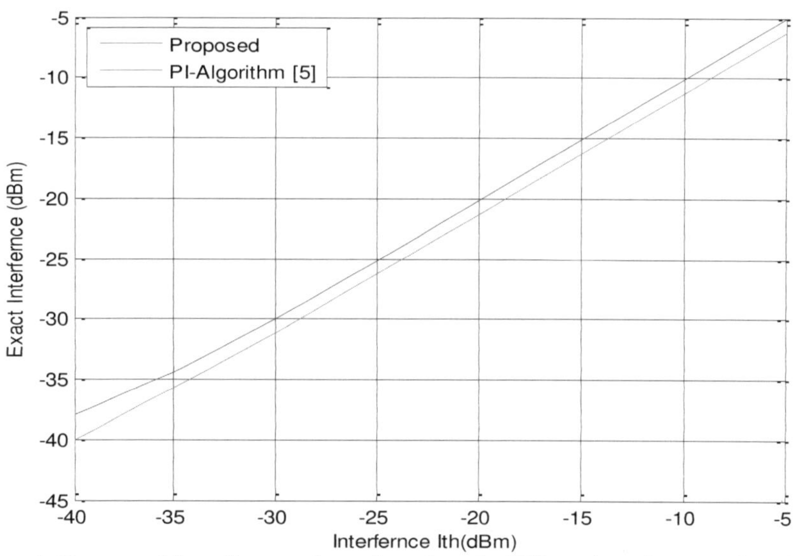

Figure 4. Generated interference introduced to the PU vs. interference threshold

Note that the net interference induced by **MPA** exhibits more flexibility in terms of interference management compared to **PI algorithm** and satisfies the predefined interference threshold, which makes **MPA** algorithm more suitable for practical scenarios. The strict management of interference induced to PU band resulted from varying the pricing factor in the adopted pricing scheme, which can be varied until the interference constraint is affordable to PU band. In contrast, in **PI algorithm**, the parameters are pre-defined in the power algorithm, which resulted in a fixed interference component that does take advantage of all of allowable interference to PU.

5.3. Computational Complexity

Note that by adopting pricing scheme, the computational complexity of the proposed **MPA** algorithm is reduced compared to **PI-algorithm** presented in [5]. This is because the proposed **MPA** algorithm utilises only one water-filling algorithm based on pricing scheme; therefore, the computational complexity is $O(N\log(N))$, where N is the number of subcarriers. In contrast, the **PI-algorithm** performs three complex steps, i.e., two water-filling algorithms and power level readjustment step; therefore, the computational complexity is $O(N\log(N))+O(L)$, where L in the number of PU. Moreover, the proposed **MAP** algorithm exhibits a much lower computational complexity compared to that of the optimal algorithm, which is $O(N^3)$.

6. Conclusion

A market-based power allocation method in an underlay downlink-cognitive radio network termed as MPA has been proposed in this paper. The pricing scheme has been used to provide flexibility in allocating the power budget to available subcarriers. Hence, CRs can utilise all allowable interference produced by PU. The existence and uniqueness of the NE have been proven mathematically. Moreover, the proposed algorithm offers a higher capacity compared to the previously tested algorithm in the context of OFDM-CRNs. Furthermore, the computational complexity of the proposed algorithm is reduced from $O(N\log(N)+\eta N)+O(L)$ to $O(N\log(N))$, which makes the proposed algorithm more suitable for practical implementation.

References

[1] FCC 2003 Notice of inquiry and notice of proposed Rulemaking *ET Docket No 03-237*

[2] Budiarjo I, Nikookar H and Ligthart L P 2008 Cognitive radio modulation techniques *IEEE Signal Processing Magazine* **25** 24

[3] Shaowei W, Mengyao G and Wentao Z 2013 Energy-efficient resource allocation for ofdm-based cognitive radio networks *IEEE Transactions on Communications* **61** 3181

[4] Farhang-Boroujeny B and Kempter R 2008 Multicarrier communication techniques for spectrum sensing and communication in cognitive radios *IEEE Communications Magazine* **46** 80

[5] Shaat M and Bader F 2010 Computationally efficient power allocation algorithm in multicarrier-based cognitive radio networks: OFDM and FBMC systems *EURASIP Journal on Advances in Signal Processing* **2010** 528378

[6] Jianwei H, Subramanian V, Agrawal R and Berry R 2006 Downlink scheduling and resource allocation for OFDM systems *40th Annual Conference on Information Sciences and Systems* **2006** 1272

[7] Jiho J and Kwang Bok L 2003 Transmit power adaptation for multiuser OFDM systems *IEEE Journal on Selected Areas in Communications* **21** 171

[8] Sadr S, Anpalagan A and Raahemifar K 2009 Radio resource allocation algorithms for the downlink of multiuser ofdm communication systems *IEEE Communications Surveys & Tutorials* **11** 92

[9] Bansal, G, Hossain J and Bhargava V K 2008 Optimal and suboptimal power allocation schemes for ofdm-based cognitive radio systems *IEEE Transactions on Wireless Communications* **7** 4710

[10] Shaowei W, Fangjiang H and Zhi-Hua Z 2011 Fast Power Allocation Algorithm for Cognitive Radio Networks *IEEE Communications Letters* **15** 845

[11] Ekram Hossain *et al.* 2009 Dynamic Spectrum Access and Management in Cognitive Radio Networks *Cambridge University Press*

[12] Toroujeni, S M M, Sadough, S M S and Ghorashi, S A G 2013 Spectrum leasing for OFDM-based cognitive radio networks *IEEE Transactions on Vehicular Technology* **62** 2131

[13] Haijian Z, Le Ruyet D, Roviras D and Hong S 2012 Noncooperative Multicell Resource Allocation of FBMC-Based Cognitive Radio Systems *IEEE Transactions on Vehicular Technology* **61** 799
[14] Tan C K, Chuah T C and Tan S W 2011 Fair subcarrier and power allocation for multiuser orthogonal frequency-division multiple access cognitive radio networks using a colonel blotto game," *IET Communications* **5** 1607
[15] Chao X, Min S, Chungang Y, Xijun W and Liang W 2014 Pricing-based multiresource allocation in ofdma cognitive radio networks: an energy efficiency perspective *IEEE Transactions on Vehicular Technology* **63** 2336
[16] Cheong Yui W, Cheng R S, Lataief K B and Murch R D 1999 Multiuser OFDM with adaptive subcarrier, bit, and power allocation *IEEE Journal on Selected Areas in Communications* **17** 1747

Analysis of non-rainy attenuation on earth-space path in Ota, Southwest Nigeria

T V Omotosho[1] S A Akinwumi[1], M R Usikalu[1], O O Ometan[2], M O Adewusi[2] and M Abdullah [3]

[1]Department of Physics, Covenant University, PMB 1023, Ota, Ogun state, Nigeria
[2]Department of Physics, Lagos state University, Ojo, Lagos state, Nigeria
[3]Institute of Space Science, University Kebangsaan Malaysia

E-mail:oluwasayo.akinwumi@covenantuniversity.edu.ng[1], ometanfunmi@gmail.com[2],omotosho@covenantuniversity.edu.ng[1], moji.usikalu@covenantuniversity.edu.ng[1], madewusi@gmail.com[2],mardina@eng.ukm.my[3]

Abstract. Propagation effects due to atmospheric gases and tropospheric scintillation requires accurate modelling in the design of satellite communication systems. The combination of the two attenuation phenomena was observed within the period of August 2014 to December 2015. The result of this paper presents the on-going observation and data analysis of non-rainy attenuation on earth-space path in Ota, Southwest Nigeria. Results of clear-sky attenuation vary between 0 dBm and 4.85 dBm in January and February 2015 respectively. While a value of 4.23 dBm and 4.75 dBm were observed in October 2014 and 2015 respectively. The results will be useful for satellite communication system design and will be submitted to ITU-R Study group 3 Databank.

1. Introduction

As demand for satellite, services are increasing especially at higher frequencies, under the influence of microwave propagation. Earth-space path link is a major concerned in the design of satellite communication systems. Major propagation factors that might impair the signal quality and also affect availability of satellite communication service includes rain attenuation, gaseous attenuation, tropospheric scintillation, cloud attenuation, melting layer attenuation, and rain and ice depolarization. Particularly, at Ku-band (14/12 GHz) and above [1, 2].

Rain attenuation is the main impairment effect for millimetre wave signals propagating through the troposphere. Interestingly, communication satellite operating in Ku and Ka bands links only requires small size of antennas that reduce or eliminate the use of tracking method and mostly cheaper cost of earth stations with the use of Very Small Aperture Terminals (VSATs). Therefore, rain may only form a little effect of the total propagation link margin [2]. However, the frequent presence of atmospheric gases and tropospheric scintillation may cause significant fades in these higher frequency bands for a large portion of the time.

Atmospheric gases, such as nitrogen (78.09%) and oxygen (20.95%) are predominantly present in the atmosphere. Water vapour (0.25%) though very small, is the most important in gaseous attenuation and it's has a major effect on propagation of electromagnetic wave. However, absorption caused by oxygen and water vapour is fundamental in understanding how attenuation due to gases depends on electromagnetic radiation [3]. Scintillation is caused by absorption and scattering of signal in the transmission link between the earth station and satellite station. Scintillation phenomena in earth satellite link is characterised at low elevation angles and high frequencies [4].

Little or no non-rainy attenuation ground measurements has been carried out in Nigeria up till now, most of the previous works on gaseous attenuation prediction model was based on theoretical model developed for the temperate region [5, 6, 7, 8].

In tropical countries, accurate performance of signal quality has turn into a growing demand due to the fact that modern satellite communication technology applications are being increasingly

Content from this work may be used under the terms of the Creative Commons Attribution 3.0 licence. Any further distribution of this work must maintain attribution to the author(s) and the title of the work, journal citation and DOI.
Published under licence by IOP Publishing Ltd

demanded for during highly tropical weather impairments and there is a scarcity of information as regards the non-rainy attenuation such as tropospheric scintillation analysis particularly in tropical region [9,10,11,12,13]. Therefore, the prediction, measurement and modelling of atmospheric gases and tropospheric scintillation on earth-satellite path links at Ku band and above in Southwest, Nigeria during clear-sky is essential, hence form the basis of this research.

2. Non-Rainy Attenuation Measurement

The ground data experimental measurement system at our laboratory incorporates Very Small Aperture Terminal (VSAT) with 0.9 m diameter. This satellite dish VSAT antenna was installed at the rooftop of a building at College of Science and Technology (CST) of Covenant University (CU), Ota, for approximately 52 m above the ground level. On the point of latitude 6.7° N and longitude 3.23° E and is directed toward Astra satellites (Astra 2E/2F/2G) on the geostationary orbit of longitude 28.2 ° E and the elevation angle of the receiver antenna is 59.9°.

The measurement setup at CU in Figure 1 shows VSAT transmitting data and simultaneously receiving returned data from Astra satellite. The beacon fixed-frequency of 12.245 GHz is transmitted by the satellite for reception on the ground and is vertically polarized. The output signal of the low noise block (LNB), at the dish was connected to a spectrum analyzer, which was interface to a computer at CU, Ota as shown in Fig. 2.

The 9103 handheld Spectrum Analyser (Aeroflex) is a multi-channel remote sensing active device that has many applications such as logging of attenuation received signal level data frequency every seconds in decibel meters (dBm), troubleshooting of antenna and cable installations among others. The averaged channel power attenuation (in dBm) at one-minute interval is recorded and stored on a personnel computer.

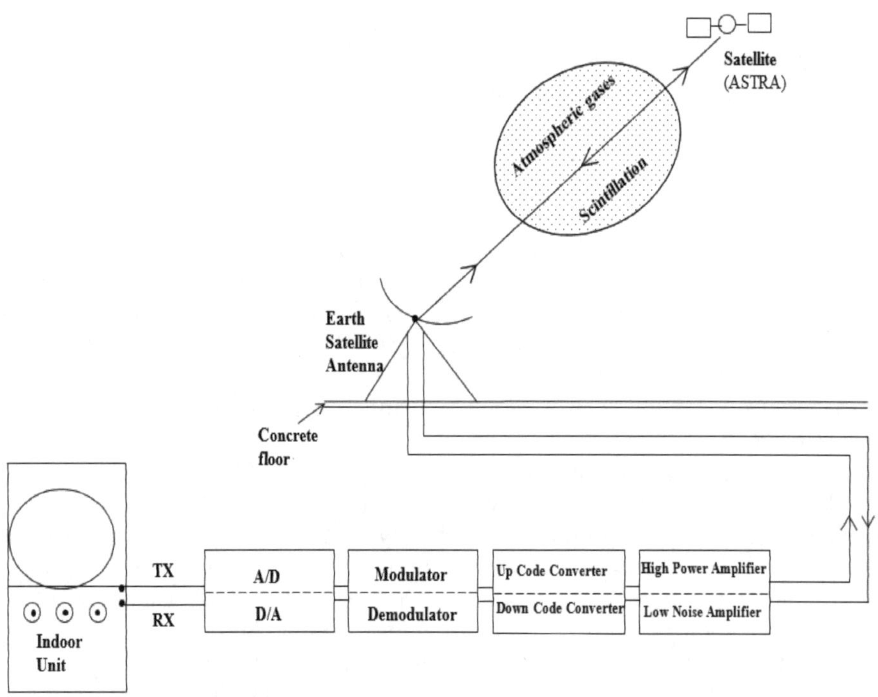

Figure 1. Experimental setup and illustration

The separation of gaseous attenuation and scintillation and other attenuation phenomena were done by filtering. Although, scintillation occur during rainy as well as non-rainy periods but rain events are excluded in this report. Also, from Davis weather station the rain rate at 0 mm/h were observed for non-rainy days while rain rate above 0 mm/h were filtered away from the corresponding days and time data from spectrum analyser within the period of August 2014 to December 2015.

The non-rainy attenuation is obtained by subtracting a reference signal level data from measured received signal level data on every one minutes for each clear-sky day. The reference level data is therefore, obtained by averaging the entire received signal level data on each month and at each place during non-rainy term.

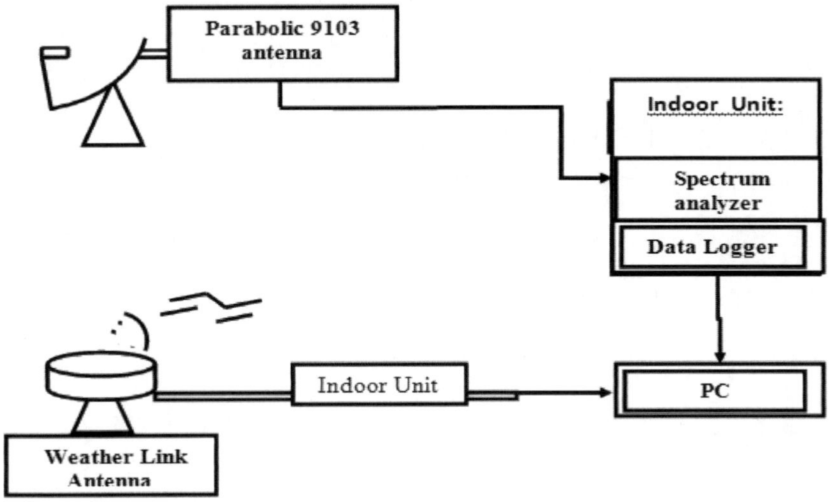

Figure 2. Block diagram of the VSAT setup in CU

3. Results and Discussion

Gaseous attenuation and tropospheric scintillation fade distributions were predicted using Spectrum Analyser and Davis weather station for a period of two years (August, 2014 to December, 2015). The non-rainy attenuation for different percent (0.001%, 0.01%, 0.1% and 1%) of time of the year determined by percentage of ordinate exceedance curves is shown in Figures 3 to 6 for the year 2014 and 2015 respectively. The results obtained showed the time portion gets smaller with an increase received clear-sky attenuation distribution, the reason for this is attributed to little effects of gases and scintillation on the total attenuation at small percentage of time.

Table 1 show summary of results of clear-sky attenuation minimum values of 0 dBm experience in January and March, 2015 respectively. While, Maximum value of 4.85 dBm was observed in February 2015 been a dry month, this may be due to the presence of cumulonimbus cloud which cause scintillation (presence of high water vapour density) contribute to the results. However, maximum values of 4.23 dBm and 4.75 dBm was recorded in October of 2014 and 2015 respectively, this is in conformity with high attenuation during rainy season.

Table 1: Summary of the result

YEAR	2014		2015	
MONTH	Min. (dBm)	Max. (dBm)	Min. (dBm)	Max. (dBm)
JAN	-	-	0	4.00
FEB	-	-	0.05	4.85
MAR	-	-	0	4.50

APR	-	-	0.02	3.92
MAY	-	-	0.05	3.85
JUN	-	-	0.04	4.06
JUL	-	-	0.05	4.75
AUG	0.03	3.63	0.02	4.48
SEP	0.04	4.04	0.05	3.65
OCT	0.03	4.23	0.05	4.75
NOV	0.03	3.77	0.01	3.91
DEC	0.05	3.95	0.03	4.23

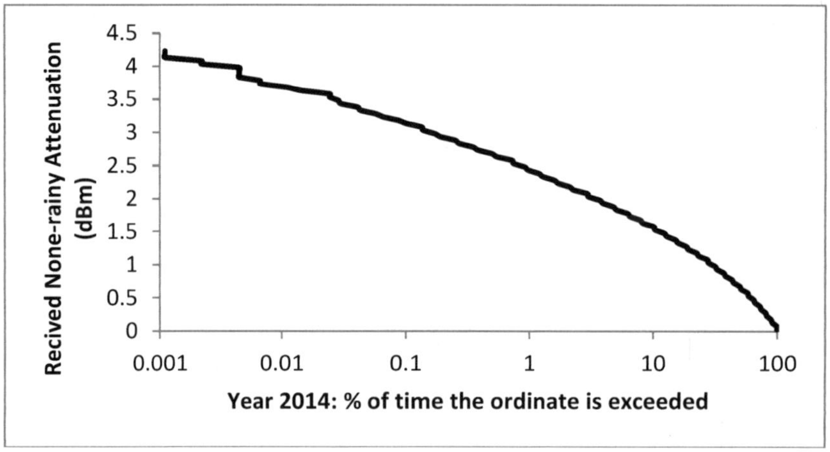

Figure 3. Total measured cumulative distributions for non-rainy Attenuation in 2014

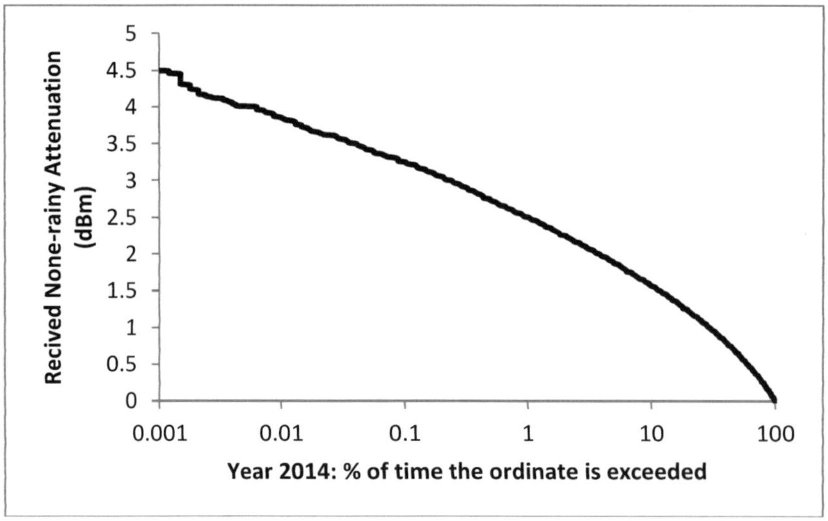

Figure 4. Total measured cumulative distributions for non-rainy Attenuation in 2015

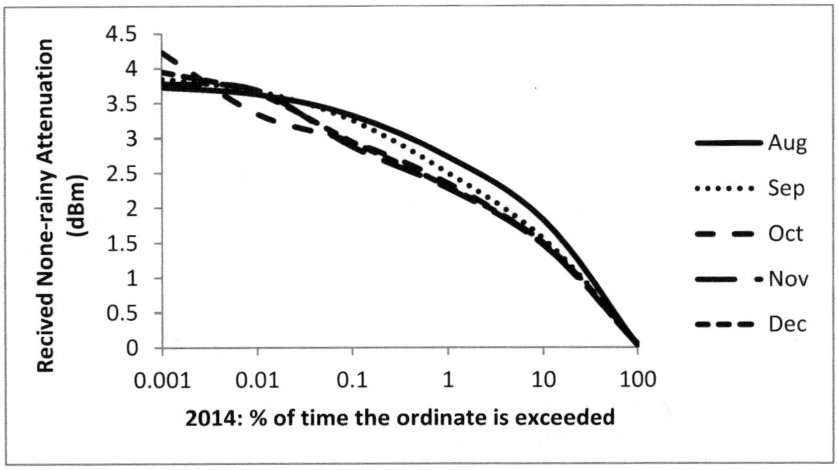

Figure 5. Cumulative distributions of non-rainy Attenuation in CU, Ota from August to December 2014

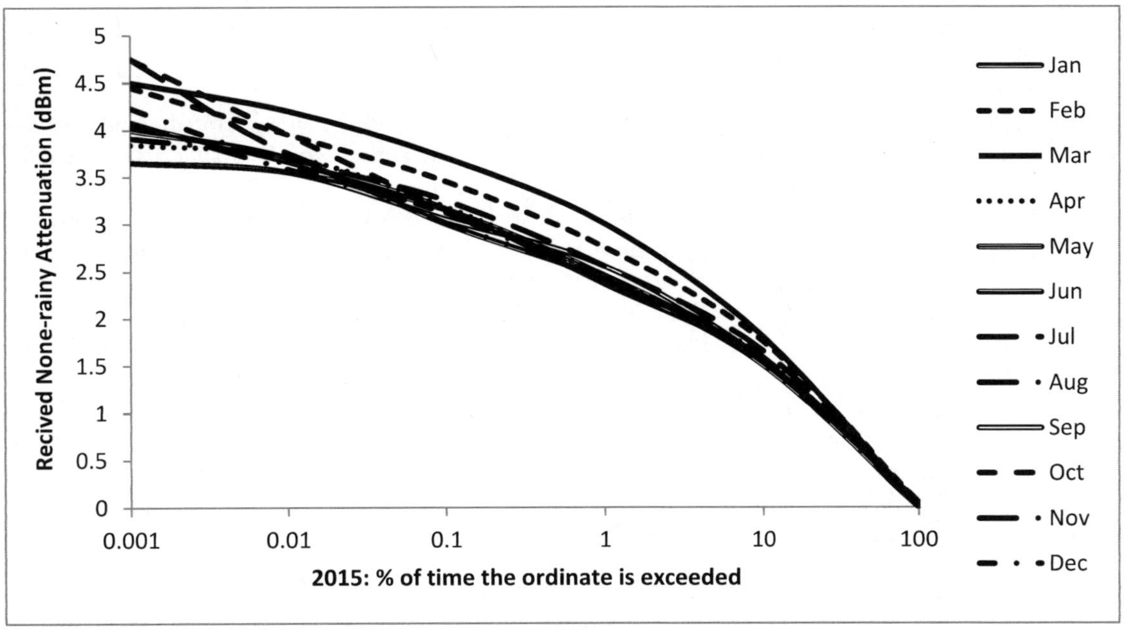

Figure 6. Cumulative distributions of non-rainy Attenuation in CU, Ota from August to December 2015

4. Conclusion
The results from Covenant University show that atmospheric gases and tropospheric scintillation can be combined accurately with the fact that is statistically independent of each other. However, the results non-rainy attenuation may be significant for low fade margin system designs at higher frequency bands. The model and data analysis process is ongoing in the study area and the results may be useful for satellite communication system design and submitted to ITU-R for future recommendations.

Acknowledgement
The authors thankfully acknowledged Covenant University for sponsorship of the research work.

References

[1] Zulfajri B H, Kiyotaka F, Kenichi I and Mitsuo T 2002 Measurement of Ku-band rain attenuation using several VSATs in Kyushu Island, Japan. *IEEE antennas and wireless propagation letters* **1**

[2] Mattioli V, Basili P, Bonafoni S, Ciotti P and Westwater E R 2009 Analysis and improvements of cloud models for propagation studies. *Radio Science* **44**

[3] Karasawa Y and Matsudo T 1991 Characteristics of fading on low-elevation angle earth-space paths with concurrent rain attenuation and scintillation. *IEEE Trans. on Antennas and Propagation* V**39** 657

[4] Gustavo A S, Jose M R and Pedro G 2015 Atmospheric Attenuation in Wireless Communication Systems at Millimeter and THz Frequencies. *IEEE Antennas and Propagation Magazine* **57**

[5] Ajayi G O, Feng S R and Reddy B M 1996 Handbook on Radio Propagation Related to Satellite in Tropical and Subtropical Countries 58

[6] Adenugba D A 2000 Microwavea absorption by oxygen and water vapour at tropical latitudes *M.TECH Thesis* The Federal University of Technology, Akure

[7] Adenugba D A and Kolawole L B 2006 Study of Gaseous Attenuation at Tropical Location. *Nigerian Journal of Physics* **18** (1)

[8] Akinwumi S A, Omotosho T V, Usikalu M R, Ometan O O and Adewusi M O 2016 Atmospheric gases attenuation in West Africa *IEEE radio and Antenna Days of the Indian Ocean 2016 Conference, Reunion Island*

[9] Ali M A, Aduwati S, Mandeep J S, and Alyani I 2013 Extracted atmospheric impairments on earth-sky signal quality in tropical regions at Ku-band *Journal of Atmospheric and Solar-Terrestrial Physics* **104** 96

[10] Ramachandran V and Kumar V 2007 Modified rain attenuation model for tropical regions for Ku-band signals *International Journal of Satellite Communications and Networking* **25**(1) 53

[11] Bryant G H, Adimula I, Riva C, Brussaard G 2001 Rain attenuation statistics from rain cell diameters and heights *International Journal of Satellite Communications* **19**(3) 263

[12] Dissanayake A, Allnutt J E, Haidara F 1997 A Prediction Model that Combines Rain Attenuation and Other Propagation Impairments Along Earth-Satellite Paths *IEEE Tran. On antennas and propagation* **45**

[13] Yamada M and Karasawa Y 1994 Increase of reliability for satellite communication circuits by use of long-distance site diversity *In: Proceedings of the 24th European Microwave Conference* 1581

IOP Conf. Series: Journal of Physics: Conf. Series **852** (2017) 012040 doi:10.1088/1742-6596/852/1/012040

Installing the earth station of Ka-band satellite frequency in Malaysia: conceptual framework for site decision

M R Mahmud[1,2], M N M Reba[1,2], S W Jaw [1,2], A Arsyad[2] and M A M Ibrahim[2]

[1]Department of Geoinformation, Faculty of Geoinformation& Real Estate, UniversitiTeknologi Malaysia, 81310, Skudai, Johor Bahru, Malaysia

[2]Geoscience & Digital Earth Centre, Research Institute for Sustainable Environment, Universiti Teknologi Malaysia, 81310, Skudai, Johor Bahru, Malaysia

E-mail : rizaludin@utm.my

Abstract. This paper developed a conceptual framework in determining the suitable location in installing the earth station for Ka-band satellite communication in Malaysia. This current evolution of high throughput satellites experienced major challenge due to Malaysian climate. Because Ka-band frequency is highly attenuated by the rainfall; it is an enormous challenge to define the most appropriate site for the static communication. Site diversity, a measure to anticipate this conflict by choosing less attenuated region and geographically change the transmission strategy on season basis require accurate spatio-temporal information on the geographical, environmental and hydro-climatology at local scale. Prior to that request, this study developed a conceptual framework to cater the needs. By using the digital spatial data, acquired from site measurement and remote sensing, the proposed framework applied a multiple criteria analysis to perform the tasks of site selection. With the advancement of high resolution remotely sensed data, site determination can be conducted as in Malaysia; accommodating a new, fast, and effective satellite communication. The output of this study is one of the pioneer contributions to create a high tech-society.

1. Introduction

The evolution of broadcast satellites using Ka-band frequency instead of the existing Ku-band is an emerging trend towards faster, cheaper, and efficient communication. This is because the new mode of high throughput satellites using Ka-band allowed extensive frequency reuse enabled larger amount of bandwidth support for higher transmission rates, and utilization of smaller but efficient antennas [1,2]. This advancement qualities would be strategic in addressing the growing needs of the increasing population and economy of Malaysia (~30 million, [3]); one of the rapidly developing semi industrial-agriculture country in Southeast Asia. However, the major drawback of this Ka-band in this region is their strong attenuation effects from large size water droplet of rainfall from atmospheric to ground.

Scientifically, the increasing frequency of Ka-band (26 – 40 GHz) would increase the speed of signal transmission and shortened the wavelength. The aftermath of this condition in humid tropics is the limitation to penetrate over intense rainfall with larger water droplets [4]. Literature has shown that the diameter of the rainfall in the tropics is relatively larger [5,6]. One of the effective strategy to overcome this constraint is via site diversity; selecting the area with less rainfall for satellite communication and diversifying the selection according to the local-scale wet seasons [7]. In Malaysian context, it is wise to decide based on a fine resolution of climatology data due to its high

Content from this work may be used under the terms of the Creative Commons Attribution 3.0 licence. Any further distribution of this work must maintain attribution to the author(s) and the title of the work, journal citation and DOI.
Published under licence by IOP Publishing Ltd

dynamic climatological pattern [8]. In addition, we also should consider the environmental aspect of the sites, its logistics and strategic condition.

Achieving both objectives can be materialized with the availability of the innovated climatology products and usage of Geographical Information System (GIS). A proper conceptual framework would be a pioneer effort in promoting this potential technology to a local scene. The design conceptual framework should be adhered a right product, approach and results which fit the local scale environmental and climatology characteristics. This study tends to occupy that effort by developing the conceptual framework based on integrated satellite-improved climatology input data that operated on GIS environment.

2. Dealing with the attenuation factors by hydro-climatology factors in Peninsular Malaysia

Peninsular Malaysia categorized as humid tropical region with perature of 26° Celcius throughout the year and annual rainfall of 2500mm. With no dry season, the hydro-climatology is driven by monsoon flows. The northeast monsoon (Nov.-Jan.) constitutes the heaviest rainfall especially to the east part (200-500mm/month). The southwest and inter-monsoon period (Jul.-Oct.) resulted second heavy rainfall which concentrated on the west region. Added the effect from high elevation areas of the Titiwangsa fringe, the local rainfall pattern can be highly heterogeneous especially during the non-monsoon season [9] (Figure 1).

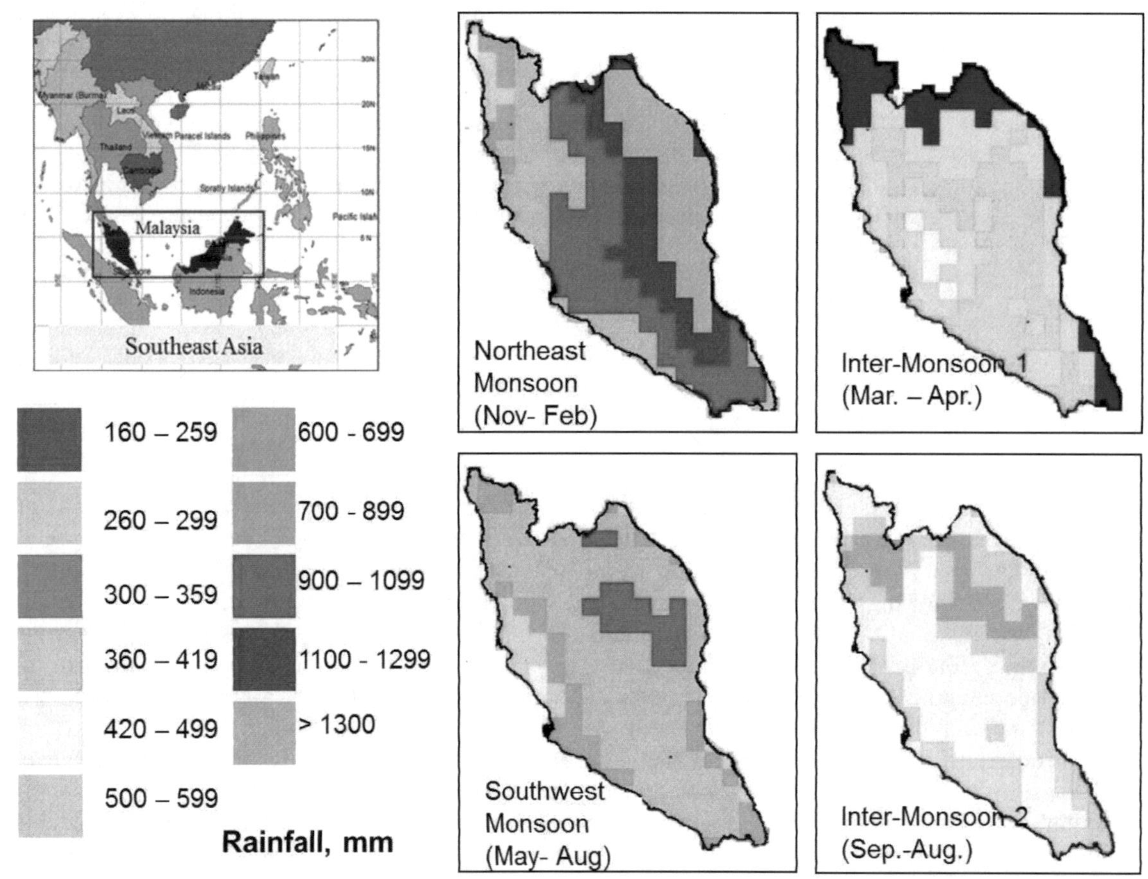

Figure 1. Peninsular Malaysia and its seasonal rainfall variation

Due to this humid tropics and dynamic monsoon seasons; the attenuation of Ka-band frequency not only come from rain, but also the precipitable water in the atmosphere [10]. Considering the sky-free condition is depending on seasonality of monsoon flows, the wise option is to employ a selection with seasonal weightage. This can only be materialized by using high resolution hydro-climatology data; which now available to be obtained from satellites and publicly accessed.

3. High resolution of satellite hydro-climatology data & geographic computing

The advancement of remote sensor, retrieval algorithm and data sharing has significantly improved the representation of hydro-meteorological variables at both local and global scale. The high availability of such data especially one from microwave scanning [11,12] provides a perfect opportunities to tropical region, where large vapors converged, cloud formed, and rainfall downpours were indicated; a useful option to our framework.

Integration of those data with series of logical condition, programmed by geographic computing or GIS, enabled us to precisely determine the appropriate site. Given that all the data were correctly projected and the resolution is fine enough, successful data manipulation can be obtained. Those two criteria is our core to this conceptual framework.

4. Conceptual framework for site selection

The developed conceptual framework emphasize the following logic in sequential manner; (a) rainfall, precipitable water, & water vapor effects to Ka-band frequency, (b) sensitive areas which restricted for developments, and (c) strategic and logistics factors. Figure 2 summarized the design framework.

This conceptual framework utilized spatial-based decision making, where the site would be determine as multi-criteria analysis, followed the design condition in Table 1. A geographical information system usage is optimized where all the queries, display, analysis and storage of large databases, were implemented.

The pixels which fit the design condition will be given a higher score and vice versa. This method is chosen because it is relative; so relative description is the best option. At the end of the process (a), (b), and (c), is the map with selected pixels with different score; how suitable is the pixels for the earth station installation.

4.1 Attenuation from rainfall, precipitable water, & water vapor

For the first conditions, the primary goal is to determine the attenuation level contributed by rainfall, precipitable water, & water vapor. Because of the dominant effects of monsoon flows, it is best to quantify it temporally according to the monsoon season. Regarding the attenuation model, the utilization of the approach from International Telecommunication Union (ITU) was recommended. The following attenuation computation procedures for rain, cloud and water vapor from [13], [14], and [15] respectively can be utilized. This is to ensure that the attenuation was computed following the standard operation procedures.

The input of this part should be able to represent an effective spatio-temporal dimension to represent its corresponding subject of interest; at least ~5km grid. For rainfall, input from dense rain gauge network should be the first priority. Substitute or support information from other apparatus, such as precipitation radar or satellites can be considered. For precipitable water and water vapor, an input from the atmospheric & meteorological satellites or fine climatological model is the best option. In measuring total column of precipitable water and vapor, among the data that fit our objective are from The Special Sensor Microwave Imager (SSM/I), Special Sensor Microwave Imager Sounder (SSMI), TRMM Microwave Imager (TMI), The Advanced Microwave Scanning Radiometer (AMSR) or from the web-host that supply integrated products from various satellites such as the operational Microwave Integrated Retrieval System (MIRS) and the Morphed Integrated Microwave Imagery (MIMIC) (16).

In term of frequency selection, because the effective frequency Ka-band (~26-40 GHz) in tropical region is varies via location [17], selecting samples according to the constant interval is suggested.

After the attenuation rates for each parameter were obtained, the areas which have higher attenuation is ranked low and vice versa. In addition, the Ka-band attenuation effects were strongly influenced by larger water droplets, therefore, it is suggested that the weightage is assign to give emphasis in the following orders; rain, precipitable water, and vapor.

Table 1. Summary of the input, condition and ranking system of the developed conceptual framework

<table>
<tr><th rowspan="3">Phase</th><th rowspan="3">Input</th><th rowspan="3">Condition</th><th colspan="5">Ranks</th></tr>
<tr><th>1</th><th>2</th><th>3</th><th>4</th><th>5</th></tr>
<tr><th>Least suitable</th><th>Moderately suitable</th><th>Fairly suitable</th><th>Suitable</th><th>Most suitable</th></tr>
<tr><td>Phase 1 - Attenuation from rainfall, precipitable water & atmospheric water vapor</td><td><u>Monthly maps:</u>
Rainfall
Precipitable water
Water vapor
<u>Attenuation model:</u>
Attenuation co-efficient
Viewing angle
Selective frequency from 26-40GHz</td><td>Attenuation rate</td><td colspan="5">Maximum attenuation rate ----> Minimum attenuation rate</td></tr>
<tr><td rowspan="3">Phase 2 - Sensitive areas restriction</td><td>(i) Natural Protected areas
- Forest reserve maps</td><td>Boolean logic, Yes or No</td><td colspan="2">No</td><td>-</td><td colspan="2">Yes</td></tr>
<tr><td>(ii) Natural Sensitive areas
- Rivers, swamp, and peatland</td><td>Boolean logic, Yes or No</td><td colspan="2">No</td><td>-</td><td colspan="2">Yes</td></tr>
<tr><td>(iii) Private entities
- Building, roads, expressways, railways</td><td>Boolean logic, Yes or No</td><td colspan="2">No</td><td>-</td><td colspan="2">Yes</td></tr>
<tr><td rowspan="4">Phase 3 - Strategic & logistics</td><td rowspan="4">High resolution aerial images</td><td>Non-occupied land</td><td colspan="2">No</td><td>-</td><td colspan="2">Yes</td></tr>
<tr><td>Non-residential areas</td><td colspan="2">No</td><td>-</td><td colspan="2">Yes</td></tr>
<tr><td>Appropriate slope</td><td colspan="2">No</td><td>-</td><td colspan="2">Yes</td></tr>
<tr><td>Accessible</td><td colspan="2">No</td><td>-</td><td colspan="2">Yes</td></tr>
</table>

4.2 Sensitive areas restriction

The output from the attenuation rate will undergo further screening of sensitive areas. In Malaysia context, environmental sensitive areas were divided into three broad category, (i) natural protected areas, (ii) private entities, and (iii) natural reserve areas. Development or commercialization of these areas were prohibited by law and also destabilizing the ecosystem.

Environmental sensitive and protected areas include forest reserve; both inland and mangrove, and their corresponding adjacent areas (depending on context). Next, the natural reserve areas are natural features which not gazette as protected by law but contribute major services for ecosystem; rivers, swamp, peatland and their buffers. For private entities, accounted features are major road and expressways, railways, power grid & transmission areas, and private land or building.

To conduct this analysis, maps of land use, transportation, utilities and sensitive areas are used. Areas which located in this sensitive condition are excluded from the selection. A series of Boolean logic is applied for the operation.

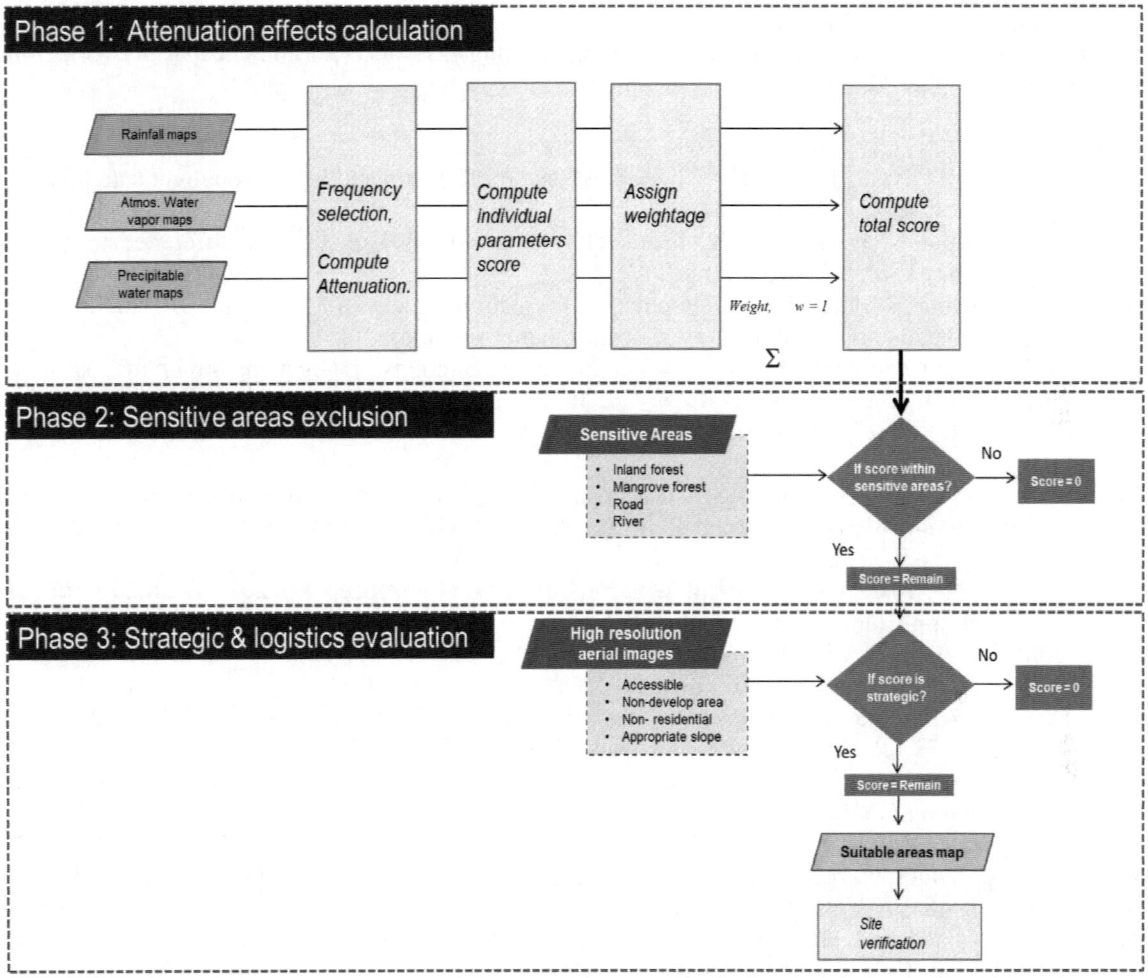

Figure 2. Conceptual framework

4.3 Strategic & logistics factors

The third phase of the framework is determining the strategic & logistics factors. This factor represents the elements which associated with the logistics aspect the earth station. That includes non-occupied land, non-residential areas, low to middle topographic slope and logistically accessible via vehicles (~200-300m).

A high resolution images are utilized. Basic filtering started with separation between densely built up areas and vegetated and non-developed areas. The use of vegetation indices is encouraged for fast processing. There are wide array of data selection for this process; depending on the quality of the desired output and cost. A low altitude and small unmanned vehicle or an aircraft with high resolution cameras can obtain the most precise resolution (1-2 meter). Meanwhile, high resolution data from satellite can offered between 10m – 15m accuracy results.

5. Conclusion

This paper developed a conceptual framework in precise determination of the suitable location for the earth station of Ka-band frequency signal. The conceptual framework emphasized the Ka-band ability to obtain effective transmission in high resolution area scale (~1km); with considering the attenuated by water droplets from rainfall, cloud and atmospheric water vapor. The site selection then being conducted using high resolution satellite imageries to potentially identify the non-sensitive areas, strategic location, meets the logistics demand and targeted community. The realization of the conceptual framework would be the pioneer step in creating fast, efficient and cheaper network from the high throughput satellites.

References

[1] Butash, T and Evans, B 2016 IJSCN Special Issue on Ka band and high throughput satellites *Int. J. Satell. Commun. Network* **34** 459
[2] Gilat Satellite Network 2011 Ka-band versus Ku-band, Whatmakesthe difference in VSAT technology? *Satellite Evolution Magazine*
[3] Dept of Statistics Malaysia 2017 Population Projection (Revised) Malaysia, 2010-2040 Access online (30/01/2017) at https://www.dosm.gov.my/v1/index.
[4] Ellis S M and Vivekanandan, J 2011 Liquid water content estimates using simultaneous S and Ka band radar measurements *Radio Sci*. **46** RS2021
[5] Maciel L R, Assis M S 1990 Tropical rainfall drop-size distribution *Int. J. Satell. Commun*. **8** 181
[6] Tenorio R S, Moraes M C D S and Sauvageot H 2011 Raindrop size distribution and radar parameters in Coastal Tropical Rain Systems of Northeastern Brazil *J. App. Met. Clim.* **51** 1960
[7] Semire F A, Mokhtar R M, Ismail W, Mohamad N and Mandeep J S 2014 Evaluation of site diversity rain attenuation mitigation technique in South-East Asia *Acta Astronautica* **96** 303
[8] Wong C L, Venneker R, Uhlenbrook S, Jamil A B M and Zhou Y 2009 Variability of rainfall in Peninsular Malaysia, Hydrol *Earth Syst. Sci. Discuss.* **6** 5471
[9] Cheang B K 1980 Some aspects of winter monsoon and its characteristics in Malaysia.In M.M. Service (Ed.) *Research Publication* Malaysian Meteorological Service
[10] Saurabh D, Maitra A and Shukla A 2013 Diurnal variation of slant path Ka-band rain attenuation at four tropical locations in India *Indian J. of Radio & Space Physics* **42** 34
[11] Boukabara S -A, Garrett K and Wanchun C 2010 Global coverage of total precipitable water using a microwave variational algorithm *IEEE Transactions on Geoscience and Remote Sensing* **48**(10) 3608
[12] Petty G W and Li K 2013 Improved passive microwave retrievals of rain rate over land and ocean. Part II: Validation and intercomparison *J. Atmos. Oceanic Technol.* **30** 2509
[13] Recommendation ITU-R P 840-6 2013 Attenuation due to clouds and fog, P-Series *Radiowave Propagation*
[14] Recommendation ITU-R. P. 676-3 2013 Attenuation due to atmospheric gases, P-Series, *Radiowave Propagation*
[15] Recommendation ITU-R P 676-3 2013 Specific attenuation for rain for use in prediction methods, P-Series *Radiowave Propagation*
[16] Wimmers A J and Velden C S 2011 Seamless advective blending of total precipitation water retrievals from Polar-Orbiting Satellites *J. Appl. Meteor. Climatol.* **50** 1024
[17] Dasgupta K S, Charania A, Shukla, A and Acharya R 2009 A new propagation campaign in tropical areas: The Ka-band propagation experiment over India with the GSAT-4 satellite. Antennas and Propagation 2009 *EuCAP 2009 3rd European Conference on Antennas and Propagation* 23-27 March 2009

Estimation water vapor content using the mixing ratio method and validated with the ANFIS PWV model

W Suparta[1,*], K M Alhasa[1], M S J Singh[2]

[1]Space Science Centre (ANGKASA) Institute of Climate Change Universiti Kebangsaan Malaysia 43600 Bangi, Selangor Darul Ehsan, Malaysia
[2]Dept of Electronic and System Engineering Universiti Kebangsaan Malaysia 43600 Bangi, Selangor Darul Ehsan, Malaysia

*Email: wayan@ukm.edu.my

Abstract. This study reported the comparison between water vapor content, the surface meteorological data (pressure, temperature, and relative humidity), and precipitable water vapor (PWV) produced by PWV from adaptive neuro fuzzy inference system (ANFIS) for areas in the Universiti Kebangsaan Malaysia Bangi (UKMB) station. The water vapor content value was estimated with mixing ratio method and the surface meteorological data as the parameter inputs. The accuracy of water vapor content was validated with PWV from ANFIS PWV model for the period of 20-23 December 2016. The result showed that the water vapor content has a similar trend with the PWV which produced by ANFIS PWV model ($r = 0.975$ at the 99% confidence level). This indicates that the water vapor content that obtained with mixing ratio agreed very well with the ANFIS PWV model. In addition, this study also found, the pattern of water vapor content and PWV have more influenced by the relative humidity.

1. Introduction

Our Earth is made liveable by the existence of gases as called greenhouse gases that allowed trap long-wave radiation emitted from the Earth's surface. Greenhouse gasses originated from anthropogenic and their effect on atmospheric temperature have been widely studied (e.g., [1], [2]). Without these gasses, the average of surface temperature in Earth's would be about −18 °C (0 °F) and vice versa with existence these gases the present average temperature of Earth's surface of 15 °C (59 °F). This phenomenon is called as the greenhouse effect. Although water vapor is not come from anthropogenic, it has impact to greenhouse effect about 60% in clear skies compared to the other greenhouse gases such as the carbon dioxide, ozone and trace gases including methane and nitrous di oxide [2].

Atmospheric water vapour is one of the most important components in determining and understanding of global climate change. The content and distribution of water vapour are critical variables for description of the state and evolution of many physical processes in the Earth's atmosphere. A long time ago, Manabe and Wetherald [3] have described the water vapour feedback show an important role in the behaviour of the climate and weather system. For example, water vapour plays important role in dynamic of spread and in radiation exchange within the atmosphere. This process is fundamental to transfer and formation energy in the atmosphere, and propagation of weather [4].

In addition, the water vapour is not only limited role in absorbing and radiating energy from the sun, but also its influence to formation of clouds, aerosols, and chemistry of the lower atmosphere [5]. On the other hand, the distribution of water vapour is closely related to the event of clouds, rainfall, convection and storm, as noted by many authors [6, 7]. Thus, the understanding of amount water vapor concentration is primarily important to monitor weather and climate changes in our Earth.

The purpose of this study is to attempt the calculate of water vapor concentration in our atmosphere using the mixing ratio formula and then validate its value using the total column water vapor in term of precipitable water vapor (PWV) that derived by adaptive neuro fuzzy inference system (ANFIS) PWV model. Observation one-week data from Universiti Kebangsaan Malaysia Bangi (UKMB) station was provided to compare and estimate water vapor concentration and PWV.

Content from this work may be used under the terms of the Creative Commons Attribution 3.0 licence. Any further distribution of this work must maintain attribution to the author(s) and the title of the work, journal citation and DOI.
Published under licence by IOP Publishing Ltd

Later on, this formula will be used to calculate concentration of water vapor in peatland area, since the water vapor is one of primary component greenhouse gasses that to contribute to greenhouse effect. In the next study, the concentration of water vapor content which calculates with the mixing ratio formula will be compared with the other component greenhouse gasses such as the carbon monoxide (CO_2) and methane (CH_4) to know the relevance of them.

2. Methodology

2.1. The available data

In this study, three parameters of surface meteorological data such as surface pressure, surface temperature and relative humidity were used to determine of water vapor concentration and PWV. The data were taken from the space science centre (ANGKASA), UKMB station (2°55'23"N, 101°46'23"E). All the surface meteorological data were collected directly from the PTU300 meteorological sensor. The PTU300 is a one of meteorology sensor that manufactured by Vaisala. This sensor consists of three parameters measurement which is surface pressure (in mbar), surface temperature (in ^{O}C) and relative humidity (in percent). All the surface meteorological data were collected at 1-minute interval.

2.2. ANFIS PWV model

Since, the distribution of water vapor content has primary role in weathers and climate changes such as the determinant success of the weather forecast. Many researchers have developed technique to measure water vapor content in terms of PWV, such as a traditional measurement (radiosonde and water vapor radiometry), and ground-based global positioning system (GPS) [8]. Compared to traditional measurement, the GPS technique has an advantage in the measurement of water vapor content as thought high accuracy, large capacity, and quick variation. However, in the current state, the GPS data have a lacking due to the incomplete data or missing in 24 hour, especially in remote area or area without GPS receiver provider.

To solve this shortcoming, Suparta and Alhasa 2016 [9] was developed the model to estimate the PWV without GPS data, which is only use the surface meteorology data as well as pressure, temperature, and relative humidity. The model have constructed as the multi-layer feedforward neural network. It consists of five layers, where the first layer and fourth layer have four membership functions and rules, respectively, as shown in Figure 1. To achieve the high accuracy, the model has trained with the hybrid learning algorithm. This model was validated with the other techniques such as GPS measurement. The validation result showed that the PWV value that provided by ANFIS PWV model has a strong relationship with the GPS PWV value. For more detail about the physical of this model, please refer this paper [9]

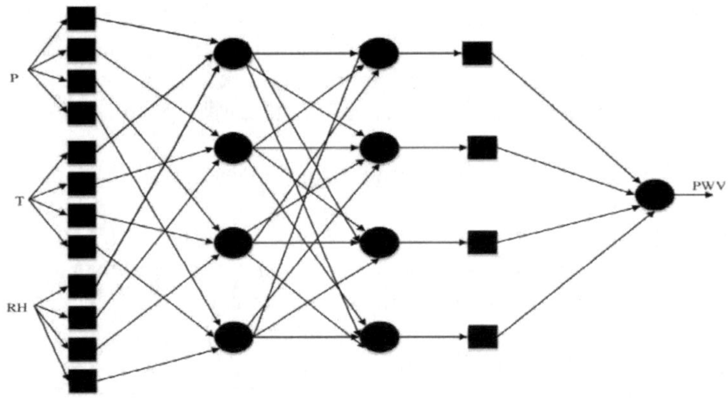

Figure 1. Structure of ANFIS PWV model

Figure 2 shows the workflow of ANFIS PWV model. Before the model started, the input surface meteorological data such as pressure (P), temperature (T), and relative humidity (RH) that coming to the model should be normalized to scale -1 until 1. After the scaled process, the input value will be entered to first layer and transform their with membership function into fuzzy value. This stage is called as fuzzification process. The membership function that usually used is in the form of a triangular, trapezoid, gauss and generalized bell. For this model the generalized bell membership function was used to transform the input value to fuzzy value. In the next stage, the fuzzy value that have range 0-1, also known as degree membership function will be evaluated by knowledge and rules base in second and third layers. There are four rules that were used in those layers. A linear equation also known as a first-sugeno model in fourth layer was used to calculate PWV value for each rule. This stage is known as the defuzzification process, where the fuzzy values that were evaluated in the previous layer are transforming to crips value (real value). Finally, in the fifth layer, the PWV from ANFIS model were obtained by summation and averaged all PWV value from each rule with use the weight average.

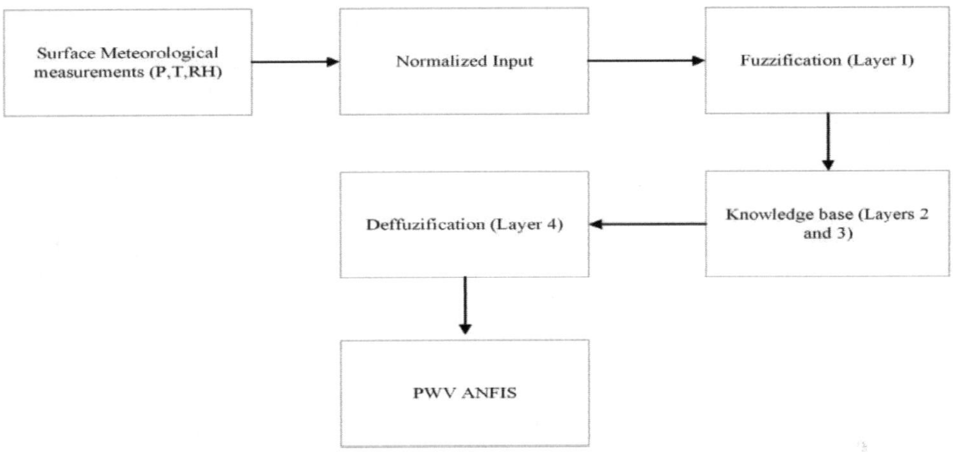

Figure 2. Flow diagram showing PWV estimation using ANFIS PWV

2.3. Water vapor mixing ratio

In the atmospheric science, the mixing ratio is designed as the ratio of amount mass in a constituent that in a given volume to the total amount of mass from all elements of air in that volume [10]. In this case both of air and water vapor show gaseous constituents, but they are not in condensed phase water or particulate matter. The mixing ratio has advantage in the atmospheric chemical applications which it unchanged by the differences of altitude associated in pressure and temperature or with meteorological variance. In addition, according to equation of state, their concentration is more influence on pressure and temperature.

Related to water vapor, it can be measured with the mixing ratio as called water vapor mixing ratio. It has a definition as the mass of water vapor per mass of dry air, typically expressed in kilogram per kilogram. The equation (1) is describing how to calculate the water vapor mixing ratio (w) [11].

$$w = \varepsilon \frac{P_w}{P - P_w} \tag{1}$$

where ε is the molecular weight ratio of water to dry air (= 0.622), and P_w is the partial pressure of water vapor that is obtained In the above equation, $P_{\square}$ is obtained from the relative humidity (RH) and saturation vapor pressure (e_w) as recommended from World Meteorological Organization (WMO), Guide to Meteorological Instruments and Methods of Observation (CIMO Guide)[12] and given by

$$P_w = \frac{RH}{100} e_w \tag{2}$$

$$e_w = 6.112\, e^{\left(17.62\,T/(243.12+T)\right)} \tag{3}$$

Since the unit of water vapor content is kilogram per kilogram and mixing ratio of trace gases are commonly given in units of part per million (ppm), then its unit should be converted to ppm unit. In term of mass per mass, 1 ppm is equal to 1 milligram (mg) per kg. From this conversion, it will obtain 1 kg/kg is equal to 1000000 ppm.

3. Result and Discussion

Figure 3 shows the trend all the parameters which measured by the system from 20 to 23 December 2016. The alphabetic letter in this figure presents the sub figure for each parameter. The *a,b,c,d,e* is represented for pressure (P) in mbar, temperature (T) in degree Celsius, relative humidity (RH) in percent, precipitable water vapor (PWV) in millimeter, and water vapor content (wp) in part per million (ppm), respectively. All parameter that shows in this figure has interval of one-minute measurement. In addition, the empty value was shown in this figure due to the system have loss of the power supply as of the data is not available at that time.

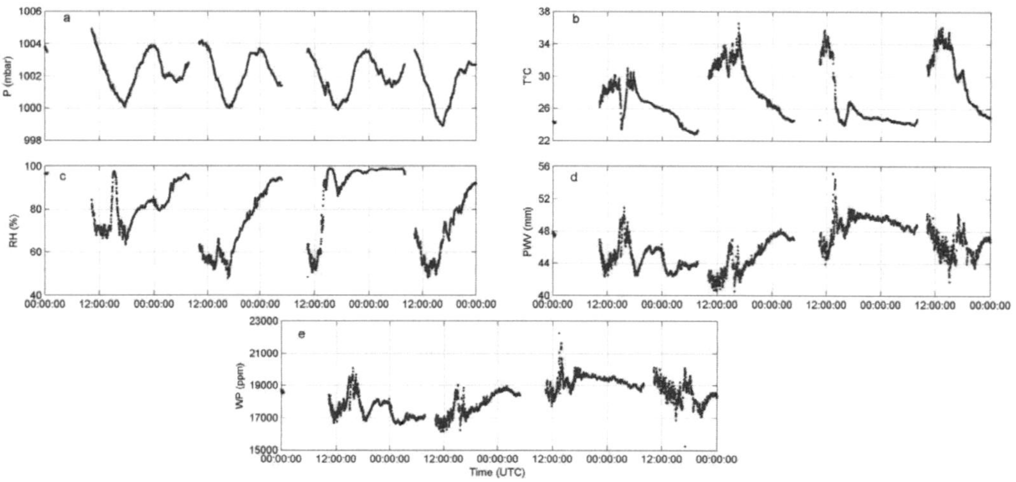

Figure 3. Variation (a) Surface pressure, (b) temperature, (c) relative humidity, (d) water vapor content, and (e) precipitable water vapor variations for the period of 20 – 23 December 2016 at UKMB station

Look into the pattern of P, T, RH, PWV and wp, they have an interrelationship to each other. According to Figure 3, the temperature have trend at a mid-day will be higher as compared during the night time. In addition, during the four days measurement, the temperature has a reached maximum value in the second day at about 36.55 °C. Opposite trend was showed by relative humidity, where it will be lower at a mid-day and rise in the night. Since the moisture will decrease in conjunction with the rising of temperature, and vice versa. On the other hand, the pattern of pressure is more influence on the height of site position. The higher of sensors installed, the air pressure in the air will be higher since the air layer stretched and thinned. From the measurement, the pattern of pressure at a mid-day time will be decreasing and rising at night time. It will be happened due to the Coriolis Effect.

Related to the other result as thought the PWV and water vapor content, it can be shown in Figures 3d and 3e. Both of the PWV and wp have a trend is closely correlated to each other. As can be seen, the trend of PWV showed a similar pattern to the wp. It can identify that the distribution of wp varies from time-to-time follows the PWV variations, only difference in varying of both ranges. During the four days measurement the range of PWV for this area is varied about 40.54 mm to 55.13. At the same time, the water vapor content has varying from 15240 ppm to 2240 ppm. From the result, it also able to indicate the PWV and wp variation during the last day measurement have a close variation between night time and day time. It could be happen due to light precipitation occurred in this area which was caused a slightly increase in the water vapor during nighttime. Moreover, both of wp and PWV have more influenced by relative humidity parameter compared to the other surface meteorological parameters. By comparing Figures 3c, 3d, and 3e, it can be noticed from their pattern that either wp or PWV is closely correlated with the relative humidity. It can be seen at the trend of relative humidity rise, either pattern of PWV or water vapor content also rise and vice versa. To clarify the measurement of water vapor content, the relationship between water vapor content and PWV is plotted in Figure 4. The result demonstrated that either water vapor content or PWV have a good relationship, where all data points from them are roughly fall into the line trend. In addition, both water vapor content and PWV have a strong linear correlation with r = 0.976 at the 99% confidence level.

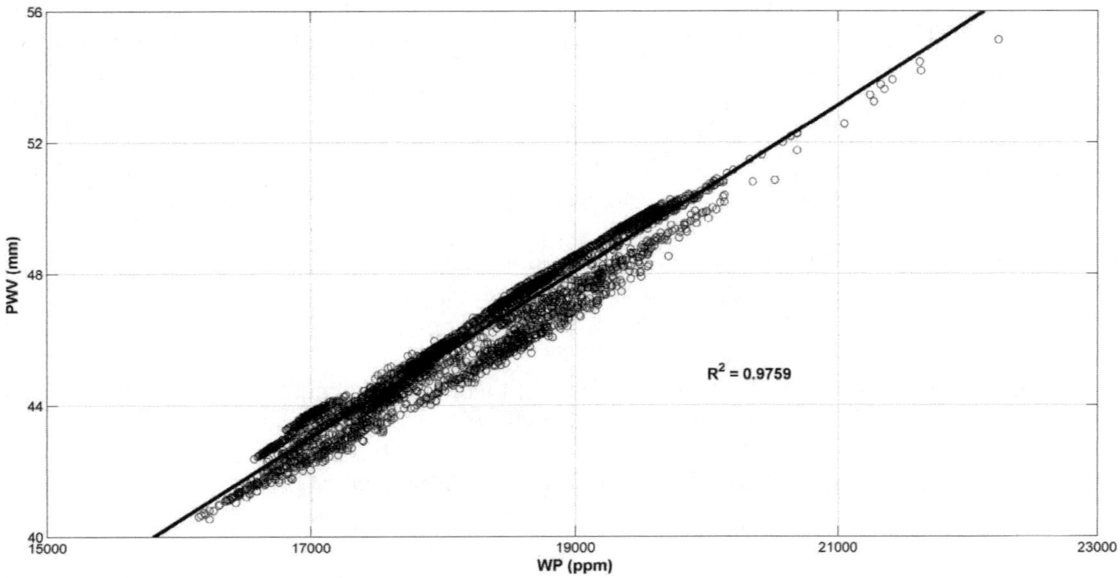

Figure 4. Relationship between water vapor content and PWV in UKMB station for the period of 20 – 23 December 2016

4. Conclusion

For the first time, this study reports the successful of estimation of water vapor content with the mixing ratio method in the area of UKMB station. The water vapor content data was compared with PWV ANFIS and found that the water vapor content was a good relationship with the GPS PWV. In addition, either of water vapor content or PWV was found their values are influenced by the surface relative humidity. The patterns of them are following the surface relative humidity. During the four days measurement, the reading of PWV and water vapor content values in the normal air is 46.22 mm and 18263 ppm on average, respectively. By this result, the water vapor mixing ratio will be implemented in the microcontroller to monitor greenhouse gasses variation in Pekan, Pahang peatland.

Acknowledgments

The authors would like to thank the Ministry of Science Technology and Innovation (MOSTI), since this work is supported under Grant 06-01-02-SF1241. This paper was funded by STACLIM scheme of Institute of Climate Change, UKM: TKS 182.

References

[1] Najafi M R, Zwiers F W and Gillett N P 2015 Attribution of Arctic temperature change to greenhouse-gas and aerosol influences *Nat. Clim. Chang.* 5 **3** 246

[2] Pereira J, Figueiredo N, Goufo P, Carneiro J, Morais R, Carranca C and Trindade H 2013 Effects of elevated temperature and atmospheric carbon dioxide concentration on the

emissions of methane and nitrous oxide from Portuguese flooded rice fields *Atmos. Environ.* **80** 464

[3] Manabe S and Wetherald R 1967 Thermal equilibrium of the atmosphere with a iven distribution of relative humidity *J. of Atmos. Sci.* **24** 241

[4] Rocken C, Hove T V and Ware R 1997 Near real-time GPS sensing of atmospheric water vapor *Geophys. Res. Lett.* **24** 3221

[5] Raatikainen T, Nenes A, Seinfeld J H, Morales R, Moore R H, Lathem T L and Bougiatioti A 2013 Worldwide data sets constrain the water vapor uptake coefficient in cloud formation *Proc* of the National Academy of Sciences vol 110(10) p 3760-3764.

[6] Adams D K., Fernandes R M, Holub K L, Gutman, S I, Barbosa H M, Machado L A, and DeMets C 2015 The Amazon Dense GNSS Meteorological Network: a new approach for examining water vapor and deep convection interactions in the tropics *Bull. Am. Meteorol. Soc.* 96 **12** 2151

[7] You Q, Jiang Z, Bao Y, Pepin N and Fraedrich K 2016 Trends in upper tropospheric water vapour over the Tibetan Plateau from remote sensing *Int. J. Climatol.* **36** 4862

[8] Wang J, Dai A and Mears C 2016 Global water vapor trend from 1988 to 2011 and its diurnal asymmetry based on gps, radiosonde, and microwave satellite measurements *J. Clim.* **14** 5205

[9] Suparta W and K M Alhasa 2016 Modeling of precipitable water vapor using an adaptive neuro-fuzzy inference system in the absence of the GPS network *J. Appl. Meteorol. Clim.* **55** 2283

[10] Williams J 2013 *The AMS weather book: the ultimate guide to America's weather* (Springer Science & Business Media)

[11] Moteki N and Kondo Y 2013 A new theoretical method for calculating temperature and water vapor saturation ratio in an expansion cloud chamber *J. Geophys. Res. Atmos.* 118 **12** 6633

[12] World Meteorological Organization 2008 *Guide to Meteorological Instruments and Methods of Observation* (Geneva: CIMO Guide) Appendix 4B WMO-No. 8

Measurements of radioactivity levels in part of Ota Southwestern Nigeria: Implications for radiological hazards indices and excess lifetime cancer-risks

K D Oyeyemi[1], M R Usikalu[1], A P Aizebeokhai[1], J A Achuka[1] and O Jonathan[2]

[1]Department of Physics, Covenant University, Nigeria
[2]Department of Computer and Information Science, Covenant University, Nigeria

Email: kdoyeyemi@yahoo.com, kehinde.oyeyemi@covenantuniversity.edu.ng

Abstract. Super SPEC RS-125 radiation detector with large 2.0 x 2.0 NaI crystal and linear energy ranging from 0.80 MeV to 1.2 MeV was used to measure the activities of primordial nuclides and the radiation dose exposures rate in Iyana-Iyesi, Ota, southwestern Nigeria. The measured activities vary from 17±0.02 $Bqkg^{-1}$ to 30.49 ±0.01 $Bqkg^{-1}$, 50.01 ±0.16 $Bqkg^{-1}$ to 158.49±0.17 $Bqkg^{-1}$, and 406.9±0.42 $Bqkg^{-1}$ to 1275.48±0.82 $Bqkg^{-1}$ for ^{238}U, ^{232}Th and ^{40}K respectively. The acquired gamma radiation dose rate range from 138.696 ±2.06 ($nGyh^{-1}$) to 350.103±7.21 ($nGyh^{-1}$) with mean value of 148.22 ($nGyh^{-1}$), almost three times higher than the recommended safe limit of 55 ($nGyh^{-1}$). The measured activities and radiation dose rate were engaged to estimate the annual outdoor effective dose, gamma index, excess lifetime cancer risks and annual gonadal dose equivalent. It was observed from all the estimated parameters, those values in the study area are well above the recommended safe limit for normal background radiation. This suggest that the dwellers and those using the excavated geomaterials from this area for construction purposes are exposed to very high radiation from natural radionuclides. Further research to evaluate the mineralogy and geochemistry of the clay deposits in the area is highly recommended.

1. Introduction

Sources of radiation exposure for humans are the natural radionuclides within the environment constituting the background radiation level. Terrestrial components of this background radiation level consist of those radionuclides localizing within the soil, water, air and materials for building constructions whose quantitative abundance in an area depends significantly on the physiography, localized rock types and the regional geological setting. Man-made sources such as nuclear activities and accidents have also been reported to contribute immensely to the background radiation levels [1] [2] [3] [4]. Measurements of radionuclides distribution is a critical prerequisite for estimation of radiological health hazards and risks caused by radiation exposure. Lung cancer, hepatic skin, leukaemia and atrophy of the kidney are among the health hazards usually attributed to the long-term exposure either by inhalation or otherwise. Prolong exposure to non- ionizing radiation can also result to acute leucopoenia, sterility, anaemia and ultimately death; children whose mothers were exposed to radiations during pregnancy suffer the risk of mental retardation [5]. It is of prime importance to evaluate the natural environmental radiation level and compare the measurements with the standard dose limits of public exposures. Estimation of the distribution pattern of the natural occurring

Content from this work may be used under the terms of the Creative Commons Attribution 3.0 licence. Any further distribution of this work must maintain attribution to the author(s) and the title of the work, journal citation and DOI.
Published under licence by IOP Publishing Ltd

radionuclides together with their consequent radiological hazard risks is equally essential in providing some sense of control on prevailing radiation levels. Such study would also establish a baseline for future research work within a particular region. The objective of this work therefore, was to measure the activity concentration level in selected locations within Ota, southwestern Nigeria with the aim of assessing the potential outdoor radiation dose and consequent health risks their exposure pose to people within Ota and its environs.

2. Methodology

2.1. Study Area

The study site is located within the Iyana-Iyesi, Ota, southwestern Nigeria. The physiographic setting is characterized by gentle sloping low area with mean altitude of about 65 m with respect to the sea level. Ota is situated in a humid region characterized by both dry and raining seasons ranging from November to March and April to October respectively. Though intermittent rainfalls are often experienced and reported during dry season due to its closeness to the Atlantic Ocean. The regional geology as presented in Figure 1 is that of Dahomey basin [6], breaking apart from the eastern part of the Niger Delta basin by the continental extension of chain fracture zone [7]. The stratigraphy comprises of Ise, Afowo, Araromi, Ewekoro, Akinbo, Oshosun, Ilaro, and Benin Formations with the first three formations belonging to the Abeokuta Group. However, the local geological setting within Ota consists of the clayey sand, sandy clay, lateritic clay, kaolinitic clay lens, and unconsolidated sand units with the last rock type serving as the main groundwater yield aquifer in Ota and in the area.

2.2. Field Survey

Super SPEC RS-125 spectrometer with large 2.0 x 2.0 NaI crystal was used to measure both the activities of natural occurring radionuclides and the radiation dose exposures in ten location points within Ota, southwestern Nigeria. The data acquisition was maintained about 1 metre above the topsoil. The equipment was calibrated using 5 minutes spectral accumulation on Thorium, Uranium and Potassium pads with 10 minutes accumulation on the background (BG) pad according to Canadian Geophysical Institute. RS-125 portable handheld radiation detector was choosing for this research due to its high accuracy with probable error of about five percent. It also offers good integrated design with large detector, large data storage, high sensitivity and ease of use. Five readings at the interval of 80 seconds were acquired with average result recorded at each station point. The linear energy of the detector ranges approximately between 0.80 MeV and 1.2 MeV which cover majority of the radiation emissions from the terrestrial sources. The mode of data presentation by the equipment was parts per million for both ^{238}U and ^{232}Th, while ^{40}K was in percentage; the data were later converted to $Bqkg^{-1}$ using [8] conversion factor.

2.3. Estimation of radiation hazards parameters

The outdoor absorbed dose rate of 1 metre height can be estimated using UNSCEAR [1] guidelines which is based on the assumption that the naturally occurring radionuclides will have a uniform distribution at that height. The outdoor annual effective dose in units of $mSvy^{-1}$ was computed using the equation (1) based on [9] with conversion coefficient of 0.7 $SvGy^{-1}$and outdoor occupancy of 0.2 denoting that people around the world spend 20% of their time outdoors on the average. The gamma Index (I_γ) was estimated using the equation (2) according to [10] for the radiation risks resulting from the exposure to the naturally occurring radionuclides. It determines the level of radiation associated with the activity concentrations K_U, K_{Th} and K_K of Uranium, Thorium and Potassium.

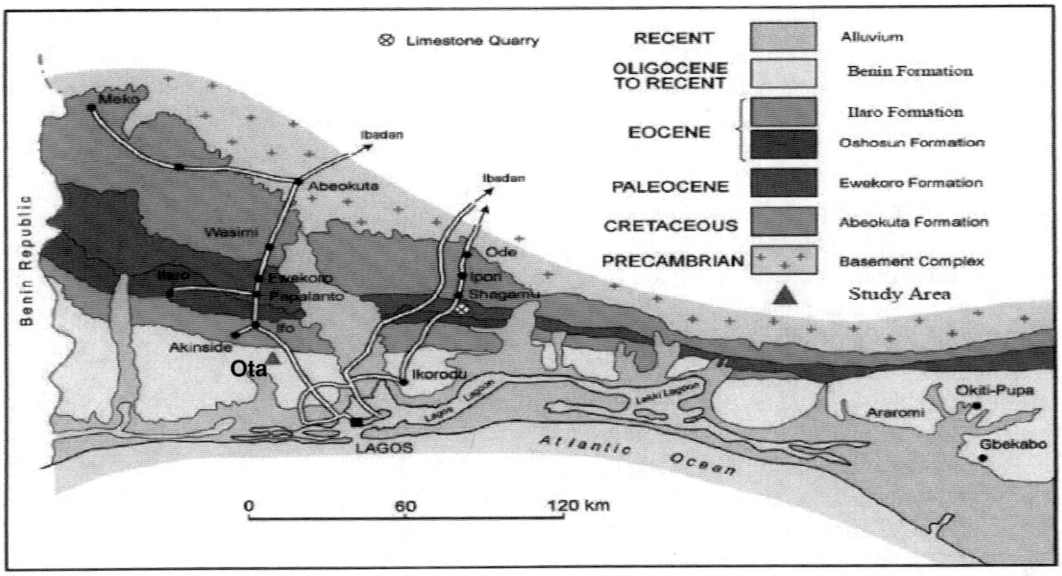

Figure 1. Geological map of the Nigerian section of Dahomey basin (modified after [6])

Excess lifetime cancer risk was calculated according to equation (3) [11]. This determines the possible chance a person exposed to radiation (from birth till death) has over the lifetime risk of cancer development or being diagnosed with cancer. E is the outdoors annual effective dose, LE is the lifetime expectance of people in Nigeria (average of 54.5 years) according to the latest data published by world health statistics [12]; life expectancy for male is 53.5 years, while that of the female is 55.6 years. Cancer risk factor per Sievert (RF) was stochastically calculated as 0.05 for the general public [13]. Howlader [14] employed this same radiological tool (ECLR) to evaluate the widespread of cancer in the United States. He reported the possibility that a man living in US will develop cancer in his lifetime is 43.3% whereas that of a woman is 37.8%. In this study, the calculated annual gonadal dose equivalent (AGDE), which evaluate the possible effects of the specific activities of ^{238}U, ^{232}Th and ^{40}K on the bone marrow activities and bone surface cells was determined using the equation (4) after [15] with all symbols having their usual meanings.

$$H_E(mSvy^{-1}) = 1.23\times10^{-3} \times \text{ Dose rate} \quad (1)$$

$$I_\gamma = {K_U}/{150} + {K_{Th}}/{100} + {K_K}/{1500} \quad (2)$$

$$\text{ECLR} = \text{E x LE x RF} \quad (3)$$

$$\text{AGDE } (\mu Svy^{-1}) = 3.09\text{K}_\text{K} + 4.18\text{K}3.09K_K + 4.18K_{Th} + 0.314K_K \quad (4)$$

3. Results and Discussion

The measured activities and the radiation dose across the entire ten location points are displayed in Table 1. The activity concentration for ^{238}U, ^{232}Th and ^{40}K radionuclides are within the ranges 17±0.02 $Bqkg^{-1}$ to 30.49±0.01 $Bqkg^{-1}$, 50.01±0.16 $Bqkg^{-1}$ to 158.49±0.17 $Bqkg^{-1}$, and 406.9±0.42 $Bqkg^{-1}$ to 1275.48±0.82 $Bqkg^{-1}$ respectively in the entire area of study. The corresponding mean values of measured activity concentrations are 25.498 $Bqkg^{-1}$, 77.772 $Bqkg^{-1}$, and 710.704 $Bqkg^{-1}$ for ^{238}U, ^{232}Th and ^{40}K respectively. These values were compared to the worldwide standard mean activity concentrations of 32, 45 and 420 $Bqkg^{-1}$ for ^{238}U, ^{232}Th and ^{40}K respectively according to [1]. It was observed that the measured activity concentrations of ^{238}U are lower and acceptable in all the locations. However, the mean activity concentrations for ^{232}Th and ^{40}K across all the location points in the area are higher than the worldwide average standard. There are also strong positive correlations of

0.999 and 0.747 between the later estimated excess lifetime cancer risk and the activities of both ^{232}Th and ^{40}K respectively as shown in Figure 2 (a and b). These high activity concentrations of Thorium and Potassium may be connected to the local geology with clay as the dominant lithology in the area. The stratigraphy of the area comprises of the clayey sand, lateritic clay and kaolinite clay and unconsolidated sand units. The mineralogy composition of the clay lithology within the area appears to be rich in K-feldspar which must have contributed directly and indirectly to the activity concentrations of the measured naturally occurring radionuclides. These observations suggest that the clay rich geomaterials from this area pose a great danger of very high radiation exposure to the dwellers, and construction of buildings and roads with such materials should be discouraged.

Evaluation of the eternal primordial gamma radiation on the entire study site required the measurement of the total absorbed dose rate D ($nGyh^{-1}$); the acquired values for this radiological parameter across all the station points range from 138.696 ±2.06 ($nGyh^{-1}$) to 350.103±7.21 ($nGyh^{-1}$) with mean value of 148.22 ($nGyh^{-1}$). The measured dose rates are found to be higher than the world average dose rate of 55 ($nGyh^{-1}$). The increase in the level of the acquired total γ- radiation dose rates can be linked directly to the high levels of activity concentrations of ^{232}Th and ^{40}K. The measured absorbed dose rate and the radionuclides activity of both ^{232}Th and ^{40}K are observed to be elevated for locations L8 and L9. These observations are not unconnected to the outcrops (surface expressions) of a thick lateritic clay unit close to these locations points. These clay materials are equally being used for levelling and filling during both road and building construction in Ota and its environs, thereby contributing significantly to the high level of radionuclides activity concentration and other estimated radiological hazards in the entire area.

Table 1. Measured Activity concentrations of naturally occurring radionuclides and absorbed dose rate

Location	^{238}U ($Bqkg^{-1}$)	^{232}Th ($Bqkg^{-1}$)	^{40}K ($Bqkg^{-1}$)	Dose rate ($nGyh^{-1}$)
L1	29.79±0.02	56.84±0.12	432.33±0.53	152.006±12.2
L2	27.37±0.03	59.48±0.12	425.16±0.61	152.442±5.57
L3	26.14±0.02	60.9±1.04	485.15±1.54	158.783±4.13
L4	30.41±0.02	58.51±0.12	446.02±0.72	155.471±1.74
L5	30.03±0.04	53.62±0.15	406.9±0.42	139.984±6.52
L6	20.84±0.02	50.01±0.16	627.95±0.33	138.696±2.06
L7	20.45±0.02	87.97±0.12	852.92±0.63	208.863±3.87
L8	30.49±0.01	158.49±0.17	1242.22±0.47	350.103±7.27
L9	22.46±0.02	135.94±0.13	1275.48±0.82	300.286±3.38
L10	17±0.02	55.96±0.12	912.91±0.45	144.837±4.21
Mean	**25.498**	**77.772**	**710.704**	**148.422**

Table 2 shows the results of all estimated radiological parameters in this study. The annual effective dose in air received by an adult outdoors calculated across all the station points spanned from 0.17 $mSvy^{-1}$ to 0.429 $mSvy^{-1}$ with mean value of 0.233 $mSvy^{-1}$. These values are obviously higher than the worldwide outdoors annual effective dose average of 0.007 $mSvy^{-1}$ [1], but below the set-limit of 1.0 $mSvy^{-1}$ maximum dose recommended for public exposure by the International Commission on Radiologic Protection [16]. Location L8 recorded the highest calculated annual effective dose across the entire study site as shown in the contour image map (Figure 3 (a)). The results of the estimated γ-radiation hazards index also called representative index or gamma index is displayed in Table 2. This radiological parameter determines the level of γ-radiation associated with the measured activity concentrations of the primordial nuclides. The value of the gamma index I_γ must be less than unity (1) in order to keep the radiation level insignificant, however the range of the estimated gamma index (Table 2) is from 1.008 $Bqkg^{-1}$ to 2.616 $Bqkg^{-1}$ with mean value of 1.422 $Bqkg^{-1}$. The calculated values in all the location points are greater than the recommended permissible limit of unity according to the recommendation of [1]. Figure 3(b) shows the spread of gamma index in the study area with location L8 having the highest and also a strong relationship in terms high positive correlation of 0.976 was observed between the levels of estimated γ-radiation index and ELCR (Figure. 4(a)).This further confirms that prolong exposure to the high level of γ-radiation in this area will pose lifetime cancer risks to the dwellers.

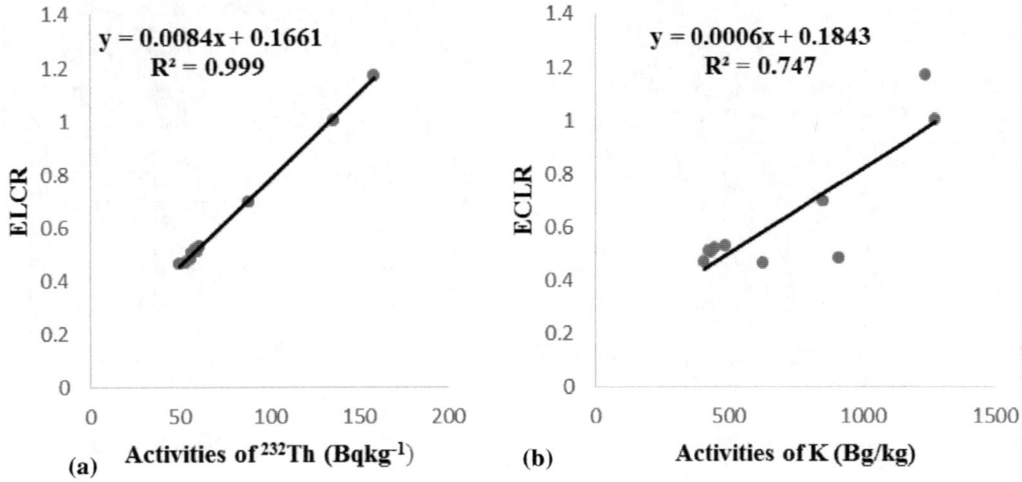

Figure 2. Correlation of ECLR and activities of (a) ^{232}Th and (b) ^{40}K

Table 2. Estimated radiological parameters

Location	H_E ($mSvy^{-1}$)	I_γ	ELCR ($mSvy^{-1}$)	AGDE (μSvy^{-1})
L1	0.186	1.055	0.508	465.39
L2	0.187	1.061	0.509	466.7
L3	0.195	1.107	0.531	487.67
L4	0.191	1.085	0.52	478.59
L5	0.172	1.008	0.468	444.69
L6	0.17	1.058	0.464	470.61
L7	0.256	1.585	0.698	698.72
L8	0.429	2.616	1.17	1146.76
L9	0.368	2.359	1.004	1038.13
L10	0.178	1.282	0.484	573.1
Mean	**0.233**	**1.422**	**0.635**	**627.04**

Excess lifetime cancer risks factor was directly determined using the annual effective dose radiation which is solely dependent on the measured radiation dose rate in the area of study. ELCR is usually used to quantitatively evaluate the effects of prolong exposure to the γ – radiation associated with the measured natural occurring radionuclides. The estimated ELCR values as shown in Table 2 and Figure 5(a) ranging from 0.464 $mSvy^{-1}$ to 1.17 $mSvy^{-1}$ with mean value of 0.635 $mSvy^{-1}$ are higher than the world permissible standard of 0.29 $mSvy^{-1}$ [10].

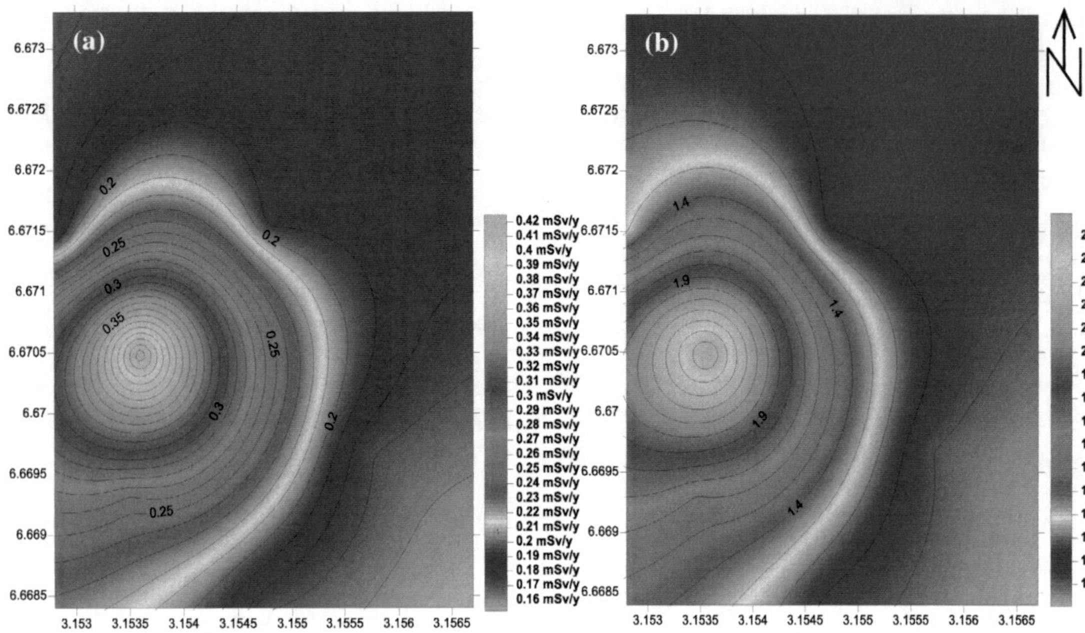

Figure 3. Location contour maps for estimated (a) Outdoor annual effective dose and (b) Gamma Index

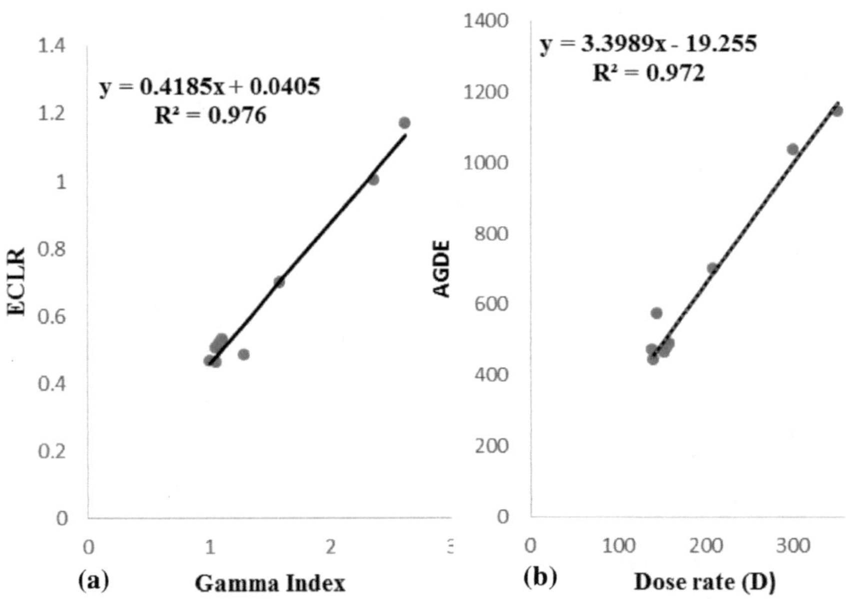

Figure 4. Correlation of (a) ECLR and Gamma Index I_γ (b) AGDE and γ − radiation dose rate

The United Nations Scientific Committee on Effects of Atomic Radiation (UNSCEAR) have so much interests on the activity of the bone marrow, this is because gonads are essential organs that can be affected by radiation exposure [17]. Exposure to radiation with high annual gonadal dose equivalent can result to leukaemia in the bone marrow. The calculated AGDE (μSvy^{-1}) in this research spanned from 444.69 (μSvy^{-1}) to 1146.76 (μSvy^{-1}) with mean value of 627.04 (μSvy^{-1}). These values are two times above the average $300(\mu Svy^{-1})$ annual gonadal dose equivalent values around the world. The elevation in the estimated values for AGDE can be linked to the high levels of activity concentrations of ^{232}Th and ^{40}K nuclides which may not be unconnected to the local geology. Several geologic materials such as granites, silt, and clay have been reported to be rich in naturally occurring radionuclides of ^{232}Th and ^{40}K [18]. Figure 5 (b) shows the contour image map of the estimated AGDE in the study area and the strong connection between the AGDE and the measured $\gamma-$radiation dose rate with positive correlation of 0.972 (Figure 4 (b)). This reveals that both measured radionuclides activity concentrations and dose rate within the study area are pointer to the possibility of severe AGDE risks associated with prolong exposure to $\gamma-$radiation exposure.

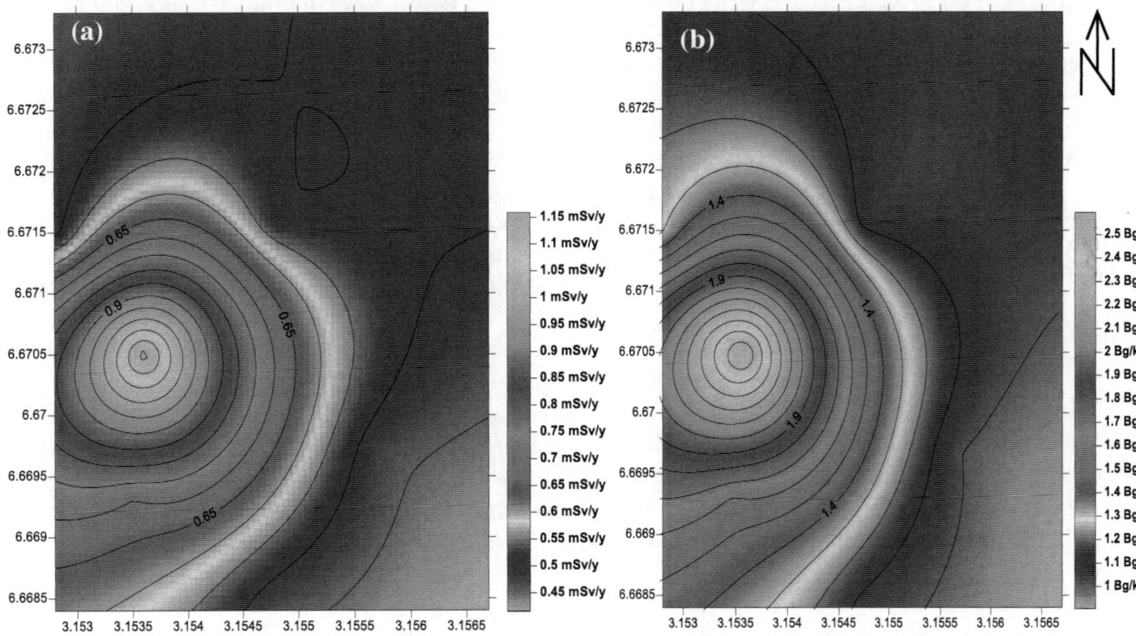

Figure 5. Location contour maps for estimated (a) ECLR and (b) AGDE

4. Conclusion

The activity concentrations of radionuclides and $\gamma-$radiation dose rate were measured in ten locations within Iyana-Iyesi Ota, southwestern Nigeria using Super SPEC RS-125 spectrometer. The measured activity concentration of ^{238}U were observed to be below the worldwide average of 32 $Bqkg^{-1}$. However, the rest of the measured radionuclides including ^{232}Th and ^{40}K were observed to be higher than the worldwide average of 45 and 420 $Bqkg^{-1}$ which may be interpreted to be from the lateritic clay deposits close to the site and the geological setting of Ota which is predominantly silt, clay, clayey sand and unconsolidated sand. The average measured radiation dose rate in the area is 148.422 $(nGyh^{-1})$which is equally higher than permissible limit of $55(nGyh^{-1})$. Consequently, all the estimated radiological parameters in the area such as outdoor annual effective dose, gamma index, excess lifetime cancer risks and annual gonadal dose equivalent are above the permissible limits of $0.007\,mSvy^{-1}$, Unity (1), 0.29 $mSvy^{-1}$and $300(\mu Svy^{-1})$ respectively. This suggests that people living in this community and those using the excavated geomaterials from this environment for buildings and roads construction are exposed to radiation above the worldwide recommended safe limits. Further research involving the mineralogy and geochemistry of clay with other geomaterials within this area is highly recommended so as to further understand the sources of radiation.

References

[1] United Nations Scientific Committee on the Effect of Atomic Radiation (UNSCEAR) 2000 Sources of ionizing radiation. Annex B: exposures from natural radiation. Report to the General Assembly, United Nations, New York p 156

[2] Dowdall M, Vicat K, Frearso I, Geland S, Linda B and Shaw G 2004 Assessment of the radiological impacts of historical coal mining operations in the environment of Ny- Alesund, Svalbard. *Journal of Environmental Radioactivity* **71** 101

[3] Musa M A, Funtua I I, Malam S P and Arabi A S 2011 Determination of Absorbed and Effective Dose from Natural Background Radiation around a Nuclear Research Facility *American Journal of Environmental Sciences* **7**(2) 173

[4] Sadiq A A and Agba E H 2011 Background radiation in Akwanga, Nigeria. *Working and Living Environmental Protection* **8** (1) 7

[5] National Research Council (NRC) Committee on the Biological Effects of Ionizing Radiation (BEIR V) 1990 *Health effects of exposure to low levels of ionizing radiation* The National Academy of Science, Washington, DC

[6] Gebhardt H, Adekeye O A and Akande S O 2010 Late Paleocene to initial Eocene thermal maximum foraminifera biostratigarphy and paleoecology of the Dahomey Basin, southwestern Nigeria. *Jahrb. Geol. Bundesanst.* **150** 407

[7] Onuoha K O 1999 Structural features of Nigeria's coastal margin: an assessment based on age data from wells. *Journal of African Earth Sci.* **3** 485

[8] International Atomic Energy Agency (IAEA) 1993 Safety standards: International basic safety standard for protection against ionizing radiation and for the safety of radiation sources 115 Vienna

[9] Baranwal V C, Sharma S P, Sengupta D, Sandilya M K Bhaumik B K, Guin R and Saha S K 2006 A new high background radiation area in the geothermal region of Eastern Ghats Mobile Belt (EGMB) of Orissa. *Indian Radiat. Meas.* **41** 602

[10] Tufail M, Akhtar N, Jaried S and Hamid T 2007 Natural Radioactivity Hazards of Building Bricks Fabrication from Soil of Two Districts of Pakistan *Journal of Radiological Protection* **27** 481

[11] Taskin H, Karavus M, Ay P, Topuzoglu A, Hindiroglu S and Karahan G 2009 Radionuclide concentrations in soil and lifetime cancer risk due to gamma radioactivity in Kirklareli, Turkey *J Environ Radioact* **100** 49

[12] World Health Statistics 2015 *Global health indicators* World Health Organization p 122

[13] International Commission on Radiological Protection (ICRP) 1990 *Recommendations of the ICRP* Pergamon Press, New York, ICRP Pub **60**

[14] Howlader N, Noone A M, Krapcho M, Garshell J, Miller D, Altekruse S F, Kosary C L, Yu M, Ruhl J, Tatalovich Z, Mariotto A, Lewis D R, Chen HS, Feuer E J, Cronin K A(eds) SEER Cancer Statistics Review, 1975–2011, National Cancer Institute. Bethesda, MD. http://seer.cancer.gov/csr/1975_2011/, based on November 2013 *SEER data submission* SEER website

[15] Mamont-Ciesla K, Gwiazdowski B, Biernacka M and Zak A 1982 *Radioactivity of building materials in Poland* In: Vohra G, Pillai KC, Sadavisan S (eds) Natural radiation environment Halsted Press, New York 551

[16] International Commission on Radiological Protection (ICRP) 1977 *Recommendations of the ICRP* (New York: Pergamon Press) ICRP Pub **26**

[17] United Nations Scientific Committee on the Effect of Atomic Radiation (UNSCEAR) 1988 *Sources, effects and risk of ionizing radiation* United Nations, New York

[18] Jallad K N 2016 Radiation hazard indices and excess lifetime cancer risk in sand from the northern and eastern of Kuwait *Environ. Earth Sci.* **75** 156 doi:10.1007/s12665-015-5028-9

Micro – ring resonator with variety of gap width for acid rain sensing application: preliminary study

B Mulyanti[1], H Ramza[2*], R E Pawinanto[1], J A Rahman[3], M S Ab-Rahman[4] W S Putro, L Hasanah[5], A B Pantjawati[1]

[1]Department of Electrical Engineering Education, Universitas Pendidikan Indonesia, 40154, Bandung, West Java, Indonesia
[2]Department of Electrical Engineering, Universitas Muhammadiyah Prof. DR. HAMKA, 12130, Jakarta, Indonesia
[3]Department of Electrical Engineering, Faculty of Engineering, Universiti Industri Selangor, 40000, Shah Alam, Selangor Darul Ehsan, Malaysia
[4]Department of Electronic, Electrical and System Engineering, Universiti Kebangsaan Malaysia, 43600, Bangi, Selangor Darul Ehsan, Malaysia
[5]Department of Physics Education, Universitas Pendidikan Indonesia 40154, Bandung, west Java, Indonesia

E-mail: ramza.harry@gmail.com

Abstract. The acid rain is an environmental disaster that it will be intimidates human life. The development micro-ring resonator sensor created from SOI (Silicon on insulator) and it used to detect acid rain index. In this study, the LUMERICAL software was used to simulate SOI material micro-ring resonator. The result shows the optimum values of fixed parameters from ring resonator have dependent variable in gap width. The layers under ring resonator with silicone (Si) and wafer layer of silicone material (Si) were added to seen three conditions of capability model. Model – 3 is an additional of bottom layer that gives the significant effect on the factor of quality. The optimum value is a peak value that given by the FSR calculation. FSR = 0, it means that is not shows the light propagation in the ring resonator and none of the light coming out on the bus – line.

Keywords: Micro-ring resonator, gap width, quality factor, and Free Spectral Range (FSR)

1. Introduction

Acid rain is categorized an environmental hazard due to chemical condition in lower atmosphere. The chemical material involved hydrocarbon oxides and nitrogen (NO_x) it's very dangerous for human life [1]. In current study of acid rain chemical material, the NO_x gases sensor were used to detect chemical emission index and calculated with precipitation index [2]. However, the result of calculation NO_x and precipitation have weakness due to lack of data measurement. Furthermore, to improve the calculation result Cai *et al.* [3] were proposed to develop new sensor using stabilized zirconia and MoO_3–In_2O_3 Nano composites. The highly sensitive using stabilized zirconia and MoO_3–In_2O_3 were improve NO_x gases sensor and calculation result. In this study, the basic sensor design of ring resonator were developed based optical field [4, 6]. The discrete wavelength, the optical field will mix constructively with another optical fields while both of light running and requiring of extraction conditions from optical representation of optical phase [4, 5]. The optical field from the sensor has highly capability to

Content from this work may be used under the terms of the Creative Commons Attribution 3.0 licence. Any further distribution of this work must maintain attribution to the author(s) and the title of the work, journal citation and DOI.
Published under licence by IOP Publishing Ltd

transmit the information NO_x gases in near future. In this study, we analyze Free Spectral Range (FSR) of micro-ring resonator sensor design by wafer Silicon layer (Si), Silicon Oxide (SiO2) Layer and Silicon layer (Si) using LUMERICAL software.

2. Methodology

2.1. Micro-ring Parameters

Basic of ring resonator can be learnt by considering a dielectric ring that embedded on the dielectric material with smallest refractive index as seen in Figure 1. An optical field is coupled into the ring that it is circumstances and reflection light will transmit continuously on the ring boundaries with smallest refractive index that close to the material [5, 6].

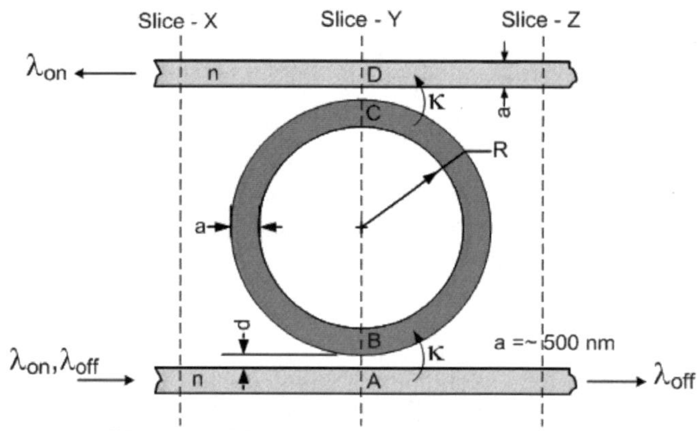

Figure 1. Ring dielectric optical resonator

The parameters from a micro-ring resonator is free spectral range (FSR), it is defined as a part of frequency between two sequences of resonance. This case can be approached by following amount of m and non-dispersive material expressed as;

$$\lambda_m = \frac{2\pi R n_{eff}}{m} \tag{1}$$

$$\Delta f = \frac{c}{2\pi R n_{eff}} \tag{2}$$

In the wavelength part, it is given by,

$$\Delta\lambda = \frac{\lambda^2}{2\pi R n_{eff}} \tag{3}$$

where λ is wavelength at the vacuum space.

The single ring optical resonator with pairing waveguide signal (see Figure 1) in bottom side of signal wave called input bus and the top side is named by output bus. The resonance bandwidth is one of critical parameter for micro-ring and it gives equation,

$$F = \frac{FSR}{\Delta\lambda_{FWHM}} \tag{4}$$

$\Delta\lambda$ is defined as resonance bandwidth (full width half maximum) and it was proved to be,

$$\Delta\lambda = \frac{2\kappa^2\lambda^2 v_g}{(2\pi)^2 RC} = \frac{2\kappa^2\lambda^2}{(2\pi)^2 R n_{eff}} \tag{5}$$

κ is coupling efficiency that will discussed further by applying the couple mode theory, and last expression on the equation (5) produces group of velocity approach (v_g) with phase velocity (c/n_{eff}), we ignores the dispersive effects for simplifications.

2.2. *Q – Factor and Transmittance*

Quality or Q – Factor for a resonator is a size of frequency structure sensitivity. Q is given by timing of averages energy that stored in every optical path and it is divided by coupling power or output scattering from resonator. Q is limited with radiation and scattering lost. A multi-device functioned from 1.3 μm until 1.5 μm that requires boundary radiation (Q) amount 1500 or more for achievement of a width of spectral line is `1 nm or lowest. By inserted equation (5) into the Q – definition $(Q = \lambda/\Delta\lambda)$causes [5, 6],

$$Q = \frac{2\,\pi^2\, R\, n_{eff}}{\lambda\, \kappa^2} \tag{6}$$

In fact, Q is an important variable for gives of maximum performance and it will be discussed from alternative at the next part.

2.3. *Propagation Loss*

The different material interface scatters light and causes loss on the each materials and it cannot particularly apply for bent plane waveguide. The wall roughness of waveguide scatters light and loss coefficient is given by Tien's model for a plane waveguide and random roughness distribution expressed [5]

$$\alpha_{st} = 2 \cdot \sigma^2 \left(n_{eff}^{\ 2} - n_0^{\ 2}\right) \cdot k_0^{\ 2} \cdot \frac{k_x}{\beta} \cdot E_s^{\ 2} \tag{7}$$

where n_{eff} is effective core refractive index, n_0 is cladding refractive index and k_0 is propagation constant in the air, $k_0 = 2\pi/\lambda$. k_x and β are transverse and modal constant propagation respectively, $k_x = \sqrt{n^2 k_0^{\ 2} - \beta^2}$, then E_s defined as electric field at the interface between core and cladding.

Generally, when the existing of waveguide on the substrate has equal or greater value of the refractive index, this problem causes leakage of substrate. The bent waveguide, additional leakage occurred cause the existing of radiation loss. Using conformal mapping for planar waveguide solved this problem. The leakage loss α is expressed by:

$$\alpha = k \cdot e^{-\frac{R}{R_c}} \tag{8}$$

where k is a constant depend waveguide characteristic, R is radius of bent waveguide and R_c is critical radius that is shown with,

$$R_c = \frac{1}{\beta}\left(\frac{n_{eff}}{2 \cdot \Delta n_{eff}}\right)^{3/2} \tag{9}$$

β is propagation constant and Δn_{eff} is the different of effective refractive index and effective refractive index of cladding waveguide. In principle, the loss mechanisms can be shown independently.

2.4. *Micro-ring Parameters*

The micro-ring resonator consists of three layers for silicon on insulator (SOI), namely, wafer Silicon layer (Si), Silicon Oxide (SiO_2) Layer and Silicon layer (Si). This structure is fabricated by initial producing the waveguides through lithography and etching process, then wafer bonding an initial mechanical separation, second and third wafer containing layers that ultimately create the micro-ring resonator (see Figure 2). After wafer bonding, the substrate is detached from the third wafer, and lithography and etching process are used to create the ring resonator [5, 6].

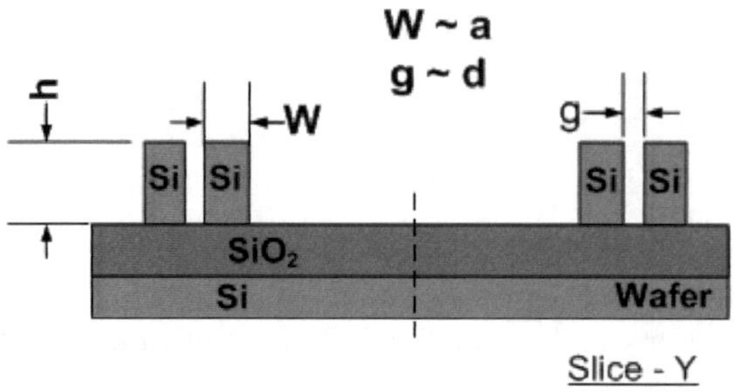

Figure 2. Cross section of micro-ring resonator base on slice-y

The parameter of micro ring resonator is determined to look at the beam propagation of signal waveguide. In this simulation study, there are many parameters that are filled to find the best of quality factor (Q). This simulation analysis of micro-ring resonator waveguide uses Lumerical software. As a preliminary study, we were divided into 3-model analysis of micro-ring resonator as seen as Table 1 [7].

Table 1. Assuming values of initial parameters for micro-ring resonator

	Symbol	Unit	Model 1	Model 2	Model 3	Remark
Design Expectation						
Wavelength	λ	nm	1500	1500	1500	For optical communication
Width of Ring	W	nm	500	500	500	
High of Ring	h	nm	220	220	220	
Total of Upper layer	H	nm	0	340	340	
Silicon bottom layer		nm	0	0	500	
Free Spectral Range	FSR	nm	20	20	20	
Result Expectation						
Factor of Quality	Q		5000	5000	5000	Sensor application
Factor of Quality	Q		10000	10000	10000	Better Condition
Free of variable						
Length of gap	g	nm	5 – 200	5 – 200	5 – 200	

These models will analyze the optimum design parameter from LUMERICAL simulator. The simulation result is calculated base on the sweep of time drop from the variety of gap. Three models shows the difference of layer dimension at the upper layer between Figure 3(a) and 3(b), then the additional layer at the bottom part can be seen at Figure 3(c) [5, 7].

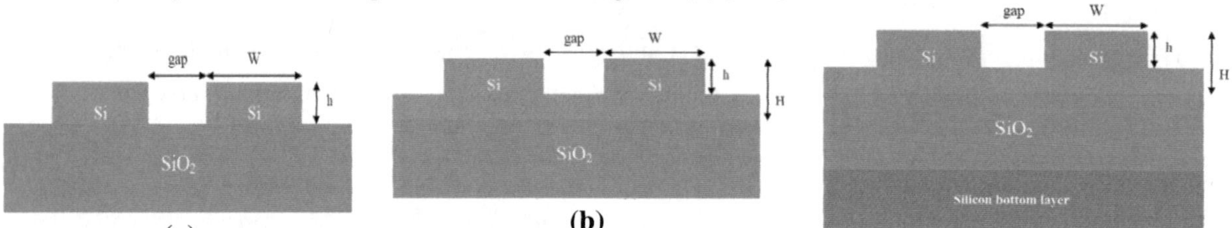

(a) (b) (c)

Figure 3. Three models of micro – ring resonator with the difference of dimension layer (a) Model 1, Ring resonator and signal waveguide are separate part, (b) Model 2, The ultimate layer connecting ring resonator and signal waveguide, and (c) Model 3, addition layer at the bottom part.

2.5. Micro-ring resonator for acid rain sensing application

The applications of micro-ring resonator to detect acid rain were developed in this study. The combination designs of micro-ring resonator using SOI with three layers were analyzed using LUMERICAL software. Figure 4 shows flowchart of micro-ring resonator mechanism for acid rain sensing in near future.

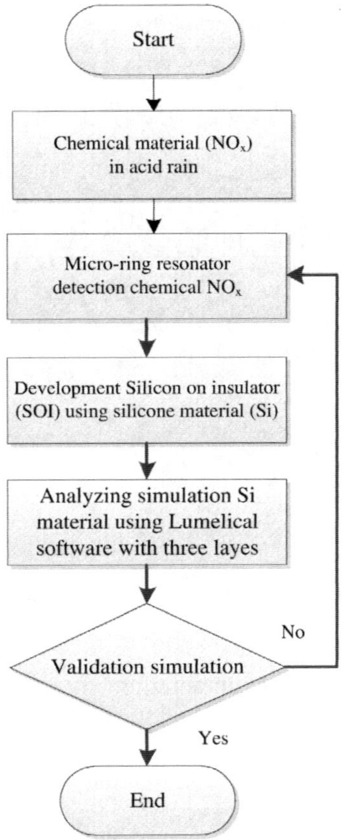

Figure 4. Flowchart mechanism micro-ring resonator for acid rain sensing

Simulation of Micro Ring Resonator for sensing application was compared between the distance of sliced input and the ring resonator part. The bottom part between the ring part and the sliced output were similarly. Furthermore, the two of gap distances must be symmetrical size for getting the result of sweep time and the quality factor.

Three conditions of gap will be seen base on the model 1, 2 and 3, while simulation result will provide the optimum condition. We changed the variables referred on the actual circumstances or conditions that we expect in the measurements. Simulation result will shows the calculation using FDTD analysis, as shown in Figure 5.

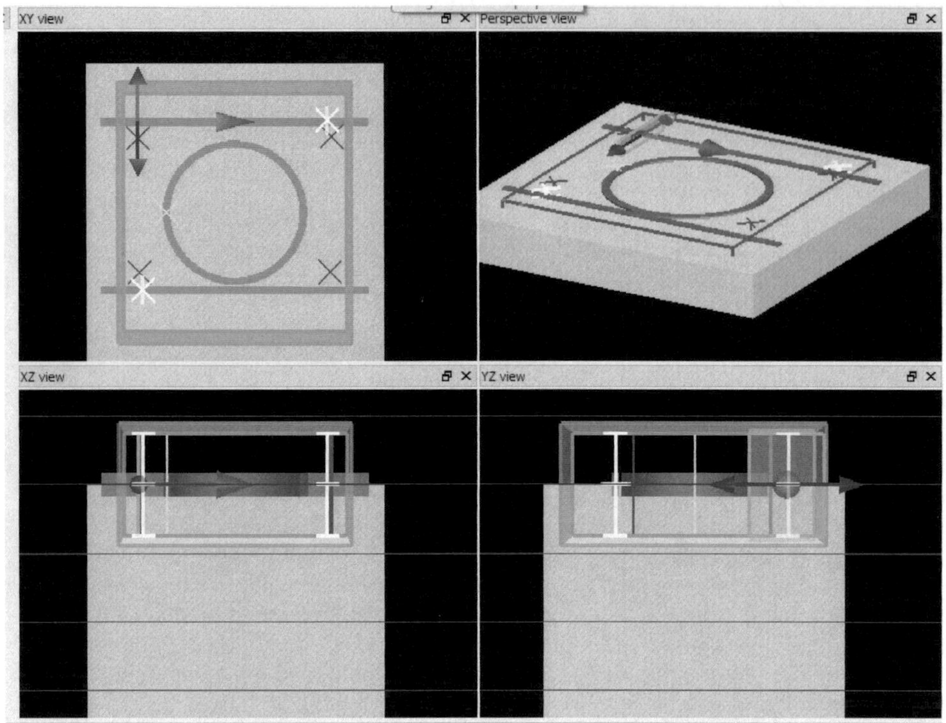

Figure 5. Simulation modelling setup of Micro Ring Resonator for Acid Sensor

3. Results and Discussion

Base on Q – factor, this micro – ring resonator can determine the amplitude of the light field that refers to the sweep of time drop. It is started from 0 until 5000 second with the variation of gap 10 nm, 50 nm and 200 nm. The optimum design of model 1 that was chosen is 50 nm for gap width based on table 2. Figure 6 shows the spectrum result that was affected by background noise. Sweeping time is started from 0 second until 5000 second with increment step is 500 second. The optimum value is defined as lowest amplitude of the light field. Comparing between Figure 4 and Figure 5, it can be seen that alteration intensity of light field along the sweep of time – drop. Figure 6 gives the spectral information that higher of gap width, it will perform low amplitude of the light field and will occur the background noise effects.

Here, Figure 7 shows the quality factor (Q) with variety of the wavelength of the light source. The range of the wavelength is determined from calculation range of FSR in order to be able to compare the quality of light source changes issued as transmitted information. In optimum condition with the gap width of 50 nm, it can be stated that great slope lines indicate good quality factor. If the condition of the quality factor is fixed throughout the wavelength range, it can be stated that the wavelength of light source is mixed with the external disturbances. This condition still not meets qualification of sensor application caused the result of quality shows below 5000.

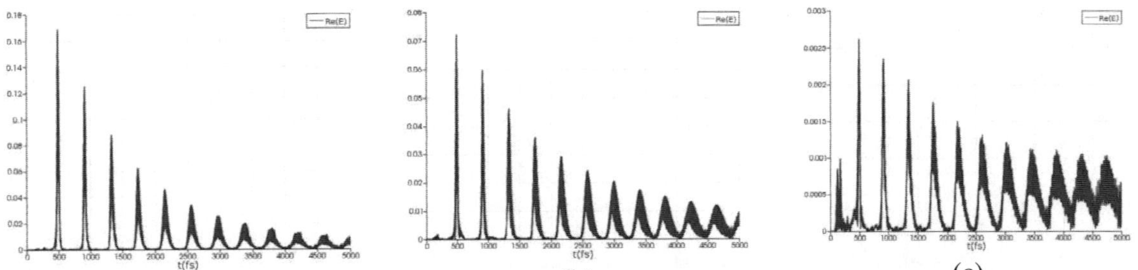

(a) (b) (c)

Figure 6. The sweep of time – drop spectral result base on gap width at model 1 for (a) gap 10 nm, (b) 50 nm and (c) 200 nm.

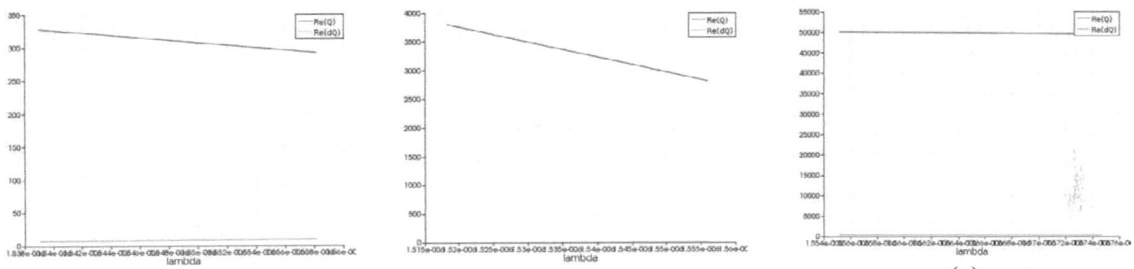

(a) (b) (c)

Figure 7. The quality factor (Q) of micro-ring resonator based on the wavelength of the light source for (a) gap 10 nm, (b) 50 nm and (c) 200 nm.

Furthermore, Figure 8 is the spectrum results of light field amplitude according the variety of gap. The variety of value is started from 80 nm to 200 nm that shown in below. These graphs are results of simulation that refers to model – 2 from ring resonator in Figure 3. The variation of simulation gives the assumption to add the ultimate layer. The additional value is given as a constant that will compared with the variety of gap width. The optimum parameter of model – 2 is gap width amounted 120 nm, it is a result simulation that sufficient with design expectation of Q = 7670.5 (sufficient for sensor) and FSR = 20.69 nm. Figure 9 are the graphs of the quality factors that fitted by the gap width 80 nm, 120 nm and 200 nm. This model shows different of Q where the optimum condition creates the significant value that near to the parameters of sensor application in micro-ring resonator. The optimum condition of this model can fix up the quality of the signal that is injected into the ring resonator because the external disturbances or as a sensor can affect the refractive index on the circular ring resonator area.

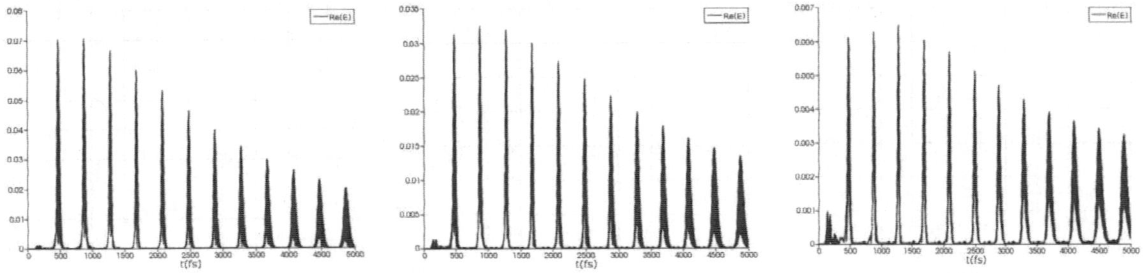

(a) (b) (c)

Figure 8. The sweep of time – drop spectral result at model 2 for (a) gap 80 nm, (b) gap 120 nm and (c) gap 200 nm

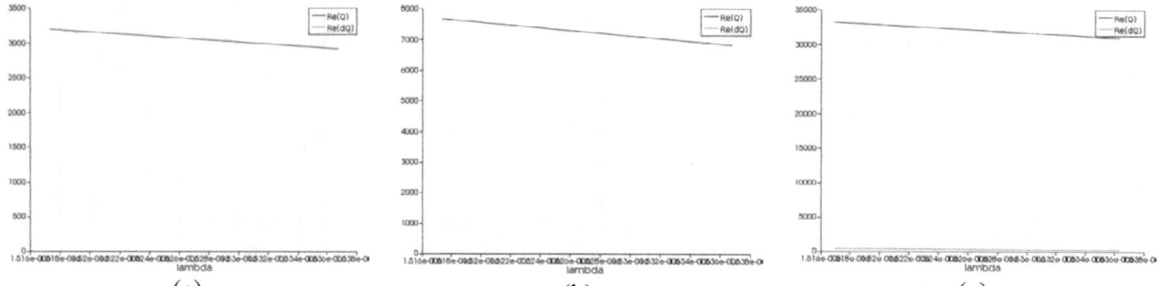

(a) (b) (c)

Figure 9. The quality factor (Q) of micro-ring resonator based on the wavelength of the light source at model 2 for (a) gap 80 nm, (b) 120 nm and (c) 200 nm.

Figure 10 are the spectrum results of light field amplitude according the variety of gap at model 3. These variations are started from 50 nm to 200 nm where a simulation result shows the assumption to add of bottom layer from ring resonator. The additional of bottom layer is given as a constant that will affect the factor quality. The optimum design parameter of gap width for model 3 is 130 nm and it is meet sufficient with design expectation of Q around 9599.85 (sufficient for sensor) and FSR = 20.74 nm.

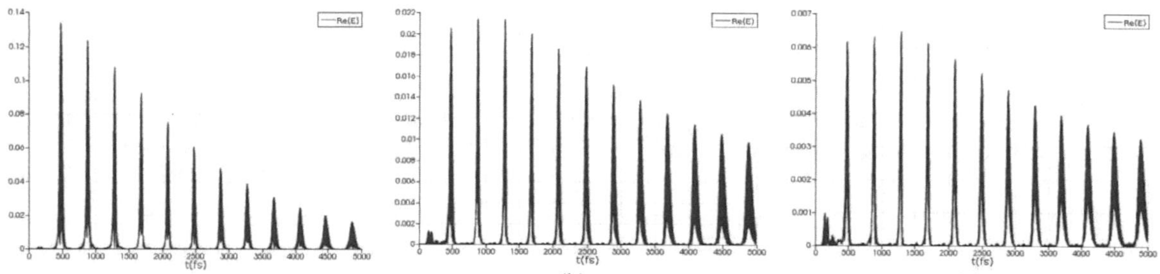

(a) (b) (c)

Figure 10. The sweep of time – drop spectral result at model 3 for (a) gap 50 nm, (b) gap 140 nm and (c) gap 200 nm

Figure 11 are the graphs of the quality factors that fitted by the gap width 50 nm, 140 nm and 200 nm. Q factor of model 3 creates the difference of slope line and Q values. The additional effects of bottom layer from ring resonator was performed the quality factor that it is close to initial assumption.

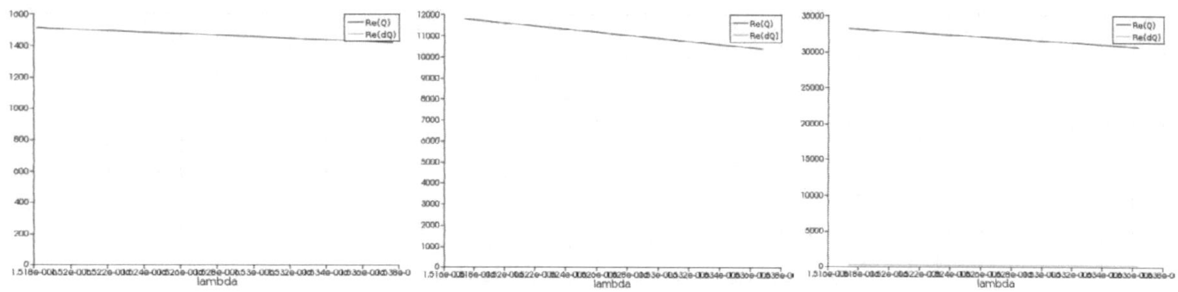

(a) (b) (c)

Figure 11. The quality factor (Q) of micro-ring resonator based on the wavelength of the light source at model 3 for (a) gap 50 nm, (b) gap 140 nm and (c) 200 nm

4. Conclusion

The development micro-ring resonator with three layers using wafer Silicon layer (Si), Silicon Oxide (SiO_2) Layer and Silicon layer (Si) to detect acid rain was successful. The highest value of the width gap will provide the effect of the amplitude value and also low coefficient results. So as, the high value of the ring resonator will supply the spectral disturbance effects and present the slope line and tend to be flat along the predetermined of the FSR value. This simulation shows the real conditions base on the assumed parameter caused the optimum value is a maximum value for the FSR condition. When the FSR value gives the 0 point, it means that no occurrence of light propagation in the ring resonator and the absence of light coupling process from input bus – line to the ring resonator and vice versa.

Acknowledgments
The authors would like thanks to Professor Dr. Sahbuddin Shaari and the other parts were supporting this research activity and all staff from Institute of Microelectronic and Nano Electronic (IMEN), Universiti Kebangsaan Malaysia (UKM).

References

[1] Seinfeld J H and Pandis S N 2016 *John Wiley & Sons* **1** 3
[2] Seppälä A, Verronen P T, Clilverd M A, Randall C E, Tamminen J, Sofieva V, Backman L and Kyrölä E 2007 *Geophysical Research Letters* **34** L12810
[3] Cai H, Sun R, Yang X, Liang X, Wang C, Sun P, Liu F, Zhao C, Sun Y, and Lu G 2016 *Ceramics International* **42** 12503
[4] Chin M K and Ho S T 1998 *Journal of Lightwave Technology* **16** 1433
[5] D G Rabus 2007 *Springer-Verlag Berlin Heidelberg* **1** 41
[6] Mulyanti B, Menon P S, Shaari S, Hariyadi T, Hasanah L and Haroon H (2014) *Sains Malaysiana* **43** 247
[7] Mulyanti B, Hasanah L, Hariyadi T, Novitasari R, Pantjawati A B and Yuwono H (2015) *Advanced Materials Research* **1112** 32

IOP Conf. Series: Journal of Physics: Conf. Series **852** (2017) 012044 doi:10.1088/1742-6596/852/1/012044

Development of a solar charged laboratory bench power supply

W A Ayara[1,1], T V Omotosho[1,2], M R Usikalu[1,3], M S J Singh[2] and W Suparta[2]

[1]Department of Physics, Covenant University, Ota, Nigeria
[2]Space Science Centre, Institute of Climate Change, Universiti Kebangsaan Malaysia, Malaysia

[*]E-mail: [1]williams.ayara@covenantuniversity.edu.ng,[2]omotosho@covenantuniversity.edu.ng, [3]moji.usikalu@covenantuniversity.edu.ng,[2]wayan@ukm.edu.my

Abstract. This product is an improvement on available DC laboratory bench power supply. It is capable of delivering low voltage Alternating Current (AC) and Direct Current (DC) to carry out basic laboratory experiment for both secondary schools and also at higher education institutions. The power supply is capable of delivering fixed DC voltages of 5V, 9V, 12V, the variable voltage of between 1.25–30V and a 12V AC voltage. Also Incorporated is a USB port that allows for charging cell phones and other mobile devices, and a dedicated 12V DC output to power 5-7 Watt LED bulb to provide illumination in the laboratory for the instructor who may need to work at night in the absence of utility power.

1. Introduction

Erratic power supply is one huge factor limiting advancement in education in the developing world especially in rural environments, taking its toll in the area of basic experiments in sciences, engineering, and technology. Electric power supply is the most important commodity for national development. With electrical energy, the people are empowered to work from the domestic level and the cottage industries, through the small-scale and medium industries to employment in the large-scale manufacturing complexes [1].

Laboratory Bench Power supply is a common electrical/electronic power unit found in laboratories. These available units have only one source of input power by which they can be put to use. Erratic mains power or the complete mains outage renders this all important laboratory power unit non-useful appliance. Hence basic laboratory experiment requiring constant low power Alternating Current (AC) or Direct Current (DC) voltages are forced to be put on hold.

The unit is designed such that the internal energy storage bank can be charged using solar energy with the help of an internal charge controller circuitry. Hence it becomes very important as it is conveniently used to meet the need of low DC and AC power requirement for basic experiments in schools located in places where there are habitual power outages or total power failure from the utility supply. As a result, the usual postponement or outright cancellation of experiments in practical classes due to power outage will be a thing of the past.

2. Basic Overview of The Entire Unit

The functionality of the device is based on regulated DC power supply (fixed and variable) in conjunction with fixed AC supply. The primary requirement considered in the design of the regulated

Content from this work may be used under the terms of the Creative Commons Attribution 3.0 licence. Any further distribution of this work must maintain attribution to the author(s) and the title of the work, journal citation and DOI.
Published under licence by IOP Publishing Ltd

DC power supply are the output voltage V_{dc}, maximum current I_{max} required by the load, percentage load and line regulation [2, 3].

2.1 The Transformer and Full-wave rectifier
This is a multiple output transformer that receives an input AC mains supply voltage of between 220-230V and steps it down to lower voltages of 2x15V, 2x12V, and 2x7V all center-tapped. The AC voltage from the transformer is rectified using a full-wave rectifier made up of four P-N junction diodes to form a rectifying bridge [3] providing DC supply. Depending on the configuration of the transformer, the full-wave rectifier could give single positive, or equal positive and negative voltages that are needed by several DC operated circuits such as operational amplifiers [2, 3].

2.2 The Smoothing Circuit and Voltage Regulation
The ripples are smoothened using a low pass filter in the form of a shunt capacitor [3]. The effect of continuous charging and partial discharging causes ripples on the DC level. Hence, depth of the ripple depends on the capacity of the reservoir capacitor and also the amount of current drawn by the load. To attain better smoothing and minimize ripple, large value capacitor (1,000μF of more) is used, so that the potential difference across its terminals falls more slowly [3 - 5].

The voltage regulation stage of supply provides a stabilized DC voltage to the output interface as set by the integrated circuit regulator (IC) [4]. The regulators provide both fixed and variable output DC voltages of 5V, 9V, and 12V both of positive and negative voltages.

2.3 The Solar Panel
The sun supplies energy in the form of radiation, without which life on Earth could not exist [6]. The energy derived from the sun is converted to DC voltage by means of holes and electron exchange. Energy derived from the sun by the solar panel is used for energy storage in the battery bank. The solar panel in addition to the battery bank forms the core of this product.

2.4 The Charger, Battery bank and Inverter section
This charger is meant to deposit electrical energy in the battery bank for storage that will be useful whenever there is a power outage. As stated earlier, energy derived from the sun through the solar panel is stored in the battery bank by means of the charger. This is a charger controller that regulates the charging rate of the battery for best performance.

The battery bank is a combination of individual cells in series and parallel arrangement. Batteries are available in different sizes, voltage, amp-hour rating, vented or non-vented. The battery bank's size determines the length of time the inverter can supply AC output power. The larger the bank, the longer the inverter can run and also the longer the time to recharge [7].

The section of the inverter takes energy stored in the form of DC voltage from the battery bank and converts it to AC voltage whenever AC voltage is required when power from the mains fails.

2.5 The Mobile Device Charger and DC output for LED lighting
Mobile devices are now our everyday companion. They operate on battery and to be useful their internal power source must always be available for use. This part of the unit is reserved to supply enough energy to the battery of mobile devices like Tabs, iPod and other mobile devices whose power may be low or completely depleted.

The DC output for LED lighting is a 12V DC dedicated output for lighting up light emitting diode LED. Its usefulness is harnessed mostly at night whenever there is a power outage and the instructor still intends to stay back in the laboratory.

3. Design considerations
To fully deploy the Solar Charged Laboratory Bench Power Supply; several sub-units within the entire unit were put into consideration. A few of them are considered below. It is worthy to note that the

device is low cost when compared with other commercially available DC power supply for laboratory work. Its durability is further enhanced by the fact that energy from the sun which will always be available is favorably harnessed for energy storage in the power bank.

The core focus on the design consideration is targeted at places where school laboratories are situated, having an erratic power supply or power not available for use and cost of using petrol or diesel to power generator set is high. In addition, lighting as an important factor was also considered. Therefore, the device is meant to be able to provide light at night via Light Emitting Diode (LED BULB) for any teacher or instructor who may which to work in the laboratory for some hours at night. Figure 1 shows the block diagram of a solar charged laboratory bench for a power supply.

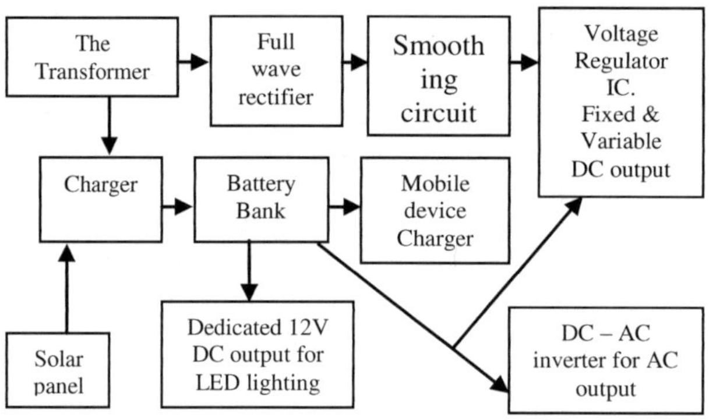

Figure 1. Block diagram of a solar charged laboratory bench power supply

4. Hardware Considerations

The entire system consists of several units of hardware put together for its proper functioning. Areas to be considered include voltage transformation and rectification section, DC voltage smoothing section, DC voltage regulation section, solar charge controller section etc.

4.1 Voltage transformation and rectification

The transformer used is a $50H_z$ $230V_{ac}$ multiple low voltage output transformers capable of delivering 30V, 24v and 14V all centered tapped. The importance of the transformer it has the ability to completely isolate the power supply circuitry and the user from the mains hence safety is ensured. The operation of the transformer is guided by the relation:

$$\frac{N_p}{N_s} = \frac{I_s}{I_p} \tag{1}$$

The rectification of the step-down AC voltage to DC was achieved by using four (IN4007) P-N junction diodes that can handle as much as 2 Ampere.

The output of a rectifier consists of a DC component and an AC component known as the ripple. The smaller the AC component in the DC output, the better the rectifier [8]. This factor K is a ratio of the r.m.s value of the AC component to the DC component in the rectifier output. The effective r.m.s value of total load current is given by:

$$I_{rms} = \sqrt{I^2{}_{dc} - I^2{}_{ac}} \tag{2}$$

or

$$I_{ac} = \sqrt{I^2{}_{rms} - I^2{}_{dc}} \tag{3}$$

By dividing through with I_{dc}, we get,

$$\frac{I_{ac}}{I_{dc}} = \frac{1}{I_{dc}}\sqrt{I^2{}_{rms} - I^2{}_{dc}} \tag{4}$$

But the ratio I_{ac}/I_{dc} is the ripple factor. The ripple factor (K) is defined as,

$$(K) = \frac{1}{I_{dc}}\sqrt{I^2{}_{rms} - I^2{}_{dc}} = \sqrt{\left(\frac{I_{rms}}{I_{dc}}\right)^2 - 1} \tag{5}$$

In half wave rectification, $I_{rms} = I_m/2$ and $I_{dc} = I_m/\pi$, Therefore,

$$K = \frac{1}{I_{dc}}\sqrt{I^2{}_{rms} - I^2{}_{dc}} = \sqrt{\left(\frac{I_m/2}{I_m/\pi}\right)^2 - 1} \tag{6}$$

In full-wave rectification, $I_{rms} = I_m/\sqrt{2}$ and $I_{dc} = 2I_m/\pi$, Therefore,

$$K = \frac{1}{I_{dc}}\sqrt{I^2{}_{rms} - I^2{}_{dc}} = \sqrt{\left(\frac{I_m/\sqrt{2}}{2I_m/\pi}\right)^2 - 1} \tag{7}$$

This is equal to a value of 0.48 [8], which shows that the pulsations in the output of the full-wave rectification are less than that of the half-wave. Hence the full-wave rectification is used in the design.

4.2 DC voltage smoothing and Dual Rail Power Supply section

In order to clean up the residue of AC component found in the DC voltage after rectification, smoothing is carried out using appropriate filtering capacitor. In a well-designed power supply unit, the fall in output voltage is small. Before the voltage has had time to fall far, the next pulse from the other half circle arrives and the capacitor is recharged to its peak level. This continuous charging and partial discharging causes ripple on the DC level of the power supply. This ripple can be sufficiently minimized by the use of a large value capacitor [3,4].

The dual rail power supply section is responsible for providing the required DC output. The regulators LM317, LM339, and LM78xx & LM79xx series were employed to determine the + or – voltages required at the output section as shown in Figure 2. The voltages provided by this section include fixed voltages of 5, 9 and 12V and a variable voltage of 1.25-30V.

A series regulator configuration is considered for the design. Here the regulator ifs placed in series with the load [8].The variable voltage from the LM317 was determined using the formula:

$$V_{out} = 1.25V\left(1 + \frac{V_R}{R}\right) + I_{adj}V_R \tag{8}$$

where V_R is a variable resistor (pot). Note that I_{adj} is mostly controlled to less than 100μA, the error that is related to this term is negligible of many applications [9].

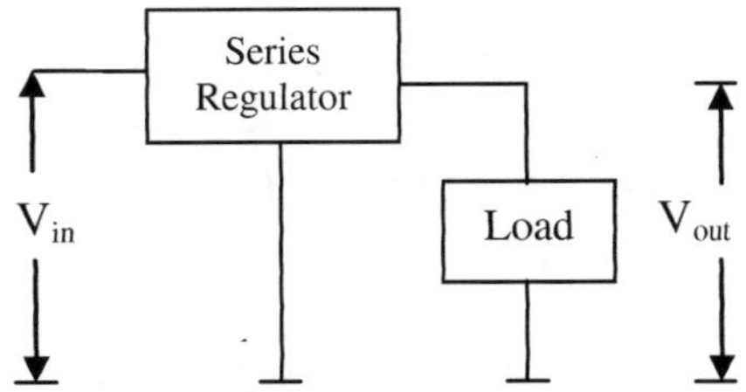

Figure 2. A regulator placed in series with a load

IOP Conf. Series: Journal of Physics: Conf. Series **852** (2017) 012044 doi:10.1088/1742-6596/852/1/012044

4.3 Charger and DC-AC Inverter Section

Two charger sections exist, one to provide storage energy to the battery. Battery bank comprises of two 12V banks connected in series. Hence the LM317 employed is configured such that it produces a charging voltage of 26V. The second charger which is for mobile devices is designed also using the LM317 to attain a maximum output voltage of 5V at 1A using equation (8) in both cases.

The consideration for the inverter section is to provide a 12Vac voltage in the event of mains failure. Since some experiment will require AC voltage for their operation, 12V transformer wound with a turn ratio of 1:1 connected to a low power square wave inverter is used to provide the desired $12V_{ac}$ output.

5. Testing Result and Discussion

The power supply having being calibrated was tested for voltage regulation by applying different load condition in the form of resistance to different values. Voltage regulation was carried out to determine the stability of the power supply between when the supply is on-load and when not on load. To obtain this result, two fixed voltage were set and the resistive load was connected in turn. The percentage regulation is given by the formula:

$$VR = \frac{V_{NL} - V_{FL}}{V_{FL}} x100\% \quad (9)$$

where V_{NL} is the DC output at no load, and V_{FL} is the DC output at full load [8]. Table 1 shows the values obtained when a test of voltage regulation is carried out with applied load of between 100-1500Ω at a test voltage of 12.25V. A very close observation reveals that the power supply exhibit a stable voltage regulation between when a load of 400 - 1,350Ω is applied at a set voltage of 12.25V. At load exceeding 1,350Ω, voltage regulation is no longer stable. This implies that higher amount of current is being drawn. The trend of this increase is plotted in Figure 3.

Table 1. Percentage regulation at 12.25V with varying resistive load

Load (Ω)	Positive voltage (V) V_{NL}	Positive voltage (V) V_{FL}	Regulation (%)
100	12.25	12.21	0.33
200	12.25	12.19	0.49
300	12.25	12.17	0.65
400	12.25	12.17	0.65
500	12.25	12.17	0.65
600	12.25	12.17	0.65
700	12.25	12.17	0.65
800	12.25	12.17	0.65
900	12.25	12.17	0.65
1000	12.25	12.17	0.65
1100	12.25	12.17	0.65
1200	12.25	12.17	0.65
1300	12.25	12.17	0.65
1400	12.25	12.16	0.65
1500	12.25	12.16	0.74

Similar to Table 1, Table 2 shows values obtained from a test of voltage regulation at a test voltage of 5V when load of 100-1,500Ω is applied. As shown in the table, two distinct stable voltage regulation states and a drift are observed. At load of 900-1,000Ω a sudden upward surge in current drawn by the load occurred showing instability of 0.2% in regulation. Observably, with increase in applied load from 1,050Ω and above better voltage regulation was observed all at a test voltage of 5V.

The trend of this increased is plotted in Figure 4(a), while the voltage drop at around 5V is plotted in Figure 4(b).

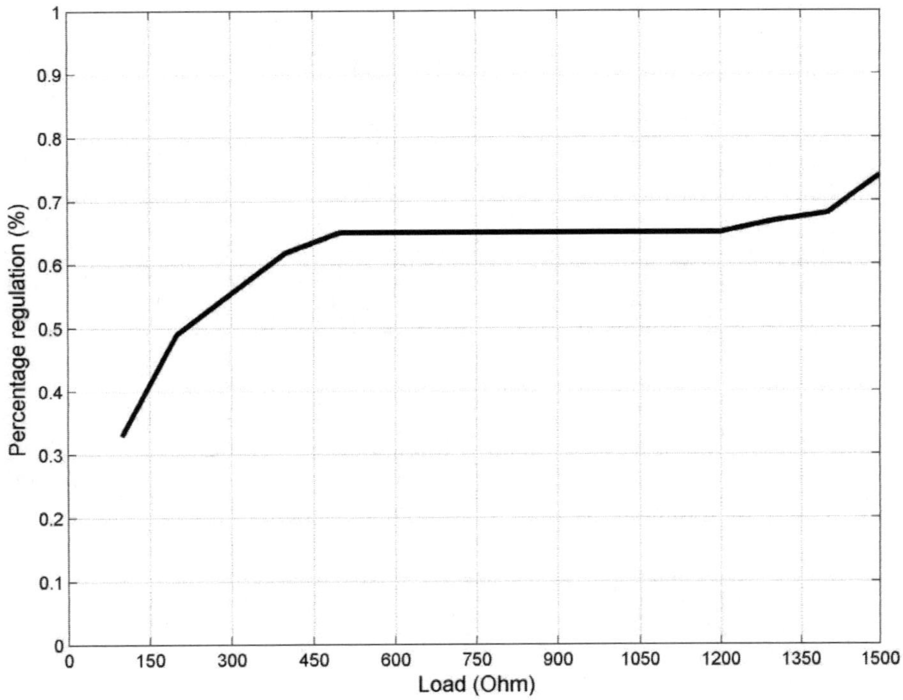

Figure 3. Graph of plot showing percentage regulation against load

Table 2. Percentage regulation at 5V with varying resistive load

Load (Ω)	Positive voltage (V) V_{NL}	Positive voltage (V) V_{FL}	Regulation (%)
100	5.00	4.96	0.81
200	5.00	4.96	0.81
300	5.00	4.96	0.81
400	5.00	4.96	0.81
500	5.00	4.96	0.81
600	5.00	4.96	0.81
700	5.00	4.96	0.81
800	5.00	4.96	0.81
900	5.00	4.96	0.81
1000	5.00	4.95	1.01
1100	5.00	4.95	1.01
1200	5.00	4.95	1.01
1300	5.00	4.95	1.01
1400	5.00	4.95	1.01
1500	5.00	4.95	1.01

In order to see the voltage drop during the operation, we have tested the circuit at 12V with a 5W LED bulb as load as presented in Table 3. It was observed that at the fixed voltage of 12V the connected LED bulb remained bright providing illumination for up to 3 hours (180 minutes) with minimum voltage drop. Beyond 3 hours, the brightness of the bulb dropped as a result of the high voltage drop observed. The testing system is shown in Figure 5. From the table, increasing of time operation of the power supply will significantly increasing the voltage drop which is a possibly by dissipation energy.

Table 3. Time - Voltage drop at 12V with 5Watt LED bulb as load

Time (min)	Voltage (V)	Voltage at time t (V)	Voltage drop (V)
0	12.00	12.00	0.00
30	12.00	11.95	0.05
60	12.00	11.90	0.10
90	12.00	11.91	0.09
120	12.00	11.86	0.14
150	12.00	11.82	0.18
180	12.00	11.76	0.24
210	12.00	10.66	1.34

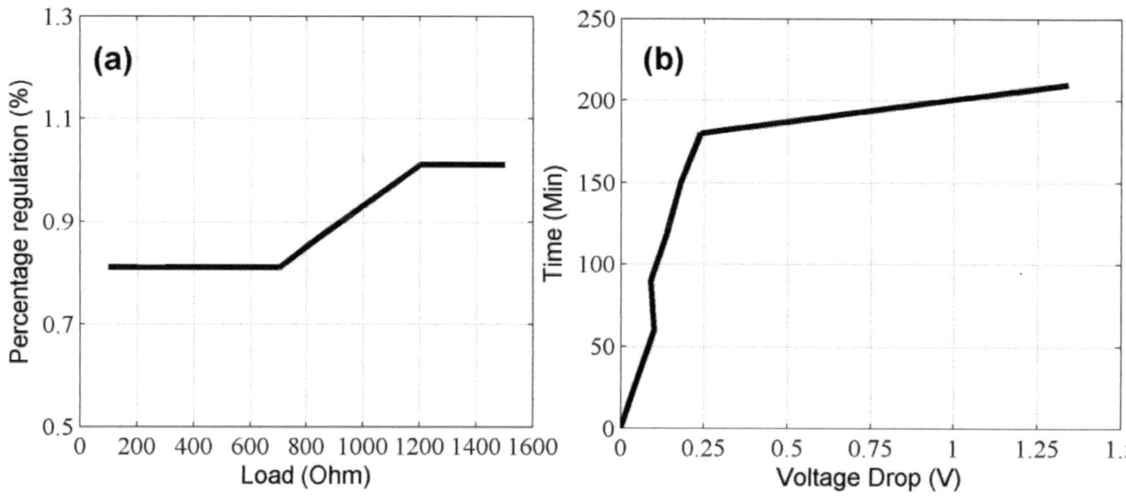

Figure 4. (a) Graph of plot showing the percentage regulation against load and (b) Voltage drop at 12V with 5 Watt LED bulb as load

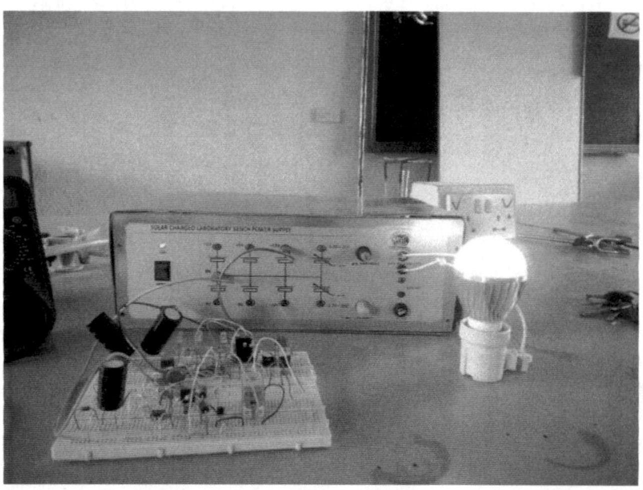

Figure 5. Picture of the power supply being tested in the laboratory to power a 7W LED bulb and a circuit on bread board

As can be seen from Figure 5, the device is not plugged into mains supply, yet from the energy stored in the power bank using a solar panel, it is being used to power a 7 Watt LED bulb and a circuit under test. This has eventually corrected the usual need to either stop or postpone an experiment in the laboratory due to an erratic power supply or total power outage from the mains. The observed result obtained from Figures 3 and 4, it shows that the power supply has better voltage regulation at a lower voltage as compared with the result obtained at a fairly increased voltage i.e. 12V. The power supply showed appreciable voltage regulation of over 80%. Figure 6 shows the complete circuit diagram for the development of a solar charged laboratory bench power supply.

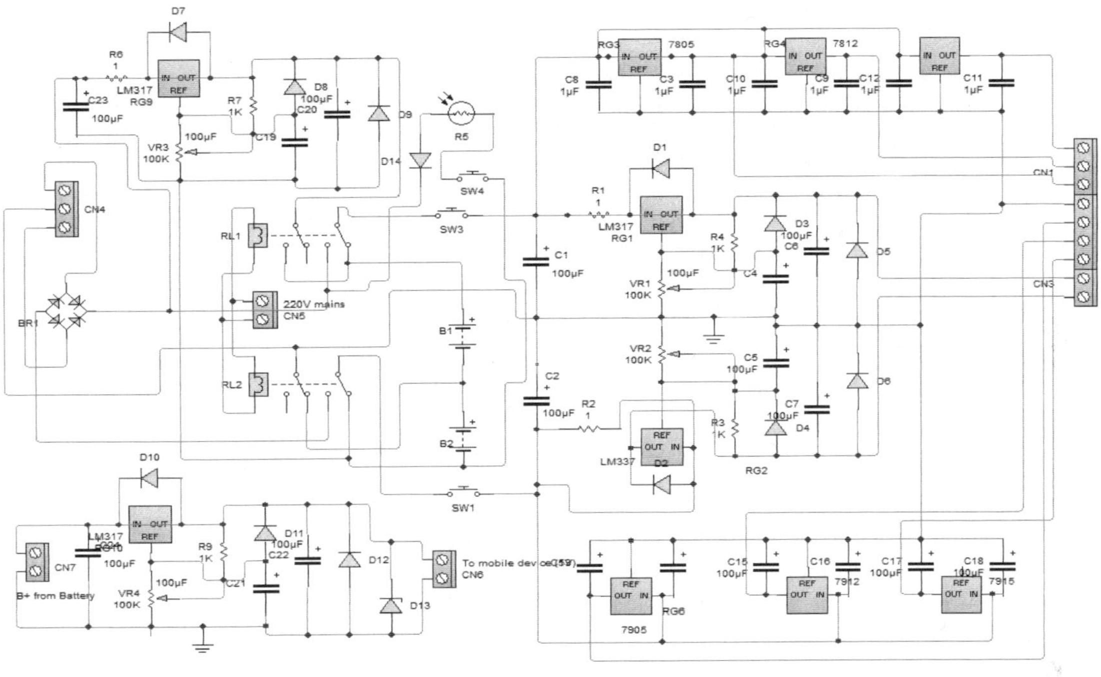

Figure 6. Complete circuit diagram

6. Conclusion

This work has successfully presented a power supply that is functional, low cost and durable. The test carried out shows that the power supply can support basic experiments in electricity and electronic class hence ensuring that no practical class is suspended in the absence of supply from the mains.

Acknowledgement

I sincerely do express my profound appreciation to the Department of Physics and the covenant University Center for Research innovation and development (CUCRID) for sponsoring this work.

Reference

[1] www.gamji.com/article9000/news9075.htm
[2] Shoewu O, Olaniyi O M and Ogunleye O M 2011 *Design and development of an intelligent variable power supply*, the pacific journal of science and technology, **12** p 30
[3] Owen Bishop 2011 *Electronics circuits and systems*, fourth edition p 141
[4] Nigel P Cook *Practical Electronics*, second edition, p 72
[5] Horowitz P and Winfield H 1995 *The art of electronics*, second edition p 307
[6] *Planning and installing Photovoltaic system*, second edition, The German Energy society
[7] Xantrex 2006 *Battery bank for inverter systems, application note* Xantrex Inverter/Chargers 976-0114-01-01 Rev A (April 2006) p 1
[8] Mehta V K 1998 *Principle of electronics*, sixth edition p 408
[9] ON Semiconductor, LM317 datasheet, January, 2016 – Rev 14

IOP Conf. Series: Journal of Physics: Conf. Series **852** (2017) 012045 doi:10.1088/1742-6596/852/1/012045

Background radiation dose of dumpsites in Ota and Environs

M R Usikalu, O O Ola, J A Achuka, I O Babarimisa and W A Ayara

Department of Physics, Covenant University, PMB 1024, Ota, Ogun State, Nigeria

Email: moji.usikalu@covenantuniversity.edu.ng

Abstract. In-situ measurement of background radiation dose from selected dumpsites in Ota and its environs was done using Radialert Nuclear Radiation Monitor (Digilert 200). Ten measurements were taken from each dumpsite. The measured background radiation range between 0.015 $mRhr^{-1}$ for AOD and 0.028 $mRhr^{-1}$ for SUS dumpsites. The calculated annual equivalent doses vary between 1.31 $mSvyr^{-1}$ for AOD and 2.28 mSv/yr for SUS dumpsites. The air absorbed dose calculated ranged from 150 $nGyhr^{-1}$ to 280 nGy/hr for AOD and SUS dumpsites respectively with an average value of 217 $nGyhr^{-1}$ for all the locations. All the estimated parameters were higher than permissible limit set for background radiation for the general public. Conclusively, the associated challenge and radiation burden posed by the wastes on the studied locations and scavengers is high. Therefore, there is need by the regulatory authorities to look into the way and how waste can be properly managed so as to alleviate the effects on the populace leaving and working in the dumpsites vicinity.

1. Introduction

General environmental radiation and the outward vulnerability from gamma radiation, depend mainly on the earthly and geographical condition, and appear at various levels in the soils of each region in the world [1]. In the biosphere, the propagation of radionuclides depends on the transportation of the geological materials from which they are made from and the processes which establish them at a given location in specific area. The major way to understand these distributions, therefore, is to study the transportation of the source materials and the physical and geochemical processes that lead to increased concentrations of radionuclides under given conditions [2].

Man is continually exposed to different levels of ionizing radiation dose, due to the nature of his environment [3, 4]. This is mainly due to the radionuclides found in different rock types propagated around our environment, with little addition from cosmic rays (cosmogenics). Radionuclides enclosed in rocks form a key part of the Naturally Occurring Radioactive Material (NORM) in the environment, which when not properly audited and controlled, are capable of causing harm to man and the environment.

In the last few decades, the natural index of radionuclides has been elevated due to human activities. Waste generation and disposals have also added in no small quantity to the increased levels of human radiation exposures. Human radiation exposure could be from either external sourses from ^{40}K, ^{226}Ra and ^{232}Th in soil or internal sources from inhalation of radon and its daughter particles in dust and fumes from waste disposal sites. The measurement of radiation exposure levels from waste dump-sites will provide information on any possible radiological risk that can arise from waste generation and disposal to human health and environment.

Radioactive dose obtained from various waste dumpsites have been studied by many researchers in Nigeria [5 – 11]. They reported that dumpsites show no significant radiological health hazards to the people living or working around the dumpsites. But in most studies, dumpsites doses were higher when compared with sites that are free from waste disposal. This is the first study on radiation dose from dumpsites in Ota and Environs. The background radiation dose measurement was carried out in some dump-sites so as to estimate the radiological challenges the dumpsite pose on the population

Content from this work may be used under the terms of the Creative Commons Attribution 3.0 licence. Any further distribution of this work must maintain attribution to the author(s) and the title of the work, journal citation and DOI.
Published under licence by IOP Publishing Ltd

around them. Therefore, this study will be useful in providing baseline data on radiation dose incurred in Ota and its environment from dumpsites.

2. Materials and Method

The assessment was carried out between March and May 2016 in 13 locations within and around Ota, a commercial town in Ogun state (Figure 1). Ota is located in 6° 41'N and 3° 41'E. It is the local government headquarter of Ado-odo and has an estimated 163,783 population as adopted by the last headcount in Nigeria.

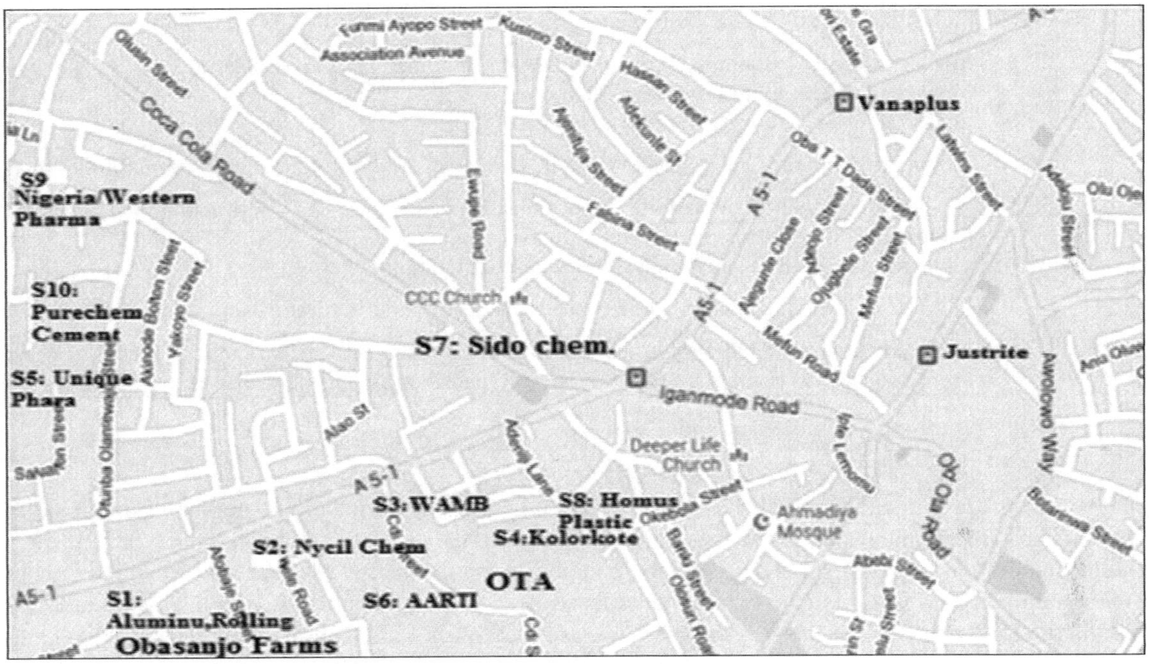

Figure 1. Location Map of the study area

In–situ method of measurement of radiation dose was employed in some selected dumpsites in Ota, Iju and Atan area of Ogun state. Table 1 showed different dumpsites used for this study and their associated waste. The measurement was done using a portable Geiger Muller Digilert 200 (S.E. Inspector) handheld digital radiation detector. The detector is a portable, survey meter based on microprocessor working principle and it has high sensitivity to measure low levels of alpha, beta, x-rays, gamma, for common isotopes to measure the dose rate or exposure in mixed field. It detects Alpha down to 2 MeV, Beta down to 0.16 MeV; typical detection efficiency at 1 MeV is approximately 25% and detects Gamma low to 10 KeV through the end window. Measurement was made randomly at 10 points from each thirteen dumpsites and average calculated for each location. For effective detection, the survey meter was positioned at gonad level i.e. at about one meter above ground level. It was switched on to absorb radiation for some seconds and stable value was recorded. The procedure was repeated at each location and readings were taken in micro sievert per hour.

Table 1. Dumpsites and the associated wastes.

S/N	NAME	ASSOCIATED WASTE
1	KNL	Construction and iron steel wastes: irons, steels, alluminium etc
2	KN	Industrial waste and commercial wastes:mostly paper and wood
3	AOD	Market, residential and commercial wastes: rotten food, glass, plastics, ashes, paper, nylon, tin cans and an abattior site etc.
4	LPH	Mixture of medical and pharmaceuticals with industrial wastes: tin cans, nylons, paper, bottles, damaged electronics.
5	SNL	Wastes from industrial action: production of paper
6	TSM	Industrial wastes: Aluminium wastes, papers and nylons etc.
7	CUD	Different wastes including market, household, commercial and agricultural wastes dumped: vegetables, fruits, nylons, packaging materials and ashes etc.
8	VGP	Industrial and office wastes: plastics, nylons, paper
9	KD	Commercial, market and household waste: vegetables, fruits, nylons and unfinished foods
10	ARM	Waste from construction activities and industrial waste: building construction, welding metal scraps, and other commercial wastes around the area.
11	SUS	Construction and iron steel wastes: irons, steels, aluminium etc
12	IDA	Market, residential and commercial wastes: spoilt food, glass, clothes, smoke and ashes
13	ID	Different wastes including market, household, commercial and agricultural wastes dumped there includes vegetables, fruits, nylons, paper, packaging materials, ashes etc.

3. Results and Discussion

The average dose rate and standard deviation measured from all the locations are shown in Table 2. The background radiation measured from the dumpsites range from 0.015 $mRhr^{-1}$ to 0.026 $mRhr^{-1}$. SUS, a steel producing company has the highest radiation, while the lowest dose rate was observed in AOD (Atan Oja dumpsite). Highest radiation observed in SUS may be attributed to the different types of construction waste such as steels and iron waste dumped in this site. It has been noted that steel and metals generally emits a high radiation dose than all other household waste and agricultural waste. In AOD it could be as a result of the composition of residential, market and commercial wastes that do not emit much radiation other sites. The background radiation exposure rates obtained in this study are slightly higher than the global background radiation standard of 0.013 $mRhr^{-1}$. The annual equivalent dose was estimated and conversion to absorbed dose was done using equation [1]

$$1\, radh^{-1} = 1.0 * 10^{-2}\, Gyh^{-1} \tag{1}$$

The calculated absorbed dose and annual effective dose equivalent is presented in Table 3. This was done in order to estimate the quantity of energy (radionuclides) delivered by ionization radiation to the human body in a given period. This was done to prevent any somatic, epidemiological, and radiological health implication recommended so as to set the maximum permissible limit for non–radionuclide industrial worker and the public [12]. The estimated annual equivalent dose rate (mSv/yr) ranged between 1.31 $mSvyr^{-1}$ and 2.28 $mSvyr^{-1}$ for AOD and SUS dumpsites respectively. Figure 2 shows the calculated annual equivalent dose rate for each location compared with [1] threshold. The estimated equivalent dose rates for all the thirteen locations studied were higher than the permissible limit of 1.3 $mSvyr^{-1}$ recommended [1].

The calculated absorbed dose rate ($nGyhr^{-1}$) ranged between 150 $nGyhr^{-1}$ to 280 $nGyhr^{-1}$ for AOD and SUS dumpsites respectively with an average of 217 $nGyhr^{-1}$. Figure 3 shows that the average air absorbed dose rates ($nGyhr^{-1}$) for the locations were higher than the world's average value of 60 $nGyhr^{-1}$ [1]. The higher values recorded in this work may be attributed to chemical, medical and other hazardous materials dumped together on the dumpsites. This will probably affect the surface water, soil and underground water sources of the environment. It may serve as source of contamination due to accumulation of radionuclide in the atmosphere, seepage and precipitation from the dumpsites Leachate. Therefore, people living and working around the dumpsites areas are exposed to different doses of radiation; these may result in health problems such as cancer, radiation poisoning and cell mutation. This calls for a serious concern and detailed studies of all the dumpsites in the state to ascertain the level of radiological impact of the sites workers, communities and the environment.

Table 2. Average radiation dose for all locations.

S/N	SAMPLE	Average Radition ($mRhr^{-1}$)
1	KNL	0.021 ± 0.0033
2	KN	0.019 ± 0.0022
3	AOD	0.015 ± 0.0027
4	LPH	0.021 ± 0.0030
5	SNL	0.022 ± 0.0039
6	TSM	0.022 ± 0.0039
7	CUD	0.023 ± 0.0037
8	VGP	0.022 ± 0.0039
9	KD	0.026 ± 0.0057
10	ARM	0.017 ± 0.0027
11	SUS	0.028 ± 0.0084
12	IDA	0.024 ± 0.0069
13	**ID**	**0.025 ± 0.0069**

Table 3. Air absorbed dose rate and equivalent dose rates.

S/N	Location	Average Dose Rate ($mRhr^{-1}$)	Annual Equivalent Dose Rate, E($mSvyr^{-1}$)	Average Air Absorbed Dose Rate ($nGyhr^{-1}$)
1	KNL	0.021 ± 0.0033	1.84	210
2	KN	0.019 ± 0.0022	1.67	190
3	OD	0.015 ± 0.0027	1.31	150
4	LPH	0.021 ± 0.0030	1.84	210
5	SNL	0.022 ± 0.0039	1.93	220
6	TSM	0.022 ± 0.0039	1.93	220
7	UD	0.023 ± 0.0037	2.02	230
8	VGP	0.022 ± 0.0039	1.93	220
9	KD	0.026 ± 0.0027	2.28	260
10	ARM	0.017 ± 0.0027	1.49	170
11	SUS	0.026 ± 0.0084	2.28	260
12	IDA	0.024 ± 0.0069	2.10	240
13	ID	0.025 ± 0.0069	2.19	250

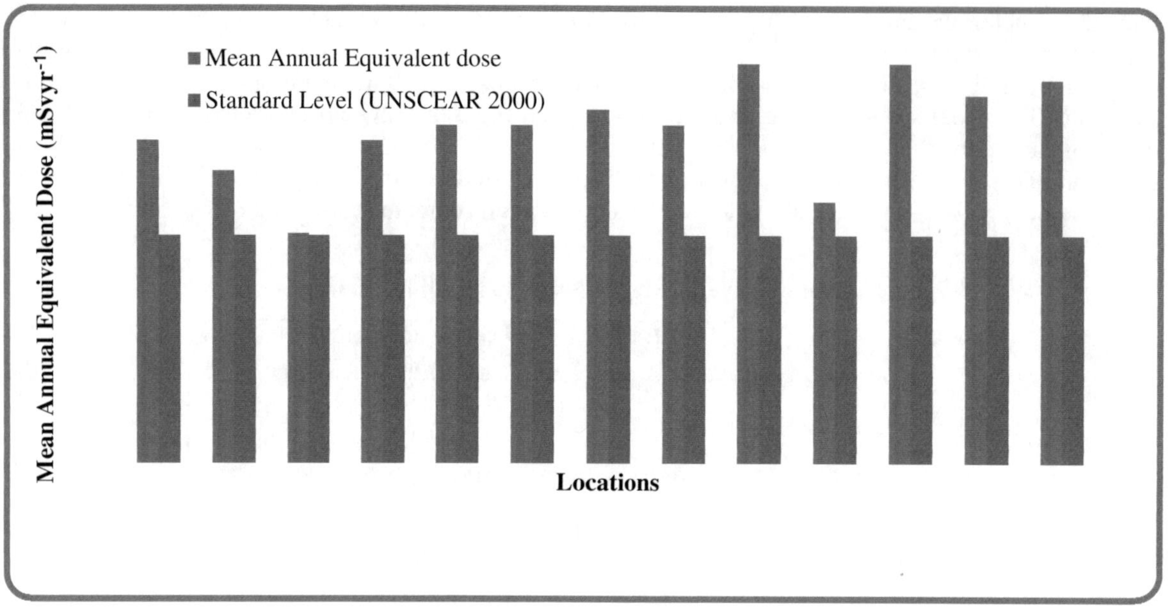

Figure 2. Comparison of average equivalent dose ($mSvyr^{-1}$) measured with the UNSCEAR (2000) Threshold.

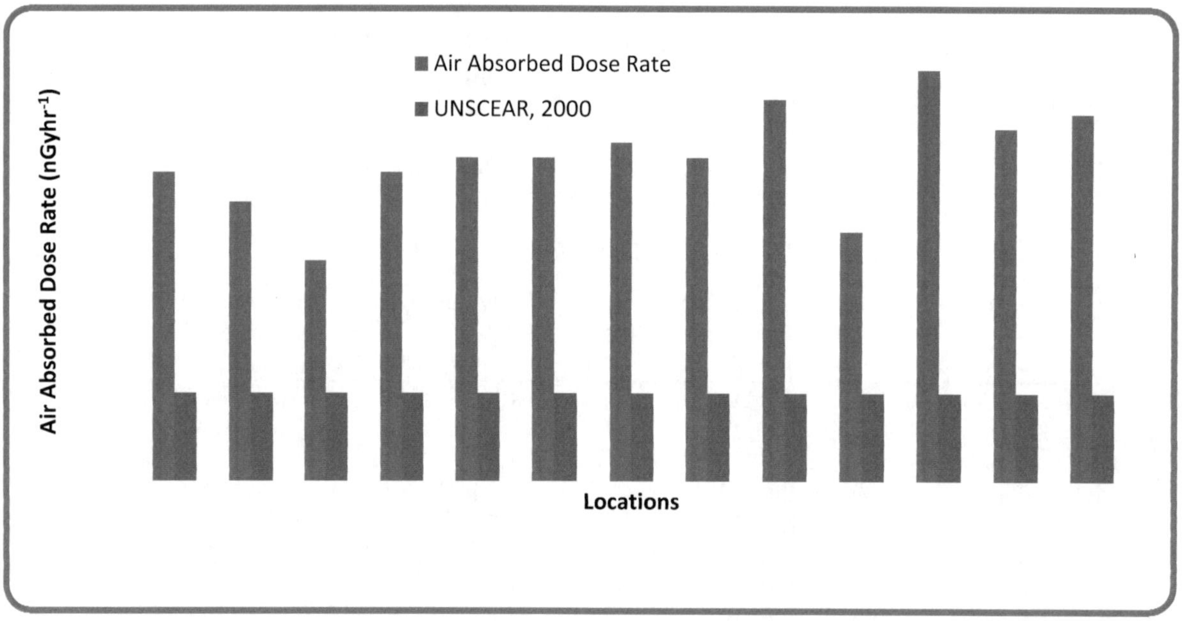

Figure 3. Comparison of average air absorbed dose ($nGyhr^{-1}$) measured with the UNSCEAR (2000) Threshold.

4. Conclusion

Radiaton dose from thirteen dumpsites in Ota and environs have been measured using the Digital alert meter. The estimated mean equivalent dose for all the dumpsites in all the locations was all higher than the permissible limits for background radiation for general public but lower than the 20 $mSvyr^{-1}$ for radiation workers. Based on these findings, the potential risk posed by wastes in most of the study locations to the environment (human, plants and animals) is minimal for now but may be a great concern with time. From the result of the study, we suggest that a perimeter fence at 10 m limits to dumpsite be constructed to serve as demarcation from building and other human activities in order to provide protection for the public. Therefore it is recommended for the local relevant authorities to provide perimeter fencing in order to restrict people from getting close to the dumpsites within the Metropolis while carrying out their daily activities.

References

[1]. United Nation Scientific Committee on the Effect of Atomic Radiation Sources, Effects and Risk of ionizing radiation UNSCEAR 2000 Report of UNSCEAR to the General Assembly UN-New York

[2] IAEA 2003 Extent of Environmental Contamination by Naturally Occurring Radioactive Material (NORM) and Technological Options for Mitigation (*Technical Reports Series No.419*) IAEA in Austria

[3]. Eyebiokin M R, Arogunjo, Oboh A, Balogun M G, F A and Rabiu A B 2005 Activity concentrations and absorbed dose equivalent of commonly consumed vegetable in Ondo State *Nigeria Nig. J. Phys.* **17S** 187

[4]. Farai I P and Vincent V E 2006 Outdoor radiation level measurement in Abeokuta, Nigeria by thermoluminescent sosimeter *Nigerian Journal of Physics* **18**(1) 121

[5]. Ademola A K, Babalola A I, Alabi F O, Onuh O D and Enyenihi E E 2014 Assessments of natural radioactivity and determination of heavy metals in soil around industrial dumpsites in Sango-Ota, Ogun state, Nigeria *Journal of Medical Physics* **39** 106

[6]. Avwiri G O and Olatubosun S A 2014 Assessment of environment radioactivity in selected dumpsites in Portharcourt, Rivers State, Nigeria *International Journal of Scientific & Technology Research* **3** 263

[7]. Emelue H U Eke B C Ogbome P and Ejiogu B C 2013 Evaluation of radiation emission from refuse dump sites in Owerri, Nigeria *IOSR Journal of Applied physics* **4** 1

[8]. Faweya E B, Babalola A I 2010 Radiological safety assessment and occurrence of heavy metals in soil from designated waste dumpsites used for building and composting in Southwestern Nigeria *The Arabian Journal for Science and Engineering* **35(2A)** 219

[9]. Isinkaye M O and Faweya E B 2006 Occurrence of natural radionuclides in refuse dumpsites within the city of Ado-Ekiti, southwestern Nigeria. Central Europe *Jourrnal of Occupational and Environment Medicine* **12**(1) 9

[10]. Odunaike R K, Laoye J A, Alausa, S K, Ijeoma G C and Adelaja A D 2008 Radiation emission characterization of waste dumpsites in the city of Ibadan in Oyo State of Nigeria *Research Journal of Environment Toxicology* **2** 38

[11]. Oladapo O O, Oni E A, Olawoyin A A, Akerele O O and Tijani S A 2012 Assessment of Natural Radionuclides Level in Wasteland Soils around Olusosun Dumpsite, Lagos, Nigeria, *IOSR Journal of Applied Physics (IOSR-JAP)* **2**(3) 38

[12]. International Commission on Radiological Protection (ICRP) 1999 *The 1995 – 99 recommendation of the International Commission on Radiological Protection Publication 76* Pergamon Press

Impacts of ionospheric electric fields on the GPS tropospheric delays during geomagnetic storms in Antarctica

W Suparta

Space Science Centre (ANGKASA), Institute of Climate Change, Universiti Kebangsaan Malaysia, 43600 Bangi, Selangor Darul Ehsan, Malaysia.

*Email: wayan@ukm.edu.my

Abstract. This paper aimed to overview the interaction of the thunderstorm with the ionospheric electric fields during major geomagnetic storms in Antarctica through the GPS tropospheric delays. For the purpose of study, geomagnetic activity and electric fields data for the period from 13 to 21 March 2015 representing the St. Patrick's Day storm is analyzed. To strengthen the analysis, data for the period of 27 October to 1st November 2003 representing for the Halloween storm is also compared. Our analysis showed that both geomagnetic storms were severe ($Ap \geq 100$ nT), where the intensity of Halloween storm is double compared to St. Patrick's Day storm. For the ionospheric electric field, the peaks were dropped to -1.63 mV/m and -2.564 mV/m for St. Patrick and Halloween storms, respectively. At this time, the interplanetary magnetic field Bz component was significantly dropped to -17.31 nT with $Ap >$ 150 nT (17 March 2015 at 19:20 UT) and -26.51 nT with $Ap = 300$ nT (29 October 2003 at 19:40 UT). For both geomagnetic storms, the electric field was correlated well with the ionospheric activity where tropospheric delays show a different characteristic.

Keywords: Ionospheric electric fields, Tropospheric delays, Geomagnetic storms, Antarctica

1. Introduction

Characterization the cause-effect mechanisms driving the formation and evolution of middle and the coupling between the upper and the lower levels of the atmosphere is challenging task. The physical mechanism on how the ionospheric activities interact directly or indirectly to troposphere is still not clear. It is well known theoretically that the high latitude ionosphere-troposphere contains the footprints of processes that have their origin in the planetary space. Many different techniques are now available for probing the ionosphere-troposphere, from radar measurements to the analysis of radio propagation noise. Among them the use of Global Navigation Satellite System (GNSS) measurements allows to describe the 3D plus time evolution of the ionospheric plasma and water vapor over restricted regions [1,2]. Mathematical techniques combined with experimental observations provide the ability to study the ionosphere from high in the F-region to the lower atmosphere. Thus the coupling processes between the magnetosphere and the neutral atmosphere can be approximated. Therefore, the possible linkages between solar phenomena, solar-induced interplanetary disturbances, the magnetospheric state and the chemistry of the middle and lower terrestrial atmosphere have been explored over the years with intent to separate natural variations from the anthropogenic forcing [3]. However, a solar influence on water vapor over Antarctica during the intense magnetospheric disturbance [2] found that the coupling between the upper and lower levels of the atmosphere is less sensitive enough to sense the coupling process. In fact, there are many phenomena occurs from the ionosphere to the lower atmosphere such as precipitation events, including water vapor distribution, formation of clouds and aerosols and the chemistry of the lower atmosphere [3], formation of polar stratospheric clouds, lightning and atmospheric electricity [4], and a long-lived trace gas in the mesosphere provides a possibility atmospheric motion and drags as well as hazard material detection during active weather. In this sense, global electric circuit is identified as a possible physical mechanism for the coupling process.

Lightning and thunderstorm activity, as related to global electric circuit plays an important role in the coupling between Earth's lower atmosphere and the ionosphere [5]. Thunderstorm activity has an effect on the E-sporadic layer and the global electric circuits are shown to be organized by lightning phenomena over the geomagnetic equator. Global thunderstorm charge ionosphere and current returns

Content from this work may be used under the terms of the Creative Commons Attribution 3.0 licence. Any further distribution of this work must maintain attribution to the author(s) and the title of the work, journal citation and DOI.
Published under licence by IOP Publishing Ltd

to ground through fair weather conduction. The effects thunderstorms on atmospheric circulation and their role in the lower atmosphere are very important to forecast weather and ultimately find out how the climate is changing. Measurement of lightning is usually conducted by Very Low Frequency (VLF) band (2~200 KHz) and thus lightning will release a large amount of energy into that propagates for long distances. However, the exact role of lightning and other variables in the frequency band still not well understood. In this study, an electric field which may drive the ionospheric phenomena is employed as a media or wire to transfer solar energy from space to Earth's surface for the coupling process. To realize the investigation, the geomagnetic activity during St. Patrick's Day storm (17 March 2015) and the Halloween storm (29 October 2003) is analyzed.
Therefore, the study of ionospheric effects of geomagnetic storms and their connection to the lower atmosphere remains one of the most important scientific tasks. Measurements of an electric field for determining atmospheric parameters will act as a better basis for understanding the electrical coupling between the ionosphere and the atmosphere. The most importance is how electric field from space interacts through ionosphere and affect thunderstorms in the upper troposphere.

2. Data and Methods

To study the response of electric field connecting by space weather, data for the period of 15~20 March 2015 during St. Patrick's geomagnetic storm is compared with Halloween storm that has occurred between 27 October and 1 November 2003. The geomagnetic index called planetary index (*Ap*) and interplanetary magnetic field in the southward (z) component (IMF Bz) in Geocentric Solar Magnetospheric (GSM) rather than Geocentric Solar Earth (GSE) were used to indicate the level of the geomagnetic disturbance. Data for *Ap* index and IMF Bz are obtained from the OmniWeb website (http://omniweb.gsfc.nasa.gov) of the NASA Goddard Space Flight Center (GSFC). Both parameters are in units of Nanotesla (nT) and the data was averaged from 1-min to 1-hour.

To analyze the response of the ionospheric phenomena on the Global Positioning System (GPS) propagation at the lower atmosphere, the zenith path delay (ZPD) data at Bernardo O'Higgins (OHI3: 50°S, 74°W and ellipsoidal height 32.15 m) and Palmer station (PALM: 64.77°S, 64.05°W and ellipsoidal height 31.0427 m) was used. The observations are focused in the Antarctic Peninsula where the region is expected to have experienced global warming. The ZPD for St. Patrick's storm was taken at a five-minute interval from the Crustal Dynamics Data Information System (CDDIS) NASA (ftp://cddis.gsfc.nasa.gov/pub/gps/products/troposphere/zpd/). However, for the Halloween storm, the ZPD data was obtained with an interval of 2 hours. Since no ZPD data for PALM during the Halloween storm, the ZPD is estimated using adaptive neural-fuzzy interference system (ANFIS), which using surface pressure, temperature, and relative humidity as inputs. The meteorology data is obtained at the respective station that was managed by British Antarctic Survey (BAS) website (https://legacy.bas.ac.uk/met/metlog/). The estimation approach is called the ANFIS ZTD. More in-depth about the ANFIS ZTD concept and the implementation can be found in Suparta and Alhasa [6].

Then, to link the ionospheric phenomena to the case of lower atmosphere such as thunderstorm, interplanetary electric field (IEF) obtained from OmniWeb of the NASA GSFC is employed. Since the IEF data at certain days was missing (or no data recorded), the equatorial ionospheric eastward electric field (EEF) from Cooperative Institute for Research in Environmental Sciences (CIRES) of University of Colorado (https://geomag.colorado.edu/) is employed. The real-time variation of EEF data was calculated with transfer function model from solar wind data and a climatological model to account for the quiet day variations of EEF [7]. In the equatorial ionosphere of the Earth, the wind-driven currents coupled with the Earth's primarily horizontal magnetic field produce the EEF. On the other hand, EEF is a result of solar wind electric fields penetrating from high latitudes to the equator. Based on this concept, EEF is predicted on a real-time basis using the solar wind data from the Advanced Composition Explorer (ACE) satellite. Detail of the model can be referred to Manoj et al. [8].

3. Results and Discussion

Figure 1 shows the indication of severe geomagnetic storms as represented by *Ap* index and IMF Bz in hourly basis. As shown in Figure 1(a), the severe geomagnetic storm was occurred with $Ap > 100$ nT on 17 March 2015 which so-called the St. Patrick's day storm. The 2015 St. Patrick's Day storm was the first storm of solar cycle 24 with the G4 level based on the NOAA scale. The sudden storm commencement (SSC) was registered at ~04:45 UT on 17 March 2015 and then there was a quick drop of the IMF Bz to the value of −18.1 nT, observed at ~16:00 UT. This commencement caused by a coronal mass ejection (CME) at 02:00-02:30 UT on 15 March 2015. This geomagnetic storm was first reported by Kamide and Kusano [9]. As seen in the figure, prior to severe level, the event was initiated with a minor storm in the early morning of 17 March 2015 and then the IMF Bz component sharply turned southward and reached the minimum of -16.1 nT at 08:20 UT with *Ap* < 75 nT, but became positive (northward) at 11:50 UT with IMF Bz ~10 nT. After that the IMF Bz is largely southward to -18.10 nT for ~30 min of time at ~14:32 UT when the *Ap* > 150 nT (severe level). About 3 hours later, it has northward direction to -11 nT when the *Ap* = 150 nT and three hours later reached IMF Bz of -17.3 nT at 19:20 UT when the *Ap* reached more than 150 nT. Finally, it became positive (northward) at 24:00 UT on 17 March 2015 and subsided in the noon on 18 March 2015 (with the minor storm due to the effect of CME on 15 March 2015) and the IMF Bz still maintained in the northward direction. During the severe geomagnetic storm (~6 hours), the observed periods with largely negative IMF Bz led to the interconnection between the IMF and the Earth's magnetic field lines. This makes the storm of the 2015 St. Patrick's Day storm of 17–18 March 2015 is the first largest in the solar cycle 24 [10].

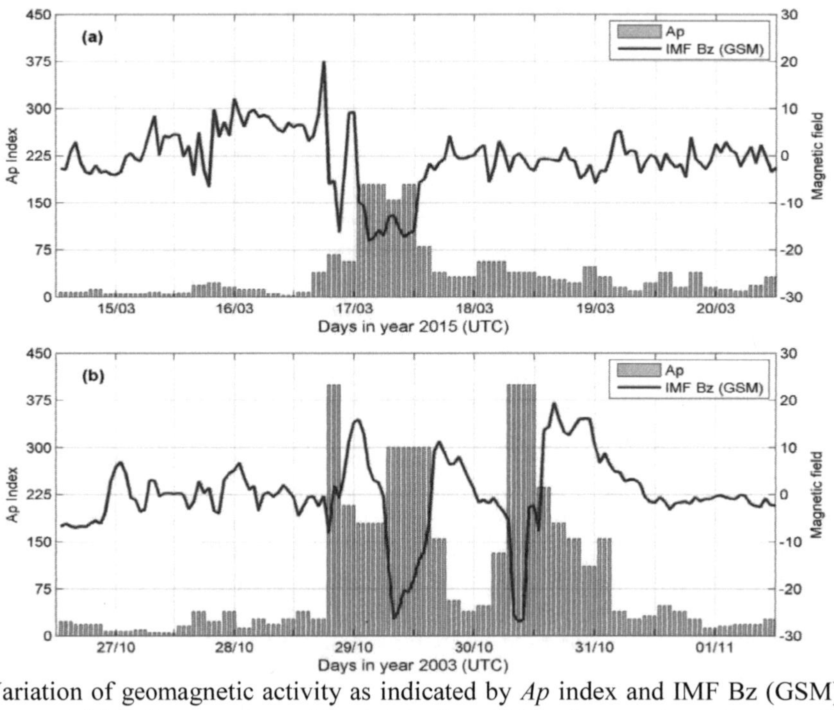

Figure 1. Variation of geomagnetic activity as indicated by *Ap* index and IMF Bz (GSM) for (a) St. Patrick's storm and (b) Halloween storm. The date in the middle of graph shows the noon UT time.

Figure 1(b) shows the super geomagnetic storm which was well known as the Halloween storm for comparison with the St. Patrick's Day storm. These storms reached *Ap* > 375 on 29 and 30 October 2003. The most significant impact was to have four negative peaks with two the largest of IMF Bz with -26.5 nT (*Ap* = 300 nT) at 19:40 UT and -27.1 nT at 20:36 UT, respectively. However, prior to the first largest negative peak, it has IMF Bz of -8.2 nT where *Ap* > 375 at 06:27 UT and the last one after 2.5 hours is -8 nT in the morning of 31 October 2003. The Halloween storm shows a double strength in the intensity of geomagnetic activity as compared with the St. Patrick's storm. However, the pattern of both storms is almost similar with a W shape. Details character of this storm has been published in many scientific journals by several authors [e.g., 2, 11, 12]. The difference between the two storms are the occurrence of flare activity, where Halloween storm recorded X17.2/4B and St. Patrick's storm was recorded more only in C and M classes (or no major solar flares). Although the G4 storm possibly will not affecting the astronauts, it could affect the communication system (GNSS and radio signals on Earth). More than that how it propagates and finally, how it affects the weather on the ground (Earth surface) or lower atmosphere phenomena like a thunderstorm are still a challenging task to be answered.

The idea behind the use of ZPD to connect it to the upper atmosphere during geomagnetic storms is a possible cause of GPS disturbances at high latitudes. Patches with containing small-scale plasma structures is one of the phenomena which related to the distribution of scintillations at high latitudes and clearly disturbed the GPS signals [e.g., 13,14]. Jacobsen and Andalsvik [15] have studied the impact of the disturbances on the network RTK and Precise Point Positioning (PPP) techniques and found that the positioning errors is increased rapidly for both techniques. In other word, with the same GPS signals, ionospheric TEC is strongly influenced by geomagnetic storms. Figure 2 shows the variation of zenith tropospheric delay (ZTD) or zenith path delay (ZPD) from CDDIS NASA derived from GPS measurements over the Antarctic Peninsula during (a) St. Patrick's storm and (b) Halloween storm, respectively. As shown in Fig. 2a, the GPS signals more experienced with an irregular variation on 16 March 2015 after the CME was subsided. It takes approximately 8 hours after CME subsided and reach the lower atmosphere with ZPD is 2.30 m (PALM) and 2.34 m (OHI3) where PALM leads ~3 hours compared to OHI3. Both ZPD shows a similar variation and the ZPD peak different a few hours which is expected due to the geographical variations between the stations. Looking at the St. Patrick's storm on 17 March 2015, the ZPD was decreased to ~2.25 m which is 0.74% lower compared to their average value. The ZPD will increase again ~1.2% at midnight of 18 March 2015. Comparing to Fig. 1a, during a major storm occurs (St. Patrick's Day), it can be said there is no direct response to the lower atmosphere. This indicates that the effects of the St. Patrick's storm for the selected GPS station in Antarctic Peninsula are somewhat less pronounced. Looking at the time difference in upper atmosphere and on the ground (travel ~8-10 hours), the positive response of ZPD is after the CME shocked on 15 March 2015. The response by increasing peak of ZPD is on 16 March 2015 at ~06:00 - 14:00 UT with an increase of 5.7% compared to their mean average. However, the response of ZPD with duration time at post the CME event and to make a firm suggestion regarding this needs further investigation.

When looked at the Halloween storm in Fig. 2b and compared with Fig. 1b, it can be seen that two stations (OHI3 and MARA) shown a positive response of Halloween storm to lower atmosphere. Noted in Fig. 2 that since meteorological data is only available daily at PALM and two days were no ZPD data at OHI3, then ZPD estimated using ANFIS ZTD for Rothera (ROTH: 67°33'57.1"S, 68°07'43.5"W) and Base Marambio (MARA: 64°14'41"S, 56°39'25"W) is plotted for the comparison. Selection of these stations is only based on the data available for both storms and in the scope of Antarctic Peninsula. From Fig. 2b, PALM and ROTH showed a drop peak on 29 October 2003 as compared to other two stations. ROTH station is located at Rothera point, Adelaide Island with the crushed rock surface. The geographical distance between ROTH and MARA is about 640 km where ROTH is on the west side of OHI3. The high response of lower atmosphere during this Halloween storm is possibly due to the enormous of solar flares compared to St. Patrick's storm. The solar flare was correlated well with the occurrence of the ionospheric storm and the GPS signals were strongly delayed through the irregular variation of total electron content (TEC) in the upper atmosphere [10]. By comparing the ZPD average between both geomagnetic storms, the ZPD during St. Patrick's Day is 2.6% higher compared to Halloween storm. However, the high of ZPD observed possibly due to a different in the meteorological season is still needs to be investigated further. In addition, whether solar flares have a connection to the lower atmosphere is still a big challenge to prove. The ZPD response from two effects of geomagnetic storms disturbances on the GPS receiver on the ground demonstrated very dynamically. Their impact on tropospheric delays changing very fast and seems to be random. Propagation signals from the upper atmosphere seems never reach the troposphere, it has been diminished or disappeared due to the attenuation and a possible occurrence of convection and absorption of waves. On the other hand, the shielding of the earthward region is very strong to break by the electric field current from the magnetosphere. This is in contrast with the upper atmosphere that patches associated with scintillation can be clearly observed at high latitudes due to precipitation of energetic particles.

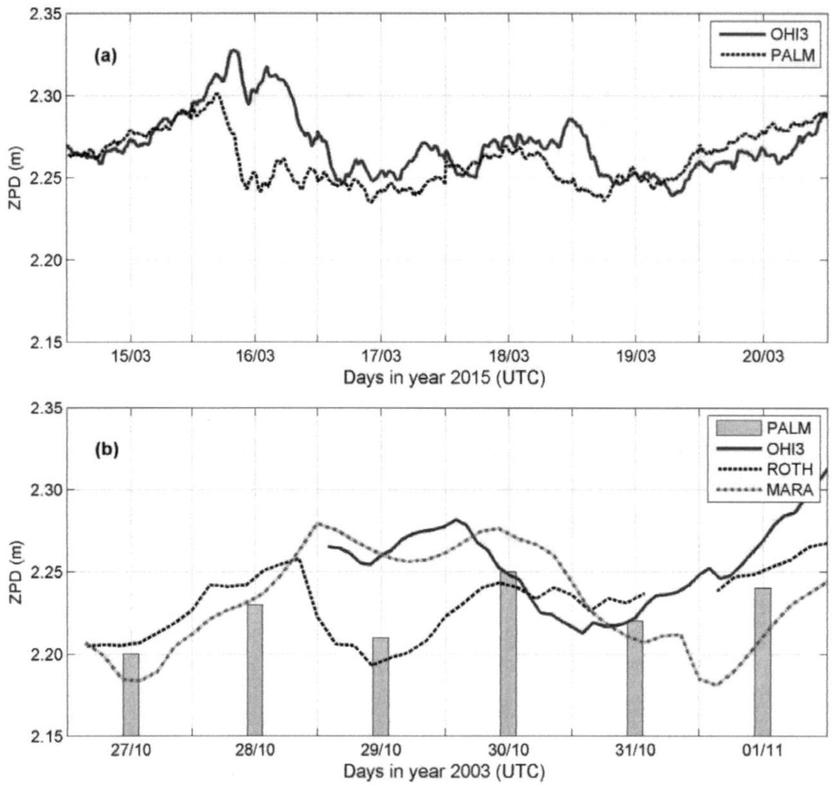

Figure 2. Variation of zenith tropospheric delay from GPS signals during (a) St. Patrick's storm and (b) Halloween storm, respectively

Another way to study the connection between the upper atmosphere and the lower levels of the atmosphere is analyzing the penetration of electric field as presented in Fig. 3. The magnitude of penetration electric field drive by the solar wind/IMF and supported by the strength of the magnetospheric convection will possibly bring the electric field current to the Earth's surface. As introduced in Section 2, EEF is predicted on a real-time basis using the solar wind data from ACE satellite. As shown in Fig. 3a, Ey observed from NASA representing the east-west component of the electric field was plotted due to its induced on the ground by geomagnetic activity. The use of Ey assuming that the solar wind is due to the ExB drift of the interplanetary magnetic and electric fields, IEF Ey is calculated based on the Ey = Vx × Bz, where Ey be the dawn-dusk component of the IMF, Vx be the solar wind velocity, and Bz be the vertical component of the interplanetary magnetic field. The vectors are relative to the geocentric solar magnetospheric coordinate system [8]. On the other hand, the x-axis is along the Sun-Earth line with the origin of the Earth, and is negative toward the Earth. The y-axis is the cross product of the Earth's dipole axis and is positive toward dusk. The z-axis is perpendicular to the xy plane and is positive to the Earth's magnetic north. Figure 3a shows the IEF variation during quiet day times from CIRES. It shows two peaks every day at around 14:00 and 22:00 UT which possibly due to the action of the atmospheric dynamo. Penetration of electric field is conducted by solar wind in the inner edge of the ring current in the shield region of equatorward. This electric field penetrating the ionospheric region with produces a net dusk-to-dawn electric field during geomagnetic storms. This can be depicted from the figure that the strongest disturbance for the quiet + penetration of EEF variation was a quick drop to a -1.46 mV/m at ~04:50 UT. This peaked drop was confirmed by IEF to a -1.0 mV/m (red color) with led about one-hour. On the other hand, the irregular variation of both EEF and IEF confirmed that it was strongly influenced by the St. Patrick's Day storm. This means that the IEF was successfully penetrated to the middle- and low-latitude ionosphere.

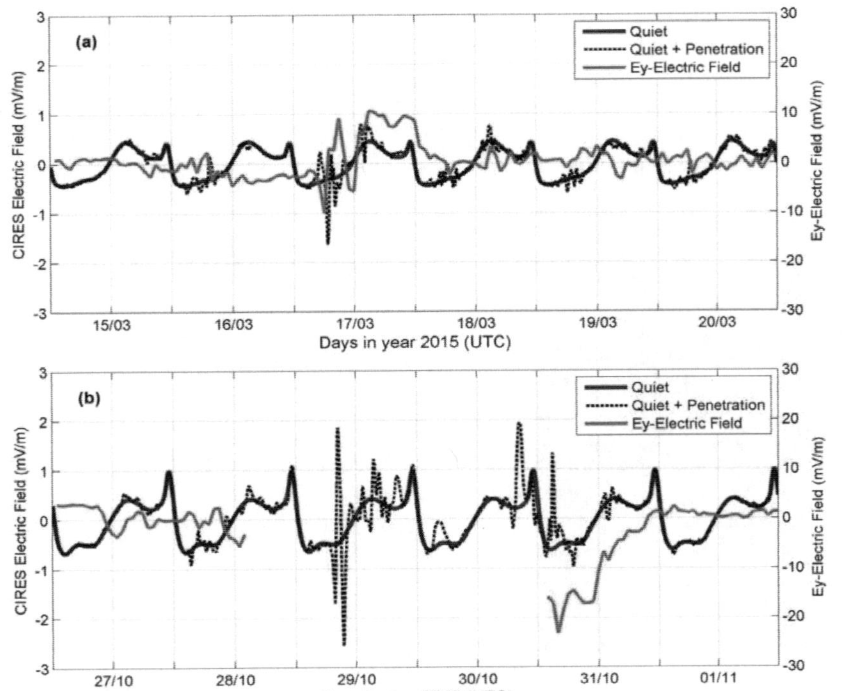

Figure 3. Variation of Equatorial Electric Field (EEF) to indicate ionospheric electric field disturbance for (a) St. Patrick's storm and (b) Halloween storm, respectively

For Fig. 3b, the CIRES data (EEF) initially first drop to a -1.7 mV/m at 18:00 UT on 29 October 2003 and two hours later become a positive at 20:00 UT to a 1.8 mV/m. The significant drop of EEF was observed at 22:00 UT with a -2.6 mV/m and fluctuated between -0.25 and 1.4 mV/m. The high positive of EEF was observed at 20:00 UT on 30 October 2003 with a value of 1.9 mV/m and reduced to 1.4 mV/m at 05:00 UT on 31 October 2003. Unfortunately, during the super intense of a geomagnetic storm, it was no observation data for the Ey-electric field from late afternoon 28 until 30 October 2003. The negative peak of Ey has recorded ~-24 mV/m at 04:00 UT on 31 October 2003. Thus the high fluctuation during 29 to 30 October 2003 indicated that the geomagnetic activity was strongly influenced by space weather as well as the satellite facilities. No data during this period indicates that there has been a power outage (blackout) in which super geomagnetic storms causes damage to electric transmission lines.

4. Proposed a connection between the upper and the lower level of the atmosphere

Based on the response of GPS tropospheric delays on the geomagnetic storms, the different in intensity of the geomagnetic storm will affirm to have a different characteristic of ZPD. Looking at the onset of geomagnetic storms, the Halloween storm is more pronounced as compared to the St. Patrick's storm. The delay of GPS signals represented by ZPD is increased during the major storms. In contrast, the ZPD during St. Patrick's Day has been reduced, or the energy from space can be said to have been lost before reaching the neutral atmosphere. The interaction of upper-lower atmosphere from space perspective can be illustrated in Fig. 4. In general, solar activity has clearly direct influence to the upper atmosphere, so-called the space weather. ZPD derived from GPS signals from the neutral atmosphere is very important for weather forecasts, numerical weather prediction as well as climate when the ZPD value is transformed into the precipitable water vapor (PWV). PWV is an indicator of climate change which can be useful for synoptic use, prediction of thunderstorms, etc.

If the connection between the upper-lower atmospheres happening, some physical mechanisms of how solar activity influences terrestrial weather/climate changes can be explained. For example in Fig. 4a, the occurrence of thunderstorms will generate lightning. A lightning discharge will create large earth currents in the atmosphere and this initial charge separation will create an electric field through the global electric circuit (GEC). Thunderstorm is the major source of DC electric field. Rycroft [16] reviewed that thunderstorm can connecting to the magnetosphere through electromagnetic radiations generated from lightning discharges. These effects look like from below that the ionospheric equipotential surface above the thundercloud is lowered when the energetic electrons is precipitated. Now, this work is to propose how the space weather has a direct or indirect connection to earth's climate and climatic changes with lightning activity as illustrated in Fig. 4b.

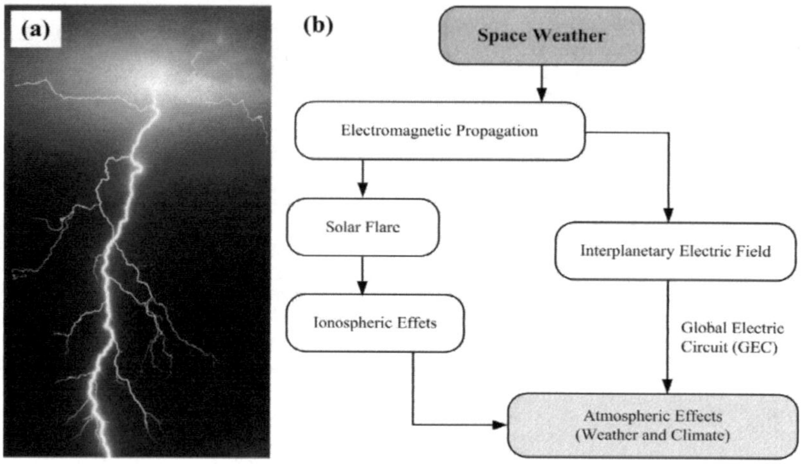

Figure 4. (a) The occurrence of thunderstorm with generates an electric field and (b) proposed a simple mechanism between the upper and the lower levels of the atmosphere

As shown in the Fig. 4b, the space weather clearly affecting the upper atmosphere by the disturbance of radio signals. The propagation delays of this signals can be indicated by TEC fluctuation and the strong effects of space environment is also characterized by how big the occurrence of solar flare. The direct effect of ionosphere to the lower atmosphere in the current research during short-term study is still difficult to prove. The effect maybe indirectly to GPS water vapor variability. The ionosphere is most contained of electron content, while the Earth's surface (ground) can be assumed as chassis or negative charge. In DC role, the current should be flow from positive to negative charges (or, ground to negative charge). This connection will be from below and form a GEC. Another way from right side of Fig. 4, the interaction from space to the ground can be thought thundercloud that can also form a GEC. GEC plays an important role in the coupling between the ionosphere and the Earth's lower atmosphere. The global atmospheric electric field acts as a medium to transfer energy from the Sun to Earth's surface. This situation can be measured through IEF/EEF. IEF/EEF will affect the atmosphere through the interaction between solar wind and thunderstorm, where more active of thunderstorm in the upper level of the troposphere the lightning activity will more generate current. Despite of thunderstorm, the variations in the galactic cosmic ray intensity, modulated by solar activity, are also manifest in changes the atmospheric ionization; however there is some uncertainty that these have significant potential to affect the lower atmosphere [17]. These mechanisms suggest that the temporal distribution and strength of solar activity as indicated by geomagnetic disturbances are important inputs for GPS tropospheric delays.

5. Conclusion

The use of interplanetary electric field (IEF) to connecting the upper and the lower levels of the atmosphere during a major geomagnetic storm based on GPS propagation were successfully proposed. GPS signals were clearly disturbed during the major geomagnetic storm. The electric field variation was also correlated well with the ionospheric activity. It is noted that the tropospheric delays during St. Patrick's Day are slightly affected by the electric field as compared to the Halloween storm. A small to moderate solar flare that occurred during St. Patrick's Day may hard to affect the region of the lower atmosphere. The energetic particle from electromagnetic emission is insufficient to ionize the atmosphere. Finally, the response of the lower atmosphere to the space weather can be said very small, or delayed in few hours. In contrast, the Halloween storm shows more energy transferred to the lower atmosphere as indicated by the blackout of electric transmission lines during the high event.

The coupling process proposed is an increase in the electric field of the upper atmosphere indicates substantial disturbances to the GEC through an increase in the incident density of energetic particles in the atmosphere. This increase is from magnetosphere by dynamo activity. In addition, the major change in the conductivity from the ionospheric E-region (mesosphere) to the lower atmosphere with changes of cloud structure and cloud base temperature is possibly by cosmic ray modulation (accompanied by a change in downward longwave radiation). Thus, GEC may constitute a variable physical mechanism linking space weather and the earth's weather and climate. Since the ZPD to have different response on the level of geomagnetic storms as well electric field, each of GPS satellite needs to be investigated in future work. The links of the ionospheric electric field to the lower atmosphere through lightning activity also needs to be validated, especially for the Antarctic Peninsula whether they have a direct or indirect response to weather and climate.

Acknowledgements
This work is supported by MOSTI Flagship Program under ZF-2014-016 grant and the National Antarctica Research Centre (NARC). The interplanetary magnetic field data and planetary index data was provided by NASA, and Equatorial Electric Field (EEF) data provided by Cooperative Institute for Research in Environmental Sciences (CIRES) of University of Colorado. Author would like to thank again to NASA and British Antarctic Survey (BAS) for providing the ZPD and meteorological data, respectively.

References

[1] Fedrizzi M, de Paula E R, Langley R B, Komjathy A, Batista I S and Kantor I J 2005 Study of the March 31, 2001 magnetic storm effects on the ionospheric GPS data *Adv. Space Res.* **36**(3), 534–545
[1] Suparta W, Abdul Rashid Z A, Mohd Ali M A, Yatim B and Fraser G J 2008 Observation of Antarctic precipitable water vapor and its response to the solar activity based on GPS sensing *J Atmos Sol-Terr. Phys.* **70**,1419-1447
[2] Rycroft M J, Israelsson and Price C 2000 The global atmospheric electric circuit, solar activity and climate change *J. Atmos. Sol-Terr. Phys.* **62**(17-18), 1563-1576
[3] Stokes G M and Schwartz S E 1994 The atmospheric radiation measurement (ARM) program: Programmatic background and design of the cloud and radiation test bed *Bul. Amer. Meteorol. Soc.* **75**(7), 1201-1221
[4] Lakhina G S 1993 Electrodynamic coupling between different regions of the atmosphere *Current Science* **64**(9): 660-686
[5] Suparta W and Alhasa K M 2015 Modeling of zenith path delay over Antarctica using an adaptive neuro fuzzy inference system technique *Expert Systems with Applications* **42**(3), 1050-1064
[6] Manoj C and Maus S 2012 A real-time forecast service for the ionospheric equatorial zonal electric field *Space Weather* **10**, S09002, doi:10.1029/2012SW000825
[7] Manoj C, Maus, S, Lühr H and Alken P 2008 Penetration characteristics of the interplanetary electric field to the daytime equatorial ionosphere *J. Geophys. Res.* **113**, A12310, doi:10.1029/2008JA013381
[8] Kamide Y and Kusano K 2015 No major solar flares but the largest geomagnetic storm in the present solar cycle *Space Weather* **13** (6), 365–367, doi: 10.1002/2015SW001213
[9] Astafyeva E, Zakharenkova I and Förster M 2015 Ionospheric response to the 2015 St. Patrick's Day storm: A global multi-instrumental overview *J. Geophys. Res.* **120**(10), 9023-9037, doi: 10.1002/2015JA021629.
[10] Skoug R M, Gosling J T, Steinberg J T, McComas J T, Smith C W, Ness N F, Hu Q and Burlaga L F (2004) Extremely high solar wind: 29 – 30 October 2003 *J. Geophys. Res.* **109**, A09102, doi:10.1029/2004JA010494
[11] Tsurutani B T, Mannucci A J, Iijima B, Guarnieri F L, Gonzalez W D, Judge D L, Gangopadhyay P and Pap J 2006 The extreme 2003 solar flares (and Bastille Day, 2000

Flare), ICMEs and resultant extreme ionospheric effect: A review *Adv. Space Res.* **37**(8): 1583-1588

[12] Kintner P M, Ledvina B M and de Paula E R 2007 GPS and ionospheric scintillations *Space Weather* **5**(9), S09003, doi: 10.1029/2006SW000260

[13] Jin Y, Moen J I and Miloch W J 2015 On the collocation of the cusp aurora and the GPS phase scintillation: a statistical study *J. Geophys. Res.* **120**(10), 9176–9191, doi: 10.1002/2015JA021449

[14] Jacobsen K S and Andalsvik Y L 2016 Overview of the 2015 St. Patrick's day storm and its consequences for RTK and PPP positioning in Norway *J. Space Weather Space Clim.* **6**, A9, doi: 10.1051/swsc/2016004

[15] Rycroft M J 2006 Electrical processes coupling the atmosphere and ionosphere: an overview *J Atmos. Solar-Terr. Phys.* **68**, 445-456

[16] Erlykin A D, Gyalai G, Kudela K, Sloan T and Wolfendale A W 2009 Some aspects of ionization and the cloud cover, cosmic ray correlation problem *J. Atmos. Sol-Terr. Phys.* **71**, 823-829

Institute of Physics
Dirac House, Temple Back
Bristol BS1 6BE UK

ISSN:1742-6588
ISBN 978-1-5108-4243-4

9 781510 842434

CURRAN ASSOCIATES INC.
proceedings.com